Principles of
Electronic Media

SECOND EDITION

Principles of Electronic Media

William R. Davie
University of Louisiana

James R. Upshaw
University of Oregon

PEARSON

Boston   New York   San Francisco
Mexico City   Montreal   Toronto   London   Madrid   Munich   Paris
Hong Kong   Singapore   Tokyo   Cape Town   Sydney

Series Editor: *Molly Taylor*
Series Editorial Assistant: *Michael Kish* and *Suzanne Stradley*
Senior Development Editor: *Cheryl de Jong-Lambert*
Senior Marketing Manager: *Mandee Eckersley*
Production Editor: *Janet Domingo*
Editorial Production Service: *Lynda Griffiths*
Composition Buyer: *Linda Cox*
Manufacturing Buyer: *Megan Cochran*
Electronic Composition: *Omegatype Typography, Inc.*
Interior Design: *Ellen Pettengell*
Photo Researcher: *Katharine S. Cook* and *Larissa Tierney*
Cover Administrator: *Joel Gendron*

For related titles and support materials, visit our online catalog at www.ablongman.com.

Between the time website information is gathered and then published, it is not unusual for some sites to have closed. Also, the transcription of URLs can result in typographical errors. The publisher would appreciate notification where these errors occur so that they may be corrected in subsequent editions.

Cataloging-in-Publication data is not available at this time.

ISBN 0-205-44975-1

Text Credit: Page 45: "57 Channels (And Nothin' On)" by Bruce Springsteen. Copyright © 1990 Bruce Springsteen (ASCAP). Reprinted by Permission.

Photo Credits: Photo credits appear on pages 408 and 409, which constitute a continuation of the copyright page.

Printed in the United States of America

10   9   8   7   6   5   4   3   2   1   VHP   09   08   07   06   05

Brief Contents

Contents

CHAPTER 3

Cable and Satellite 44

CHAPTER 4

Radio and Television Technology 66

PART TWO
The Business of Electronic Media

CHAPTER 5
Digital Domains   92

CHAPTER 6
The Industry   114

Over the Top! *Democratic candidate for president Howard Dean was skewered by late-night talk-show hosts for his overexhuberant "I Have a Scream" speech. The unforgiving eye of the TV camera can magnify one moment in time and amplify it a thousand times.*

Industry Gambles. *Retired anchorman Walter Cronkite found young viewers doing political analysis for MTV. Sirius Satellite Radio wooed shock-jock Howard Stern away from earthbound channel. Wife Swap, a TV "reality" experiment, mixed families for fun and profit.*

CHAPTER 7

Programming and Distribution   142

CHAPTER 8

Broadcast News   174

CHAPTER 9

The Audience   196

CHAPTER 10

Advertising and Promotions   218

Man + Audience. The Bernie Mac Show *lost some ratings punch when its star suffered health problems, but the unusual comedy series has regained and held on to strong backing from family viewers.*

Not-So-Hidden-Persuaders. *Product placement is a subtle yet powerful form of advertising. Stars of* Million Dollar Baby, *Clint Eastwood and Morgan Freeman, appear to implicitly endorse Coca-Cola in this scene.*

PART THREE

Electronic Media: A Broader View

CHAPTER 11

Law 244

CHAPTER 12

Professional Ethics 278

CHAPTER 13

Theory and Research 304

CHAPTER 14

Public Broadcasting 330

Red and Blue States. *National pollsters guide presidential stragegists by charting desirable prey—undecided voters. Pollsters often sample about 1,000 voters, asking, "If the election were held today, who would you prefer?"*

Funny Bits. *Using a primitive communication method, Carl Kasell and Peter Sagal of public radio's comedy news-quiz show* Wait Wait . . . Don't Tell Me! *clown for the camera.*

CHAPTER 15
The World   362

Preface

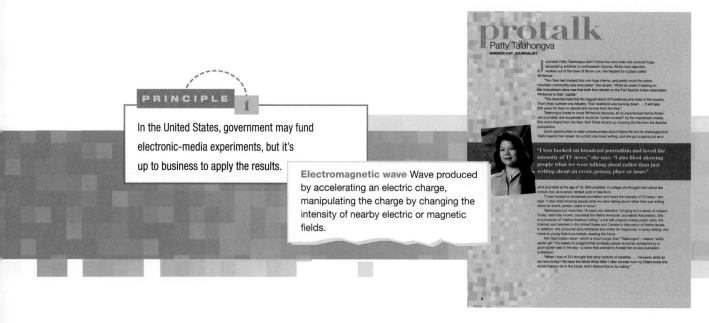

PRINCIPLE 1

In the United States, government may fund electronic-media experiments, but it's up to business to apply the results.

Electromagnetic wave Wave produced by accelerating an electric charge, manipulating the charge by changing the intensity of nearby electric or magnetic fields.

TO THE STUDENT

Welcome to the exciting, demanding, often controversial, and occasionally baffling world of electronic media.

If you suspect you already know this world—you do, after all, inhabit it and are in fact a major reason for its existence—then prepare to be surprised.

Through *Principles of Electronic Media*, you're about to tour a vast territory whose music and messages have the power to penetrate your private moments and to enliven many of your public ones. You'll confront people, ambitions, practices, institutions, and social causes and consequences that have shaped this field since the first glow of a vacuum tube.

As the tour unfolds, we'll supply some important signposts, including:

▶ *Principles*—not just theories but guidelines that almost seem to function as natural laws on which modern professionals in electronic media base decisions and do their work. You'll find **Principles** in the text of each chapter.

▶ *Lingo*—that is used in every area of electronic media. Learn all the language—from jargon and job titles, to names of concepts and equipment—by reading the **glossary terms** that are highlighted in the text, defined in the page margins, and listed in the Index at the end of the book.

▶ *Voices*—of current professionals whose lives every day reflect the rewards and shortcomings of these media, not just the voices of inventors and industry magnates who have helped to build them. These voices emanate from the **ProTalk** features sprinkled through the book.

▶ *In-depth looks*—at media conglomerates and at great moments in media history. Organizational structures and statistical analyses can be found in the Sidebar features that appear in the pages of every chapter.

▶ *Signposts*—toward your own successful pursuit of jobs and careers in electronic media. You'll discover these in the **Career Focus** passages at the end of most chapters.

► *Fuel for reflection*—on the astonishing influence of radio, television, and the Internet, transcending borders, oceans, cultures, values, political systems, and the landscape of our collective imagination. There's fuel aplenty in the **Food for Thought** that ends each chapter. Its central aim is to provoke critical thinking about the forces that conspire or collide as we use and observe electronic media in our daily lives.

CHARTING THE TOUR

Like a small country divided into three distinct but related provinces, *Principles of Electronic Media* is organized into three parts. Our basic tour will roam through all the regions of a single province, then move on to the next. You may notice repetition or overlap, but that's deliberate: Some lessons need reinforcement. Besides, you (or your instructor) may see fit to navigate this book's chapters in a different sequence from ours, for your own purposes.

Part One provides a foundation for understanding how electronic media came into being and how they grew into open-ended media with a flair for the digital. Recent events animate *Chapter 1*, an overview, and connect the media's pervasiveness today to the founding sparks that pushed them toward the future, as outlined in *Chapter 2*, a history. *Chapter 3* charts the rise of cable TV and the role of satellites, and *Chapter 4* explains the technology on which broadcasting is based. This leads naturally into the newer phenomenon of digital convergence—where all electronic media seem to meet—as described in *Chapter 5*. Here, you'll read of the computer's influence and its ever-useful ties to the Internet.

In **Part Two**, we take on the business of electronic media. *Chapter 6* focuses on the organization of the industry, which encompasses all for-profit enterprises, great and small. Next, *Chapter 7* charts the development and production of programming with which the industry attracts audiences, and thus advertisers. A special and influential category of programming—news, in which we are well steeped—occupies *Chapter 8*. *Chapter 9* decodes the art and craft of identifying and measuring audiences. Finally in this "province" of the book, *Chapter 10* examines the people and strategies of advertising and promotion; they compose what the ancient Greeks might have called the siren song of electronic media.

Part Three takes a broader look at the field, ranging well beyond its most obvious features. The laws and regulations that have defined the limits and possibilities of electronic media are spelled out in *Chapter 11*. *Chapter 12* delves into the ethics that do or should guide media professionals in their socially important work. In *Chapter 13*, we discover research and theories that help explain how the sounds and images of electronic media influence our lives. *Chapter 14* moves the expedition through *non*commercial media zones where so-called public broadcasting and its grass-roots relatives fight the good fight. *Principles of Electronic Media* closes with *Chapter 15*, tracing the increasingly global reach of broadcasting and the Internet, the international media systems, and how the U.S. industry functions around the world.

Along the way, as we've noted, this book highlights important working patterns and guidelines, showcases personal views, supplies fuel for making connections, and suggests what it all means. If you finish a chapter without having studied its *Principles* and *glossary terms*, its *ProTalk* and *Sidebar* features, its *Career Focus* section, and its probing *Food for Thought* questions—you haven't really finished the chapter.

The key now is to start, travel the territory, and move on, gradually absorbing a clearer, richer understanding of today's electronic media.

TO THE INSTRUCTOR

First, since you may well be familiar with the founding edition of *Principles of Electronic Media*, here are some important improvements in this second edition:

▶ *Chapter 1*, an overview of how electronic media meet social needs and are tied to historical contexts, not only introduces the usefulness of **Principles** but it also includes a few of them to get the critical thinking under way.

▶ Throughout the book, **ProTalk** features have been updated or replaced with fresher ones, introducing new people and stories to this series of brief professional profiles. Among the latest individuals are:
  - *Steve Harris*, vice president of programming for XM Radio—one of the cutting-edge architects of a new media pathway, satellite radio
  - *Patty Talahongva*, a Native American broadcast journalist with a uniquely purposeful slant on covering the news
  - Cable executive *Anne Doris*, of Cox Communications, with many insights into the role of new-media "bundles"
  - *Robin Sloan*, who sharpens our understanding of "digital domains" in electronic media through his work at IndTV
  - NBC/Telemundo executive *Don Browne*, drawn to television by its civil rights coverage in the 1960s and now courting Hispanic audiences for a giant network
  - Newspaper-editor-turned-dot-com-journalist *Ernest Sotomayor* of Newsday. com, intent on diversifying online journalism

▶ *Principles of Electronic Media*'s second edition examines media types, technologies, and developments that have emerged or grown more important since the first edition was published. Among these are:
  - The spreading impact of consolidation on a fractured media landscape
  - The continuing decline of profits for local television news
  - What has become the routine migration of broadcast news stories to the Web and other platforms

- Many TV viewers' eagerness to use digital devices to shuffle their daily menus
- "Reality shows" exploding—and in some cases, imploding—across the broadcast schedule
- The staggering blows delivered to the recording industry and FM radio by online music delivery
- Accelerating efforts to measure and analyze new-media audiences and do a more precise job of gauging traditional-media viewers and listeners
- An experimental foray of TV programs in which gays play leading roles
- Federal Communications Commission's attention to bursts of "indecency" on the airwaves
- The arrival of satellite radio as a potentially powerful competitor to earthbound broadcasting
- The burgeoning digitization of electronic media in general

Media change is always on the morning breeze, and of special note in this edition of *Principles of Electronic Media* are some sections we've added to the Public Broadcasting chapter (*Chapter 14*). They're intended to do justice to some interesting alternatives—some old, some new—to noncommercial radio and television. These include "community" radio stations; public-access TV channels; college broadcasting on campuses (still the home of many public radio and public television stations); and the embryonic but already fascinating citizen experiments in low-power FM and TV.

All of these noncommercial options now receive, deservedly, more extensive coverage than our first edition supplied. And we believe you'll find that true of our narratives and features throughout all chapters. They are still history-conscious but are infused with the new horizons, grounding this textbook both in the formative past and in the present with an eye toward the future. We have used our studies, experiences, and professional contacts to mine information about electronic-media structures, aims, and practices—noting how practice refracts against principles.

Beyond the results of our own efforts, the value and success of *Principles of Electronic Media* rest with you and your students.

SUPPLEMENTS

Not content merely to supply this revised textbook, we have developed four supplementary tools. We hope you and your students will find them valuable in organizing a collective learning experience.

For the Instructor

▶ *The Instructor's Manual and Test Bank*—This manual provides a suggested syllabus with a course outline, as well as ideas for research-paper assignments. It also includes supplemental lecture topics, lists of videotapes, and other resources. Also in the manual are test bank questions, including multiple-choice, true-false, and essay questions to test critical-thinking skills.

▶ *Computerized Test Bank*—This provides test questions electronically through our computerized testing system, TestGen EQ. The fully networkable test-generating software is now available in a multiplatform CD-ROM. The user-friendly interface enables instructors to view, edit, and add questions; transfer questions to tests; and print tests in a variety of fonts. Search and sort features allow instructors to locate questions quickly and arrange them in a preferred order.

▶ *PowerPoint Presentation*—This provides sets of key points for all chapters that can be displayed with strong visual emphasis to your classes. It can be found at the Instructor Resource Center at www.ablongman.com.

For the Student

▶ *Companion Website*—This is intended as an essential companion to the book. We outline each chapter, adding to it news, analysis, and further questions. Here is a multipurpose resource, available online around the clock to stimulate thought and discussion and to provide self-testing opportunities. Check it out at www. ablongman.com/davie2e.

Allyn & Bacon Supplements

▶ *Allyn & Bacon's Research Navigator™ Guide for Mass Communication*—This booklet by Ronald Roat, Southern Indiana University, includes tips, resources, activities, and URLs to help students. The first part introduces students to the basics of the Internet and the World Wide Web. Part Two includes over 30 Net activities that tie into the content of the text. Part Three lists hundreds of web resources for mass communication. The guide also includes information on how to correctly cite research and a guide to building an online glossary. In addition, the Research Navigator™ Guide contains a student access code for the Research Navigator™ database, offering students free, unlimited access to a collection of more than 25,000 discipline-specific articles from top-tier academic publications and peer-reviewed journals, as well as popular news publications and *The New York Times*. The guide is available packaged with new copies of the text.

▶ *Communication Tutor Center*—Our Tutor Center (access code required) provides free, one-on-one tutoring for students who purchase a new copy of participating Allyn & Bacon textbooks. Qualified instructors tutor students on all material covered in the texts. The approach is highly interactive, providing both knowledge of the academic discipline and methods for study of that discipline. Tutoring assistance is offered by phone, fax, Internet, and e-mail during Tutor Center hours. For more details and ordering information, please contact your Allyn & Bacon representative.

▶ *The Career Center*—This service is available to students using a selected Allyn & Bacon textbook. Qualified career specialists assist students who register for our service. Once registered, students are entitled to eight 30-minute sessions, a total of four hours of career consultant time. The Career Center is designed to address the wide range of preparation and life stages of individuals attempting to develop their career through education. Making the jump from being a student to becoming a professional is challenging for many reasons. Initiating this process is sometimes the most challenging step, and the Career Center is dedicated to making this transition as smooth as possible.

ACKNOWLEDGMENTS

We continue to be grateful for the wise counsel and constant support of our associates at this book's publisher, Allyn and Bacon. We thank our steadfast series editor, Molly Taylor (in it for the long haul); our patient and persistent development editor, Cheryl de Jong-Lambert; the warmly helpful production editor, Janet Domingo; editorial assistant Michael Kish, a man of many answers; the resourceful photo scout Kate Cook; the witty and merciful copyeditor Lynda Griffiths; Mandee Eckersley, a cheery friend in charge of marketing the book; Meaghan Minnick, who helped to round up new media professionals for us to interview; production technology supervisor David Ershun, who taught us about photo capturing online; and Larissa Tierney, who added her touch to the selection of photos.

Also of immense value to the publication and currency of this book have been a long series of reviewers who helped us stay true to the realities of electronic media and the needs of educators and students. They are Kenneth A. Fischer, Southern Illinois University at Carbondale; John L. MacKerron, Towson University; Antone J. Silvia, University of Rhode Island; and Ron Stotyn of William Paterson University.

Many, many other people in media professions and academe have provided photos, illustrations, permissions to publish, statistics, and other materials that proved helpful and even essential to us in telling the story of electronic media. On this list of partners are Carolyn Aguayo and Debbie Davis of KTLA; Steve Behrens of *Current* magazine; Bob Bruck of the *Owensboro* (KY) *Messenger-Inquirer;* Paul Capelli and Jonathan Klein of CNBC; Gayle Chisholm of KLCC-FM; Sarah Colley of Turner Broadcasting; Matt Heffernan of WTVR-TV; Kelly Hudgens of General Electric; Bojana Jevtic of Radio B92; Dennise M. Kowalczyk of KBOO-FM; Jo LaVerde of Nielsen Media Research; Amy Menard at KATC-TV; Wacuka Mungai of the Committee to Protect Journalists; Bob Papper of Ball State University; Vanessa Ruiz of WGBH-TV; Paul Sturlaugson of KXGN-TV; Al Tompkins and Bob Steele of the Poynter Institute; Richard Wiley, former Federal Communications Commission chair and now managing partner of Wiley, Rein & Fielding LLP; and Rick Chessen of the FCC.

Colleagues aplenty at our two academic homes gave us great boosts when we needed them as we revised this edition of *Principles of Electronic Media.* At the University of Louisiana at Lafayette, Professor Davie wishes to thank Department Head T. Michael Maher, Dean A. David Barry for his support, and Michael Gervais, chief engineer, for his technical advice and photographic eye. At the University of Oregon, Professor Upshaw thanks Dean Tim Gleason of the School of Journalism and Communication for continuing to accommodate his occasional need for time off to work on the textbook.

Topping all in their personal importance to us are family members and friends who excused us from social events, listened patiently to (and often improved on) our ideas, and voiced their support for this project at every turn. Without the love and patience of these dear and special people, the textbook we sometimes refer to by the acronym of *POEM2e* would not have come into existence.

Comments? Questions? Contact Us!

William R. Davie
wrdavie@louisiana.edu
(337) 482-6140

James R. Upshaw
Jupshaw@uoregon.edu
(541) 346-3745

Principles of
Electronic Media

Overview

We thought, because we had power, we had wisdom.

—STEPHEN VINCENT BENET, *American poet and novelist*

Chapter 1

It was a frightening war, largely because no one could be sure when it would end—or what it might be starting.

When U.S. forces attacked Iraq in March 2003—to remove dictator Saddam Hussein as part of a campaign against world terrorism—national controversy raged over whether it should be happening. Debate continued throughout the six-week "combat phase" and into a postwar occupation that triggered bloody guerilla violence.

Rich in action and personal risk and national import, all of this seemed ready-made for modern electronic media. And as the war dragged on, the media not only gathered news and transmitted it but they also *became* news:

- ▶ NBC correspondent David Bloom added high-tech video gear to an Army vehicle, placed himself in front of a remotely operated camera, and sent back vivid, rolling battleground reports, before his tragic death.
- ▶ Fifty-one journalists and media workers gave their lives covering the war, according to the Reporters Without Borders website.

- ▶ Arab media, led by Al-Jazeera TV, showed a different dimension of the conflict—close-ups of dead Iraqis, including civilians—that fanned anti-U.S. anger.
- ▶ Clear Channel, the largest owner of U.S. radio stations, hired a public-relations firm to fend off criticism that its top executives were drumming up pro-war demonstrations.
- ▶ High-speed Internet links brought millions of people heavily visual war reporting—"the coming of age of the broadband news medium," said one web-news executive.
- ▶ Nontraditional reports came from web logs, or "blogs," a new Internet phenomenon that some war journalists used to convey personal insights.

Flowing into our lives by electronic means—into TV sets, car radios, home computers, even cell phones—news reports, background stories, and political claims and arguments helped Americans reach conclusions about the war. However, while this national drama played out, the largest electronic media were waging a different battle among themselves—over prominence and consumer support.

■ Media Battles

Television and radio enterprises in the United States are mostly commercial businesses trading in *entertainment*. While they supply *information* as well, the great bulk of effort in this industry is aimed at diverting us and engaging us, for profit. This has not been as easy in recent years as it once was. New media channels are slicing the "mass" audience into smaller and smaller fragments, thinning the payoff for all companies.

Radio has fought back chiefly by hiring talk-show hosts who drive up listener temperatures. TV has sprouted dozens of unscripted "reality" shows. Still, audience loyalty is elusive, and ratings keep sliding downward. ABC apologized to its affiliated stations for coming in a distant third in reporting the launch of the Iraq war, but the network showed much deeper remorse over failing to support stations with strong prime-time entertainment programs. NBC's president Jeff Zucker admitted that, during TV's all-important fall season in 2003, "some of the programming just sucked."

Electronic media are a colorful, often fascinating field of study, as these facts and issues suggest. Today's electronic media both create and communicate much of our *popular culture*—the stories, ideas, and images that pass among us, fueling conversations, shaping tastes, feeding (or titillating) appetites.

For some people, modern media are irresistible, and for most, they remain inescapable. And because business is behind most of these enterprises, they push their messages on us continuously and relentlessly. We can't even reach all of the "off" switches needed to take electronic media out of our lives.

That alone makes it worthwhile for us to study these media closely and carefully—and there are many other reasons to begin.

STRICKEN STAR. NBC correspondent David Bloom was rushed from an Iraqi battle zone—and later died of a pulmonary embolism. Bloom had gained fame reporting from a specially equipped Army vehicle he helped refurbish. Long days cramped in it evidently hastened his death.

How Principles Matter

Our study will travel a road bounded by *principles*—what amount to guardrails that should help us stay on the main track.

One major dictionary defines a principle as "an accepted or professed rule of action or conduct." Thus, a principle is either an accepted rule—that is, it's in practice all or most of the time—or it's merely something that someone professes, or claims, to follow. The same dictionary notes that principles, taken together, can be simply a "basis of conduct" or a "guiding sense" of what's the right thing to do.

In short, there's wiggle room around principles (as we all surely know from the personal dilemmas we encounter). But if our electronic media profess a principle, do we want them to adhere to it—to obey the rules that follow from it?

We'll suggest a few principles in each of the chapters of this book. The first of these principles is that the complexity and subtlety of electronic media and their effects require us, most of all, to *think critically*. This sort of effort doesn't have to be especially difficult, but it must be focused and must remain at the center of the learning project.

Educators use the phrase *critical thinking* frequently and almost automatically, knowing how easy it is for anyone to slip into a read-and-memorize habit. And any business—in fact, almost any human organization—runs on some assumed principles that rarely are probed or questioned, much less followed faithfully.

A good example showed up in 1990 in a speech by the then-chairman of the Federal Communications Commission (FCC), which oversees U.S. broadcasting. He told an audience that "one of the most fundamental American broadcasting principles . . . is **localism.**"

The principle of localism? What's that? The chairman was referring to an understanding, encoded in the Communications Act of 1934, that broadcasting's first obligation is to its *local* community—the people that stations are licensed to serve. That idea may not seem controversial, but in fact whether it is a *principle* of broadcasting must be questioned, given the lessons of the past seven decades.

Critics long have complained that locally oriented radio and TV content is diminishing. One study found that less than one-half of 1 percent of broadcast time goes to local public affairs programming. (Local entertainment programming is skimpier still.) Broadcasters sometimes admit privately that they produce little local content because it draws small audiences and thus costs much more than it earns. That leaves hometown news as the only program-generating function in many TV stations. Radio has largely abandoned most local reporting. So, it has been argued, these media can profit from free use of public-owned airwaves with little regard to supporting their communities.

Is localism, then, really a principle of the business? And if it is, then, given that it evidently fails to guide some broadcasters' behavior, is the principle accepted as part of a basis of conduct, or is it merely professed?

We've tried to identify other accepted or professed principles of electronic media. Still, we think it is useful to remember that the existence and validity of any principle deserves to be questioned, not just in theory but in the real world we all inhabit.

PRINCIPLE 1

Understanding the work and effects of electronic media requires *critical thinking.*

PRINCIPLE 2

The power of any principle stems from the commitment of its users.

Localism Principle putting broadcasters' primary emphasis on serving local viewers or listeners.

■ The Basics

Let's start by breaking our central phrase, *electronic media*, in two and by reexamining both parts briefly. *Electronic*, as we'll use the term in this book, refers to the human manipulation of electronic energy—the flow of electrons, which can be found everywhere in the universe. We've harnessed them to create radio, television, the Internet, and more. Other electronic possibilities may lie ahead, adding to the variety of paths over which people communicate.

Media, for our purposes here, are conduits or carriers of human communication that can reach great numbers of people. Remember that *media* is the plural of *medium*. For example, the newspaper is just one medium; add in other mass communication forms, such as TV and the Internet, and you have media. We also apply the term *media* to the companies and enterprises that use and operate those paths and, usually, profit from generating content.

MESSAGES AND POWER

Electronic media basically are intermediaries; their primary functions are to record, clarify, and transmit messages among people. However, electronic media also may *create* messages. This can complicate our task of distinguishing reality from fantasy. For example, when broadcast journalists relay news to us, we usually process those messages as real and respond accordingly. Reversing this, when TV producers or radio disc jockeys invent outrageous skits or horror tales and present them to us as such, we usually process them accurately, as fantasy. This, though, can be a shaky undertaking, because electronic media deliver both factual and fictional material and they sometimes sound and look very similar.

The most famous example of this phenomenon occurred back in 1938, when actor-producer Orson Welles electrified a national radio audience with a play about Martians landing on earth. Welles said more than once that it was a fictional yarn, but some people packed their cars and raced to remote areas to wait out the alien invasion that hadn't really happened. Since then, television has presented similarly terrifying and realistic stories, and invariably some viewers miss the disclaimers and become frightened. All of this amounts to a warning: Electronic media can be terribly powerful, sometimes in unintended ways.

OBSERVATIONS

The breakdown of the term *electronic media* leads us to three observations:

1. There's nothing especially new about human use of electrons. People had been thinking about them long before the legend of Ben Franklin, his kite, and a bolt of lightning. Most modern uses of electrons to transmit pictures and sounds are derived from basic discoveries made in centuries past, as this book will explain.
2. The messages (all sorts of "content") we receive through electronic media seem to have been shaped, sometimes in subtle ways, by the nature of the media themselves. Why, after all, do some of us prefer to read the news on paper or the Internet, whereas others prefer to hear it on radio or see it on TV? Here are three possible reasons:

   ▶ Because each form of media has a built-in limitation. *Space*-limited newspapers and magazines can provide only so many details of the news; *time*-limited broadcast media can give us only so much understanding of it.
   ▶ Because personal tastes and situations vary. Some people like to see or hear news in brief, relatively dramatic form. Others like to read it, processing its meaning at their leisure.

▶ Because our brains distinguish among different media. *Hearing* words and sounds on the radio, we make "pictures" in our heads to represent reality or fiction.[1] *Seeing* television lets us fuse both audio and video in our minds to construct a sense of reality or fiction. *Reading* print media is yet a third mental exercise, in which we intellectually "make meaning" that we believe decodes fact or fiction.

3. An observation central to this book: How we interact with electronic media helps to give them great influence on our lives and culture. Even if we doubt critics who claim that media and their advertisers are out to bend us to their will, we each need to ask: How many minutes or hours today did I spend with radio, TV, or the Internet?

IDENTITIES

It's hard not to have a strong identity when your hometown is Glendive, Montana. Glendive likes to call attention to its charms and its history: Lewis and Clark stopped there in 1806, as did General Custer before meeting his end at the Little Big Horn. The surroundings are rugged and picturesque; the Yellowstone River flows through town. Some 6,000 people live in Glendive, often joined by visiting anglers. However, the town is shrinking. Located at the edge of the Badlands, it's an old hub for grain growers and doesn't draw many new settlers. Perhaps appropriately, one of its chief claims to fame is the paddlefish, a spoon-billed creature that thrives in waters nearby and happens to be prehistoric.

KXGN-TV (GLENDIVE, MT). You can't expect architectural grandeur from #210 on Nielsen's list of 210 U.S. television markets. Little KXGN-TV brings both NBC and CBS programs to viewers in rural eastern Montana.

The "Market" Factor

Glendive is the smallest television **market** in the United States. That means it's the smallest distinct population of potential consumers of advertising—ordinary people—served by any U.S. television station, according to Nielsen Media Research. The town and a few smaller ones nearby add up to market number 210 of 210,

Market In broadcasting, a geographically discrete potential audience, usually identified by the name of its town, city, or metro area.

PULLING POWER. A single radio set has the power to draw many people together. The resulting audience can hear not only talk, music, or information but also the advertising that supports commercial media.

TABLE 1.1	Sample of U.S. Television Markets (2004–2005)		
RANK	**DESIGNATED MARKET AREA (DMA)**	**TV HOMES**	**% OF US**
Large Market Areas			
1	New York, NY	7,355,710	6.712
2	Los Angeles, CA	5,431,140	4.956
3	Chicago, IL	3,417,330	3.118
4	Philadelphia, PA	2,919,410	2.664
5	Boston, MA (Manchester, NH)	2,391,840	2.183
Middle Market Areas			
71	Honolulu, HI	417,120	0.381
72	Tuscon (Sierra Vista), AZ	417,070	0.381
73	Des Moines–Ames, IA	412,230	0.376
74	Portland–Auburn, ME	409,060	0.373
75	Rochester, NY	396,880	0.362
Small Market Areas			
206	Helena, MT	25,360	0.023
207	Juneau, AK	25,070	0.023
208	Alpena, MI	17,930	0.016
209	North Platte, NE	15,590	0.014
210	Glendive, MT	5,150	0.005

Source: Nielsen Media Research.

LOCAL REPORTER-PRODUCER.
Darcie Fisher of Boston's WCVB-TV prepares "Healthwatch" segments for the newscasts—an example of service to local audiences, a traditional duty of broadcasters.

Consolidation The act of bringing together different entities or parts into a single or unified whole, as when media owners buy or take control of competing or complementary companies.

and are unlikely to move up the list (see Table 1.1). The lone TV station, KXGN, brings in network programming and also broadcasts local news. Never mind that far-off media corporations and big-city stations dwarf it; little KXGN serves the people of Glendive and thus plays a part in the colorful tapestry of U.S. electronic media.

Wherever they are, in large cities or small, electronic media strive for strong identities. That's because all of them have one common objective: to catch and hold our attention. Most channels are commercial; they need to gather us into measurable audiences to "sell" to advertisers, by far the predominant source of electronic-media revenue. To achieve this, traditional broadcast networks (ABC, CBS, NBC, and others) and some cable channels and radio stations seek the widest possible audience appeal—that's why we call them *mass* media.

That term is being redefined, however, as the media marketplace changes from one into many. In recent years, with the growth of newer networks, specialized channels, and Internet sites, the mass audience has been breaking into fragments. Each of these clusters has its own characteristics and interests, making it an identifiable target for tightly focused advertising—and a magnet to entrepreneurs who may even tailor a new channel to fit the niche.

Interestingly, despite their diverse programming in recent times, electronic-media companies retain a monolithic image among many Americans. It's easier to praise or condemn stations, networks, or Internet media enterprises as just "the media"—indeed, it's easier to *remember* them that way—than as individual trees in an electronic forest. After all, many companies have changed their identities through mergers with other companies, complex deals that can be difficult to sort out. These moves toward media **consolidation** have allowed some relatively small companies to reach broader—even global—advertising markets, and have created efficiencies that increase profits for stockholders. They also have left a nucleus of mammoth corporations at the core of U.S. media control.

BRIAN WILLIAMS. This NBC News anchor succeeded chief anchor Tom Brokaw in 2004. Williams had spent years in lesser roles for NBC and its cable cousins MSNBC and CNBC.

Electronic-Media Roles

In a process known loosely as **globalization,** U.S.-based electronic media are helping to construct a complex—and often controversial—web around the earth. To use transoceanic telephone lines or transcontinental cables, of course, is nothing new. However, for growing media giants to bring a rainbow of radio, TV, and Internet programming to dozens of countries, via satellite and with digital services to boot, is new indeed.

Electronic media never could be described very precisely in terms of *place;* their signals tend to travel too far to confine their impact to a single neighborhood or even a town.[2] Now, geography has become almost immaterial, as broadcast programming flies into previously unserved or underserved areas of the globe. *Time* is a different matter; it still both marks and limits not only messages and programs but also industry developments. Electronic media haven't been around forever and they didn't all arrive at once. Each came into being in an appropriate moment in time because people figured out how to make them and because they met pent-up personal needs.

FILLING OUR NEEDS

The main importance of our media hinges on their ability to fill demands that have existed throughout history, but that only in the past century or so have been seriously addressed. Electronic technology made that possible.

Instant News

The September 11, 2001, attacks on the World Trade Center and the Pentagon were unprecedented; the country never had experienced large-scale terrorism. The attacks jerked the media into action and prompted Americans to use their media as never before. Fear for their own families' safety and for the security of everyday life was paramount. Panic was rare but shock and concern were almost universal.

Most Americans were working, commuting, in school, or just waking up when the carnage occurred, and they learned of it on the radio. Then came wall-to-wall

Globalization Movement to deliver programming and other media services to many countries, often through international corporate partnerships or mergers.

protalk
Patty Talahongva
BROADCAST JOURNALIST

Journalist Patty Talahongva didn't follow the herd when she covered huge, devastating wildfires in northwestern Arizona. While most reporters worked out of the town of Show Low, she headed for a place called Whiteriver.

"Two fires had merged into one huge inferno, and pretty much the whole mountain community was evacuated," she recalls. "What we weren't hearing on the mainstream news was that both fires started on the Fort Apache Indian reservation. Whiteriver is their 'capital.'

"The Apaches have the the biggest stand of Ponderosa pine trees in the country. That's their number-one industry. Their livelihood was burning down . . . it will take 200 years for them to rebuild and recover from the fires."

Talahongva chose to cover Whiteriver because, as an experienced Native American journalist, she suspected it would be "under-covered" by the mainstream media. She and a friend from the *New York Times* wound up covering the fire from the Apache perspective.

Such opportunities to raise consciousness about Native life and its challenges bind Talahongva to her career. As a child, she loved writing, and she got a paying job as a

"I was hooked on broadcast journalism and loved the intensity of TV news," she says. "I also liked *showing* people what we were talking about rather than just writing about an event, person, place or issue."

print journalist at the age of 16. Still unsettled, in college she thought hard about law school, but, as a senior, landed a job in television.

"I was hooked on broadcast journalism and loved the intensity of TV news," she says. "I also liked *showing* people what we were talking about rather than just writing about an event, person, place or issue."

Talahongva put more than 16 years into television, bringing to it a sense of mission. Today, nationally known, she leads the Native American Journalists Association. She is a producer of "Native America Calling," a live talk program linking public radio, the Internet, and listeners in the United States and Canada in discussion of Native issues. In addition, she produces documentaries and writes for magazines. In every setting, she mentors young Native journalists, seeding the future.

Her Hopi Indian name—which is much longer than "Talahongva"—means "white spider girl." It's based on a legend that someday people would be connected by a giant spider web in the sky—a name that seemed to foretell her on-line journalism profession:

"When I was in TV I thought that story foretold of satellites. . . . However, what do we have today? We have the World Wide *Web!* I often wonder how my Elders knew this would happen far in the future. And I believe this is my calling."

TV news coverage, shoving Tuesday soap operas and then evening comedies and dramas off the air. All the while, millions of people pounded their computer keyboards, searching out details of the disaster according to their own practical and psychological needs.

These responses came as no surprise to media researchers in the electronic age. People always rush desperately to find news of a disaster in which their loved ones may have been injured or worse. What's more, the peculiar directness of electronic media—plus what seems to be a perceived intimacy—draws people to radio and TV in particular, for various deep-seated reasons. Families of air-crash victims, for example, often find healing in viewing news reports repeatedly.

Even in normal times, we crave some types of news, defined broadly as fresh information about human experience and the ways in which people pass their days. **Cultural** programs that send cameras roaming the world are part of this fabric, as are thoughtful **documentary** treatments—long-form, in-depth examinations of ordinary high schools and hospitals, for example—and replays of history. Media companies have created **voyeuristic** Internet news sites, radio talk shows on which we can hear and argue with others about social issues, TV newsmagazines that trace real human stories, and "reality" shows that pit eager contestants against nature.

By now it's clear that many of today's offerings from the electronic media spill across old boundaries between news and entertainment. In fact, some have dropped boundaries and become blenders. Newsmagazines such as *Dateline NBC* use Hollywood-style visual and aural techniques to heighten their viewers' emotions over stories of true crime and personal tragedy. Popular radio hosts mix serious issues with colorful sexual references and personal attacks. On CNN's *Crossfire*, and similar talkshows, political analysts engage their opponents in loud, arm-waving arguments that discourage contemplation.

NEWS HELICOPTERS. When the terrorism of September 11, 2001, raised security concerns about the skies over the nation's cities, the government grounded one television news tool. The ban on "chopper" flights lasted almost three months.

Diversion and Enrichment

Recent trends may trouble journalists, but broadcast managers see them as necessary concessions to pressure for **ratings.** However, more significant than these dry head-count numbers are the needs and urges of the people behind them: **"Infotainment"** draws millions of fans who evidently distill both the serious and the frivolous from these programs—or at least don't allow one to obscure the other.

Many people do not seem especially concerned with placing content in neat categories. Rather than focusing on news and information needs, they seek one or both of two qualities that aren't as easily boxed in: diversion and enrichment. The great majority of electronic-media content can be seen as meeting at least one of these two needs.

Diversion may be described as the process of having our attention moved from whatever's bothering us to something more pleasant. News and analysis often bother people as badly as budgets, bosses, and boyfriends or girlfriends do—and they turn it off. That helps explain why radio offers many hours a day of music as diversion while TV presents scores of comedies, game shows, and cop dramas.

Enrichment, the addition of value or significance to our lives,[3] is a more elusive concept than diversion. Commercial broadcasting is structured mainly to profit from light entertainment, not from the enrichment offered by profound music, extended and probing news treatments, arts displays, and sociopolitical documentaries. Public broadcasting favors such fare but has been pushed toward airy content,

Cultural Pertaining to the culture—the sum total of behaviors, customs, and beliefs—of a particular society or group.

Documentary Based on or relating to an actual person, era, or event and presented factually and without fictional content.

Voyeuristic Pertaining to the practice of self-gratification by viewing certain acts or situations intended to be private; traditionally applied to sexual acts, but recently broadened in media contexts to include other invasions of privacy.

Rating Size of the audience (usually in households) for a particular program.

Infotainment Blend of news (or information) and entertainment.

Diversion The function that electronic media perform when they distract viewers from unpleasant or bothersome subjects.

Enrichment The process of adding value or significance to one's life, a function sometimes performed by electronic media.

DAN RATHER. "Discovered" covering a hurricane for a Houston station in 1961, Rather built a bright network career. He resigned as top CBS anchor after broadcasting unsubstantiated negative reports about President George W. Bush's National Guard service.

including British comedies and *Car Talk*-style radio banter, by constantly growing pressures for bigger audiences and more donations.

However, to halt discussion here would mean taking a narrow—even elitist—view of enrichment. Whether we've received added value is determined not just by where we end up but by where we *begin*. Most of what we see and hear about worlds beyond our usual routines can add substance to our lives. People with few contacts outside the home may find all broadcast programs enriching, and those of us who have "surfed" onto unfamiliar TV terrain have found that simply allowing exotic messages to roll across our brains can broaden our knowledge.

Books may be best at imparting enrichment, but it can also be found in broadcasting and is readily accessible through the Internet media, with their click-click interactivity and bottomless information pits. Their potential to connect us to troves of learning has only started to develop. Meanwhile, the older electronic media provide diversion and some measure of enrichment to a rushed society accustomed to *multitasking*—doing many things at once.

A Special Need

Media scholars have found evidence that we all have a built-in **surveillance** instinct, a need to keep an eye on our surroundings. Communications scholar Pamela J. Shoemaker links this to human biological and cultural evolution and calls it being "hardwired" for news.[4] Surveillance requires us to go beyond what we usually define as news to a broader combination of information and impressions of the world we inhabit. To keep our balance in that world, so full of strangers and unexpected events, we constantly scan the people, things, and spaces around us. We are looking for reassurance, for a zone of safety, often not realizing, or only half-realizing, that we're doing so.

Surveillance Observation or monitoring (of the human environment).

ENTERTAINMENT SHOWS. Light TV fare such as comedies has helped people escape their cares, a valuable function in times of war and terrorism. The Emmy-winning *Arrested Development* treats tense situations in an eccentric family with doses of edgy humor.

Within this sphere of truth, rumor, and hope, electronic media hold an important place. Their directness and emotional power have helped them excel at filling our surveillance needs quickly, at presenting and sometimes even helping us sort out and store away the meanings of each day's events. Historic newspaper headlines are fascinating relics, but more often it's radio voices and television images that have burned incidents into our memory. Profiles of government officials, entrepreneurs, teachers, and firefighters—people against whom we can assess our own lives—all contribute to our intellectual-emotional loading.

Television often catches us at our peak levels of fear or concern and shows us vividly what has happened; that is its assumed mission. In extreme cases, this can be alarming and even damaging, as when too much local TV crime news frightens some people into hiding in their homes.[5] Most of the time, however, if our minds are otherwise in order, we take in stride the fruits of our surveillance and use them to keep living successfully.

PRINCIPLE 3

Electronic media supply clues to where we stand in our "surveillance" society.

An Industry "Morphs"

Today's media occupy such a large share of our popular culture—the ongoing stream of public messages, images, and social symbols in which all of us swim through our days—that each of their decisions ripples far across the cultural landscape. The way we flocked to the media in the aftermath of the 2001 terrorism onslaught helped to demonstrate this; we interact with these institutions almost automatically. What they do, for their own commercial or other purposes, has consequences for us. When one company or group takes an action that works, others are sure to imitate it, multiplying the impact on society. For example, Tribune Company's acquisitions in television, cable, newspapers, and other enterprises help Tribune draw revenue from dozens of communities—large and small—and spur rival corporations to make similar moves (see Figure 1.1).

Every day the business and entertainment pages of newspapers (and websites) ripple with changes brought about by the rush of technology, economic

Broadcast and Cable

Television

WPIX	New York
KTLA	Los Angeles
WGN	Chicago
WPHL	Philadelphia
WLVI	Boston
KDAF	Dallas
WATL	Atlanta
KHWB	Houston
KCPQ	Seattle
KTWB	Seattle
WBZL	Miami/Ft. Lauderdale
KWGN	Denver
KTXL	Sacramento
WXIN	Indianapolis
WTTV	Indianapolis
KSWB	San Diego
WTIC	Hartford/New Haven
WTXX	Hartford
WXMI	Grand Rapids

WGNO	New Orleans
WNOL	New Orleans
WPMT	Harrisburg
WBDC	Washington
WEWB	Albany
KPLR	St. Louis
KWBP	Portland
WB Network (partial stake)	
Tribune Entertainment	

Cable
WGN-TV
Chicago Land Television
Central Florida News

Radio
WGN-AM (Chicago)

Newspapers
Newsday (Long Island, NY)
Los Angeles Times
Chicago Tribune
Baltimore Sun

South Florida Sun-Sentinel
Orlando Sentinel
The Hartford Courant
The Morning Call (Allentown, PA)
Daily Press (Hampton Roads, VA)
The Advocate (Stamford, CT)
Greenwich Time (CT)
La Opinion (Los Angeles)
Exito (Chicago)
Hoy (New York)
El Sentinel (Orlando)

Other
Chicago Cubs
Tribune Media Services
Classified Ventures (partial)
Brass Ring
Zap 2 It
BlackVoices.com
Chicago magazine

FIGURE 1.1 *One Corporation's Holdings: Tribune Company (July 2003)*

Sources: Columbia Journalism Review and "Who Owns What" website, Aaron J. Moore, creator.

opportunities, individual ambitions, and sociocultural flux. We can track changes on several fronts, including:

- *Audiences.* With Hispanic immigrants pouring into major U.S. cities, increasingly important Spanish-language radio stations compete furiously, some of them pushing limits on sexual and profane speech to grab audiences.
- *Choices.* Internet news sites go through a "shakeout," with some sites shutting down, but overall they enjoy healthy growth; they do especially well among viewers in their thirties, and have pushed video **streaming** (playback of on-line video onto computer screens) into the limelight since the 2001 terrorist attacks.
- *Corporate reach.* Mogul Rupert Murdoch's News Corporation owns *two* television stations in 9 of the top 16 U.S. markets. It also owns the DirecTV satellite service, Fox Motion Pictures, British Sky Broadcasting, and television systems in Australia, Italy, Japan, and Latin America. Power over multiple voices in broadcasting continues to gravitate toward a few of the largest corporations, upsetting critics of "big media."

▐ The Book and the Quest

Important questions arise from the riddles and gaps of uncharted territories exposed to one's critical thinking. That is a process this book is intended to launch. Some questions involve not just media functions but something more human—the communities and people affected by them. What follows is an example of this, a real situation from the television industry.

MOTIVES AND A COLD DAY

Will Wright was news director of WWOR-TV in Secaucus, New Jersey, when, one winter day, a woman fell through the ice. News crews raced to the scene. The woman was rescued. She spoke only Spanish. When video of the incident arrived at the station, a writer-producer was assigned to supervise its editing. In doing so, the writer *permitted other sound to cover the rescued woman's Spanish words.* The news director saw this on the air and was displeased.

"I was bemoaning the fact that the writer didn't let the Spanish portion pass through," Wright said the next day. "That, to me, showed disrespect for that woman's culture. Are we so tight (on air time) tonight that we can't let at least 10 seconds of that language go on the air?"

So, a well-meaning manager was irked because one story on one news day omitted a touch of Spanish. That's what we learn from the anecdote—but what more do we need to learn? Here are some questions and answers to consider:

- *What is the station's audience?* WWOR-TV, based in New Jersey, competes for viewers in the massive New York City–area television market.
- *OK, but what are the **demographics** (the key characteristics of the audience)?* The market includes more than one million Hispanic households, about 11 percent of the total.[6]
- *Did Will Wright care about that audience?* Restate the question more precisely, please.
- *Did Wright care personally about the Hispanic community?* No one could read his mind, but he was a well-known contributor to minority causes around New York and an advocate for better minority coverage by the broadcast industry. After the rescued-woman incident, he exclaimed, "Surely the entire Spanish-Latino culture in New York is worth 10 seconds!"
- *Could other factors also have been on Wright's mind?* Yes, ratings, for one; market positioning, for another. Wright was building WWOR-TV's image as a champion of minority news and issues. It was a gamble, but he believed in it, and WWOR-TV did need to distinguish itself from its competitors.

Streaming Digital transmission of audio or video to computer, using software decoding tool; user sees or hears transmission during download, rather than waiting for playback.

Demographics Audience characteristics (age, gender, race, etc.) used to identify "target" audiences or consumers at which advertising can be aimed.

▶ *Still, why would a news director react so strongly to a simple editing decision?* The station and the group that owned it were to be acquired soon by Fox, which was sure to cut jobs while merging operations. Consequently, every daily decision bearing on audience perceptions and support could affect the fate of Wright and his staff. At least he hardly could be blamed for being nervous.

Indeed, Will Wright later left his job after Fox bought the station and brought in its own executives. Wright had earned credit, according to industry sources, for "attracting to the newscast the young and largely urban minority audience," and soon became an executive news producer at Black Entertainment Television (BET).

PRINCIPLE 4

This textbook's value rests on the *new* questions it inspires in its readers.

Summary

Electronic media are far more to us than suppliers of entertainment and news. They meet social and psychological needs of which we barely are aware but which are fundamental to our lives, individually and collectively. Radio, television, and the Internet both mediate and originate communication. No longer merely witnesses or storytellers, they are massive producers of culture and they are economic players; they are *active* in shaping society.

These media are connected to us through their principles—some merely professed, some usually followed; some historically grounded and some invented to suit the times. Whether and how a tele-vision station should report on violent crime, for example, are among the questions best answered through the application of principles. Media responsibility and accountability are never guaranteed.

None of us can measure the full influence of electronic media simply by consuming their products. Because media largely are businesses and adapt their behaviors to what their markets require, we can't expect to understand and keep up with them unless we examine them closely. Beyond the basic act of study—and ultimately far more important—lie daily monitoring, probing inquiry, and constant critical thinking.

Food for Thought

1. Is confusion of media fact with fantasy a problem? Do streams of true and false information sometimes merge? What solutions do you see?
2. What kinds of principles should we expect in our electronic media? List a few and explain their potential importance as we consume media content.
3. Which of three content types—news, diversion, or enrichment—dominate your use of elec-tronic media? Does this help in balancing your life? Explain.
4. Do you agree with Will Wright regarding the rescue story about the Hispanic woman? How would you address viewers' ethnic sensitivities? Explain your answer.

History *of* Radio *and* Television

> *The human brain must continue to frame the problems for the electronic machine to solve.*
>
> —DAVID SARNOFF, *broadcasting pioneer*

> *What we are doing is satisfying the American public. That's our job. I always say we have to give most of the people what they want most of the time. That's what they expect from us.*
>
> —WILLIAM S. PALEY, *builder of CBS*

Chapter 2

It seemed like a magician's trick at first: Two telegraph keys were clicking coded messages that pulsated through a wire linking distant operators, but what if the wire were removed? Could you make the two telegraphs still talk to each other? That was the big question a number of late–nineteenth-century "magicians" hoped to answer. How these inventors tried to solve the mystery was by reading about each other's experiments, trying out each other's ideas, and adding their own secret solutions. One of them experimented in the Blue Ridge Mountains, another one tinkered with the machines of New York's electricity companies, while a third labored in the Tuscan hills of Italy.

Mahlon Loomis was a dentist by trade, and he became fascinated by electricity. After bouncing radio waves between two flying kites in the hills of West Virginia, Loomis won an early patent for his wireless telegraphy in 1872. Later, a Serbian inventor from Croatia came to the United States and designed a "four-tuned radio circuit." Nikola Tesla applied his ideas about electricity while working with Thomas Edison and later at Westinghouse.

Meanwhile, in Italy, Guglielmo Marconi experimented in "wireless telegraphy," seeking to send messages between a station on shore and ships at sea. It was first in Great Britain and then in the United States that Marconi became a business success. All three men contributed to what we now take for granted: *wireless communication.* Whether it is between cell phones, computers, or a satellite in space and a home television set, the principles of electromagnetic energy are the same. It just took a number of brilliant minds to unlock the secrets.

No single experiment or inventor held the key to the wonders of wireless electricity; there were many doors and many keys. The birth of our electronic media was gradual, disjointed, and messy. What humans brought into being—the power to send sounds, pictures, and more across oceans and continents and then around the globe—would transform human society. Government was there to encourage some of the experiments, but it took American enterprise to capitalize on the inventions.

Once broadcasting became viable, it experienced periods of explosive growth requiring important decisions that would influence the progress of innovation. In content, broadcasting at first experimented, then it exploited its popularity; today, surrounded by formidable rivals, it tends to follow rather than lead social trends.

The 1800s: Broadcasting's Prelife

A critical event occurred on May 24, 1844, in the chambers of the U.S. Supreme Court. That's when a preacher's son from Massachusetts clicked a verse of scripture on a telegraph key: "What hath God wrought?" The question seemed appropriate as Samuel Finley Breese Morse opened and closed his circuit breaker to transmit those words from Washington, DC, to Baltimore's Mount Clare Railroad Station.

Morse had hoped the government would then take over and operate his invention as a national system. Instead, entrepreneurs rushed in to capitalize on this early form of "dot-com" business. Western Union and American Telegraph Company launched the telecommunications industry—while fortune again eluded Morse. He later told a friend that the telegraph brought him only "litigation, litigation, litigation."[1] Even though the U.S. government is willing to bankroll experiments for the good of science and commerce, Morse's story shows how private industry is needed to build the enterprise.

PRINCIPLE 1

In the United States, government may fund electronic-media experiments, but it's up to business to apply the results.

RADIO

The transmission of Samuel Morse's language of dots and dashes was a startling and path-breaking development. It foreshadowed radio and later forms of digital media, but rather distantly in time. More than 30 years would pass before a speech teacher for the deaf—Alexander Graham Bell—attempted the next leap forward: a machine to amplify the human voice. First, some crucial discoveries were needed.

Maxwell's "Ether"

An early revelation came from one of Bell's fellow Scots, a man named *James Clerk Maxwell*. He was investigating **electromagnetic waves**, which seemed promising as an invisible means of telecommunication. Maxwell wondered how an electric current could act like a magnet on a compass, and how these currents could spread out in fields. Could it be, Maxwell wondered, that electromagnetic waves undulated in an invisible ocean of currents—"a luminiferous ether"? His "ether," it seems, was just a metaphor, but the idea provoked other scientists to investigate electromagnetic waves in order to uncover their true identity and powers.

Bell's Tele-Tinkering

If *Alexander Graham Bell* was born to be a great tinkerer, it wasn't apparent in his early years. Bell wished to follow in the footsteps of his father—an educator in Scotland renowned for teaching the deaf how to use visible speech. As Bell read about electromagnetic experiments in telegraphy, he thought about the possibilities of transmitting *sounds* by telegraph. Suppose a harmonic machine could be built to vibrate in response to a voice, as the human eardrum does; could that help the deaf to hear?

In 1875, Bell, his future father-in-law, and a local businessman pooled their resources to build such a machine. Thomas A. Watson signed on to help solve mechanical problems. On March 7, 1876, Bell secured a patent for a blueprint to transmit vocal sounds telegraphically. Three days later, he recorded in his diary the moment when he first exclaimed by phone, "Mr. Watson, come here, I want you." Bell showed off his device at the Philadelphia Centennial Exposition, where it was ignored until the emperor of Brazil paused by the telephone to observe, "My word, it talks! "[2]

Electromagnetic wave Wave produced by accelerating an electric charge, manipulating the charge by changing the intensity of nearby electric or magnetic fields.

ALEXANDER GRAHAM BELL. Bell never lost interest in helping the deaf to hear, which led him to create something of immense social impact—the telephone, in 1876. Bell showed that voices could be transmitted through electrical wires.

Bell and his partners formed a company in 1877, introducing the name *telephone*. Seven years later, they created a subsidiary to handle long-distance communications: American Telephone & Telegraph Company. However, once the telephone drew attention as another method of making money, Bell lost interest in his invention and resumed teaching the deaf. It was others who developed the business of "harmonic telegraphy," and turned the telephone into the basis of a national communication system.

Hertz Makes Waves

Researchers continued work that eventually would tie together the strands of future electronic media. In Germany, *Heinrich Rudolph Hertz* decided to make the invisible visible, permitting people to see electromagnetic waves. Hertz used Maxwell's formulas to create something called a **spark-gap detector.** He would use it along with **oscillating** currents to turn his ideas to reality.

In a corner of his Berlin classroom in 1888, Hertz electrified a pair of metal rods. When the rods took on opposite charges, they sparked, and waves oscillated back and forth. Hertz had verified Maxwell's theorems. Electromagnetic waves existed, usually unseen, in about the form Maxwell had predicted. Interestingly, the scientist seemingly was unaware of his experiment's practical implications—and thus its historical importance. His name would endure, however, as the label for one cycle of electromagnetic energy. Broadcast engineers today still call one cycle a **hertz.**[3] This professor revealed how electromagnetic energy exists in the atmosphere as an invisible force, and that it can be bounced with precision between two points.

Message from Marconi

Meanwhile, in Bologna, Italy, a physics professor was delving into the mysteries Maxwell had calculated and Hertz had demonstrated with his own spark-gap

PRINCIPLE **2**

The key to telecommunications is that electromagnetic energy can be transmitted between two points.

Spark-gap detector Device that revealed electromagnetic radiation by making a spark jump a gap, emitting waves that triggered another spark-jump some distance away.

Oscillating Swinging back and forth, like energy in an alternating current.

Hertz Cycle of a radio wave; honors Heinrich Hertz for demonstrating waves' existence. (One *kiloHertz* = 1,000 waves per second; one *megaHertz* = 1,000,000 waves per second.)

The role of a twenty-first-century media device linked to the inventor of the telephone dispatched news around the globe. Correspondents in Afghanistan in 2001, covering the fall of the Taliban regime, sent back odd-looking video—pictures that broke into digital fragments off and on, words falling out of synch with a speaker's lips. Its quality was less than desirable, but the content was *news* that otherwise would not have "made air." Later, after the 9/11 terrorist attacks staggered telecommunications in Manhattan, the same technology kept some reporters on our screens. It popped up again in Iraq before, during, and after the 2003 U.S. military strikes.

This was the **videophone,** a great journalistic tool when the scene of a news story was distant or in chaos. It stemmed from a device created in 1880 by Alexander Graham Bell to help the deaf hear. Bell's instrument—the "photophone"—transmitted sound by way of vibrations in a beam of light. Much later, engineers perfected it and made it portable. In 1999, Kyocera Corporation started selling the videophone (essentially a TV camera linked to a computerized satellite-phone system) and, two years later, its transmissions danced across our TV screens.

The videophone's greatest asset was its ability to transmit *something* visual out of remote crannies where battles raged. The system was battery operated, quick to put up or tear down, compact, and easy to hide—a reassuring quality to journalists in hot spots. Besides that, it worked "live" and looked digital in an age when digital was trendy. Best of all to broadcast executives, a videophone unit costs a fraction of what they paid to buy, equip, and operate most conventional satellite "earth stations."

GUGLIELMO MARCONI. Marconi envisioned the use of radio waves for wireless telegraphy and soon spanned the ocean with messages in Morse Code, the first of them merely letters of the alphabet.

detector. While academics are often absorbed by the excitement of theoretical discovery, their students may see more practical applications.

Guglielmo Marconi had read about electromagnetic experiments, and even dropped in on university lectures. He began prodding professors to answer what we might call "real-world" questions. Marconi considered whether and how a *wireless telegraph* could send messages to warn ships of threatening seas, or to relay reports from distant battlefields. These large notions propelled Marconi, and he made the family granary his laboratory to begin trying out his ideas.

On a summer day in 1895, Marconi sent his older brother and a tenant farmer to carry a radio receiver to the edge of the family vineyard. They were to respond with a rifle blast if they heard the telegraph receiver. Marconi sent a simple message and was rewarded with the boom of a gunshot. That sharp sound from a Tuscan hillside marked the true beginning of radio, and thus of all its descendants.

Marconi formed his first wireless radio company in London in 1897. Two years later, seeking North American sponsors for new investments, he arrived in the United States and incorporated the American Marconi Wireless Telegraph Company. Driven to transmit over greater distances, he sent the letter S across the Atlantic in Morse code by wireless in 1901, and began building a stronger amplifier for future transmissions.[4]

TELEVISION

The 1800s found dreams of television drawing strength from the wireless and other inventions. Efforts to reproduce images from life had begun long before; by the

1830s, French artist and chemist Louis-Jacque-Mandé Daguerre was perfecting photography, and those early photos became known as *daguerrotypes.* Over the next 50 years, other advances that foreshadowed *moving* images would emerge in Europe and the United States.

Scanning

Some innovations, such as a way of sending telegraph signals that would reproduce pictures or symbols, showed little potential. Others were more tantalizing. A German group discovered how the **cathode ray**—a focused beam of energy from the electrode inside a **vacuum tube**—could project shadows on the tube. As Thomas Edison's phonograph made sound recording possible, telephone creator Alexander Graham Bell and others labored to transmit not just sounds but pictures.

A vital contribution came, like the daguerrotype, from France: Maurice Leblanc's 1880 revelation of **scanning.** It was a technique, still crucial today, through which an image could be broken systematically into lines that then were translated into energy. Scientists and engineers were realizing that only by taking an image apart could they translate it into forms that could be transmitted to another place, where it then could be reassembled.

A German innovator, Paul Nipkow, soon employed scanning with encouraging results. In 1884, he patented a system of perforated disks, rotating at constant speeds, through which images could be transferred from their source to a viewing device. The mechanical approach embodied by the "Nipkow disk" lent great momentum to TV's development.

De Forest's Tube

Mechanical means of giving the magic of mobility to pictures eventually would give way to the *electronic* transmission of images. As the nineteenth century ended, Lee De Forest was advancing that cause—first by way of radio. This new medium needed some way to amplify a signal as it entered the receiver inside a radio set. De Forest, an Alabama minister's son and engineering graduate, had an answer: a glass tube containing a new type of **electrode.** He proudly dubbed his "triode" bulb the **Audion,** and patented it in 1906.

De Forest faced a fight over his claim to the invention—among the first of his many legal battles. One was a dispute with Columbia University Professor Edwin Howard Armstrong over the critical circuit that increased radio signal strength. That fight would rage on until 1934, when the U.S. Supreme Court awarded patent rights to De Forest.

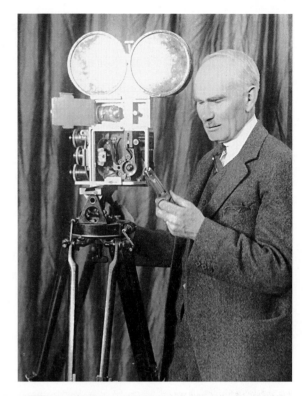

LEE DE FOREST. This American inventor pursued research on the vacuum tube—crucial to TV—by adding a third element to a glass tube, creating a "triode." It amplified and controlled electromagnetic signals.

Videophone Portable camera-computer unit that can transmit news via satellite.

Cathode ray Focused beam of energy from the electrode inside a vacuum tube that projects shadows (images) on the tube.

Vacuum tube Electron tube used to amplify signals; made obsolete by transistor for most uses (the cathode ray tube in TV is a vacuum tube).

Scanning Left-to-right motion of an electron beam across the face of the cathode ray tube; modulating the color and intensity of the beam produces TV images.

Electrode Conductor through which an electrical current enters or leaves a medium, such as a vacuum tube.

Audion Innovation by Lee De Forest using three conducting elements (triode) in a bulb to amplify and oscillate radio waves.

■ 1900–1930: Birth of an Industry

While De Forest experimented with boosting the sounds radio transmitted through his Audion, other inventors were tinkering with the cathode ray tube, which became

the glowing heart of the future television set. By the third decade of the century, the excitement radiating from television labs was drawing brilliant minds into an international race to realize the dream of "seeing radio." Scientists and engineers were hard at work in the United States, Europe, and Japan.

RADIO

This community of scientists and engineers scarcely paused as radio stepped out of the laboratory to make public history: The first stations went on the air. Thereby hangs a debate that has kept broadcasters and scholars enthralled for nearly a century: Which station should be entitled to call itself first in the nation in terms of **broadcasting**?

Who Was First in the Nation?

Station WWJ in Detroit laid its claim by documenting its initial broadcast on August 31, 1920, which announced returns from Michigan political primaries. However, in San Jose, California, an ancestor of the West Coast powerhouse KCBS had gone on the air in 1909. It broadcast Charles Herrold's farm reports to owners of **crystal sets.** These crude radio receivers were invented around 1906. Each featured an adjusting arm, two binding posts, and a string of wires called a "cat's whisker," which tickled a crystal, passing along a lot of static and occasionally radio frequencies.

In Madison, Wisconsin, experimental station 9XM (later WHA) signed on at the state university's campus in 1909 as a physics department experiment, and reached a few crystal sets. Some of the most vigorous claims to "first" status in radio point to KDKA in Pittsburgh, based on its uninterrupted service since 1920. The station's story begins when a Westinghouse engineer, Dr. Frank Conrad, built a workshop over his garage in 1916 and began transmitting. Listener letters in response to that station, then called 8XK, impressed Westinghouse Electric Corporation, which decided it could use Conrad's station to help sell its surplus of wireless radio sets.

KDKA STATION OPENING. Claiming a "first" guarantees controversy—but KDKA in 1920 launched the first continuous, scheduled radio program service licensed in the United States.

Westinghouse built a new 100-watt transmitter on top of its plant in east Pittsburgh and made plans for an inaugural broadcast. A commercial license arrived from the U.S. Department of Commerce just in time for KDKA to "sign on" on November 2, 1920, and report the presidential-election returns. Vote totals were phoned to the station from the *Pittsburgh Post,* and KDKA announced that Warren G. Harding had defeated James M. Cox (who, ironically, would leave politics to begin what would become the Cox Broadcasting Corporation).

It's true that other stations preceded KDKA on the air, but that 1920 event marked the *first continuous, scheduled program service by a U.S.-licensed station.*[5] Still, others note that WBZ in Springfield, Massachusetts, another Westinghouse station, received its license in September of that year, nearly two months before KDKA. Anyway, the news quickly spread: Radio was here. Enthusiasts lined up at electrical shops to buy receivers. They could at last begin to hear life in motion—even if they were still years away from being able to watch it in their living rooms.

Broadcasting Originally from agriculture (distribution of seed); redefined by the Communications Act of 1934 as dissemination of radio communications to public.

Crystal set Early radio receiver; with a silicon crystal connected to a wire coil, antenna, and headset, hobbyists could tune in programs.

RCA's Sarnoff Gamble

As radio entered the Roaring Twenties, the Radio Corporation of America (RCA) gambled on the commercial potential of a mass audience—mainly because of one man, David Sarnoff. He had moved upward quickly, and his famed "music box memo" imagined a future in which the radio set would be as common in homes as the piano or phonograph.[6]

As RCA general manager, Sarnoff proposed to run a new company devoted mainly to the art and science of broadcasting (it would become NBC, the National Broadcasting Company). Board members wanted evidence that commercial radio would pay off—and got it when Sarnoff's broadcast of the 1922 World Series brought raves from the New York regional audience. By 1924, local stations were linking to generate full coverage of political party conventions. Clearly, it was time for permanent networks to blossom.

Continuing disputes over technology patents and over leasing telephone lines among stations barely slowed Sarnoff's fierce ambitions. Soon NBC gave him two networks to supervise: "NBC Red," connecting stations owned by the American Telephone and Telegraph Company (AT&T) to RCA's station WEAF, and "NBC Blue," tying a half-dozen stations to WJZ in New York. To celebrate all this, NBC on September 9, 1926, broadcast live from New York's Waldorf-Astoria Hotel, Chicago's Drake Hotel, and the Kansas City dressing room of America's favorite comic-philosopher, Will Rogers.

DAVID SARNOFF. Sarnoff sparked the birth of NBC as an offspring of the Radio Corporation of America (RCA). Skilled at self-promotion, he moved from commercial manager at RCA to chairman of the corporation.

Local radio stations were proliferating rapidly. In 1927, Congress created the Federal Radio Commission (FRC) to license them in a way that would prevent signal interference among them. The government was watching the new medium as stations colonized the public's airwaves and planted the flags of their **call letters.** The FRC and its 1934 successor, the Federal Communications Commission (FCC), would stand watch over radio **frequencies** and increase regulation during the next few decades.

CBS and Paley

The story of CBS begins in 1926. That's when an orchestra manager and a promoter with the New York Philharmonic and Philadelphia's symphony formed their own radio program corporation. They invited a third partner, the Columbia Phonograph Company, on board to shore up their finances; in exchange, they would name the network the Columbia Phonograph Broadcasting System. The **network** grew to 16 "affiliated" stations, but it kept losing money. In 1927, the phonograph company bailed out—leaving behind, simply, the Columbia Broadcasting System (CBS).

At about the same time, William S. Paley started thinking big. He had been infatuated with radio ever since his family's cigar company rose in sales after sponsoring a show on WCAU in Philadelphia. Hearing that the Columbia network was up for sale, Paley and his father took over a majority of its stock to see if they could turn the company around. In 1928, CBS lost more than $330,000; yet after Paley became president he moved the headquarters to New York and balanced the network's books.

Paley developed affiliate-friendly ideas, giving stations the shows they needed plus money to air network advertising "spots" (commercials). He hired a professor from Princeton for research and recruited leading newspaper and wire service journalists to build CBS News. The network pushed away advertising-agency influence, taking control of its own shows. Paley—with strong CBS business ability and an instinct for program quality—began to build what would become known as the "Tiffany Network," after a New York jewelry company known for elegance. Paley had the brains for the business, and he found the talent to entertain the nation.[7]

Call letters Broadcast-station identifiers; stations west of Mississippi River have call letters beginning with *K*; those east of the river begin with *W*.

Frequencies Means for counting electromagnetic waves by counting the number of times per second (frequency) they pass a particular point.

Network Group of radio or television stations (affiliates) connected by contract to a central source of programming.

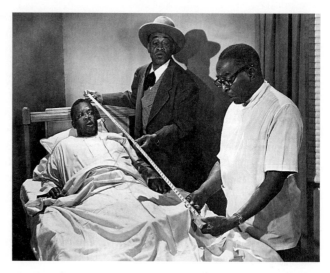

THE COLOR OF CHANGE. On NBC radio, *Amos 'n' Andy*—its two white creators at the microphones—presented comic African American stereotypes and drew huge ratings. But on TV in 1951, with civil rights advocacy stirring, the show used black actors and polarized viewers. It lasted only two years on CBS.

Programming the Twenties

In the decade that preceded the Great Depression, two stage performers from Illinois and Virginia kept the national mood light. Charles Correll and Freeman Gosden were white, but developed a radio serial in black dialect, *Sam and Henry*, on Chicago's WGN in 1926, and it became a hit. In 1929, NBC Red gambled on the renamed *Amos 'n' Andy*—and put it on the network schedule. It quickly attracted a national audience that would follow the humorous pair for decades. Presidents Truman and Eisenhower would confess to being fans of the show. Although it had listeners of all races, and was popular among African Americans, it eventually sparked criticism for its racial stereotyping.

Elsewhere on the air, *The Eveready Hour* set the standard for musical variety in radio's early days. The National Carbon Company premiered the program on New York's WEAF in 1923. It cost up to $6,000 a week to produce—the money required to bring in telephone lines, a 16-piece orchestra, singers, and guest celebrities. Other musical variety and comedy shows would pop up on radio during the Depression, including *Rudy Vallee's Variety Hour, Kay Kyser's Kollege of Musical Knowledge, Jack Benny,* and *George Burns & Gracie Allen.*

PRINCIPLE   4

History shows that broadcasting's nature is to follow social trends, leading them only rarely.

TELEVISION

Television was moving along with deliberate speed. Systems employing the Nipkow disk were patented in the 1920s. In England, John Logie Baird fabricated the first working TV set on that mechanical principle, and earned accolades as the British "father of television." The Baird system had a limited capacity to produce large, high-resolution pictures, and international brainpower increasingly focused on achieving visual quality by electrical means.

Westinghouse Electric Corporation hired Vladimir Zworykin, and, in 1920, he invented the *iconoscope*, which could scan pictures and break them into electronic signals. He conceived and created the first practical television "picture tube." Zworykin's work transformed tiny laboratory images into large, bright pictures that could be viewed from across the room.

Americans of color performed in more dramatic and substantive settings than the minstrel-style misadventures of *Amos 'n' Andy.* CBS began a dramatic series, *John Henry, Black River Giant,* in 1933. An African American writer in the U.S. Office of Education produced a series for NBC entitled *Freedom's People,* featuring stories about such prominent figures in American culture as boxer Joe Louis and actor/singer Paul Robeson.

New York City radio stations WMCA and WJZ broadcast African American dramas in the 1930s. In 1944, WMCA began airing a series of programs devoted to issues in the black community, titled *New World a Comin',* written by Roi Ottley and based on his book of the same name.

In Chicago, Richard Durham developed the first African American soap opera for radio, *Here Comes Tomorrow.* Durham also produced a weekly series on Chicago's WMAQ, *Destination Freedom,* spotlighting a range of African American history makers, including baseball's Jackie Robinson, journalist Ida B. Wells, and surgeon Dr. Daniel Hale Williams.

Another star of TV development was ascending. Philo T. Farnsworth was a Utah-born farm boy who, at age 15, realized how an image might be scanned electronically, and at age 21 he transmitted the first wholly electronic television picture. Farnsworth formed his own company, won a long patent dispute with Zworykin, and continued innovative work—but never secured a major financial stake in television. His creations, notably the "image dissector" tube that improved picture resolution, greatly speeded the improvement of TV.

PHILO FARNSWORTH. This former farm boy invented the image dissector tube, which brought high resolution to TV pictures, helping to ensure the medium's future.

23

Broadcasting's development always has turned on collaboration, and often on happy accident. Start with radio: When Swedish immigrant Ernst Alexanderson went to General Electric and met another inventor, Reginald Fessenden, they clicked. On Christmas Eve 1906, Alexanderson used his new high-frequency **alternator** to help Fessenden transmit songs and instrumental music—the world's first radio program.

Years of obscure research and development followed, but in 1924, Alexanderson sent a wireless-telegraph *picture* across the Atlantic. Four years later, he inaugurated the world's first TV station—just a big mechanical device with a tiny screen and a perforated, rotating scanning disk. It was a station because it sent out a signal—for just 15 or 20 miles, enough to reach an experimental TV set in Alexanderson's home in Schenectady, New York. That marked the first successful home-TV reception.

Shortly after that—May 18, 1928—came the first program from the inventor's TV station, WGY. It soon broadcast the first remote TV news report, from the state capitol as Governor Al Smith announced his candidacy for the presidency. And WGY broadcast *The Queen's Messenger,* the nation's first live TV drama. Television screens of that day were so small that the viewer could see an actor's hands or face—but not both at once. Today, as WRGB-TV, Alexanderson's pioneering engine of TV innovation serves New York's capital region as a CBS affiliate.

Still, the medium lagged well behind radio in reaching mass audiences. Those who watched early television programs did so mainly at special showings in public halls. Such venues scattered and limited TV's impact and, at that point in history, it was just as well: The U.S. economy was sinking in the Great Depression. Only a minority of families could afford any of the few TV sets in existence, and many technical issues still stood unresolved.

■ 1930–1945: Growth, Regulation, and War

To broadcasters, the period from 1930 to 1945 would prove pivotal as well as suspenseful. Both radio and the fledgling television medium came out of the Roaring Twenties on a technological roll that would sometimes falter but headed inexorably forward.

RADIO

Radio so far had been transmitted through a technology called **AM** (amplitude modulation), and would remain AM-dominated for decades. However, a new radio form with wide future appeal was about to emerge.

From FM to FDR

In 1933, Edwin Howard Armstrong obtained his first patents for radio based on **FM** (frequency modulation), which minimized noise and distortion. This mode especially delighted music listeners. De Forest demonstrated his system for RCA's David Sarnoff and, by 1936, was on the air with an experimental FM station. This invention would prove critical decades later, but for now, it was obscured by bigger events.

During the Great Depression that had descended in 1929, with more than one-third of the work force jobless, a growing number of Americans tuned in their radios for relief and psychological support. An estimated 12 million U.S. households invested in radio sets in 1930. President Franklin D. Roosevelt gave radio an

Alternator Machine that converts mechanical energy into electrical energy.

AM Amplitude modulation; a way of adjusting radio waves to carry sound by changing their height and depth but not their width.

FM Band of 100 channels in very high frequency range, 88 to 108 megaHertz.

PRESIDENT FRANKLIN D. ROOSEVELT. The thirty-second president of the United States, FDR was a master of broadcasting who lifted the nation's spirits during the Great Depression and used radio to strengthen the country's will to fight World War II.

even more powerful role: He began his famous series of "fireside chats" on March 12, 1933, with a talk aimed at soothing the nation's fears as a remedy to the banking crisis. Over the next 11 years, FDR comforted and informed the nation 298 times in that way, using an intimate medium perfectly suited to his warmth and charm. At the end of the 1930s, radios were in 51 million homes.

Mystery dramas and serialized dramas proved popular as the networks widened their hold on audiences. Hits included *Ma Perkins, Charlie Chan,* and *The Shadow.* In 1934, *Lux Radio Theater* (Lux was a brand of soap) introduced adaptations of films, with movie stars performing their roles on radio. *The Lone Ranger* formed the cornerstone of a fourth network in 1934: The Mutual Broadcasting System started as a consortium of stations in Detroit, Chicago, Cincinnati, and Newark—all selling advertising cooperatively.[9]

Radio gained such a grip on the nation that millions of listeners fell hard—dangerously hard—for a trick-or-treat prank. Actor Orson Welles narrated a 1938 Halloween tale of Martians landing in New Jersey, based on H. G. Welles's *War of the Worlds.* The show's reality disclaimers failed to register with Americans already primed for disaster on the eve of world war; instead, the radio show's realistic use of news flashes sent many frightened listeners scurrying for the hills. [10]

War News

Soon, as TV waited out a wartime license "freeze," radio began to show its highest potential. Onto the national stage stepped reporters who had been dispatched to war posts overseas—notably CBS's Robert Trout, William Shirer, and Edward R. Murrow who now filled U.S. living rooms with battle bulletins, word portraits, and strategic analyses. (Men ruled this age of news, but women were rising: Pauline Frederick would broadcast from China in 1945, and covered the United Nations from its founding in 1948.)

For many families, these war reports were a link to their sons, brothers, and fathers on ships and in trenches. Radio stars such as Bob Hope and Jack Benny comforted families at home by visiting soldiers abroad. The medium was launching relationships with millions of Americans that would continue into the postwar adolescence of television.

Reining in the Networks

Before the war, CBS, along with NBC's Red and Blue networks, had tough contracts with big-city radio stations. The contracts required each station to pull its own programming aside ("pre-empt" it) if a network had something it wanted to put on the air. In addition, musicians and actors hoping for network careers had to join "artists' bureaus" first, giving CBS or NBC broad rights as their agents and employers.

Such power plays by networks were staggered in May 1941 by the Federal Communications Commission's *Report on **Chain Broadcasting***. It prevented any new licensee from affiliating with the two NBC networks. Also, the FCC gave stations the right to refuse any network program, and questioned the networks' total power over performers. The report underscored the importance of localism—the stations' mandate to serve their local communities first. NBC and CBS fought it to the U.S. Supreme Court and in 1943 lost the case.

As a result, NBC sold its smaller NBC Blue network for $7 million to the chairman of Life Savers, a candy company. The network would change its name to the American Broadcasting Company (ABC) and also would take Paramount studios as a partner to help it become financially stable. CBS and NBC closed their artists' bureaus and gave affiliates the "right of first refusal" to *not* air network shows if they so chose.

TELEVISION

In 1935, RCA leader David Sarnoff announced plans for the first modern U.S. television station and for the manufacture of TV receivers for everyone. The country was about to enter an age in which its living rooms would become audiovisual theaters, with social consequences that no one could yet fathom.

A Slow Start

Within the next couple of years, the British were telecasting certain events—indeed, they already had a TV chef on the air! In 1939, Sarnoff's RCA used an NBC camera to transmit pictures from the opening of the World's Fair in New York. However, long after TV had become technically sound, it still was not routinely available across the United States.

It didn't help that commercial and legal squabbles echoed in the boardrooms of New York and the official chambers of Washington. In one major dispute, Philo Farnsworth refused to sell RCA the manufacturing rights to some of his most important advances. In 1940, to calm such disturbances and to fashion a national TV infrastructure, the FCC stepped in—first authorizing limited transmission over the public's airwaves and then slowing TV's growth to hold technical hearings.

The following year brought a breakthrough: The National Television Standards Committee, formed by manufacturers, produced an agreement setting standards for home picture quality, which had been a sticking point. So, despite a lingering dispute with NBC's hard-driving David Sarnoff, the FCC gave the medium its final go-ahead. The age of commercial television in the United States dawned on July 1, 1941, when WNBT (now WNBC) in New York began broadcasting. CBS went on the air that same month.

There was little time for euphoria, however, much less growth: Soon after the Pearl Harbor attack touched off World War II in 1941, television's development came to a halt. Government blocked both the manufacture of sets and construction of stations; materials for both would be funneled into the war effort. Mass access to the delights of this new medium would have to wait.

Chain broadcasting At first, simultaneous broadcasting of the same program by two or more connected stations; later used to describe group ownership.

1945–1960: A "Golden Age" Dawns

World War II ended in the defeat of Germany, Italy, and Japan in 1945, and broadcasters gratefully returned to the airwaves. A British announcer marked his return with "As I was saying before I was so rudely interrupted. . . . " Broadcasting, a key to American morale during the international conflict, was poised to extend its influence through a more peaceful era.

TELEVISION

As waves of veterans returned to the United States, the young TV industry maneuvered, setting the stage for a postwar television boom. CBS raided the other networks for stars, especially those whose name recognition would give new television programs a boost. Network contracts that helped celebrities become "corporations" and benefit from a tax loophole appealed to Jack Benny, Red Skelton, Bing Crosby, and other radio luminaries.

MILTON BERLE. The broad humor of the rubber-faced Berle amused huge audiences during national TV's infancy and turned him into "Mr. Television."

HOWDY DOODY. The TV puppet Howdy Doody, guided by "Buffalo Bob" (Smith), starred in one of many early shows aimed at children. Howdy emerged on NBC in 1947 and his show lasted until 1960.

RCA chief David Sarnoff had been a communications aide to General Dwight D. Eisenhower during the war, and emerged as a brigadier general. Now he marshaled big manufacturers (some of them also programmers) in lowering prices on TV sets. Television crept, but more often swept, through towns and cities, quickly initiating new viewers. (Virtually left out of the process was the seminal genius Philo Farnsworth; his major patents had expired before the boom and he would die largely unrewarded for his breakthroughs.)

TV's pictures were black and white, but efforts toward color programming moved rapidly. CBS in 1946 demonstrated a color system to the FCC, and after RCA entered the arena, development accelerated. CBS was promoting a "color wheel" that spun red, green, and blue filters to create colored images. Another company, Dumont, pushed a similar wheel.

By now, TV was creating unprecedented excitement as an adjunct to postwar national optimism. Returning veterans helped launch a baby boom, goods became plentiful and cheaper again, and a new kind of community called the *suburb* was spreading. Similarly, television spread through the nation's leisure hours. Baseball was televised starting in 1946, and the first TV World Series triggered a rush to buy home sets. By 1948, the upstart ABC network had joined NBC, CBS, and Dumont to compete for viewers. As added bait, a cable to carry programs nationwide was installed

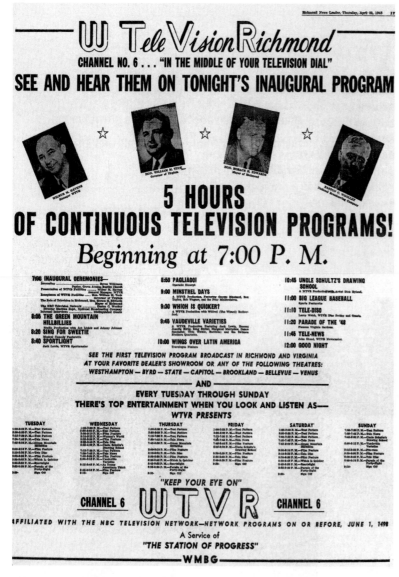

"FIVE HOURS OF CONTINUOUS TELEVISION PROGRAMS!" In 1948, WTVR in Richmond, Virginia, became the first TV station in the South to broadcast and ran a splashy newspaper ad promising to fill weekday evenings with local programs.

Very high frequency (VHF) Originally referred to TV channels 2 through 13.

Spectrum The array of electromagnetic "airwaves" (identified by wavelength) that broadcasters harness to transmit radio signals.

Ultra high frequency (UHF) Originally referred to TV channels 14 through 69. Some of those channels have now been reassigned to make way for DTV.

between New York and Los Angeles. The "tube" was about to go coast to coast.

Another Freeze

In 1948, 108 TV stations were operating. That's when the FCC once again froze the issuance of new television licenses—this time to study video and color standards, interference, frequency allocation, and educational uses. That took four years, after which the commission's *Sixth Report and Order* officially thawed the freeze, assigning TV frequencies to about 1,300 communities. The new channels were not just in the existing **VHF** (very high frequency) portion of the **spectrum** (the whole range of bandwidths in which radio and TV can operate). Many newly allocated channels were in the **UHF** (ultra high frequency) band.

A particularly thorny issue for the FCC had been whether to hold aside part of the spectrum for noncommercial broadcasting. The forces of commercial television stridently opposed that idea. However, a new commissioner, Frieda Hennock, crusaded doggedly for the use of TV for education unaccompanied by advertising. She thus became a hero to critics of the general push for profits. When the freeze thawed, Hennock had prevailed; the FCC reserved 242 slots for noncommercial TV.

By the 1950s, radio's reign as the people's choice for comedy and drama was beginning to wane. Not only had top network radio shows and performers switched to television, but so had advertisers. From 1948 to 1958, radio's network revenues would slip by $18 million, dropping the average station's earnings to half of what it made in 1948. Americans were scurrying to position little antennas called "rabbit ears" over their small home TV screens.

National Impact

When U.S. industries stepped back from their war footing, people left farms in great numbers and sought better work in towns and cities. Local TV stations, always located mainly in sizable population centers, hurried to build rich and varied program schedules. WTVR in Richmond, Virginia, made its debut in 1948 as the first TV station in the South, and proudly advertised "FIVE HOURS OF CONTINUOUS TELEVISION PROGRAMS!" Almost all of the shows that first day originated locally and had strong flavors of that place and time: *Inaugural Ceremonies* (for the station), *The Green Mountain Hillbillies, Sing for Sweetie,* and *Minstrel Days.*

Excited Richmond viewers crowded around storefront TV sets along Broad Street. The medium's magnetism stemmed partly from most of the programming's *live* status. Good ways of recording programs for later airing hadn't come along yet. Stations could shoot film off the television screen—the results were called *kinescopes* or "kinnies"—but the shows lost much picture quality in the process. The 1948 Rose Bowl game was broadcast live; and the futile 1949 attempt to rescue 3-year-old Kathy Fiscus from a deep California well where she had fallen kept viewers glued to a national TV hookup for days.

By 1950, an estimated seven million TV sets were bringing programs to viewers across the country—in black and white. Experiments with color, such as a spinning color wheel fastened like a lens over the screen, were foundering. Even in black and white, though, TV was a continuous hit, transporting voters into the 1952 political conventions. Expecting this, major advertisers had signed on as sponsors, their representatives sitting right in the control booths (as NBC producer Reuven Frank would vividly recall).[11] Sig Mickelson, the CBS executive who arranged three-network coverage of the conventions, later said: "Someday, history will demonstrate that television was really made more by the national political conventions of 1952 than by any other single event."[12] Among other things, the event launched the rise of an anchor named Walter Cronkite who would dominate TV news for more than a quarter century.

TV's Variety

Television was enough of a novelty to win audiences for a wide variety of programming, including serious and even highbrow shows. CBS journalist Edward R. Murrow launched the respected *See It Now* news-documentary series in 1951. Live plays by important playwrights came to the small screen, notably in series such as *Playhouse 90*. Undemanding shows, such as *Arthur Godfrey's Talent*

KATHY FISCUS VIGIL. In 1949, studio-size cameras brought the nation one of its first live-TV events as rescuers worked in vain to rescue a 3-year-old girl from an abandoned well near Los Angeles. (Photo courtesy KTLA-Los Angeles)

Scouts, drew many loyal viewers too; entertainment ruled the airwaves from the beginning. Western dramas, a growing sports menu, quiz shows (some of them, such as *Twenty-One*, secretly rigged until a national scandal broke), and family sagas also thrived. They even changed the way viewers talked. People wisecracked, "Just the facts, ma'am," in light conversations, echoing the tough-cop series *Dragnet*.

Comedy flourished well beyond what "Uncle Milty" Berle was offering. The unique Groucho Marx and his cigar dominated the modest quiz show *You Bet Your Life*. In *The Jackie Gleason Show*, Gleason and an ensemble cast created an often darkly realistic comedy series called "The Honeymooners" that was funny as well as poignant. Other comic greats, including Sid Caesar and Carl Reiner, were hugely popular. *I Love Lucy* arrived in 1951, focusing on a Cuban American bandleader and his ditzy, manipulative wife; the show was gentle but tart, touched on familiar marital themes, and would become one of the most beloved series in TV history. It also was shot on film, making editing possible—a safety net for comic timing.

That same year, notably, TV's first all-black series premiered. *The Amos 'n' Andy Show*, a crossover from radio, featured light comic situations and stereotyped characters that raised the ire of the NAACP. Protests by black leaders failed to stop the show's production, however. African Americans were divided on the controversy, since actors of color needed the work and some assumed the show's fumbling of the English language and character buffoonery would not be mistaken for real by white audiences. It was losing viewers that eventually took *Amos 'n' Andy* off the air after three seasons despite an Emmy nomination for its team of writers, who would later script such hits as *Leave It to Beaver*.[13] Another important import from radio, the soap opera, premiered on television in 1952. *The Guiding Light* would

I LOVE LUCY. The 1950s CBS-TV series, a domestic comedy featuring Lucille Ball and her bandleader-husband Desi Arnaz, created viewer loyalty that survived into the twenty-first century via reruns.

enjoy a run that (for television) was almost breathtaking, extending all the way into the twenty-first century, and even some of its rivals spanned decades.

From Murrow to "Dobie"

When RCA at last put a small-screen color set on the market, it cost one thousand precious 1954 dollars and sold very slowly nationwide. Of course, black-and-white TV was appropriate for watching white men in suits run for office, or for watching journalists, including CBS's Ed Murrow, who tended to challenge viewers' moral values. Murrow, more than any other figure in broadcast history, made television an essential conveyor of news as vital information to the American people. His long-form reports, such as *Harvest of Shame*, a searing close-up of migrant workers' poverty, brought startling social truths into innocently unaware living rooms.

Murrow's televised assault on Senator Joseph McCarthy's shaky grip on power and decorum took much of the steam out of early Cold War anticommunist witch-hunts. Murrow became a towering icon of broadcast journalism—yes, advocacy journalism, and a shining model for many of his professional descendants.

Elsewhere on the TV dial, the values of the Eisenhower age were reflected and sometimes gently lampooned by growing numbers of situation comedies, known as *sitcoms*. *Ozzie and Harriet* depicted an idyllic and amusing family life for 14 seasons. *Father Knows Best* brought all problems under firm but benign paternal control. *Leave It to Beaver* featured the antics of the family's youngest son in its title and showcased impish humor. *The Many Loves of Dobie Gillis* brought moony teenage angst to the delight of millions of baby boomers.

Elvis and Quiz Shows

One of the richest years of the decade was 1956, when videotape was introduced. This technology opened television at last to top-quality editing of high-resolution pictures, permitting a high level of visual storytelling that live broadcasts couldn't attain. The first effective wireless remote-control tuning device also surfaced in 1956. "The remote" would not become standard in U.S. homes for three decades, but producers knew better than to let their programs' pacing slacken. Among other reasons was that young audiences were warming to the new medium. Indeed, 1956 was the year of Elvis Presley: When he appeared on *The Ed Sullivan Show* on September 9, an estimated 54 million people were watching. That was evidence not only of Elvis's appeal but also of TV's massive diffusion across the country and its potential hold on all generations of viewers. Perhaps as striking was that 10 million TV-set owners did *not* tune in to see The King shake his way through rock 'n' roll songs, his hips carefully cropped out of frame.

There was a corps of businessmen behind television, and their integrity was called into question in 1958 when the great quiz-show scandal was uncovered. Some big-money shows, *The $64,000 Question* among them, had been feeding answers to contestants, largely to keep the most appealing personalities returning week after week. Critics pointed to the **Communications Act of 1934**, which required broadcasting to serve "the public interest, convenience and necessity." The quiz-show scandal set off debate as to whether television was placing its private

Communications Act of 1934 Federal law bringing most telecommunications under oversight of one agency and board of commissioners.

values first, and attracted government scrutiny that would intensify for years. A fictional defense lawyer launched his incredible courtroom career on television in 1959 in *Perry Mason*. Also that year, a talented writer named Rod Serling stepped in front of the cameras to introduce science-fiction morality tales from *The Twilight Zone*. The series would last only five seasons but paved the way for decades of TV "sci-fi" (or "s-f") ventures.

In 1950, just 9 percent of U.S. homes had television. By 1960, that figure had jumped to 87 percent—reflecting one of the biggest, fastest changes ever in American popular culture.

RADIO

The radio industry, meanwhile, had scarcely been sitting on its hands. The evolution of David Sarnoff's "music box" in post-war America had a lot to do with prosperity, physical mobility, and the flight of bored teenagers from their parents' favorite memories, styles, and tunes. In addition, however (and critically), companies that sold "records" worked steadily to develop new technologies. Their combined effect would be to free recorded music from some of its old confines while empowering youth to get out of the house and listen to *their* favorite tunes on a new kind of radio set. Social revolt was in the air, and the fates of radio and records were intertwined.

Records and Transistors

Until the early 1950s, the standard phonograph record was made largely of shellac (a product of hardened insect secretions) and rotated on an electric turntable at 78 rpm (revolutions per minute). Then Columbia Records came up with a way of making records of *vinyl*, a plastic. What's more, the company introduced a long-playing disc that would rotate at only $33\frac{1}{3}$ rpm. The result of all this was better sound quality, destined to be a major asset as young Americans became more and more infatuated with recorded music.

Columbia offered its new technical process to other record companies. Interestingly, RCA, Sarnoff's company, wanted no part of it. Instead, in 1949, RCA developed its own 45 rpm disc made of vinyl, seven inches in diameter with a big hole in the middle—soon to be the standard record for rock 'n' roll radio. Significantly larger than today's compact discs, the "45" was small, light, inexpensive, easy to store, and portable. Its microgroove technology put more and more music into the hands of young consumers, whetting their appetite to hear music wherever they went.

Rising to satisfy the youthful demand was a new kind of radio set. Texas Instruments in 1953 proposed something called a **transistor** radio. It was based on an amplifying device invented six years earlier by engineers who would win the Nobel Prize for it. It ran on batteries and fit nicely in the palm of the listener's hand—a personal portable radio for the Fabulous Fifties. The secret to its compact size was found in the tiny transistor buried inside it.

No major radio manufacturer was ready to take a big, expensive chance on the new idea, so a TV electronics company teamed with Texas Instruments and

ELVIS PRESLEY. A photo of his September 9, 1956, appearance on *The Ed Sullivan Show* shows much more of "The King's" physical style than the TV camera did. Producers were nervous—but an immense audience affirmed the reach of television.

Transistor Wafer-thin silicon crystal that amplified radio signals; as developed by William Shockley and his colleagues, transistors replaced tubes and led to the portable radio and later digital revolution.

a design firm to build the first batch of transistor radios. They arrived in stores in time for Christmas 1954, but each cost about $61 fully equipped—equivalent to $400 today—driving away even more adventurous shoppers. Major manufacturers rapidly began producing competing brands, and soon the transistor revolution was on. Prices came down through mass production and parents were happy to buy "transistors" for the kids, who then could leave the house to listen to, well, whatever music they wanted—which, by the way, was also changing rapidly.

Formats and Formulas

By this time, radio was rethinking itself. It had held audiences spellbound for years by presenting its content—drama, comedy, sports, popular and classical music, local features, and news—in rotating "block" style: 15, 30, or 60 minutes for one genre, then on to the next. Radio had been the electronic medium that supplied all that a family could want. That was because radio had been the only electronic medium most people had. Now, television was sweeping the country and whisking away radio's audiences. What could be done?

One answer came from Todd Storz, an Omaha, Nebraska, radio station owner, who (legend has it) made a simple discovery one day. He was in a bar near his station when he noticed patrons playing the same songs on the juke box again and again. After the bar closed, the waitress broke out her tip change to hear the same song one more time. Storz had a stunning revelation: What seemed to be happening when *listeners* controlled the program was not a search for variety but the comfortable repetition of a few favorite songs. From this discovery sprang the idea of "Top 40" radio.

It's a good story, but only that, according to Richard Fatherly, who was program director at a Storz station in the 1960s. Fatherly says that the undramatic truth is this: Storz acquired a study report in 1950 from the University of Omaha in which people cited music as one of their main reasons for listening to radio. He decided to fashion programming on his Omaha station, KOWH, after a national show, *Your Hit Parade,* and a radio program on WDSU-AM in New Orleans, *The Top 20 on 1280.*[14]

However he got the idea, Storz abandoned the variety format and focused on the 40 most popular tunes. He used local record sales, juke box plays, and trade magazines to learn what songs the public wanted to hear again and again. Storz supplemented the discs with news briefs and repeated station "IDs"; these were identifying messages required by the FCC but also useful as promos when delivered by disc jockeys avidly hyping the station's hits. Storz added contests and treasure hunts to excite listeners. He used the same formula on six other stations and, when it worked, started selling his promotion package to other broadcasters. Soon "Top 40" ruled radio.[15]

One major contributor to this profound change in radio programming was a former U.S. Navy intelligence officer named Gordon McLendon. He moved to Dallas, Texas, in 1947, and launched KLIF radio, licensed to serve the city's Oak Cliff neighborhood. It was to be the flagship of his Liberty Broadcasting network, and it was there that McLendon added his own twists to Top 40 radio. Through a fast-moving blend of "personality" disc jockeys, news reports, and promotional contests, KLIF became one of the highest-rated stations in the country—and Liberty grew to include 458 affiliates. Imitators appeared around the country and advertising sales soared, sending local radio as a business sector into the billion-dollar stratosphere of media enterprise.

PRINCIPLE 6

Once broadcasting experimented; then it exploded; now, surrounded by rivals, it hedges its bets.

▛ 1960–1980: Ferment and Change

It was fortunate that radio in the fifties invented new ways of operating to sustain it, because the sixties increasingly drew Americans to their TV sets. Combat footage from a far-off place called Vietnam was unsettling viewers of early-evening newscasts by 1961, well before thousands of U.S. troops became embroiled there. Domestic trouble also lit up the screen: Rebellious African Americans marched through the South, braving gauntlets of screaming whites and often violent cops, and were shown bullied and bleeding before hordes of stunned viewers.

TELEVISION

Millions of people—along with an episode of the soap opera *As the World Turns*—stopped in their tracks on November 22, 1963, when CBS anchor Walter Cronkite interrupted TV programming to announce that President Kennedy had been shot in Dallas. The next few days brought an unprecedented demonstration of broadcasting's power over Americans. Television (and radio) went everywhere and talked to everyone about the assassination, and everyone watched television in return. ABC anchor Ron Cochran later would say, "Television had actually become the window of the world so many had hoped it might be one day."[16]

At the "National Hearth"

The majestic state funeral for President Kennedy in Washington, DC, was covered live on TV. This was the first time television had etched a truly epochal event across the national consciousness. It would not be the last such occurrence, but it would be the last one conveyed entirely in black and white to a mass audience. Color television became technically feasible for network use by 1964, and color sets grew more and more affordable for much of the middle class. Some 13 months after the Kennedy assassination, all three networks broadcast in color, simultaneously, for the first time.

Viewers could now expect nearly lifelike color not only in their entertainment programs but also in the harsh hues of news reports. Coverage included the killings of President Kennedy's younger brother, Robert, and of civil rights leader Martin Luther King Jr., and the brutal handling of anti–Vietnam war demonstrators by Chicago police outside the 1968 Democratic Party convention. These galvanizing events supported the theory that television would become a national "hearth" where, now and then, we could gather to watch, worry, and recover—together.

Television's convenience and range of programming already were pulling people away from movie theaters and back into their homes. Now the availability of color—"living color," as NBC called it—enhanced this trend. The big Hollywood studios feared television's appeal to film audiences, but also saw in it a tempting opportunity: Perhaps, by cutting deals with television, they could assure themselves bigger financial returns. Before long, Hollywood studios were cranking out most of the important series on television. The head of Twentieth Century Fox, Darryl F. Zanuck, had sneered at the new medium: "People will soon get tired of staring at a plywood box every night." That viewpoint had collapsed, and the film industry had no choice but to try sharing in TV's success.

THE VIETNAM WAR. Television demonstrated its public-opinion power by covering clashes over the war, including one at Kent State University in 1970. Four students died when the Ohio National Guard opened fire on protesters.

TV and Society

Throughout this period, TV technology made events visible to the public and also seemed to play a role in bringing them about. Activists made sure that television got early warnings and clear views of the many protest marches against the Vietnam War. Communication satellites launched since the mid-1960s had made live international coverage possible, and now it began to shake up the world: When President Nixon went to China in 1972 to make diplomatic history, cameras went along, "going live" to beam the news globally. Television was on hand for the Olympics in Munich that year when Palestinian terrorists captured (and later executed) 11 Israeli hostages; that, too, was carried live worldwide.

The women's liberation movement launched in the sixties swept through the seventies in thousands of TV plot lines, characters, and dialogues. One must-see comedy was *The Mary Tyler Moore Show*, portraying a TV news worker whose feminism simmered near the surface; the show's success was ironic in that the seventies did not see much of an increase in the number of women employed in the *real* television industry.

On the other hand, an upsurge in black-identity movements did result in 1977 from *Roots*, an eight-night TV series of dramatized chapters in the lives of an African named Kunta Kinte and his descendants in America. At this writing, *Roots* still reigns as the third most-watched program in the history of television. Similarly, while leering jokes about homosexuality could still be heard, attempts to promote acceptance of gays began to make an impression. Billy Crystal's portrayal of the gay character, Jodie Dallas, in the 1977 comedy series *Soap* set a new standard of candor—two decades before *Ellen* came out as a lesbian on network television.

By this time, with color television routine and programmers taking risks, television was in the cultural foreground. Its resulting economic power helped prompt the FCC to force the industry to loosen its grip on the content of its shows, opening them to more perspectives. The FCC rules required networks to hand over program ideas to movie studios or other producers; *they* would then create programs and, as long as their shows lasted on networks or elsewhere, would receive license fees for them. The networks balked at this change because it would cost them a great deal of money, but the government in those days was opposing monopoly power and pushing diversity of control.

ROOTS. In this 1977 miniseries, author Alex Haley (played by James Earl Jones) traced his family origins to Africa. The show set an all-time viewership record, strengthened black-identity efforts, and prompted many families to seek their own roots.

This climate suited the seventies, a decade when the interracial satire of *All in the Family* and the feminism expressed by another CBS show, *Maude*, hinted at changes that moved beyond protest marches to education and economics. Robert L. Johnson in 1979 launched Black Entertainment Television, the first network almost exclusively serving—and controlled by—African Americans; it would prosper and last. Through the electronic media, the American social fabric had become a quilt in which every vivid patch asserted its desires and demands.

Money and Journalim

Everyone knew that TV news could make money, but the newsmagazine *60 Minutes*, launched in 1968 and a Sunday-night staple by 1970, proved that in-depth coverage of controversial topics could win top ratings and make a *lot* of money. That was a watershed development, opening paths to success for later news programming. It made life uncomfortable for journalistic idealists who had hoped to do their work without thinking about the corporate bottom line. It also extended major network news further into the glittering realm of evening prime time, where entertainment reigned.

In the mid-seventies, about three of every four TV homes tuned in to the "Big Three" networks. Families planned their evenings around such "appointment viewing," even for the often disturbing nightly news. When the CBS documentary *The Selling of the Pentagon* exposed government efforts to win bigger military budgets through expensive public-relations campaigns, it infuriated war supporters and widened political gaps. Fortunately, the comedy show *Laugh In* and comic Flip Wilson made silly slogans such as "Sock it to me" and "Here come da judge" a safe way to laugh off stress.

Television programmers had become more inventive; they could seize a news event and turn it into a franchise, as when ABC's coverage of the 1979–1981 Iran hostage crisis became the highly successful *Nightline* program. A new player, Cable News Network (CNN), debuted in 1980 to the jeers of traditionalists; it would grow, slowly, into a permanent fixture. Advances in **electronic newsgathering (ENG)** fused cameras with recorders and helped send them everywhere; for example, in 1981, they captured at close range the assassination attempt on President Ronald Reagan—shaky cameras telegraphing the frightful power of the event. Such technology, however, rarely helped voters understand government and politics; many depended heavily for their information on political *commercials,* which for some TV stations would become a lucrative stand-in for campaign news coverage.

RADIO

A new era for FM dawned in April 1961 with FCC approval of stereo technical standards. The government put its stamp of approval on a Zenith-General Electric stereo formula, rejecting Murray G. Crosby's competing system. The agency also ordered station owners to stop using their FM bands simply to carry what was already broadcast on AM bands. Under the nonduplication mandate, combined AM-FM stations had to air two different streams of program content for at least half of the broadcast day.[17]

FM's Effects

These improvements rapidly increased FM's popularity as a source of unique programming with high-quality sound. In 1964, the Audience Research Bureau began "counting heads" for radio and found a surge in FM listening. As the Vietnam War and race relations sharpened political and generational divisions among Americans, so-called progressive stations, also known as "underground radio," attracted many baby boomers. By providing alternative perspectives and supporting an explosion in rock 'n' roll music, FM gradually became the favorite choice of most radio fans.

Radio as a cultural medium started to break into different parts, and one big player moved to exploit the trend. In 1967, ABC targeted different audience groups with several different programming packages for its affiliate stations: ABC Contemporary, ABC Information, ABC Entertainment, and FM formats. To make this work, ABC offered radio station managers news and public affairs material designed for each format. Soon, this strategic move by one network became an industrywide campaign to identify and gather specific groups of listeners who could be reached by well-tailored programming and commercials.

Audience research helped to refine the new formats: "New Age," alternative rock, classical, urban contemporary, young country, and many more. Listeners happily adapted to new trends in music, news, and personalities, sorting themselves into various audience groups. This made decisions simpler for programmers as well as advertisers anticipating time-buying choices. The downside was that a station had

Electronic newsgathering (ENG) Applies to portable, videotape-based field equipment, and the capacity to "go live" from local news events.

to become one of the top choices on a listener's car radio or it would be forgotten altogether. The repetition impulse that Todd Storz had noticed back in the 1950s kept many drivers, especially teenagers and young adults, punching the same few radio buttons all the time.

Public Radio

Radio's family was about to grow. After the adoption of the Public Broadcasting Act of 1967, the FCC licensed more than 1,500 radio stations as "noncommercial educational," most between 88 and 92 megaHertz on the FM band. Many of these stations would receive federal money distributed through the Corporation for Public Broadcasting—a fact that annoyed commercial broadcasters envious of such subsidies. Besides local programming, the new FM stations would carry news, information, and cultural shows provided by their network source, National Public Radio (NPR). It began distributing programs to 93 member stations on April 19, 1971.

∎ 1980–Present: Challenges to Broadcasting

In the eighties, fictional cops, lawyers, doctors, and nurses filled prime-time TV, depicted in rawer, bolder terms than the mildly racy seventies culture had allowed. Some shows featured top-notch writing and acting; series such as *Hill Street Blues,* *St. Elsewhere,* and *L.A. Law* mixed trauma and tragedy with social issues and strong characters, and developed a large fan base. It wasn't enough to offset the challenges to come, however; the struggles within commercial radio and TV would become as dramatic as anything on the air.

TELEVISION

As the networks kept proving they could do high-quality fictional shows, public-affairs treatments and long-form documentaries migrated from the broadcast networks to cable, where smaller audiences were acceptable. The 1980 retirement of the magisterial Walter Cronkite as CBS's top news anchor seemed to symbolize this shift in the media terrain.

Marketplace Woes

By the eighties and early nineties, broadcast executives found themselves facing a list of troubling developments:

▶ Cable television was crowding "The Big Three" by carving out powerful market niches. MTV hit the air in 1981 and grabbed a large youth audience. Cable channels had a financial advantage: They not only sold commercial time but also charged subscriber fees.

▶ Syndication companies were selling popular dramas and sitcoms to cable as soon as they finished airing on the networks, and sometimes before then. This allowed cable to present and profit from programs that the networks themselves had once financed.

▶ The home VCR, which became commonplace in the eighties, was eating into broadcast revenues as viewers skipped TV offerings in favor of rented movies.

▶ Satellite operators had begun feeding movies and sports to cable systems, which in turn fed them to home sets. What's more, direct satellite transmission to homes was getting under way.

▶ "Fin-syn," the financial-interest-and-syndication rules Congress imposed in 1968, had become more and more troublesome. They were meant to keep networks from producing their own shows, favoring independent producers, while locking out Hollywood suppliers. To the networks, though, fin-syn was unfair—a nightmare of revenues missed.

These forces left networks ABC, CBS, and NBC weaker. Their audiences were shrinking, each company losing at least several million viewers a year and thus millions of dollars in ad revenue. This made "The Big Three" vulnerable to would-be buyers. With the *stations* they owned still making money and the FCC relaxing station-ownership limits, the networks were ripe targets. In just two years, 1985–1986, ABC merged with Capital Cities Communications, a large but relatively unknown media group; General Electric took over NBC, and Loews Corporation gained control of CBS. Disney later would purchase ABC, and CBS would become a Viacom property.

New Relationships

The networks soon found themselves among new corporate siblings. ABC shared its nest with the sports channel ESPN and the family-oriented Disney Channel, among others. In a huge radio consolidation, Westwood One acquired and combined the old NBC and Mutual Radio networks, which in turn were affiliated with more than half of the nation's commercial stations. Westwood One's chief said the megagroups gave radio more appeal to Wall Street investors—an increasingly important factor—and to Madison Avenue advertisers.

RUPERT MURDOCH. Murdoch kicked off Fox TV in the 1980s and used sensational programming to woo young-adult viewers.

Networks desperate to ensure that their TV programming reached the right audiences got into power struggles with cable systems and affiliates that had their own scheduling ideas. In 1995, the WB and UPN networks emerged with programming aimed at youth, minorities, and other underserved viewers, further slicing up the prime-time audience. Australian press magnate Rupert Murdoch had launched Fox TV in 1986 with sexy, boundary-stretching shows; now Fox had hits (including *The Simpsons* and *The X-Files*) and was buying more stations. A seventh broadcast network, PAX, went on the air in 1998 and entered a program-sharing relationship with NBC, helping both networks.

The year 2000 found much of broadcast television reduced to running after audiences and revenue, and they often got away. Cable and direct-broadcast-satellite channels, with their uncut adult programming and sports specials, lured away viewers. The Internet siphoned leisure time that might have gone to television. Networks answered with newscasts emphasizing "news you can use" for young families and playing into populist concerns about a changing America. A genre known as *newsmagazines* mined the country for crimes and family strife that would closely mimic popular prime-time fictional dramas.

Around the turn of the millennium, so-called reality programs caught fire. Notably *un*real, these shows dropped people into jungles or onto islands to test their endurance, marital fidelity, or just plain con-artistry. If TV handicappers thought such programming seemed too thin or outlandish to last long, they were in for a surprise. Reality shows cascaded through the network and cable schedules through the 2004–2005 season. Producers seemed able to come up with story line after story line, and would-be participants lined up for blocks around Hollywood studios to compete for character types to play in each new series. The handful of reality shows just five years earlier grew to about two dozen by fall 2004. A genre based largely on betrayal and humiliation, its hits were mostly planned and edited rather than left to the whims of amateurs. It had become the most fertile seedbed of profitable ratings in television. However, Magna Global, a large media-buying agency—which

NPR. Robert Siegel is senior host of the wide-ranging weekday news program *All Things Considered* on NPR.

procures commercial time and ad space for entertainment companies to hawk their wares—wondered whether the reality trend would last. Even while such shows were increasing and scripted programs were in decline, the company's analysts said that reality shows had become repetitive and their enduring popularity seemed fragile.[18]

RADIO

Some citizens wanting more in-depth treatment of issues than commercial broadcasting supplied had begun to rely on "public" radio. National Public Radio's newsmagazines, *Morning Edition* and *All Things Considered,* had grown since their founding in the 1970s into highly regarded public-affairs programs. Another program supplier, Public Radio International (PRI), was launched in 1982, partly to get heartland programming such as Garrison Keillor's *A Prairie Home Companion* on the air nationwide. The enterprise grew to encompass more than 550 member stations.

Unfortunately, people often watched or listened to noncommercial programming without sending in checks to pay for it. Because it lacks advertisers, public broadcasting came under serious financial strain in the eighties and nineties.

Opinion Radio

Commercial radio, on the other hand, had to make money on its own. Stations started searching for fresh angles back in 1978, as FM's audience eclipsed the listenership of AM. Among the relatively neglected formats was something called *news/talk* radio: It blended news segments with long, live conversations involving emotional and contoversial topics, directed by opinionated hosts, many of them politically conservative.

The news/talk format offered something few consumers had enjoyed on radio or TV—the chance to talk *back.* The genre caught fire when the vastly popular President Ronald Reagan led the country in a politically conservative direction. Americans were herded into opposing camps on gun ownership, abortion, big government, and other "wedge" issues. After the 1987 repeal of the **Fairness Doctrine,** which required broadcasters to present contrasting views of any controversial issues they covered, news/talk programming took root. Some shows were syndicated nationally, snapped up by stations seeking to build loyal engaged audiences.

The personalities emerging in talk radio knew how to stoke their listeners' peeves. Rush Limbaugh became the most popular talker of all. An ex-salesman, sports promoter, and disc jockey, the college dropout from Missouri became a specialist in reactionary monologues. Soon joining Limbaugh in national talk radio fame were a former FBI agent, G. Gordon Liddy, and ex-Marine officer Oliver North. They were followed to the microphones by Fox News Channel hosts Bill O'Reilly and Sean Hannity, who worked daily shifts in radio and television. These talk-show hosts shared a shrill dislike for viewpoints and politicians they labeled as "liberal." Debates focused on "hot-button" topics ranging from gay marriage to gun control, which featured "attack-and-destroy" commentaries earning the talkers impressive ratings.

In 2004, the first self-proclaimed liberal network went on the air with the launch of *Air America Radio.* It featured comedian Al Franken, who announced in his opening remarks that he was "broadcasting from a bunker 3,500 feet below the bunker that (Vice President) Dick Cheney is in," poking fun at the Republican administration. The radio network struggled at first, and within its first six months established a coalition of only 35 radio stations.

Fairness Doctrine FCC policy that held broadcasters responsible for covering divergent opinions on controversial issues of public importance; rescinded by FCC in 1987.

Liberalized Ownership

A major feature of the 1980s was deregulation—removal of restrictions—that encouraged broadcasters to view their outlets more as financial ventures than as permanent community assets. For years, the FCC had limited every radio owner to seven stations nationally, as a way of preserving diversity of control over an important medium. During the Reagan administration—committed to reducing government's power—the ownership limits were liberalized to 12 stations per owner. That was only the beginning. The ceiling on stations per owner was later raised to 18, and then to 20.

Federal Communications Commission and congressional actions in the nineties opened vast opportunities for expansion of the broadcasting business. The **Telecommunications Act of 1996** eliminated the national radio-ownership "cap," a limit on the number of U.S. stations any one company could own. It also cleared the way for an owner to hold up to eight radio stations in the same market and to collect 50 percent of the advertising revenue, depending on market size. This abruptly made stations more attractive and more available to big-money ownership groups. *Radio & Records*, the leading industry trade magazine, headlined: "Let the Deals Begin!" It was understood, correctly, that the industry's new battle cry would be "Buy, sell, or get out of the way."

Many statistics point to how large the major corporations in radio have grown, but one deal drives the point home. After Clear Channel Communications took over AM-FM Broadcasting's 443 stations in 2000, Clear Channel's control over U.S. radio channels topped 1,000. Today, the Texas-based corporation boasts that it owns or programs 1,376 radio stations reaching 65 countries. In the mid-1990s, the top 25 group owners controlled a little more than 7 percent of radio stations; nowadays they direct about 25 percent.

The Revenue Chase

Radio became a $19 billion business by the end of the twentieth century, exceeding the value of either the music business or the movies. For the first time, a larger group of co-owned stations could offer an advertiser more circulation than the local newspaper and a larger cumulative audience ("cume") than the leading TV station. Such a group could coordinate all of its stations' promotional activities.

About 80 percent of radio's advertising sales were based on local spots. Commercial clutter kept spreading through each hour, and the more clearly any company dominated its market, the more it got away with this tactic. Westwood One's Norm Pattiz called it one of the "negative side effects of consolidation." "When more stations in a market are owned by fewer groups," Pattiz said, "those groups can, without any formal collusion, increase the commercial load on the radio stations—without having to worry about their competitors running limited commercial hours and attracting more audiences."[19]

Maximization of profit had become more openly the overriding objective of commercial radio stations. This often meant layoffs, since a staff could consist of little more than a tiny sales force and one on-air person (acting as his or her own engineer), with most of the programs originating from distant locales. Localism seemed to diminish to such an extent that in 2004 the FCC undertook its own investigation to determine if community news coverage had become a rarity. Charges were made that radio announcers were just "ripping 'n' reading" off the state or national wires. Syndicated news-supply services were serving as a substitute for local news operations, and the government wanted to find out why.

NORM PATTIZ. Pattiz created Westwood One, which handles radio broadcasting for CBS, CNN, Fox, Metro, NBC, and Shadow Traffic. Pattiz also oversees U.S. government radio, including the Voice of America, as a member of the Broadcasting Board of Governors.

Telecommunications Act of 1996
First major piece of electronic-media legislation since 1934; liberalized radio ownership and relaxed licensing requirements.

Similar economies were applied to music programming. Satellites could relay syndicated music programs from distant studios, allowing stations to insert their own commercial spots and promotional messages. Radio talent, once found in many local stations, now was located primarily at program centers in cities such as Chicago, Los Angeles, Atlanta, and Seattle. News and talk-formatted stations had multiplied, from an initial 200 or so to 1,350 by the mid-1990s.

Glow of the Satellite

The FCC decided in 1997 to establish 20 channels for satellite radio. They weren't necessary then, but within a few years, digital progress would open a new window to relays from the sky. Whether and how it might transform U.S. radio was the hanging question.

Radio long had been a powerful, ubiquitous medium: The country has twice as many radio sets as people, and those people averaged more than 21 hours of listening per week. However, they've generally been stuck with whatever nearby stations could provide—and, as motorists passed them on long commutes, with signals fading in and out. Satellite radio could end that, making it possible to drive cross-country while listening to uninterrupted streams of programming.

The founders of this new branch of media, XM Satellite Radio and Sirius Satellite Radio, lost billions of dollars in their early development years. But they won multimillion-dollar investments from auto companies planning to offer the new receivers to car buyers. XM launched two satellites and hit the market by 2001; the following year, Sirius was in four cities and about to go national. By 2004, the two services were offering more than 100 channels of digital sound for more than two million subscribers.

Like satellite TV channels, satellite radio doesn't have to lease public airwaves the way conventional radio does—and therefore is not subject to FCC limits on content and language. That's the reason given by shock-jock Howard Stern for his decision to jump from standard radio to Sirius as of 2006. "The FCC . . . has stopped me from doing business," said Stern, whose raunchy material and comments had led the commission to fine him—and some stations to ban him—from time to time.

BOB EDWARDS. NPR pushed this respected but low-key 57-year-old newsman out of his anchor chair in a hunt for younger listeners. Edwards quickly jumped to XM Satellite Radio, one of two companies spending hundreds of millions of dollars hiring talent for the new medium.

⊤ Toward a Digital Future

Radio, TV, and cable are moving to exploit the digital revolution, which will take forms that men named Marconi, Zworykin, and Armstrong never could have conceived. Digital transmission promises to bring new programming, data services, and "interactivity" to radio and TV sets as well as to home, office, and portable computers. Not only broadcast companies with upgraded skills but some wholly Internet-based enterprises can be expected to join this movement. What no one yet knows is exactly which combinations of technology and content will win over enough users to make money year by year.

RADIO

By the middle of the new century's first decade, broadcast AM and FM radio channels continued to satisfy their audiences, but local station owners found themselves looking over their shoulders. There was new competition from MP3 players, satellite and cable music channels, and a new type of radio, low-power FM (LPFM). LPFM stations were broadcasting for nonprofit groups at less than 100 watts, continuing to grow in number despite the National Association of Broadcasters' (NAB) protests over their signal interference. The NAB began looking to "HD radio" (digital broadcasting) to move its AM and FM member stations into the twenty-first century.

Radio is the last analog medium broadcasting in the United States, and has been trying to keep listeners from defecting to a host of digital competitors from MP3 downloads to satellite radio. The digital migration was slow to make its way to American radio stations and, with so few digital stations on the air, listeners had little incentive to buy new digital radio receivers and electronics retailers had even less incentive to stock them.

The in-band-on channel (IBOC) solution was approved so that listeners could keep enjoying both digital and analog radio until the transition was made. Yet, the FCC set no firm date for the switch-over, and so no one seemed to know when analog radio would at last go digital. Once HD radio did arrive, the fidelity of AM

SATELLITE RADIO. Sirius and XM's phenomenal success has a lot to do with 120 new channels for music, news, and entertainment—most of which are commercial-free. It probably doesn't hurt subscription sales that satellite radios are portable and can be placed in a pickup truck or on a picnic table.

radio would achieve a level approaching FM sound, and it would push FM quality to that of a compact disc.

Web World: Riding in On a New Wave

Internet streaming once looked like the next big thing in radio, but the Recording Industry Association of America (RIAA) took aim at "webcasting" after a dramatic decline in the industry's CD sales was blamed on the Internet, particularly peer-to-peer file sharing. As a result, major record labels demanded restrictions on what both Internet and terrestrial radio stations could broadcast on-line. These rules required reporting the number of times a song was played: The number of web listeners clicking on to a station's "stream," with each click counting as a performance, creating higher fees that strapped broadcasters would have to pay.

In July 2002, the Copyright Arbitration Royalty Panel (CARP) announced how much it would assess Internet radio in music fees. Broadcasters cried foul, calling the performance-based criteria a "cumbersome and costly" system. Congress stepped in with a new measure on copyright royalties in 2004. It proposed to establish a panel of three judges for arbitrating royalty fees between the recording industry and webcasters.

TELEVISION

Television is well past its early tentative steps into the digital world. Digital components have replaced critical electronics inside TV sets, and the industry is well along in a movement that could have social impact. With its new clarity and screen proportions, high-definition television (HDTV) is expected to change our perceptions of that "plywood box," to use the late movie mogul Darryl F. Zanuck's dismissive term for television. By 2004, local stations in larger cities and all of the major networks were sending out programs in HDTV, and millions of homes had HDTV-ready sets in place.

However, the race by various enterprises to exploit television's digital present and future was forcing traditional TV to buy in, protect itself, or adapt. One major factor is the digital video recorder (DVR), which allows viewers to stop and replay live programming—and to skip commercials. Clearly, this threatens the elegant advertising strategies that for decades have filled broadcast TV's corporate coffers. One answer by advertisers has been to pay for "product placement" in program content. Many dramas and comedy shows now sport labels, logos, and blazing neon signs pushing real-life goods and services, and actors freely throw around brand names on the air.

On another front, satellite networks have long gone head-to-head with broadcast television, winning Emmys every year and offering commercial-free programming with digital quality. Selling subscriptions to users, as cable TV does, has kept satellite program providers relatively independent of some of TV's economic pressures.

There have been long inter-corporate squabbles and sudden business shifts, as when the DirecTV satellite company sold off its shares in the TiVo DVR company in late 2004. This seemed odd; of TiVo's hundreds of thousands of subscribers, more than half received its service through their DirecTV boxes. So what did the abrupt pullout mean? *Not* an inter-company spat, said the principals; but the digital side of electronic media clearly is a new arena for tough boardroom battles.

Against this turbulent background, conventional television broadcasters wield new tools of their own. The government in the 1990s gave them a deal: free additional space on the airwaves in return for the stations' broadcasting in digital format to help in developing HDTV. Now, in some cities, "spare" digital bandwidth is allowing broadcasters to deliver premium satellite-style programming to subscription customers, collecting new revenue for it. All the viewer needs is a digital set, a broadcast antenna on the roof, and a small set-top box—another sign of the opportunistic mix-and-match quality of change in the evolving television industry.[20]

Moreover, the world of computers, from which "digital" emerged, continues work on a parallel universe to television. Already, anyone with a personal computer can add a feature that, like the DVR, enables users to stop and replay live Internet programming. That means, in effect, that the PC can act as a home video server, feeding programs to other networked computers and providing other options that TV alone doesn't offer.

Similarly, computer users with broadband connections can order not just movies through video on demand (VOD) but documentaries from *National Geographic*—just like those seen on TV. Some major Hollywood film companies have joined in this enterprise. It could end up not only making money but also adding a new layer of competition with television. Depending on which expert you consult, such moves also could hasten a day when television and computers are virtually one and the same.

Summary

Radio and television have recorded and reported much of the lived experience of the twentieth and early twenty-first centuries. Springing from its roots in wireless telegraphy, the process of capturing and transmitting to us the story of our collective life has been invented and reinvented time and again. With that human story has come its unending music as well as the created comedy and drama that feed our hunger for entertainment.

In commercial broadcasting, reaching desirable audience demographics, rather than a mass audience, has come to be the main emphasis. Broadcasting's once-powerful unifying effect has been diffused as networks and stations entice listeners and viewers by age, gender, ethnicity, socioeconomic status, and other indicators. Large radio chains have become like ice cream shops, offering as many flavors in music as there are tastes.

History gives us an important perspective as we consider the state of commercial broadcasting today. It seems remarkable now that the long-ago founders of some giant broadcast companies were intent not on disseminating news and entertainment but on peddling radio sets. Business took over most broadcasting early. Advertisements fill more minutes per broadcast hour today than ever before. Since the 1980s, the FCC has worked to loosen ownership limits to facilitate more consolidation of broadcast companies. This has coincided with increased competition, the arrival of the Internet, and the promise of still other uses of digital technology to communicate. These forces will keep expanding the electronic media—continuously changing their roles in American life.

Food for Thought

1. Broadcasting evidently is losing the "mass" audience it began to build in the 1920s. What role do you think broadcasting will have in society in 10 years? 20 years?
2. The government struck at network power in 1941 with the Report on Chain Broadcasting. Was that move a good one? Why or why not? Is similar government action needed now? Explain.
3. FCC Commissioner Frieda Hennock fought for and won channels for public broadcasting. Could she win today? Why or why not? What factors would support or oppose her efforts?
4. Since the 1980s, competition has led to niche programming across radio and cable TV (e.g., MTV, ESPN, talk shows). Has this mattered to society? In what ways?
5. The government has been slow to set a future digital standard for radio. Should the industry be allowed to settle this for itself? Why or why not? What's at stake?

Cable *and* Satellite

Chapter 3

Cries of censorship could be heard as soon as CBS brass made the surprise move late in October 2003. *The Reagans,* a two-part miniseries, would *not* be shown in November as scheduled but would be reassigned to cable distribution instead. Was this a case of censorship, or just a media conglomerate trying to make the best of an awkward situation?

The firestorm that erupted at CBS television began routinely enough. Executives chose a familiar vehicle, the miniseries, to showcase a popular American president during an important time when national audiences are measured for the purpose of program ratings. What CBS had not taken into account was how Nancy Reagan would protest her portrayal as a first lady dependent on pharmaceutical aids, and her husband as a homophobic president suffering from Alzheimer's disease. The Republican National Committee responded by asking CBS to invite a team of historians to review its content, while talk-show hosts re-ignited familiar charges of liberal bias. A week later, CBS reversed its stance and declared that *The Reagans* would not be shown over its broadcast affiliates, but would be viewed on its cable network, Showtime. On cable, the two-part series featured a live forum with historians and journalists, and it garnered a respectable cable audience of about one million viewers, although it could have reached perhaps 10 times as many viewers over broadcast channels.

Afterward, CBS television president Les Moonves responded to allegations of censorship by saying he was surprised to discover after viewing a rough cut that the series lacked balance. It was "somewhat one-sided and it wasn't the movie I promised the public." So how could *The Reagans* be too biased for CBS, but just about right for Showtime? Moonves explained to *Time* magazine's Richard Zoglin, "When somebody's paying $30 a month, that's their decision. On a network, there's a public trust to it. . . . A cable network can be a bit more one-sided and do an opinion piece."

This episode at CBS Viacom illustrates the political pressures broadcast networks face and how cable operates under different rules. It further shows how the longstanding barriers between cable and broadcast properties have just about collapsed. In this chapter, we will see how principles of supply and demand gave rise to cable television, and how the upstart medium found its way to success beyond advertising revenues. We also will examine the nature of cable television competition and ideas about how media choices may vary, but the overall investment remains the same.

Cable television evolved from small-town systems hooked up to community antennas to multibillion-dollar enterprises. More than 9,300 systems weave their way across the national landscape, but their numbers are shrinking from a peak of more than 11,000 systems reached in the mid-1990s. Mergers and acquisitions are responsible for much of the decline—so many mergers, in fact, that today's top 10 conglomerates serve more than 80 percent of basic cable subscribers in the United States.

What factors gave rise to such an industry? The first systems were designed to escape the freeze imposed on new TV stations in 1948. The FCC needed time to sort out its channel allocation map while selecting the color technology to replace black-and-white TV. The 108 stations on the air simply whetted the appetite of viewers wanting "radio with pictures." Some could not wait for the thaw and began yanking distant TV signals from the sky. Their stories form the genesis of cable television, or, as it was known then, **community antenna television (CATV).**

ED PARSONS, CATV ENGINEER OF ASTORIA.

Early Days of CATV

At first, appliance shopkeepers began to erect TV antennas in high places, hoping to capture distant programs and cascade them by wire down to neighbors in the "white areas" where TV screens turned to snow. Who would've bet that these early "mom and pop" systems would one day emerge to become lucrative television enterprises? Some people did and made a fortune; others simply lost their shirts.

FIRST MOM AND POP SYSTEM

The original parents of community antenna television lived in Astoria, Oregon. E. L. "Ed" Parsons was a radio station owner and engineer who ran a marine radio service for the Columbia River's fishing boats. He and his wife, Grace, were impressed by a 1947 TV demonstration at a Chicago broadcasters' convention. She had heard a new Seattle TV station would be on the air in 1948 televising a high school football game. Grace asked her husband, Ed, to figure out a way to close the distance between Astoria and KRSC (now KING-TV) in Seattle. Ed began scouting around for a point where that signal could be tuned in, and a location where he could place his antenna. He found that spot atop the Astoria Hotel across the street from his and Grace's apartment, where they invited friends and family to watch the Thanksgiving Day sports event.

Soon, just about everyone had heard the news and wanted to see the furniture with pictures. The Parsonses discovered that if they were to have any peace, Ed would have to drop a cable down the elevator shaft to the hotel lobby and install a TV set there for others to watch. Soon, he had strung wire to local bars and about 25 "subscribing" neighbors, charging them $125 for the installation costs but without monthly fees. That was the beginning of community antenna television.[1]

APPLIANCE STORE OWNERS

The prospect of rooting for football teams on autumn afternoons inspired a TV appliance shopkeeper in Tuckerman, Arkansas, to build a 100-foot tower on his store's roof. Jim Y. Davidson thought he could capture the TV signal from Memphis, Tennessee, 90 miles away. He carefully strung cable down to the American Legion Hall so that he and his friends could watch Ole Miss play Tennessee. Davidson thought he might like this new line of work, and so he created Davidco to begin shipping ready-made cable systems to towns and cities around the country.[2]

Community antenna television (CATV) Term used for local cable systems until satellite and microwave distribution prompted a change to simply "cable."

Microwave Radio signals of at least 1,000 MHz carrying audio and video over long distances, either to satellites or to terrestrial relay towers.

Headend Technical center of a cable system where all programming is received, amplified, and retransmitted.

Meanwhile, back east, families in the hills of Pennsylvania naturally wanted to get their first peek at "seeing radio." An appliance store owner and lineman for Pennsylvania Power and Light, John Walson, hooked 700 homes to his antenna in Mahanoy City. In Lansford, Pennsylvania, another electronics appliance store owner, Robert Tarlton, was trying to figure out how to bring a signal in from Philadelphia's WCAU-TV across Summit Hill. He began rounding up appliance shopkeepers to see if they would invest in his idea. They did and, in 1950, Panther Valley CATV was born.[3]

JERROLD: CATV'S PIONEER EQUIPMENT MANUFACTURER

The excitement surrounding this CATV system sparked the interest of a Philadelphia lawyer and future governor of Pennsylvania, Milton Jerrold Shapp. Shapp watched the Panther Valley cable workers stringing lines during the Thanksgiving holiday of 1950 and decided to get involved. He named his firm Jerrold Electronics and began supplying cable wire to the Lansford system. Community antenna television was no longer an experiment now; it was a growing enterprise. At first, CATV made money based on home-installation fees of between $100 and $200 rather than monthly cable bills, which then amounted to only a few dollars a month for system maintenance.[4]

For other CATV enterprises to get started, Shapp invited Wall Street investors to come on board. Three venture capital firms in 1952 put up money for a system in Williamsport, Pennsylvania. Within two years, the Williamsport franchise grew to become the largest in the country. Shapp and Bob Tarlton took to the road, spreading the good news of community antenna television.

The first CATV systems carried only 2 or 3 channels, but in larger towns that "fill-in" service was a welcome relief. When one or more network affiliates were missing, CATV operators would import channels from distant cities so viewers could see the Big Three network channels (ABC, CBS, NBC). They relied on **microwave** companies to bounce the signals back to the distribution center that became known as the **headend.** New amplifiers and other technical improvements allowed CATV operators to increase the number

MILTON JERROLD SHAPP, FOUNDER OF JERROLD ELECTRONICS.

PRINCIPLE 1

When public demand exceeds the supply of media products, private enterprise generates more media products.

The growth of the cable industry drew on the enthusiasm and entrepreneurial spirit of big deal makers such as Bill Daniels. This ex-fighter pilot and former Golden Gloves champ was looking for work in Wyoming's oil field when, at the age of 37, he noticed something new in Denver—television. Daniels, along with other customers, came to Murphy's Restaurant every Wednesday night to watch the boxing matches. He drove in each week from Casper, hours away, where he was trying to make a go of it selling insurance in the oil industry. After reading about CATV in the newspaper, Daniels asked himself, Why not bring it to Wyoming? With an

sidebar

CABLE'S EARLY WRANGLER

investment from AT&T and training from Marty Malarkey at Jerrold Electronics, he became the first cable operator to relay TV signals 120 miles from Denver to Casper.

Daniels's salesmanship inspired others to enter the business, or at least to move forward in expanding their cable operations. In just five years, he made 147 deals for cable systems, involving $1.5 billion. Early cable giants such as TelePrompTer, Time Warner Cable, Cox Communications, Tele-Communications, Inc., and Sammons all give Daniels credit for either inspiring their start-ups or forging deals for their companies' expansion.

MICROWAVE RELAY TOWERS. Microwave towers are used to capture and transmit audio and video signals on earth. Spaced about 40 miles apart, they may be used to repeat signals across the landscape.

LAUNCH OF TELECOM SATELLITE. A rocket lifts off from a base in French Guyana carrying two telecommunication satellites to send video images to viewers in India and France.

of channels to 5, and later to 12 choices. These early entrepreneurs responded to public demand by supplying new media products through the innovation of CATV, a new delivery system.

◼ Open Skies

In the middle of the twentieth century, communication satellites were launched. Broadcast signals bounced off orbiting spheres focused on huge expanses of the earth. Satellite technology helped lift geographic barriers and became a key ingredient in cable television's development. In 1972, the FCC encouraged **cable networks** to relay their signals via satellites by adopting an *open-sky* policy. This action gave the green light for cable companies to enter the domestic communication satellite business so long as they had the technical expertise and financial support to do so.

HBO'S STAR IS BORN

A cable operator serving Lower Manhattan was having financial difficulties and imagined a way to make pay television work in the late 1960s. Charles F. (Chuck) Dolan's

Cable network A closed-circuit channel offering television programming via satellite to local cable systems, for delivery to its subscribers.

scheme came to him while aboard the *Queen Elizabeth II* on vacation; he was struck by an idea to save his struggling Sterling Manhattan Cable system—a pay TV channel devoted to movies and sports. He would label it the "Green Channel," and would offer subscribers uncut movies, New York Knicks basketball, New York Rangers hockey, and live boxing matches from Madison Square Garden. Needing financial backing, Dolan presented his idea to Time Inc.'s board of directors in 1971, and they seemed to like it. Dolan told Time's board of directors that one day HBO would become "the Macy's of television [and it would] . . . use whatever efficient transmission systems become available, from microwave to satellite, to sell television programs worldwide." He brought on board a lawyer, Gerald Levin, and a marketing specialist, Tony Thompson, and they decided a better name might be Home Box Office (HBO).[5]

On November 8, 1972, HBO aired its first film, *Sometimes a Great Notion*, at what proved to be an inauspicious occasion. The mayor of Wilkes-Barre, Pennsylvania, where HBO's debut was held, was forced to cancel his appearance, and the local newspaper thought it unworthy of coverage. Yet, the event marked the inauguration of premium cable television, ushering in both a new format and a new stream of revenue. Satellites began feeding HBO's programs in 1975. Its first satellite telecast to subscribers was "The Thrilla in Manila," a live boxing match between Muhammad Ali and Joe Frazier.

Soon, HBO had competition. Viacom had been cobbled together in 1971 from some of CBS's cable operations and syndication properties. Aspiring to become the nation's leading cable programmer, Viacom began feeding its content by videotape and microwave until domestic satellites, or "domsats," created a new system for networking. In 1978, Viacom launched Showtime as a competitor to HBO. The big winners were the telecommunication firms that auctioned off the **transponders,** satellite circuits that were leased to carry programming from earth to space for millions.

Domsats also provided another innovation for cable television—the **superstation,** a local TV station based in a major city but distributed nationwide on multiple cable systems. The same "bird" carrying the HBO signal would start carrying other passengers, including a struggling TV station in Atlanta, Georgia. When WTBS, an independent UHF station, was fed via satellite to cable systems in 1976, it became America's first superstation. In turn, WTBS inspired other independent TV stations to seek a cable channel, while other programmers planned for niche networks of news, sports, and music to fill the cable menu of channels. The guiding spirit of this early activity was an Atlanta broadcaster by the name of Robert E. (Ted) Turner.

CABLE TIERS

Cable's bottom line was secured not only by its power to retransmit local or distant channels but also by its programming options, which include premium channels and **pay-per-view.** Pay cable enables subscribers to buy programs by charging them more than the basic monthly fee. This fee structure is based on **tiering,** which gives the cable operator a menu of channels to sell based on levels or tiers of service. The basic service features local TV channels, **public access,** one or more distant superstations, and advertiser-supported networks, such as ESPN and CNN. (The average monthly fee for basic service was $36.59 in 2003.[6])

The cable operator also offers an expanded basic tier that includes channels such as Nickelodeon, A&E, and the Weather Channel. Above that tier is a premium offering with clusters or packages of commercial-free, pay-cable channels such as HBO and Showtime. Premium channels cost more, but a majority of the profits

Transponder The term is a conflation of *transmitter* and *responder.* The receiver/transmitter unit on a satellite that picks up signals on one channel and bounces them back to earth on another one.

Superstation A commercial broadcasting station whose signal is retransmitted by satellite and cable systems to a national audience of subscribers.

Pay-per-view Programs sold to customers and priced according to the show requested.

Tiering Marketing strategy for selling levels of cable service, based on over-the-air signals at the low end and premium channels at the high end.

Public access Dedicated channels allowing residents to produce and televise programs over a community cable system.

PRINCIPLE 2

To become viable, new media create new ways of making money.

TED TURNER

JOHN MALONE AND BOB MAGNESS

A New York business writer said the man looked like Clark Gable, but sounded like Huey P. Long. Perhaps an apt comparison, considering the impact Ted Turner made on both entertainment and politics. Turner's life did not always hold that promise, though. When he inherited his father's billboard business at the age of 24, he was—like his first TV station—a struggling independent, prone to make trouble. Turner had been tossed out of Brown University for having a woman in his room, and when he won the America's Cup title in yacht racing, he showed up drunk to accept his team's trophy. But whatever Turner lacked in self-control, he more than made up for it in imagination, drive, and bravado.

In 1970, Turner renamed WTCG-TV and reprogrammed it as WTBS (Turner Broadcasting System) with a lineup of sports and movies. He bought two sports franchises, the Atlanta Hawks and the Atlanta Braves, and his superstation shined a light on their seasons. Southern Satellite Systems beamed WTBS-TV programming to cable systems around the nation, and Turner became a billionaire. His genius also brought forth CNN, Headline News, Turner Network Television, and a host of enterprises from the Goodwill Games to the Turner Foundation that dispensed millions of dollars to charity and environmental causes.

In 1987, Turner began to turn over the keys of his empire to others. First, he invited investors to buy about one-third of TBS so he could pay off $1.5 billion in debt to MGM/United Artists. In 1995, one of Turner's rivals, Warner CEO Gerald Levin, assumed control of TBS and most of Turner's cable enterprises. In exchange, Turner became $2 billion richer and was placed in charge of the Atlanta operations, but he was vulnerable. Within five years, Turner saw himself "reorganized" out of his job with the company. Only weeks later, like a Shakespearean tragedy, Gerald Levin, who oversaw Turner's fall, retired from Time Warner.

Not even the briefest overview of cable history is complete without noting the contrasting but intertwined lives of John C. Malone and his mentor, Bob Magness. Malone, a Yale graduate with two master's degrees and a doctorate, made his way from Ivy League halls to the top of Jerrold Industries, a major seller of cable equipment. In the early 1970s, Malone's ambitions converged with the interests of a cottonseed salesman and part-time rancher in Texas. Bob Magness with his wife, Betty, had turned a small cable system in the Texas panhandle into the nation's tenth largest **MSO (multiple system operator),** Tele-Communications, Inc. (TCI), which had moved to headquarters in Denver, Colorado.

In 1972, Magness was looking for a partner to help manage a business that was financially overextended, and Malone was looking for a new business opportunity. The two joined forces at TCI, and over the next two decades built the largest cable enterprise in the nation. Magness was the heart and soul of TCI, a folksy sort who loved to stay in touch with his employees, clients, and customers, regardless of their rank or station. Malone, on the other hand, was the hard-driving engineer and businessman who knew how to deal with Wall Street. By the time of his death in 1996, Magness had become a billionaire and had returned to his first love, raising Arabian horses and Limousin cattle. Two years later, Malone was ready to turn the business over to someone else, and AT&T (now Comcast) stepped forward. Meanwhile, Malone moved on to one of TCI's spin-offs, Liberty Media, where he served as chairman of the board.

go to the cable operator through fee-splitting deals that may include promotional opportunities. Pay channels are devoted to movies, sports, or special events. For example, a film distributor will license a motion picture for a specific number of showings during a set period of time over the channel, or a promoter will license a sporting event or concert. Subscribers also can take a cluster of premium channels called mini-pays that are less expensive.

Multiple system operators (MSOs) Corporations that own and operate more than one cable system.

NEW CHOICES

Cable television companies create new choices of media in order to stay viable, and these options often relate to technical innovations. In 2004, for example, a consumer study showed that two-thirds of American adults preferred "à la carte" pricing and about 80 percent did not want to pay for cable tiers that included channels they didn't watch. A group of "Concerned Women for America" sponsored the poll, which was both criticized and praised. The cable industry challenged its methodology but members of Congress requested further study of à la carte pricing by the FCC. The economic structure of the industry is based on tiers, and pay-per-channel pricing would radically alter that system.

FORMER TIME WARNER CHAIRMAN AND HBO PIONEER GERALD LEVIN.

Meanwhile, the rollout of digital cable proved to be making promising financial gains. It not only expanded the number of channels to around 40 for the basic digital tier but it also gave rise to interactive television services such as digital video recorders (DVRs), video on demand (VOD), and interactive program guides (IPGs). The DVR is a tapeless video recorder that digitally records, stores, and plays back programming for cable customers. Video on demand is another digital innovation that resembles pay-per-view programming. It allows subscribers to order and watch movies on demand, as well as pausing, rewinding, or fast-forwarding them. Subscription video on demand (SVOD) charges viewers a flat rate for unlimited access to a library of video offerings. Some have predicted that SVOD will become the model for the future of television using interactive program guides, which help viewers sort and select their TV shows on screen by title, topic, and time.

Using cable modems for high-speed Internet access has become the choice of about one in five cable subscribers and an estimated two-thirds of all broadband high-capacity digital customers. By mid-2004, there were 16.1 million cable modem users in the United States. Another innovation attracting interest is cable **telephony,** or voice-over Internet protocol (VoIP), serving about 2.5 million cable subscribers. Time Warner Cable partnered with MCI and Sprint to offer digital voice services, and Cablevision began offering New York City customers unlimited local and long-distance phone service for $35 a month. Within three years of its start-up, the number of digital cable customers in the United States surpassed 22.2 million.

RELATIVE CONSTANCY

These developments underscore another principle based on relative constancy. This idea holds that consumers will spend a relatively stable amount of their income on media activities, but those choices will vary. In other words, when viewers choose video on demand, they may quit using Blockbuster for video rentals, which is why the cable industry has been promoting **bundling,** combining media choices into one pay package per month. The principle of relative constancy implies that media must maximize their chances of persuading the consumers to make a favorable decision, which is achieved by offering more media options, greater control, and optimum convenience.

PRINCIPLE 3

Media consumers spend a relatively constant amount of their income on media activities, but their media choices will vary.

◧ Cable's Costs and Benefits

Cable television companies principally rely on two sources of revenue: subscriptions and advertising. Basic channels, such as VH1 and CNN, carry advertising. Others, such as American Movie Classics and HBO, rely on subscriber fees. Certain channels

Telephony Technology associated with transmitting voice, fax, or data between senders and receivers.

Bundling Packaging together several telecommunications services—such as television, telephone, and the Internet—for a monthly fee.

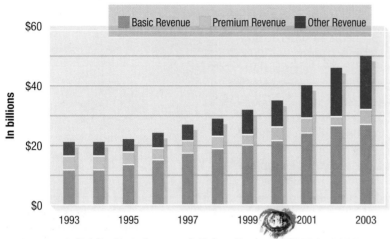

FIGURE 3.1 *Basic, Premium, and Other Revenues, 1993–2003*

Source: Kagan Research, LLC, *Broadband Cable Financial Databook, 2003*. Reprinted with permission.

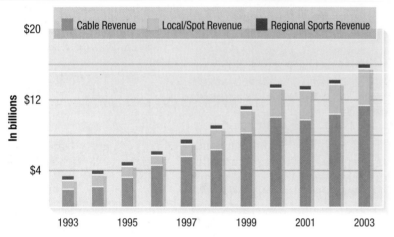

FIGURE 3.2 *Cable Network, Local/Spot, and Regional Sports Advertising Revenue: 1993–2003*

Source: Kagan Research, LLC, *Broadband Cable Financial Databook, 2003*. Reprinted with permission.

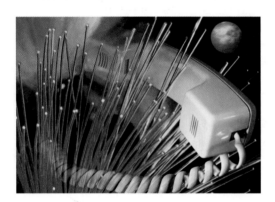

FIBER-OPTIC CABLE AND TELEPHONE. Until 1996, telephone companies could not use their lines to deliver video programs in the same communities where they provided telephone service. The Telecommunications Act of 1996 allowed phone companies to deliver television but few ventures have been successful.

give the cable system slots for selling spots to local advertisers. As Figure 3.1 indicates, the breakdown of subscriber revenue from basic cable is about 56 percent, premium cable provides another 10 percent, and the rest comes from commercial advertising or other sources.

In terms of advertising, Figure 3.2 shows how $16.4 billion in revenue was supplied from three commercial sources: *cable networks* (74%), *local/spots* (23%), and *regional sports* (3%). Other revenue streams for cable included *home shopping commissions, telephony,* and *cable modems,* which represented less than 2 percent of the total picture.

HOW IT ALL WORKS

The money invested in cable systems requires a healthy return from subscriber fees and advertising accounts. For example, a healthy cash flow is needed for satellite receivers, dishes collecting video signals, as well as amplifiers and converters to send those signals down cable lines. Also, there are expenses for laying **fiber-optic lines,** glass threads linked to **coaxial cable** drop lines. All are part of the system's *build-out* costs.

Operating expenses take another bite out of the cable budget, which is one reason systems have been merging to shoulder expenses and take advantage of economies of scale. These consolidated groups of multiple system operators factor in the same expenses but on a much larger scale. There are, for example, the local **franchise fees, copyright licenses,** satellite and microwave costs, pole attachment fees, and access channels for **public, educational, and government (PEG)** programs. In addition, there are the fees charged by about 340 cable networks, which license their programs to each system operator.

CABLE FRANCHISES

Before a cable system bids for a local contract, it must first determine what kind of return it can expect on its investment. The cost of wiring neighborhoods is calculated according to its potential revenue, based on population density and demographics. The **pass-by rate** is the number of homes passed by the cable wire that might subscribe if they could, and **penetration** is the percentage of viewers who actually do. Generally, the higher the penetration, the larger the profits. In recent years, however, cable penetration has seen a decline from more than 69 percent in 2001 to about 67 percent penetration in 2003. The rate of subscribers who buy and then cancel cable service is called **churn**, which may be the result of rate increases, technical problems, or competition from other *multichannel video program distributors (MVPDs)*, such as home satellite television.

Because cable is either strung on poles or underground, it requires right-of-way easements and public notices. Cable franchises usually are given protection from competition because accommodating more than one system poses too many challenges, both technically and economically. That's why cable has been called a **natural monopoly**. A few cable firms, however, are placed in competition in what have been described as **overbuild** situations. Whether one or more companies are involved, they must obtain permission from a local jurisdiction—usually a city council—which signs a franchise agreement that specifies fees and terms of service.

Franchising is more than just paperwork and political maneuvering. Some cities use franchise pacts to shore up public budgets. These agreements—usually lasting 10 or 15 years—cover everything from how many channels are offered to what percentage of the cable profits are paid to the government. Federal law has set the maximum franchise fee at 5 percent of gross revenues, but exacting fees for cable broadband is prohibited.

BREAKDOWN OF THE CABLE SYSTEMS

Think of the cable company as a distribution center. Programming comes into the *headend* by antenna from local channels, microwave relay antennas bring it in from more distant locations, and satellites beam down their cable networks. Programs are also produced at regional production centers, such as sports networks or at the cable company's local production studio, but not all cable systems have such facilities.

Now, step back and view one cable system in total, as depicted in Figure 3.3. Originally, **cable systems** consisted of five components: the *headend*, the *trunk* cable, the *feeder* line, the subscriber *drop*, and *terminal* equipment. The headend is where TV signals from earth and space are received, then retransmitted as video and audio channels over the trunk, feeder, and drop lines to terminal equipment in each subscriber's home. These systems relied primarily on coaxial cable (lines running parallel, or in *coaxis*, through conducting wire and a metal sheath). Coaxial signals, however, fade over distance, which is why early systems installed amplifiers every one-third of a mile to boost the signal power. In the 1990s, fiber-optic glass lines with digital equipment gave rise to higher-quality pictures and sound, greater bandwidth, and a wider array of channels. The resulting systems began to move toward something more efficient, as shown in Figure 3.3.

Star patterns of hybrid fiber/coaxial (HFC) cable have replaced most tree-and-branch systems. There are three points of transmission in the HFC system: at the *headend*, where the signal is retransmitted via fiber-optic lines to a *node*, where the video and audio signals are converted to analog for coaxial delivery to the home's *terminal equipment.*

Cable channels are assigned frequencies, the same as broadcast channels, and the VHF channels use the same wavelength as broadcast channels but without the threat of interference. Each fiber-optic strand can carry more than a million simultaneous signals through the cable without disrupting each other.

GOING DIGITAL THROUGH FIBER

Digital transmission converts light into pulses compatible with the computer's binary language of ones and zeros. These pulses travel along hair-thin strands of glass that are braided together and polished at each end to carry light and images. Fiber-optic lines not only offer greater bandwidth but they are also better suited for handling digital information. Glass fibers carry laser energy generated by light-emitted diodes (LEDs), requiring fewer amplifiers than coaxial cable, which uses

Fiber-optic lines Strands of flexible glass inside a cable transmitting pulses that carry video and audio information for cable television.

Coaxial cable Transmission line for cable television, using a center wire of aluminum or copper surrounded by a shield to prevent signal leakage.

Franchise fees Share of cable revenues dedicated to a governmental authority in exchange for an exclusive contract.

Copyright licenses Ownership rights to literary, dramatic, musical, or artistic expressions.

Public, educational, government (PEG) Government term to describe dedicated-access channels on cable systems.

Pass-by rate All homes passed by a cable feeder line, as a percentage of all homes in the area.

Penetration Percentage of customers subscribing to cable, based on all the homes passed by the cable line.

Churn Dropout of pay-cable subscribers after a short period of service.

Natural monopoly Cable systems represent a type of industry where the most efficient means of production and distribution are by a monopoly.

Overbuild Competition between two or more cable systems with lines passing the same households.

Cable system A wired network for distributing television programs on a subscription basis to homes in a single community.

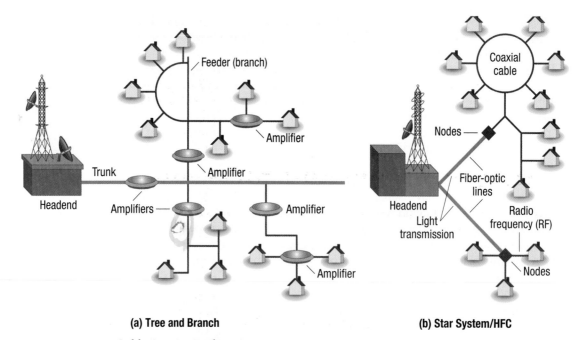

(a) Tree and Branch

(b) Star System/HFC

FIGURE 3.3 *Cable System Evolution: From Tree to Star*
Yesterday's cable systems used a tree-and-branch style of television distribution with large trunk lines branching out to smaller feeder lines and finally drop-lines to individual homes. Today's cable systems transmit from the headend over fiber-optic cable to nodes, where digital signals are converted to analog then fed by coaxial cable to homes. These systems are called *HFC* for hybrid fiber coaxial.

copper and aluminum conductors, as shown in Figure 3.4. This allows for more channels using less bandwidth. The disadvantage to this system occurs when a fiber-optic line breaks. It is difficult to splice the threads, given the size and delicate nature of glass fiber. Nonetheless, the advantage of carrying more digital channels greater distances without amplification outweighs the disadvantages.

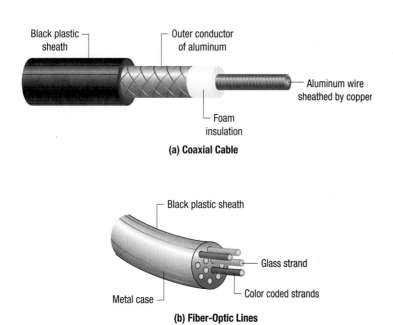

(a) Coaxial Cable

(b) Fiber-Optic Lines

FIGURE 3.4 *Fiber-Optic and Coaxial Cable*

DIGITAL HDTV

One major advantage of digital cable is high-definition television (HDTV), a format affording cinema-quality pictures with CD-quality sound. Cable's rollout of HDTV has not been without its problems, though. In 2002, the 10 largest MSOs pledged to the FCC that they would provide by 2003 a package of high-definition television services to subscribers who would pay for them. To support this promise, cable operators began ordering the necessary decoders to send HDTV signals directly to the newest digital TV sets, but the rollout was stalled by network programmers, which put cable systems in a difficult position.

Network programmers, such as ESPN and Discovery, promised to charge operators extra for HDTV because the shows were more expensive to produce. Cable owners were reluctant to pass on additional HDTV charges to subscribers after already billing them for $85-plus billion dollars in improvements to upgrade systems to fiber and to handle digital, interactive communications. Financial burdens imposed by upgrades and consolidation debt left cable systems struggling to reduce costs. Regardless, the deadline was set and the FCC made it clear that the move from analog to digital television would take place, and the government expected cable to be on board.

▐ Merger Mania and Shakeouts

How did the national cable industry grow from many small, independent systems to a few huge conglomerates? In many cases, smaller operators could not afford the equipment necessary to upgrade their cable systems, while larger firms could. When it came time to renew franchise agreements, MSOs impressed local politicians with their visionary plans to solve a host of problems, from burglar alarms to interactive billing. Older, smaller cable companies were simply overwhelmed and ended up cashing in their chips to larger firms with greater cash flow. That blue-sky era in cable is now over but not forgotten. Today, growth in the cable industry is primarily a result of two factors: horizontal and vertical integration.

HORIZONTAL INTEGRATION

At one time, thousands of cable systems with about as many individual owners served the nation; today, the owners are consolidated and fewer in number. In 1975, the top 10 multiple system operators reached 40 percent of the cable market; by 1990, the top 10 had 62 percent of the business. Today, the top 10 MSOs bill more than 80 percent of basic cable customers in the United States, and the largest one, Comcast Communications, reaches one-fifth of the nation's cable customers, or about 21.4 million subscribers.

Simply stated, *horizontal integration* occurs when two or more companies in the same line of business join forces. It is a game of economic survival and is viewed as a management strategy. It also can be a giant step toward monopoly, which is why the government has safeguards to protect fair competition through agencies such as the Federal Communications Commission. In 2002, the FCC approved a $72 billion merger of Comcast and AT&T Broadband, creating the largest cable conglomerate in the nation. The lone dissenting vote on that deal was Commissioner Michael J. Copps, who called it a "huge consolidation of commercial power [with] potential for significant harm to consumers, the industry, and the country." Copps predicted consumers would not likely see lower cable prices as a result, and this "megacombination" could use its power to deny competing multichannel video program distributors access to popular programs. What these mergers entail is often an assumption of debt and considerable downsizing that goes along with it. For example, the Comcast/AT&T merger took on an additional $30 billion in debt.

Some media mergers bear immediate consequences resulting in layoffs that sometimes reach the top of the organization. In 2001, AOL Time Warner formed the largest megamerger to date, but within two years the magic had gone out of the marriage. AOL's Chair Stephen Case and former Time Warner Chief Executive Gerald Levin had both left the company, and AOL came under investigation by the Securities and Exchange Commission for its advertising revenue and subscriber accounting practices. In 2002, the company's directors decided to remove AOL from the masthead and stock ticker, and reverted to its earlier name, Time Warner Entertainment (TWE), although it was a change in name only. Table 3.1 lists the top 10 MSOs in the nation with basic subscribers.

TABLE 3.1	Top Ten Cable System Operators in 2004	
RANK	**MULTIPLE SYSTEM OPERATOR (MSO)**	**BASIC CUSTOMERS**
1.	Comcast Cable Communications	21,468,000
2.	Time Warner Cable	10,919,000
3.	Charter Communications	6,431,300
4.	Cox Communications	6,338,300
5.	Adelphia Communications	5,469,800
6.	Cablevision Systems Corporation	2,944,000
7.	Bright House Networks	2,167,000
8.	Mediacom Communications Corporation	1,543,000
9.	Insight Communications	1,293,600
10.	CableOne	720,800

Source: NCTA.

VERTICAL INTEGRATION

Huge media conglomerates are not always interested in the day-to-day affairs of the local franchises they own. However, they are interested in the money to be made by the production interests of their companies, which is why so many acquisitions involve the takeover of program networks. Having a hand in the affairs of production, distribution, and exhibition of media content is called *vertical integration*. It defines how media conglomerates have grasped key links in the chain, such as the production studios, program networks, and system operators.

In the spring of 2004, the largest MSO in the country, Comcast, pursued a $48 billion bid for the Walt Disney Company that promised to add the ABC network, 13 cable networks, 11 television and film production studios, 10 TV stations, and a host of other properties to Comcast's stable of media. Disney rebuffed these unwanted advances; Comcast instead settled for a buyout of Tech TV.

NETWORK DEALS

The financial lure of consolidation also encouraged NBC and Vivendi Universal to merge in 2004. That deal meant the new NBC Universal would own one of the top-rated TV networks, NBC; a major movie studio; a TV production studio; seven cable channels, including USA and Sci Fi; and a group of 29 TV stations. The combined NBC and Universal studios would net profits from TV shows produced for both broadcast and cable networks, which is yet another example of vertical integration supporting the principle that the incentives of consolidation have led to larger media conglomerates.

PRINCIPLE 4

The financial lure of consolidation in the cable industry tends to create larger media conglomerates.

▐ Storm Clouds for Cable

Episodes of corruption, controversial content, and public access disputes have kept cable in the eye of the storm of public opinion. In 2004, the departure of one cable executive who helped bring television fame to Anna Nicole Smith and Howard Stern made news. E!'s chief executive, Mindy Herman, a graduate of the University of Pennsylvania's Wharton School of Business, exited her perch atop the network amid reports that she ran the cable channel as a tyrant, placed her baby shower on the company tab, and shook down her employees for gifts. Herman's contract provided for a multimillion-dollar "golden parachute" that included a percentage of the E! channel's financial gains on her watch.

More recently, cable networks have stirred controversy by programming for the gay and lesbian audience. NBC Bravo's *Queer Eye for the Straight Guy* and CBS Showtime's *Queer as Folk* began breaking ground in sexually oriented content. MTV announced its plans to program a gay-styled network called Logo. Even the City of Brotherly Love opted for a cable advertising campaign designed to lure gays to Philadelphia. These developments underscore the principle of cable's unique position with respect to controversial content. It can show more sex and violence because it is based on subscriber consent.

PRINCIPLE 5

Cable can show more sexual and violent material because it is based on subscriber consent.

protalk
Ed Bowie
EXECUTIVE DIRECTOR, ACADIANA OPEN CHANNEL

Ed Bowie runs the public access cable channel in the heart of Cajun country, Lafayette, Louisiana. Like the local cuisine, programming on Acadiana Open Channel (AOC) tends to be hot and spicy at times. The channel's call-in shows have been flashpoints for both racial and sexual tensions in the community. Bowie came on board just as the Ku Klux Klan was making a quick exit from Acadiana Open Channel; a prosecutor had brought the on-air host to trial for allowing guests to wear their hoods and masks in the studio. The Klan show was replaced by *Jabari Speaks,* a talk show that critiques the African American experience in south Louisiana. The Snake Pit finished its programming schedule by inviting vivid fantasies for cablecast from local callers.

Bowie says AOC is like the community's x-ray machine: "You don't always want to know that you have cancer but sometimes you need that x-ray machine to look inside and see that you do." Bowie, a Vietnam veteran, has to keep peace among radical factions who use the tools of his public access center. He has appeared in government chambers and before television news cameras to spread the message of tolerance that he sees embedded in the First Amendment. He believes that public access cable is vital

"We're like the community's x-ray machine."

to the community not only as an electronic soapbox but also as a means for neighbors to find out more about themselves and their community.

Acadiana Open Channel brings to cable television the civic leaders and public servants as well as the dissenters. Social service programs, produced by Goodwill and other community agencies, make up a large part of the program day at AOC, along with local government and the school board. In a media world dominated by commercial networks and homogenized content, public access is completely local and open, which is why Bowie believes it is important for cable television.

JANET JACKSON'S "WARDROBE MALFUNCTION." What was supposedly an accident exposed pop singer Janet Jackson's right breast for less than a second during the Super Bowl halftime show of 2004. It provoked critics of sexually oriented broadcasting and inspired legislation in Congress to increase indecency fines dramatically.

MORALITY IN MEDIA

Cable's freedom from censorship allows it to take programming to new areas—some would say new depths—in terms of explicit language, sexual depictions, and gratuitous violence. Even basic channels, such as MTV, seem to find new ways to violate accepted norms. In June 2003, MTV responded to the nation's growing restlessness with explicit content by censoring a music video. Calling it "livid and offensive," MTV's office of standards asked the cable channel producers to make cuts in the lyrics and imagery of Christina Aguilera's video, "Can't Keep Us Down." Aguilera cried foul, charging that a male performer would not be subject to the same censorship, but acquiesced and edited her video, nonetheless.

Seven months later, the cable music channel found itself in the center of a larger controversy that illuminated broadcast and cable's differing perspectives on indecency. MTV produced the Super Bowl half-time show for its sister network, CBS, and promoted it on the MTV website by announcing, "Janet Jackson's Super Bowl Show Promises 'Shocking Moments.'" After Kid Rock wore an American flag like a poncho and discarded it during his half-time performance, Janet Jackson took the stage with Justin Timberlake. Their duet provided the flashpoint for the Super Bowl event. Timberlake reached over to Jackson's costume and pulled at the breastplate, leaving Jackson's right breast exposed for millions to see, including parents watching the game at home with their young children.

Afterward, NFL executives announced they were "extremely disappointed by elements of the MTV-produced halftime show," adding, "it's unlikely MTV will produce another Super Bowl halftime." MTV apologized for the incident, which Timberlake at first called a "wardrobe malfunction."

ELECTRONIC SOAPBOX

Public access channels are an ideal forum for democracy, giving citizens the chance to produce TV shows on a first-come, first-served basis and make use of electronic media. These public, educational, and government (PEG) channels are periodically threatened by extremists who test the bounds of the First Amendment, and cable firms trying to reduce expenses by eliminating them. The Supreme Court, however, has stood by public access channels, declaring them to be "an important outlet for community self expression" and a "response to the increasing concentration in public discourse." Today, about 5,000 cable systems originate their own programming in studios, averaging 23 hours a week.

▌ Cable's Chronicle of Regulation

For years, cable regulation has seen a tug of war between broadcasters and cable systems at federal, state, and local levels of governments.[7] The Internal Revenue Service, for example, planned to levy an excise tax of 8 percent on the new industry in 1951. Cable leaders responded by forming the National Community Television Council at a hotel in Pottsville, Pennsylvania. A district court of appeals threw out the excise tax, and the organization endured. Today, the National Cable & Telecommunications Association (NCTA) represents cable system operators and their networks on a host of political and legal issues.

LEGAL FLIP-FLOPS

Because CATV was defined as a *common carrier,* a public communication relay that is nondiscriminatory in both content and service, the FCC took a hands-off approach to regulation at first. Independent UHF broadcasters were happy to have their channels extended via cable, but network affiliates did not wish to compete with distant channels that cable systems had harnessed by microwave relay and piped into local markets. In Wyoming, the Carter Mountain Transmission Company had relayed via microwave TV shows from Denver to three cable systems. At first, the FCC said it was okay, but when a Riverton, Wyoming, broadcaster protested, the commission turned an about-face. The agency concluded it would assess the economic consequences for local broadcasters before it approved any requests to import distant TV signals into a broadcaster's market.[8]

CABLE'S EARLY ORDERS

The federal agency was now in an awkward position. It had touched a business that was supposedly beyond its reach. The FCC attempted to clear the air by issuing the First Report and Order in 1962, asserting, among other things, that the agency indeed had jurisdiction over the medium. CATV was "ancillary" to over-the-air television, the order said, and the FCC needed to protect the public's interest in broadcasting.[9]

Anxiety among broadcasters started to rise as lobbyists for the NAB urged Congress to stop CATV systems from importing TV competition from distant cities. The FCC responded with its second report in 1966, extending its grasp over cable systems.[10] That order effectively banned distant TV signals from large markets unless a cable system could show that it was in the public interest. In *U.S. v. Southwestern Cable Co.* (1968),[11] the Supreme Court gave the final word upholding FCC jurisdiction. It could regulate cable so long as its rules were "reasonably ancillary" to broadcast services.

COPYRIGHT CONTROVERSY

Television stations were under the impression—by virtue of licensing agreements with networks, syndicators, and producers—that the shows they aired belonged to them and thus they were entitled to copyright payments. In 1976, the Copyright Act declared cable operators free to retransmit TV signals so long as they held compulsory licenses, which meant paying for programs from non-network sources. A copyright royalty panel was established to collect fees based on a share of each cable system's subscriber receipts. That money would then become royalties to the TV program's copyright holders. In 1993, Congress abolished the Copyright Royalty Tribunal and replaced it with a system based on arbitration panels appointed by the Librarian of Congress.

MUST-CARRY RULES

Certainly, the regulation that has created the most tension between broadcasters and cable companies has been **must-carry.** It holds that a cable operator must carry every TV station within a certain radius of its system. Generally, a cable system with 12 or more channels must carry local TV stations on up to one-third of its channels. Some cable owners resented the rule because it meant valuable channels had to be assigned to broadcasters instead of more profitable programming. Must-carry rules were twice rejected by federal courts for infringing on the First Amendment freedoms of cable operators. However, the Cable Television Consumer Protection and Competition Act of 1992 settled the question by providing broadcasters with the choice of either seeking payment from cable systems to retransmit their signals,

Must-carry Federal rule requiring cable systems to carry local broadcasters on a basic tier of channels.

protalk
Anne Doris
VICE PRESIDENT, COX COMMUNICATIONS

For some, cable television is a money-making machine; for others, it's about channel choices, but for Anne Doris, the initial interest in cable "stemmed from a need to understand the technology." A native of Guyana who holds degrees in journalism and business administration, she rose to become Cox Communications' vice president and system manager for southern Arizona. She has a feel for her industry and its future opportunities.

Doris is aware of the complaints about rising cable rates and customers switching to home satellite television. She agrees that "from a price standpoint, DBS [Direct Broadcast Satellite] is very competitive with cable," however, she adds that "competition leads to stable, competitive pricing and I believe that is what has happened in the cable industry."

Doris takes issue with consumer advocates who want to lower prices by giving cable customers à la carte choices. She cites the Discovery channel's prediction that it would cost at least $8 per customer per month to watch cable on that basis. If all program networks charged as much, it would drive up cable rates and actually reduce the number of channels for viewers.

Instead, Doris subscribes to a business model based on the bundling strategy, where cable combines television channels with such services as broadband Internet

> ### Doris takes issue with consumer advocates who want to lower prices by giving cable customers à la carte choices.

access and cable's telephone services, including voice-over Internet protocol (VoIP). She is wary, though, of what Washington might do to slow down this form of competition. "Regulatory uncertainty and the threat of unnecessary or overly burdensome regulation will affect whether and how VoIP services are deployed." Any new rules should provide certainty to Wall Street, she says, and serve the best interest of cable customers.

Also, the cable industry hopes to avoid stricter controls on sexual and violent content. "Any attempts to regulate indecency on television must be sensitive to the legal and technological differences between broadcasting and cable," Doris says. People don't have to buy cable, and there are ways to block channels "to limit what children or other family members can view on their televisions."

The cable industry offers multiple career paths for college graduates in communication, as well as finance, marketing, and engineering. The best way to begin, Doris says, is to "seek out internships in the cable industry," where students can fully appreciate what the digital future holds for them.

or simply accepting a guarantee for a spot on the cable dial. That law was upheld by the Supreme Court in 1997 *(Turner Broadcast System, Inc. v. FCC)*.

DIGITAL MUST-CARRY

The government began in 2001 drafting new must-carry rules to make way for digital television. The Federal Communications Commission first ruled that a cable operator was *not* required to carry more than one digital channel for each local TV broadcaster. Yet, broadcasters asked why more channels could not be created on cable, especially since one of the selling points of digital compression was its ability to multicast programming and offer more standard-definition channels for local viewers. In 2004, the FCC Media Bureau proposed a plan for obligating cable systems to give more channels to broadcasters.

The FCC said it was necessary to change the must-carry policy in order to encourage broadcast TV stations to return analog spectrum and convert to digital operations in a timely manner. Its proposal was one of a series of steps to facilitate digital broadcasting, but the NCTA protested and said that would neither help the digital transition nor improve the diversity of voices on cable systems.

▐ Alternatives to Cable

SPUTNIK. The space race began in 1957 when the former Soviet Union launched this 184-pound aluminum sphere with nitrogen gas sealed inside. Four antennas bounced signals from the "baby moon" while it was in orbit.

There are a number of alternative multichannel video program distributors (MVPDs) to cable for pay television programming. They include telephone companies' video services, satellite master antenna television (SMATV) systems, and the over-the-air system known by the curious contradiction in terms, *wireless cable*. About 10 percent of U.S. television households receive their programs from sources other than cable, known as **alternative delivery systems (ADS)**.[12] More than 22 percent of the MVPD market is made up of satellite television subscribers, and DirecTV and DISH are the only ones available in many areas.

SATELLITE COMPETITION

Satellite telecommunications is less than a half-century old, but in the last decade it has come into its own. The United States first began experimenting with the new technology after the former Soviet Union launched into space its first satellite, *Sputnik I*, on October 4, 1957. The American people became alarmed by the prospect of Communist domination in outer space and prepared to join the space race. It was 1962, however, before AT&T's *Telstar I* was able to capture microwave signals in orbit, amplify them billions of times, and bounce them back to earth. On July 10th, 1963, live scenes from France and England were relayed by television for several minutes before the orbiting *Telstar* satellite passed out of sight.

The problem of moving satellites "disappearing" from the antenna's view required a solution. One had been conceived almost 20 years earlier by a British author and futurist, Arthur C. Clarke. In 1945, Clarke proposed the geostationary orbiting belt. According to the scientist, if a satellite were propelled to a certain distance above the earth and "parked" there, it could provide uninterrupted broadcasting signals. By 1963, Clarke's vision had become a reality. The first geostationary earth orbiting (GEO) satellite was placed 22,300 miles above the planet in what is now called the Clarke belt. Figure 3.5 shows that at this

Alternative delivery systems (ADS) Distribution of video and audio content other than by broadcast or cable, including satellite, telephone, and "wireless cable."

FIGURE 3.5 *Satellite Orbit*
Telecommunications satellites can be placed in geostationary (GEO) orbit 22,300 miles in space where they reach synchronous orbit with the earth. The so-called Clarke belt has little room left in its equatorial path, so companies have also used higher elliptical orbits (HEOs), medium-earth orbits (MEOs), or lower earth orbits (LEOs) for transmissions of voice and data, including GPS-style navigation systems and global phone communications.

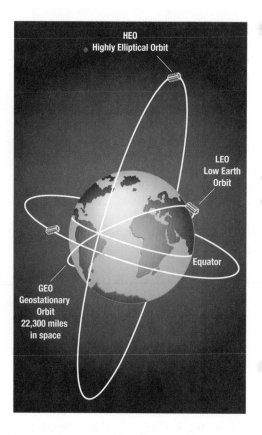

orbit, between LEO (lower earth orbiting) and HEO (higher earth orbiting) paths, the GEOs stay in phase with the earth's rotation above the equator. The number of transponders per satellite have been increasing. Prior to 2000, the average number was 31, but by 2004, some satellites had up to 49 transponders. Their beams focused on different parts of the earth's surface, with each part called a *footprint*, as shown in Figure 3.6.

SATELLITES RISING

In many respects, direct-broadcast-satellite (DBS) programming is similar to cable fare, but channels are received by a home antenna rather than through cable wire. The 18-inch dishes taking shows from DirecTV or DISH evolved from what were called *television receive only (TVRO)* antennas. These larger dishes, 10 feet in diameter or more, took about 75 channels of programming on a different wavelength, the C-band in the 3- to 6-gigaHertz (GHz) range. Even today, there are about 500,000 satellite television homes receiving TVRO programming in the United States.

The FCC-approved "Ku-band" service emerged in 1980, and today there are 23 million DBS customers. The FCC awarded construction permits to eight applicants in the beginning, but all of them failed to meet their deadline for launching. Cable systems had pressured networks to keep programming away from home satellite television. Consequently, legislation was drafted to offer a solution in 1992. The same law that clarified must-carry rules for cable also guaranteed DBS access to network programming, and prohibited discriminatory fees for satellite television programs.

In 1994, two DBS applicants, Hughes Communications and USSB, became the first to launch home satellite TV services. Hughes established a subsidiary, DirecTV, which eventually took over USSB. In 1996, EchoStar launched its Digital Sky Highway (DISH) network. Hughes Electronics Corporation wanted out of the business by 2000, and EchoStar's Charlie Ergen bid $20 billion for DirecTV, but the government blocked the merger on antitrust and public-interest grounds. Then, in 2003, Rupert Murdoch's News Corporation came forward with an offer.

By a three-to-two vote, the FCC approved the takeover of Hughes's DirecTV by News Corporation, which a dissenting commissioner said violated the public-interest standard by placing "unprecedented control over local and national media properties in

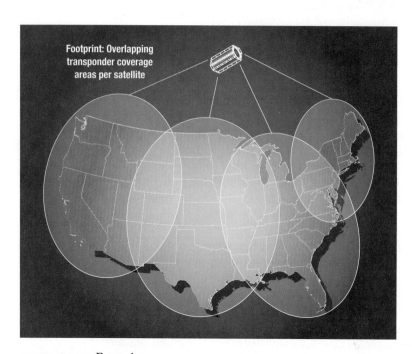

FIGURE 3.6 *Footprint*

one global media empire." DirecTV added a national video distribution platform with 12 million satellite customers in addition to News Corporation's broadcast network (Fox) and its chain of TV stations that reach more than 44 percent of the nation's television households. The megamerger deal recast the national media landscape, which Commissioner Jonathan Adelstein compared to putting a fox in charge of the hen house.

SMATV AND MMDS

Satellite delivery is also used to reach multiple dwelling units (MDUs), such as apartment complexes and hotels. It began cropping up all over the country in the 1950s and started piping television to businesses and to residential complexes. It was then called master antenna television, or MATV, and was promoted by Milton Jerrold Shapp as a way for television retailers to display their TV sets for sale on the showroom floor. The master antenna idea caught the interest of apartment managers and hotel owners who wanted such a system to feed television signals to their housing and rental units.

After the first telecommunications satellites were launched, satellite master antenna television (SMATV) services took over and began to compete with local cable companies. Motels and apartments began tacking television service to the bill and realized profits from pay movies and other special events. In 1991, the FCC gave SMATV companies a boost by designating microwave spectrum—73 channels—for them to use as links between apartment complexes and hotels.[13]

The government freed up combinations of microwave channels in 1983 to allow for another competitor in the pay-television business. Eight channels that had been designated to schools and universities, known as *instructional fixed television service (IFTS)*, and three used by businesses for point-to-point television programs called *operational fixed service (OFS)* were combined with eight more channels to form the new service. These **multichannel multipoint distribution services (MMDS)** are similar to cable but instead use a microwave transmitter to reach rooftop antennas. There are 290 wireless cable systems in the United States, serving about 200,000 subscribers.

TELEPHONE COMPANIES AND THE VIDEO BUSINESS

The breakup of AT&T in 1984 gave birth to seven regional telephone systems—sometimes called the *baby bells* or *regional Bell operating companies (RBOCs)*. These companies were required by law to stay out of the long-distance phone and video business—something that cable systems and broadcasters lobbied to prevent. However, the 1996 Telecommunications Act changed the rules to allow RBOCs and other phone companies to become MVPDs.[14] The first analog systems, known as *video dial tone service (VDTS)*, enabled the phone company to serve as a carrier of programming similar to cable channels, but digital technology and the 1996 Telecommunications Act gave rise to **open video systems (OVS).** The FCC, however, has received about 20 applications from phone companies seeking to establish OVS networks in local communities.

DIGITAL FUTURE

Cable television fought the entry of phone companies into video delivery but countered by competing for telephone service and broadband Internet access. Over the past decade, cable systems have unleashed an array of digital services bundled together in one package. Cable high-speed modems with fast download times have attracted new business, while the phone company has invested in digital service lines (DSL) to challenge cable modem sales. Satellite television is the fastest-growing MVPD, offering strong competition to cable's bread-and-butter business of specialized networks of programming.

TELSTAR 1. America's 171-pound sphere built by Bell Telephone Laboratories with AT&T funds was launched on July 10, 1962. The dawn of global satellite communications was reflected in its faceted solar cells. President Kennedy heralded Telstar as an "outstanding symbol of America's space achievements."

Multichannel multipoint distribution service (also multipoint microwave distribution systems) (MMDS) Wireless systems for delivering TV programs and Internet service over high-frequency channels to subscribers.

Open video systems Wired systems that deliver multiple channel video programming to subscribers, but are not licensed by local communities; they are required to open their systems to independent programmers.

career focus

ENTRY-LEVEL CABLE POSITIONS

Cable careers vary by their focus and by their placement in the hierarchy of the business, but they have grown substantially in number over the past 20 years, from about 40,000 in 1981 to more than 130,000 today, according to industry estimates.[15]

There are job opportunities at the local cable system level, with multiple system operators, and at cable programming networks. Generally, MSOs hire people for positions in financial and legal offices, advertising sales and marketing, corporate engineering, public affairs, and operations management.

Most MSOs are headed by a **chairman** followed by a **chief executive officer** or **president.** Below him or her are offices for the **chief operating officer** and **vice presidents** of the various departments.

The **general manager** of a local cable system supervises all aspects of the operation: policies and programming, engineering, new services, and system expansion. The **GM** is also responsible for the budget, personnel decisions, employee benefits, and planning for future growth. He or she must work effectively with managers from other offices. For example, local managers rely on an administrative staff, also called the Customer Service Department, which includes **billing clerks, accounts payable and receivable,** and **accountants.** It coordinates the technical staff's response to customer requests for new service or maintenance. Generally, a **customer service representative** handles calls about cable service and billing, and a **dispatcher** coordinates this information with the technical staff in the field.

The **chief engineer** will have working for him or her in the technical department the following personnel: **cable installers** to prepare homes for cable reception by running lines from utility poles to outlets in the home, **trunk technicians** to troubleshoot failures in the main line or feeder amplifiers, **service technicians** to respond to individual subscriber complaints, and **bench technicians** to run the cable system's repair facility.

Marketing, advertising sales, and **public affairs directors** promote the image and services of the cable system in the community. The **marketing director** is charged with building the subscriber enrollment for the cable system as well as coordinating research, advertising, and promotions. This person recruits new subscribers through promotions with homes as well as hotels and apartment complexes needing hookups to the system. **Advertising sales** or **account executives** call on businesses, which may want to run commercial spots and reach niche audiences through cable channels. **Researchers** are often hired by cable systems to determine the demographic makeup of the local audience and determine how well the system is meeting their needs.

A **public affairs director** serves as a liaison between the cable system and the community and government leaders. This person also works with the local media to get the word out on projects and events the cable system has undertaken.

Career opportunities in television production vary from system to system. The **programming and production staff** creates local programs for some systems or works with the public access center, depending on the franchise's obligations.[16]

Primarily, cable industry jobs are found in administration and clerical offices, advertising and sales, engineering/technical services, and customer services. However, more job opportunities have been created in broadband services and web operations. These include promoting and selling the cable's broadband and on-line services as well as maintaining websites for the system.

Television's digital future is making screens wider, pictures sharper, and TV sets smarter. In addition, the digital future has allowed more choices with greater convenience and has given viewers more control. In this interactive environment, the fierce competition between broadcasting, cable, and satellite is creating new rules and new players, all converging to gain the audience's favor. Not only have media converged through different delivery systems, such as cable, satellite, and other networks, so have their parent companies through mergers and acquisitions.

Summary

In its early days, community antenna television was stimulated by public demand in response to news of broadcast television. The lack of local TV stations encouraged the invention of CATV to satisfy the demand through wired networks. As the number of cable systems grew, so did broadcasters' concerns about the nature of the competition. In 1963, the FCC responded to lobbying efforts by the National Association of Broadcasters and began to exercise authority over cable television through its system of microwave relays. Those relays were employed to import distant TV signals. The FCC added to its mandate by requiring cable systems to carry television stations within the operator's area of coverage. The so-called must-carry rule was upheld in court, but nowadays cable operators and broadcasters debate new must-carry regulations for digital television (DTV).

Beyond advertising, cable enterprises found new ways to make a profit by selling subscriptions to basic, premium, and pay-per-view channels. In the future, telecommunications firms will collect revenue from bundles of digital services, including video on demand, broadband Internet access, interactive video, and telephone services.

Television innovations based on satisfying the need for speed, mobility, and interactivity require a substantial investment, which relies on the economies of scale found in large corporations. The financial lure of media combinations has led to larger and larger conglomerations, which may pose a threat to free and fair competition.

Cable and satellite media understand the principle of relative constancy, which holds that although people's choices vary in terms of media activities, the amount spent on them is about the same. Consequently, cable and satellite companies have used a strategy to encourage decisions in their favor by maximizing viewers' choices, control, and convenience in choosing media activities.

In 2004, cable television angered some by pushing the envelope in terms of sexual content. Because it is based on subscriber consent, cable programs as a rule are more daring than broadcasting channels. However, producers and cable companies find themselves facing the threat of government regulation if they continue to offend viewers.

Broadcast stations and cable are also fending off newcomers in video delivery. Multichannel video program distributors include satellite systems, telephone programs, and broadband Internet. Alternative delivery systems for video, audio, and the Internet have made the future roadmap for electronic media more intriguing. Yet, with billions of dollars and global enterprises at stake, it is not a journey for the fainthearted or underfunded.

Food for Thought

1. Early cable entrepreneurs wanted to bring distant television signals to rural customers and recover a small profit. Today, concentration of cable systems in fewer hands has meant the virtual demise of locally owned cable companies. Do you consider that a good development? Why or why not?

2. Interactive television has been an ongoing experiment with cable and satellite companies. What types of interactive services would you like to receive?

3. Cable and satellite television recover more of its revenue from subscribers than from advertising. Would you be willing to pay more if all channels were free of commercials?

4. The trend toward consolidation has meant that fewer people control both the channels we watch and how they are distributed, including cable, satellite, and broadcast channels. Do you view this as a positive or negative trend? Why?

5. The number of public access channels has declined since it became an optional requirement in 1984. Do you think that every cable system should be required to have a community access channel? Explain your reasoning.

6. Satellites are becoming more popular because of their numerous channels and digital services. What factors would determine whether you subscribe to satellite, to cable, or to video delivery by another MVPD?

7. Cable companies are now competing for telephone service using the Internet. Would you be willing to switch to a cable supplier for telephone services? Why or why not?

Radio *and* Television Technology

—R. D. LAING, The Politics of Experience *(1967)*

Chapter 4

The triumphs of invention in electronic media transcend global boundaries. Yet, there are also international conflicts as the search for common standards challenges the best minds in technology. In 1993, a South Korean working with his American colleagues broke through the thorny barriers that had frustrated technicians by digitally translating television into a higher definition. Woo Paik and Jerry Heller along with their associates at General Instrument led the way in compressing high-definition video into the computer language of ones and zeros. Paik knew the Japanese had led the race in **high-definition television (HDTV),** but had not shrunk it for digital delivery. Not only would digital compression allow "hi-def" signals to squeeze through six megaHertz channels but it would also answer complaints about ghosting of images and video snow. Paik and his associates designed a new compressed

PRINCIPLE 1

Scientific invention is based on competition and collaboration.

package that would make HDTV fit broadcast channels in the United States.[1]

Their breakthrough shook up the TV engineering world and eventually produced a pool of intellectual talent known as the Grand Alliance. After General Instrument impressed its rivals with this new, digital HDTV formula, a consortium of companies was formed that included General Instrument, AT&T/Zenith, and Philips/Thomson/Sarnoff. They all accepted the government's challenge to create a digital format of advanced television (ATV) for the nation. The Grand Alliance's system was successfully tested, and in 1996, the FCC issued its decision that would stretch the digital-television screen to cinema proportions while allowing multiple versions of screen resolution. It approved of 1,080 lines of vertical definition, which more than doubled the 525-line standard set in 1941. However, there were other versions approved, including ones with fewer lines of resolution, which were superior to NTSC only by virtue of their digital technology.

Given the choice, the world's broadcast community began to respond by moving in different directions. In the United States, NBC and CBS committed to true high-definition television, 1,080 lines, while ABC and Fox announced plans to embrace lesser standards of 720 and 480 lines. When the European Broadcasting Union convened in 2004, its members declared their intention to select a single standard for digital television, and not to accept the wide array of digital-television choices facing American broadcasters. How do these conflicts arise, and how are they resolved?

Electronic media's milestones in technology indicate how forceful the tensions of competition and collaboration can be in the creative process of innovation. Ultimately, each new invention must prove itself in the marketplace beyond the inventor's laboratory. That is where radio, television, and Internet technologies either generate sufficient demand to sustain their cost or the market ignores them until they go away. This chapter will explain how breakthroughs in technology point to the future of electronic communication.

The Big Picture: Stages in Media Production

Creating messages for radio and television is not just about writing with words; it's also about "writing" with sounds and pictures. Skilled artists and technicians know how to combine different video clips, voices, and graphics to create meaningful messages with the tools of their trade.

Before a radio or television program can wend its way through lenses and microphones to cables and home receivers, it must pass through stages in the chain of production. In technical terms, these stages are subdivided into five steps. First, *encoding* is where the raw material of pictures and sounds are recorded for production. This step calls for informed choices in audio and video recording equipment. The second step, *editing*, requires manipulating the software and hardware necessary to merge sound and images, graphics and text into meaningful content.

The third step, *storage*, places the product in its final recorded format. It may be "burned" onto a digital disc, recorded on videotape, or compressed into an MP3 file. Regardless of the storage container, the content is preserved with as little loss in quality as possible to deliver it effectively to the audience. The next step is *delivery*, when electronic media content is transmitted electronically by using wired or wireless electromagnetic energy to reach its destination. It may be transmitted by satellite, cable, or broadcast airwaves, so it is important to understand the nature of bandwidth and frequencies that serve as the vehicles.

The final step is *reception*, where electronic waves are decoded into sounds, pictures, and text appearing on a screen and heard through speakers by the audience. All of these steps require a variety of tools and machines to convey information and entertainment. By understanding the principles guiding the technology, it is easier to make informed decisions—rather than random choices—in order to produce meaningful content.

Encoding

This first stage of media production is when the messages actually begin to take shape. Once a professional learns the basic mechanics of microphones, camcorders, and the accessories, he or she can take full advantage of the means of encoding sounds and pictures. Media producers, videographers, and editors spend much of their careers learning and adapting new skills for encoding content in order to capture the audience's imagination.

High-definition television (HDTV) Increases line resolution of standard television.

TOOLS FOR SOUND ENCODING

Voice, music, or virtually any sound is first encoded by waves of pressure traveling through the air. In audio terms, these waves bounce against a microphone's diaphragm. This is the circular element common to all mics that acts as its "ears." The *diaphragm* of the mic is where the sound pressure is transformed into electrical energy by the use of a generating element or *transducer*. This process is called **transduction.**

Microphones

Mics are defined both by the way they pick up sounds and convert them into electricity and by how they are used in the field. Three mics, the dynamic, the velocity or ribbon mic, and the condenser, are the most popular, yet each serves a different purpose. The *dynamic* mic is the most durable; it uses a moving coil of wire for its transducer and is often used in the field. The *velocity* or *ribbon* mic converts sound waves into electrical energy through a thin strip of metal ribbon. It is sensitive to the subtle sounds of music and is the choice of studio artists. A small *electret* or *condenser* mic is usually powered by batteries. It holds two plates, one of which vibrates against the other to produce energy.

Microphones react to sounds in a manner similar to the human eardrum; their generating elements reproduce sound waves as electrical energy and create a **frequency response.** This is the coin of the realm; the wider a mic's range of frequency response—from low to high pitch—the better it encodes without distortion. The goal of frequency response is crystal clarity—a *flat response.*

Transduction Converts sound or light energy into electrical signals.

Frequency response The range of sounds reproduced by a microphone measured in the frequency of waves per second, or Hertz (Hz). The frequency range of human hearing usually falls between 20 and 20,000 Hz.

Pickup Patterns

Microphones also vary in terms of their *pickup patterns,* the way they "hear" sounds from different directions. Figure 4.1 shows how they absorb sounds in one of four pathways. A news reporter in the field interviews people using a *unidirectional* or

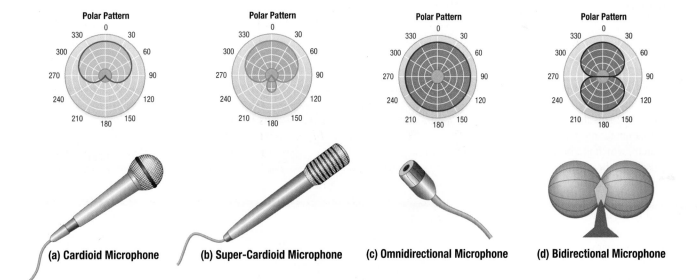

(a) Cardioid Microphone   (b) Super-Cardioid Microphone   (c) Omnidirectional Microphone   (d) Bidirectional Microphone

FIGURE 4.1 *Microphone Pickup Patterns*
Microphones are described by their mounts, transducers, and pickup patterns. Most common are cardioid microphones that pick up sounds in a heart-shaped pattern in front of the microphone. Super-cardioid patterns are found on shotgun and parabolic mics designed to pick up sounds from greater distances. Omnidirectional pickup patterns are used on lapel mics, and bidirectional mics listen from two directions (like ears).

cardioid microphone, which collects sound waves in a heart-shaped pattern. At a football game or town meeting, the same reporter might use a more focused *supercardioid* pattern. The shotgun and parabolic mics pick up sounds from long distances, and often record ambient or natural sound.

Bidirectional mics became popular during radio's Golden Age, and although rare today they pick up sounds in a figure-eight pattern. This pickup pattern was useful for dramas where two actors would speak their lines across a mic to each other. *Omnidirectional* mics uniformly register sound from all directions. *Lavalieres* (French for "pendant"), or *lavs,* are a type of omnidirectional mic often seen on the lapel of a news anchor. Some lavs are directional and may be used to reduce room noise.

Mic Mounts

Microphones are also defined by the manner in which they are mounted. A *boom* mic extends from a crane or a "fish pole"; it hovers near the speaker outside the frame of the camera's viewfinder.[2] This type of mic requires a wind screen or "blimp" to minimize noise. *Wireless* mics are popular with news crews, stage performers, and musicians. They encode signals and send them through battery-powered transmitters with tiny antennas to receivers in a console, where the sound is either recorded or broadcast.

Wireless mics, whether used as a lav, a stick, or in a headset, send radio signals over distances of between 100 and 1,000 feet. They are designed to give performers the chance to move freely (without a cable) to meet the audience or camera face to face, but they work best under controlled circumstances without obstructions—either physical or electronic ones—such as cellular phones and microwaves.

VIDEO ENCODING

The first step in *transducing* a picture for television begins in the camera lens as it captures a scene. The rays of light strike the lens and are filtered through a mirror-like prism known as a *beam splitter.* This is an optical block of glass that distributes incoming rays as three primary colors of light—**red, green, and blue (RGB)**.[3] They are transferred to silicon chips, **charge-coupled devices (CCDs)** (see Figure 4.2),

Red, Green, Blue (RGB) Primary colors of light mixed to produce a video image for television and computers.

Charge-coupled devices (CCDs) Convert light into a charge pulse and code it as a number using light-sensitive chips.

FIGURE 4.2 *TV Camera Schematics*
A television camera collects light through a lens that is filtered by the camera's internal optical system, which is a beam-splitting prism. The beam splitter of optical glass separates red, green, and blue elements of the scene and sends them to CCDs that react with photoelectron charges to the intensity of the primary light colors. These charges are then coupled with (transferred to) pixels to be scanned as a video image.

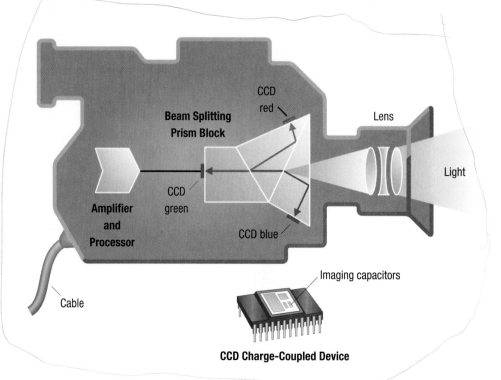

CCD red

Beam Splitting Prism Block

Lens

Light

CCD green

Amplifier and Processor

CCD blue

Imaging capacitors

Cable

CCD Charge-Coupled Device

where the light is encoded as electrical energy through the use of **pixels,** which are arranged in rows on the chip to form a mosaic of color. Professional studio cameras use three chips—one for each color of RGB. They store the light as charges to be converted line by line and pixel by pixel into a video signal that is sent to the screen. As TV cameras migrated toward digital encoding, the number of scanning lines increased along with the storage capacity of the CCD.

■ Editing

The next stage of media production, *editing,* first occurs—in a manner of speaking—in the minds of the producers, directors, audio technicians, and camera operators. Each choice about the images and sounds recorded requires professional judgment. In the postproduction process, editing takes on a somewhat different meaning. This is where all the layers of sounds and pictures are pieced together after production in the field or studio.

CHARGE-COUPLED DEVICE. This is a solid-state imaging device for converting light to electronic signals. The black square is the imaging device. CCDs were first designated for electronic-news-gathering (ENG) video cameras, and later installed in studio cameras as well.

AUDIO MIX

After the sound energy is encoded—whether it is music, voice, or special effects—it is transported by cable to an *audio console* where it is amplified and blended with other sources. The console not only serves the purpose of selecting inputs from various sound sources but it also pre-amplifies the audio, controls incoming levels, and routes multiple channels. If the audio production is in a studio, there are usually between 16 to 24 channels to mix on multitrack tape recorders. For radio stations, 8 to 12 channels may be mixed in the console before transmission through the **antenna** tower.

Digital Audio Editing

The audio industry has completed its migration from analog to digital editing, where the computer mouse has replaced marking a point on a reel of tape. The digital-audio workstation (DAW) allows producers the creative experience of editing, amplifying, and enhancing layers of sound and incorporating them all with special effects. It is more than just a matter of pointing-and-clicking at a waveform of horizontal tracks, however; today's editor must know where and how to cut, paste, and mix layers of sound so that the final production creates imagery the audience can envision as well as hear.

The challenge facing an editor involves taking layers of sound from a digital console with a variety of recording machines in the process. These sources include the recorder/players for the compact disc (CD), digital audio tape (DAT), the MiniDisc (MD), and the digital versatile disc (DVD). Audio software applications are also available for computers to control and "sweeten" the sound digitally in the editing process. Adobe Audition and Pro Tools are among the popular applications used.

VIDEO MIX

After TV cameras capture images in the studio or field, they must be routed to a switcher where a director skilled in *instantaneous editing* mixes the video sources. Each video source has a button on a switcher board as well as a row of buttons

Pixels Picture elements on a television screen's scanning line illuminate phosphors in mixtures of red, green, and blue light.

Antenna Metallic rod or wires conduct sending or receiving radio signals.

DIGITAL AUDIO EDITING.
Computer software programs
have replaced spools of tape
and splicing blocks for editing
audiotape. An audio technician
uses a waveform track to cut and
paste blocks of audio sound for
broadcast.

called a *bus*. The director punches these buttons to select a camera, a videotape, a
character generator, a live feed, or some other option. A fader bar on the switcher
enables the director to create transitions by *fading, dissolving,* or *cutting*. A picture is faded up from black, or it can be dissolved from one scene to another.
Most switchers offer special effects, including a variety of wipe techniques, borders, and backgrounds. There may be an option for a chroma-key effect, allowing
a different background to be inserted behind the talent, such as a weather map

**ANALOG VIDEOTAPE
EDITING.** A control editor
synchronizes pictures and sounds
in order to electronically transfer
scenes from a playback machine
to a recorder with analog videotape. This linear process edits the
tape from first frame to last frame
of the program.

behind a TV meteorologist. The director also may choose to flip a picture over, spin it around, or shrink it in a variety of ways using a *digital video effects (DVE)* generator.

Analog Video Editing

In the Golden Age of television, engineers would erase the mistakes of "pre-recorded" programs by cutting with scissors and splicing with adhesive tape applied across wide strips of videotape. These splices were noticeable, and soon broadcasters asked for a technology that would make edited videotape appear seamless on television.

At first, they tried to edit electronically by synchronizing the videotape playback with a second video tape recording (VTR) machine that would mix the two sources and record a final version. That solution was somewhat better, but not quite perfect. In the 1970s, U-matic videotape cassettes and recording machines linked to an edit controller made the cuts cleaner. This system of linear editing evolved through the 1990s in several formats of videotape decreasing in size from three-quarter to one-half inch, and finally to one-quarter-inch digital videotape.

Digital Video Editing

At first it was called "desktop video," but when technicians began experimenting with the computer's capacity for digitally splicing sounds and images in a random fashion, they called it **nonlinear editing.** No longer would editors assemble programs from beginning-to-end in order; rather, they could select and execute cuts on a computer. This contrasted with the analog machines where an editor joins together video and audio at the same time in *assemble* mode, or uses the *insert* editing technique and passes through the videotape at least twice—once for the pictures and again for audio. Digital video editors simply point and drag pictures and sounds on the screen. A number of popular software applications—Avid, Adobe, and Apple, for instance—lead the digital video editing field.

Nonlinear editing Manipulates scenes of video and audio in random fashion through computer software.

⌶ Storage

All program content is recorded and stored in some format, either analog or digital. The **analog** technique for audio and video recording involves converting sound and light waves in a manner that is analogous to their original shape and form. Analog recording on tape transduces waves into constellations of microscopic particles, which are engraved onto the tape's iron-oxide compounds. These recorded patterns resemble original waveforms in that they are continuous and analogous. The quality of audio recording used to depend on the speed of the tape recorder and the quality of the tape, among other variables. The greater the amount of information that could be recorded, the higher the quality of production. However, things changed radically after digital recording became the standard.

DIGITAL CONVERSION IN AUDIO

The modern process of conversion to digital electronic media began in 1983 when two recording giants, Sony Corporation of Japan and Philips Electronics of the Netherlands, introduced **compact discs (CDs)** to Japanese and European customers. Their smaller size and enhanced quality were principal selling points, while random access to digital cuts emerged as another desirable feature. The following year, CDs were sold in the United States after Ampex opened a replication plant here.

DAT RECORDING. Digital audiotape (DAT) cassettes and recorders have proven useful for certain venues such as concerts where high-quality sound is desired in the field.

SONY MINIDISC. The MD digitally records audio in the field on software that looks like a small computer disk.

DATs

Dozens of media labs soon began experimenting with **digital audiotape (DAT).** It did not use compression but digitally reproduced sound in a tape format that lacked random access capability. DAT made its way into the market in 1985, when Sony advertised to customers that they could buy expensive but smaller cassettes that eliminated the hiss and noise of standard audio cassette tapes. DAT never generated a significant consumer base, but rose to popularity as a favorite among professional studios.

MDs

In 1992, a new digital audio format, the MiniDisc (MD), was marketed by Sony and Philips. In appearance, it looked like a CD inside a floppy diskette. It was only about 2.5 inches wide—half the size of a CD—and was capable of storing 74 minutes of digital sound. However, its data compression ratio placed it below the sound quality of a CD, which was uncompressed audio. The MD used a magnetic head and a laser beam to record the digital formulas of bits and bytes. Sony experimented with several MD formats, and in 2004 shifted its emphasis to a Hi-MD format that could record up to one gigabyte of audio information, or about 347 minutes of sound.

MP3s

The digital imperatives of the Internet eventually gave way to MP3's technology for digitally compressing sound. The term is an abbreviation of Motion Picture Experts Group, Audio Layer 3. MP3 recording was developed in Germany as part of the Eureka project for **digital audio broadcasting (DAB).** Professor Dieter Seitzer of the University of Erlangen designed the algorithm necessary to compress sound for digital transfer and storage. MP3 encoders and players replaced older Walkmans, and rivaled portable CDs and MiniDisc players by offering greater storage capacity and online convenience.

Analog Television Storage

Television recording machines differ not only in terms of their tape-based or disc-based systems but also in their signal mechanics—the way they process the features of light and color. For example, a *composite* video system uses only one signal to record the TV picture's brightness, or *luminance*, and mixes it with *chrominance*, the color saturation and hue. They merge as one signal in a *composite* system, but *component* video separates the chrominance and luminance to create a superior signal.

In videotape, the trend has been toward lighter, faster, and higher-fidelity recording. Ampex introduced videotape recordings in 1956 during the Douglas Edwards newscast on CBS. Ampex used rotating heads that scanned the tape from top to bottom and recorded 15 lines of video. Television engineers soon found a more efficient way of recording called **helical scanning,** which wrapped the tape around the magnetic recording heads at a diagonal slant and compressed more picture information on less tape.

As noted, videotape widths began to shrink until eventually one-half-inch-wide cassettes of videotape were introduced by Sony's Betamax and by the Japanese Victor Corporation (JVC). Eventually, JVC's Video Home System (VHS) received wide distribution after sharing this technology as an open source. As a result, VHS and not Betamax became the standard. VHS and Super VHS (S-VHS) formats dominated the consumer market through the 1990s, while VHS camcorders became popular for making home videos.

DIGITAL CONVERSION IN VIDEO

Digital videotape recording began in the late 1980s with two three-quarter-inch tape formats (D-1 and D-2). A more portable version followed in the 1990s, when Sony introduced its Digital Betacam using a one-half-inch videotape. Panasonic's DVC Pro added its innovation, even smaller cassettes—a quarter-inch wide. Sony and Panasonic competed vigorously for commercial TV station and network sales of these digital camcorders.

"Pro-Sumer" Models

The Mini-DV camcorder was promoted as a "pro-sumer" (professional + consumer) model in 1999. This quarter-inch videotape manufactured by Canon and Panasonic

Analog Signals vary electromagnetic waves in a way that resembles the original pattern.

Compact discs (CDs) Store information on polished metal disks by optical scanning using a laser beam.

Digital audiotape (DAT) Scans digitally stored sound waves on a helical tape.

Digital audio broadcasting (DAB) Uses mathematical compression to send digital radio signals.

Helical scanning Invented in 1961 by mounting the heads at a slight angle to the tape path and wrapping the videotape in a helix around them. Rotating the heads at high speeds produced recordings in diagonal rather than longitudinal strips with higher-quality video and audio.

MINIDIGITAL VIDEO. The Mini-DV camcorder is becoming a popular item for academic, corporate, and amateur videographers. The quarter-inch tape is easily edited with computer software programs.

began to replace VHS and S-VHS camcorders and videotapes among personal video users, academic institutions, and smaller professional outlets.

The future of digital video recording will be disc-based and tapeless. Camcorders will move beyond the wear and tear of videotape and will record clips that are easily accessible by a computer server to multiple editors in a station or production house. Avid has introduced a disk-based video camera using a drive that can be removed from the camera and inserted into a computer for nonlinear editing. The industry anticipates other solid-state solutions.

DVDs

One important development in the digital recording industry has been the **digital video disc (DVD)**, which after its release in 1995 captured a major segment of the consumer market. Not only do DVDs offer an enhanced version of films and video programs but they also contain more information. A laser beam records the signals by creating "pits" on both sides of the DVD in two layers recording many times the amount of information as a CD. In 2003, more than 73 million DVD-video players had been sold in the United States, and more than 27,000 DVD movie and video titles were available for sale.

DVDs moved beyond video and began offering recorded music, still photos, games, and other media content, which called into question the significance of the abbreviation: Does *DVD* mean *digital video* or *versatile disc*? At first, the letter *V* meant *video*, but now *DVD* is a term in itself.

PRINCIPLE 2

When quality, cost, and convenience meet customer expectations, then new media succeed in the marketplace.

Digital video discs (DVDs) Record with laser video and sound with more information than CDs or digital tapes.

The popularity of computer-recording systems using DVDs increased the possibility that this format would become the standard for audio, video, and data storage—perhaps replacing CDs and CD-ROMs in computers. Clearly, the success of DVDs has been based on their versatility as well as quality and convenience. Because these factors meet customer expectations, success has followed this format's wide distribution. DVD has generated support from major electronics companies, computer hardware manufacturers, as well as music and motion picture companies. It achieved faster penetration than other software innovations, including audio cassettes and VHS tapes.

Broadcast Delivery

When a video or audio recording is finished, it must be delivered to the audience and that means another transformation. Radio and TV stations transmit on channels that use electromagnetic waves to bring sounds and pictures to life. These bursts of energy are sent from a tower in pulses that radiate outward in all directions. If they were visible, broadcast waves would resemble ripples formed by tossing a stone into a pool.

BROADCAST WAVES

Sounds and images of broadcasting ripple across a busy ocean of air called the **electromagnetic spectrum.** Radio and TV signals cruise at the speed of light—186,000 miles per second—and are measured according to their peaks and valleys, so that a single wave represents one cycle. Most humans hear sounds in a range of between 15 and 15,000 cycles, but sound waves oscillate in frequencies far below the electromagnetic spectrum broadcasters use. They generate thousands and millions of cycles (or Hertz) per second.

By international agreement, this measure was chosen to honor Heinrich Hertz, the German physicist who built a spark-gap generator to display radio waves for his students. Broadcast frequencies grow to such large numbers that metric prefixes are needed to make them manageable. For one thousand Hertz, it's *kilo*Hertz (kHz); for a million Hertz, it's *mega*Hertz (MHz); and for a billion Hertz, it's *giga*Hertz (GHz). The dial channel is the signal's **frequency** of Hertz (or cycles) per second. For example, if you tune in 820 on the AM dial, the radio station's signal is pulsating at 820,000 cycles per second, or 820 kiloHertz. These frequencies are organized into separate bandwidths, or bands.

Radio Lanes of Bandwidth

There are 12 bands recognized by the International Telecommunications Union (ITU). Channels are organized together in *low, middle* and *high* frequencies. There also are *very high (VHF), ultra high (UHF), super high (SHF),* and *extremely high (EHF)* frequency bands. Later in this chapter Figure 4.3 shows how AM and short-wave radio broadcasts in the middle frequency (MF) band, which covers signals between 300 and 3,000 kHz. FM radio stations broadcast in the **very high frequency (VHF)** band. Analog television uses both very high frequency (VHF) and ultra high frequency (UFH) bands. Digital television (DTV) will use channels in the UHF band, and eventually analog TV will be phased out, with its channels reassigned to other telecommunications.

TV Lanes of Bandwidth

Older TV stations send signals designated as channels 2 through 13. These are arbitrary numbers, since stations actually use two different frequencies for sound and pictures. In analog television, an AM signal carries the video information, and an FM carrier wave delivers the audio to the TV set.

Channels 14 through 69 are part of the **ultra high frequency (UHF)** band along with land mobile carriers, such as police and railroad radios.[4] Higher lanes of bandwidth offer frequencies for medical purposes, such as x-rays and ultraviolet radiation. As the frequencies increase in number, the waves become smaller and shorter in length, which determines their value to broadcasters.

The more radio waves pulsate, the shorter the length of each wave; thus, the higher the frequency, the shorter the wavelength. Longer waves are more durable. Energy dissipates from radio waves as they radiate outward, and as a rule, the lower the frequency, the less likely that objects in their path will block the waves. In the early days

Electromagnetic spectrum Oscillates energy from radio and television signals at the speed of light.

Frequency Wave cycles per second measured in units called Hertz (Hz).

Very high frequency (VHF) TV channels broadcasting between 30 and 300 megaHertz employed for television and FM radio transmissions.

Ultra high frequency (UHF) TV broadcasts on channels 14 through 69, but will scale back to 14 through 52 to give more room for wireless.

of radio, stations were actually identified by the length of their waves in meters and not by frequency. Even today, radio channels are given in meters in some countries.

Modulating Waves

Electrical impulses of sound produced in a radio studio are too weak and need a **carrier wave** to reach listeners and viewers. The center frequency of a radio channel is that carrier wave, which is loaded with information through the process of *modulation*. At the transmitter, the carrier waves for AM radio are modulated by amplitude. **Amplitude modulation (AM)** shapes its height and depth, but it is also more subject to interference. In **frequency modulation (FM)**, the wave's width is varied rather than its heights and depths. For that reason, FM uses more bandwidth than AM radio and broadcasts in megaHertz instead of kiloHertz (see Figure 4.3).

> **PRINCIPLE 3**
>
> The higher the frequency, the shorter the wavelength.

AM RADIO

> **PRINCIPLE 4**
>
> Lower-frequency channels are desirable in AM radio because their longer waves travel farther than the waves emitted by high-frequency channels.

AM radio is assigned a narrow corridor of spectrum; each channel has only 10 kiloHertz for broadcasting. The FCC has designated 107 channels in the AM band with just enough room on either side to protect stations from adjacent channel noise. AM radio stations are tuned in between 535 and 1,705 kHz. Lower-frequency channels, such as 590 kHz, send out longer waves with clearer signals than higher-frequency channels, a fact that has figured into the government's chart of radio station licenses.

AM Classes

There are about 4,727 AM radio stations on the air in the United States classified by their **local**, **regional**, and **clear-channel** services. The most powerful AM radio stations are clear-channel broadcasters given 50,000 watts of power. That means they have a competition-free spot on the dial. These stations fall under the *Class 1* title and include the nation's pioneers and original flagships of the radio networks. Clear-channel stations are found in larger cities, such as WLS Chicago, WABC New York, WJR Detroit, and KMOX in St. Louis.

Class II radio also broadcasts a clear-channel signal, but it is subdivided into categories based on each station's *frequency, power,* and *signal direction.* While Class I stations keep power strong day and night, Class II stations pull back their signals at sundown. Higher-frequency channels are defined as *Class III,* and broadcast at a power ranging between 500 and 5,000 watts. They extend beyond the boundaries of a single community, and are designated as *regional stations.* The smallest AM broadcasters are situated at 1230 kHz and up on the dial. *Class IV* radio stations are limited to 1,000 watts during the day and 250 watts at night, and because they serve only one community, they are classified as *local* channels. As suggested, stations with the most desirable frequencies have the lower position on the dial because their longer waves travel farther.

Radio station classifications are made not only to divide the frequencies but also to keep broadcasters from bumping into each other. Interference between stations is a special concern when the FCC grants radio licenses. Two stations sharing the same frequency must be geographically separated or vary in power enough to prevent their signals from overlapping. Wireless transmitters from cell phones or garage doors also create what engineers call "RF clutter," or radio frequency interference.

Radio Waves

The nature of AM radio allows it to send **ground waves** that radiate near the earth's surface during the day. These waves are not bounded by the curvature of the earth

Carrier Wave The signal frequency that is imprinted with information for broadcast.

Amplitude modulation (AM) Impresses sound or picture information on a carrier wave by varying its height and depth.

Frequency modulation (FM) Shapes carrier waves by varying their width and occurs in the very high frequency band.

Local services AM radio channels of limited range serving single communities.

Regional stations Serve one large community and adjacent rural areas with AM radio.

Clear-channel services Exclusive signal patterns assigned to one station.

Ground waves Propagate AM radio signals above earth's surface.

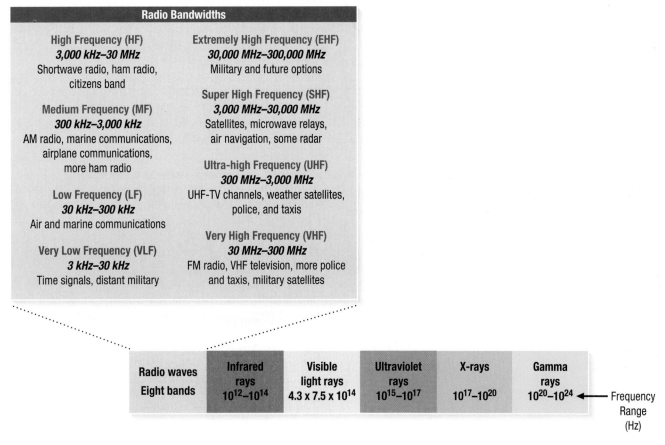

FIGURE 4.3 *Frequency Bands*
Frequency bands are grouped together by the length of their waves—the shorter their waves, the higher the frequency. Microwaves, for example, are so short that the frequency of their oscillations is measured in gigaHertz. AM radio, on the other hand, generates much longer waves and their cycles can be measured in kiloHertz.

or other obstructions. AM radio can also send signals out long distances, especially at night when **sky waves** help bounce signals off layers of the **ionosphere.** The ionosphere is an electrical layer of charged particles that also can cause interference, but it is helpful during cloudy or cold nights when sky waves extend an AM radio station's reach by hundreds of miles.

Until 1996, the FCC licensed broadcast engineers to keep each radio and TV signal centered in the middle of its channel. Even though the Telecommunications Act of 1996 took the agency out of this licensing business, engineers perform that duty. They must keep the signal properly centered and investigate all interference problems—whether produced by atmospheric changes, ground static, or signals from other stations.

FM RADIO

As noted in Figure 4.4, the frequency modulation of FM varies the wave's height and depth, and it also broadcasts in a higher spectrum band than AM. There are 100 channels designated for FM between 88 and 108 MHz in the very high frequency (VHF) band. FM not only offers higher fidelity but it also escapes much of the noise that plagues AM radio because it sends out its signals along a direct route. Figure 4.4 indicates the different patterns of modulation used in AM and FM radio.

Sky waves Refract radio waves through the ionosphere and then return to earth.

Ionosphere Holds ions in layers of atmosphere reflecting or refracting radio waves.

FIGURE 4.4 *AM/FM Wave Modulation*
Carrier waves are modulated to carry the information found in video and audio messages. The key difference between amplitude and frequency modulation is illustrated by the variations in height (AM) versus width (FM).

Unmodulated Carrier Waves are transmitted in a straight line until modulated by a video or audio signal.

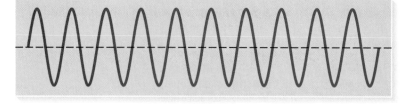

Amplitude Modulation changes the height and depth of carrier waves, but not the width.

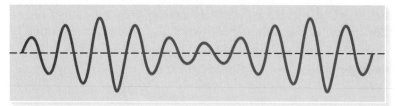

Frequency Modulation changes the width of carrier waves, but not the height or depth.

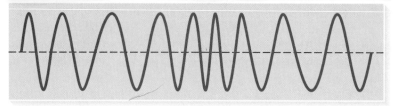

FM Direct

Unlike AM radio, FM signals are sent out in **direct waves,** and not by ground or even sky waves. As a result, FM's signal must hold to a straighter line and can reach only as far as the horizon. Since it relies on that line-of-sight path, the FM station must build a taller tower in order to avoid obstructions. Engineers for FM stations also must deal with the curvature of the earth when plotting radio coverage maps.

Stereo and SCA

Each FM channel is 200 kHz (20 MHz) wide compared to 10 kHz for AM radio, and that gives FM frequencies room to spare for additional services. The extra spectrum FM uses allows for stereo because an FM station can duplicate, or **multiplex,** its signals. **Stereo** is simply multiple sound sources transmitted through two separate channels.

In addition, FM has bandwidth available to program other services through its **subsidiary communication authorizations (SCAs).** SCAs are used to send musical atmosphere to retail outlets and to provide reading services for the blind. However, these frequencies are not found on the FM channels in your car or at home; rather, they squeeze in between those frequencies and require special receivers to tune in.

FM Classifications

FM broadcast stations are organized into categories that are different from those for AM. The FCC has distinguished FM radio stations by their *power, antenna height,* and *coverage area. Class A* stations cover the smallest terrain, about 15 miles, and are limited to 3 kilowatts (kw) of power. *Class B* stations reach twice as large an area, using 50 kilowatts of power. *Class C* stations are the strongest, covering up to 60 miles with 100 kilowatts of FM power.

Direct waves Travel in a straight line from FM radio stations or TV towers.

Multiplex Transmitting Two or more independent signals on the same channel.

Stereo Records, transmits, and plays back audio through separate channels.

Subsidiary communication authorizations (SCAs) Carry FM subcarrier signals for atmosphere music and other specialized services.

SPECTRUM MANAGEMENT

Competing interests come into play when decisions are made about what traffic will have access to what part of the spectrum. As a result, domestic and international agencies have been formed to resolve disputes. The world's airwaves grow more congested each year, and "traffic cops" are necessary to keep signals from crossing lanes, and also to make way for new, digital broadband channels.

ITU

The International Telecommunications Union (ITU) has the authority to oversee the world's spectrum traffic. This union dates back to 1865, when stringing telegraph wires across national borders created problems requiring diplomatic solutions. After the advent of radio, the ITU became the global manager for the allocation of spectrum and the arbiter of airwave disputes. Today, 150 countries comply with ITU decisions. The union's board allots bandwidth, but leaves individual channels for nations to assign and license.

FCC

The Federal Communications Commission (FCC) handles broadcast licensing in the United States. The agency assigns to each radio and TV station its broadcast power, wattage, antenna direction, and signal pattern for its coverage area. Meeting the needs of those who want to use the airwaves is not often easy, since the demand usually exceeds the channels available, which is known as *spectrum scarcity*. Just as there is unused real estate, there is unused spectrum and comparable issues in terms of location and desirability; certain frequencies are simply ill suited for broadcasting.

Low-Power Solutions

In both television and radio licensing, the government seeks to meet the needs of the "have-nots" by licensing lower-power stations. Community groups, including churches and schools, have long wanted to broadcast radio at low-power frequencies. The FCC became amenable to their needs in 2000 over the protests of commercial broadcasters who claimed **low-power FM (LPFM) radio** would threaten the integrity of their stations' signals. The issue divided engineers who testified before the FCC. Some said that interference was an unjustified concern.[5]

> **PRINCIPLE 5**
> The demand for telecommunications channels exceeds the supply controlled by the government.

The idea behind low-power FM (100 watts or less) was to set aside additional channels for radio stations that needed to broadcast only three or four miles from their transmitters. License applications came in from places such as Eureka College in Illinois and the First Presbyterian Church of New Gillette, Wyoming. Some applicants had programming ideas in mind that would not likely be found on commercial radio. An applicant in Sitka, Alaska, for example, wanted to use his low-power station to broadcast shrill, warbling whale songs.

At first, the FCC approved only 255 applicants from a pool of more than 1,200. Senator John McCain introduced a bill in 2004 giving the green light to more LPFM stations to occupy what is called *third-channel adjacency*, which would make LPFM stations next-to-next-door neighbors with full-power broadcasters. This idea did not sit well with the National Association of Broadcasters (NAB). Montana's Senator Conrad Burns, a former broadcaster himself, challenged that move toward LPFM and introduced a bill to allocate more money to further study the commercial interference question.

Low-Power Television

As in radio, there are more applicants seeking a television license than there are TV channels available. The FCC began to accommodate additional stations in

Low-power FM (LPFM) radio Broadcasts 50- to 100-watt FM stations reaching from one to three miles.

the 1970s and proposed a new solution in 1982—**low-power television (LPTV).** This move gave community broadcasters a chance to occupy vacant TV channels on both UHF and VHF bands, so long as they did not interfere with full-power stations. The FCC limited LPTV station power to 3 kilowatts (VHF) and 150 kilowatts (UHF). It was at first overwhelmed by requests for LPTV licenses, and so imposed a moratorium to establish additional guidelines. After application rules were in place, the Commission granted nearly 1,700 LPTV licenses. Today, the FCC reports more than 2,000 LPTV stations, including a statewide network of 250 LPTV stations in Alaska. Another major group owner is Trinity Broadcasting Network, and several networks feed LPTV stations their programs via satellite.

Network programming is considered by some as contradictory to the spirit of the LPTV movement—to give low-population-density areas a chance to produce and see local television. The challenge facing LPTV owners is how to find the resources necessary to make the switch to digital broadcasting—as the FCC decides where to place LPTV's channel assignments in the near future.

DIGITAL RADIO

Digital radio, or digital audio broadcasting (DAB), has been slow to arrive in the United States when compared to other countries. In the United Kingdom and Europe, listeners have been charmed by the clear reception of digital radio sound and the specialized programming they can listen to at home or in the car. In Europe, digital radio stations broadcast on Band III (221 MHz) using the Eureka 147 system. That technical plan did not suit American broadcasting, which became embroiled in a debate over whether to move radio stations with DAB to a new location on the dial. American broadcasters demanded an "in-band, on-channel" (IBOC) solution that would allow them to keep their familiar locations on the dial after making the switch to digital technology.

In 1999, three technical formats of digital radio were proposed, and eventually the major competitors merged to create the iBiquity Digital Corporation.[6] In 2002, the FCC approved of iBiquity's formula for bundling analog and digital signals using a compression technology. This consortium of companies, which included Lucent/Bell Labs, Westinghouse, CBS, and Gannett, joined in the project, which they called *HDC compression technology.* Digital radio, which iBiquity branded as "high-definition" (HD) radio, promised to reduce static, hiss, pops, and fades. It began its rollout in 2003 and slowly continues to convert stations to this day. In the United Sates, few listeners own a digital radio receiver or are even aware of the new HD radio technology.

Satellite Radio

There has been more buzz about satellite radio than HD radio ever since two companies began aggressively promoting their services. Digital signals relayed by satellite for both car and home were launched in 2001 by XM and Sirius radio. They each offered more than 100 channels of music, news, and talk from coast to coast in CD-quality sound. Both services used microwave frequencies requiring special receivers to tune in the satellite channels on the S band.

By 2005, XM and Sirius were climbing out of deep holes of start-up debt and growing a respectable base of subscribers and investors. XM had attracted 3.8 million subscribers and Sirius projected 2.5 million customers by the end of 2005. Both satellite radio firms emphasized the myriad of program channels available and the commercial-free aspect of their service. XM promised customers that the music would be commercial free but accepted advertising over the news and talk channels; however, XM charged its subscribers about $3 less per month than Sirius.

see also p 85

Low-power television (LPTV) Provides licenses to TV translators of up to 1,000 watts in VHF band.

protalk

Steve Harris

**VICE PRESIDENT OF MUSIC PROGRAMMING
XM SATELLITE RADIO**

This disc jockey from Cleveland has made a big leap toward the future. Steve Harris is on the digital frontier of a radio service that beams more than 120 channels via satellite to listeners in their cars and at home. In 2000, the seasoned professional, who played hits in cities from Chicago to Houston, began managing the music programming of XM Satellite Radio in Washington, DC. "XM is part of the entertainment landscape," says Harris. "People will think of radio in terms of AM, FM and XM."

It is a new industry, he says, and one that still needs to get the word out about what it's doing and how it's done. So far, "word of mouth" advertising has helped create most of the buzz. "When people pay for something they can get somewhere else for free, they want something special and we must always deliver a special product."

Harris placed his bets on XM for a variety of reasons but most especially *choice*. Satellite radio offers a "great content no matter what genre of music fits your tastes." There are, for example, seven Country music channels with long play lists. The same holds true for Rock, Urban, Jazz and Blues, Dance, Latin, World, and Classical channels. "Think about it," says Harris, "if you're into Blues—you now have a place to hear

> Harris placed his bets on XM for a variety of reasons but most especially *choice*. Satellite radio offers a "great content no matter what genre of music fits your tastes."

your music 24 hours a day, seven days a week." There are even programs where top recording artists discuss their careers and music.

What about satellite radio's digital competition? High-definition radio has yet to arrive for FM stations, and the impact of Internet radio is hampered, says Harris, by its lack of portability and sound quality. Compact discs have felt the impact of satellite radio, he says. "We already see evidence that people are responding to our longer play lists and are going out looking to purchase CDs they hear." There are also news and information channels on satellite radio, but those come with commercial advertising. XM's music formats are commercial-free zones.

For future radio professionals, Harris has two words: "compelling content." Learn how to produce it, and the future is yours, although good professional contacts can go a long way. He credits his mentor in Chicago, Ernest L. James, the program director who showed him how to produce "compelling content," for the overnight shift at WBMX. Whether by satellite or land, the competition in radio is more intense than ever before because there are fewer jobs in the industry now. "Experience is the key." Getting a part-time job at the local radio station while going to school can be a strong advantage, however. "Understand that radio is a business and that's what counts. Artistic successes don't mean much to Wall Street." And that's where XM Satellite Radio is watching its future dividends.

Streaming Audio

Streaming enables listeners to sample audio or video programs while they are downloading them into their computer. This technology is made possible through a process called *buffering.* By streaming audio, a file is downloaded into a buffer and then relayed to a sound card. Streaming audio servers, such as RealAudio's RealServer, tell the file to start playing shortly after the downloading begins. Interruptions can occur, and overloaded lines frustrate listeners when streaming is stalled and downloading freezes.

The popularity of Internet radio and streaming accelerated until 2002, when the U.S. Copyright Office declared that webcasters would have to pay royalty fees for all music, and that the copyright fees paid by over-the-air broadcasters did not count on the Internet. Moreover, without a broadcast license, webcasters would have to pay twice the amount assessed to radio stations for streaming copyrighted music. Streaming radio stations turned to developing listeners for news, talk, sports, and other formats where such fees are not at issue, until the matter can be resolved by Congress.

ANALOG TELEVISION DELIVERY

Each TV channel has 6 megaHertz of bandwidth to broadcast all of its audio and video information. That spectrum is about 600 times as wide as an AM radio frequency, and about two-thirds of this bandwidth is needed to carry the video signal. (Less than one-third of one MHz is required for audio signals.) Figure 4.5 illustrates how much of the 6 megaHertz is divided among video and color subcarriers, FM audio, and interference-protecting sidebands.

In order to economize on valuable spectrum, TV engineers have devised creative means for compressing the picture and sound content onto carrier waves. One method is known as *interleaving,* which alternates the video carrier's *luminance* with the subcarrier's *chrominance* signals. It is like compressing two messages into one telegram. Both of the technical features of color require scrutiny. Technicians judge the fidelity of the red, green, and blue hues by their richness or saturation when measuring the **chrominance.** They also gauge the signal's brightness, or its **luminance.**

SIGNAL INTEGRITY

Chrominance Contains color information in the TV signal.

Luminance Controls brightness levels in a color television signal.

Radio and television signals are vulnerable to deterioration, so steps must be taken to protect and stabilize them. Just as FM engineers ensure the quality of sound, TV engineers monitor their station's signals to check on quality and strength. Two different transmitters modulate the TV picture and sound. The AM transmitter generates video

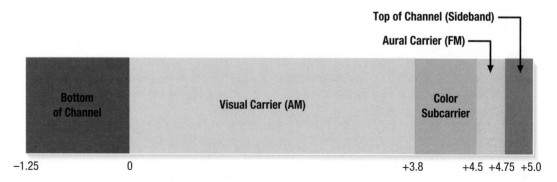

FIGURE 4.5 *Analog TV Channel*
Four of the six megaHertz in a television channel is needed to carry video information, whereas less than one-third of one megaHertz is used for the audio channel. What's left over is space for sidebands and subcarriers, including secondary audio programming (SAP).

while the FM transmitter carries the audio. A *diplexer* combines the sound and picture before they're sent to the station antenna for **propagation,** the technical term for signal distribution. Technicians look at the *amplitude* (brightness and contrast) of the picture using *waveform monitors.* They also check the color phasing through a *vectorscope.*

Once the audio and video signals have merged, they are broadcast by direct waves radiating from the TV antenna, which is positioned high to avoid obstructions in the coverage area. The **effective radiated power (ERP)** of a television station is actually a measure of its video signal strength, which is about 20 times as strong as its audio signal. Because UHF waves are smaller than VHF signals, they tend to attenuate or weaken more easily, so the FCC *affords* greater ERP for UHF stations to reach the viewers.

A broadcast station's coverage maps show just how strong the signal is through its A and B contours. The station's A contour represents its primary coverage area, while the B contour is its outer terrain. Analog television tends to fade out over great distances, and even though digital signals have a longer reach, their reception stops abruptly. This is called the *cliff effect,* because once a digital signal weakens to a certain point, it behaves as if it was falling off a cliff and disappearing.

DIGITAL TELEVISION DELIVERY

There are other differences that contrast how analog and digital TV signals are sent, but the most noticeable one is that two transmitters no longer are needed for **digital television (DTV).** The audio and video are compressed, meaning the redundant information has been eliminated so the signals are combined into one data stream.[7] This system of modulation is called **eight-level discrete amplitude vestigial sideband (8-VSB).** It fits in the 6 MHz of bandwidth and uses a pilot signal to stabilize the data stream of audio and video.

The dilemma confronting American broadcasters is not whether to broadcast digitally, but whether to do so in multiple channels and in high definition. TV stations were given one UHF channel to broadcast digitally in exchange for old analog channels to be put up for auction by the government. Some broadcasters indicated they would need to multicast on additional channels in order to create revenue and pay for new digital facilities. Washington lawmakers and citizens' groups protested that view because they thought broadcasters should act in the public's interest and use the extra spectrum to give viewers high-definition television (HDTV). As a result, some DTV stations began to do both—broadcast digitally on one high-definition channel and in standard definition on other channels. Figure 4.6 contrasts the DTV channel with the older analog format.

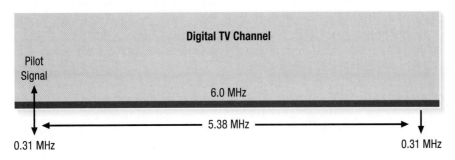

FIGURE 4.6 *Digital TV Channel*
Both analog and digital channels occupy 6 MHz of bandwidth, but the DTV channel structure differs from the analog channel. There are no partitions for visual carrier on AM; in addition, the color subcarrier and FM aural carrier are gone. Instead, the DTV channel sends a mixed data stream of video, audio, and subcarrier information, which is amplitude modulated (AM). A DTV pilot signal is added for control above the low-end of the channel during transmission.

Propagation Spreads radio waves from a transmitter over assigned areas.

Effective radiated power (ERP) Determines broadcast signal reach by the antenna gain measurement.

Digital television (DTV) Replaces analog signals by 2006 with video and audio compressed onto one channel in binary language compatible with computers.

Eight-level discrete amplitude vestigial sideband (8-VSB) Broadcasts digital television in the United States.

HDTV

To understand how HDTV began, step back to the early experiments in Japan. In 1970, the technicians of Nippon Hoso Kyokai (NHK), Japan's primary public broadcasting network, began experimenting with a new format for television. Its chief goal was to make the picture resolution richer and more detailed. It was called "high definition" because it increased the number of scanning lines on the TV screen from 525 to 1,125. The format, *multiple sub-Nyquist sample encoding (MUSE)*, also widened the picture frame, so that it resembled the cinema's wide screen.

The Europeans began studying the Japanese analog system, which required three times as much bandwidth for HDTV as standard definition television channels. The MUSE format failed to find favor with the Europeans, who came up with their own system in 1986: *high-definition multiplexed analog components (HD-MAC)*. Both early Japanese and European models used analog signals and were considered by American engineers too hungry for bandwidth to be technically efficient.

In the United States, the FCC noticed NHK's lead in high-definition television, and formed an Advisory Committee on Advanced Television Service (ACATS) in 1983. As noted early in this chapter, General Instrument's all-digital system met with favorable reviews from American engineers because of its two principal advantages: a compressible *digital* signal and compatibility with analog television. When the 8-VSB standard was selected, however, the choice met with instant controversy.

HDTV CONTROVERSY

The FCC chose 8-VSB despite protests led by the Sinclair broadcast group of Baltimore. It was convinced a better system existed. The system was called *coded orthogonal frequency division multiplexing (COFDM)*, and it had replaced HD-MAC to become the basis for digital HDTV in Europe. The European Broadcasting Union in 2004 began searching for a common standard for digital television and rejected the FCC's decision to allow scalable versions of DTV, from 1,080 to 720 and 480 lines of resolution. The U.S. government stood by its decision and allocated new spectrum space for completion of the switch from analog to digital television by 2006. By mid-2004, more than 1,233 TV stations reaching 99 percent of television households were digitally broadcasting in the United States using the 8-VSB format.

Still, a minority of American families had digital TV sets and even fewer were receiving HDTV. That fact became clear in Denver, Colorado, which was first to have a station broadcasting all of its live, local news in HDTV. KUSA-TV spent six weeks and millions of dollars completely renovating its studio, control room, and station infrastructure. On April 29, 2004, KUSA-TV presented its entire newscast in HDTV, even including a segment from its fully HD-equipped news helicopter. The only drawback of the event was that of the 1.4 million viewers in the Denver area, only about 80,000 homes had HDTV sets to enjoy this visual first in television.

In 2004, the Advanced Television Systems Committee (ATSC) announced its final approval of performance guidelines for DTV receivers. These specifications include quality control standards for everything from phase noise to signal quality indicators on smart DTV antennas. The government helped forge the last links in the DTV chain by approving "plug-and-play" rules that required consumer electronic manufacturers to have cable-ready sets available for sale that did not require a set-top box to view DTV.

▜ Reception

On your TV screen, roughly 150,000 round pixels are illuminated during the scanning process. This process, known as *scansion*, occurs so rapidly across the horizontal lines of your set that scenes appear to meld together. The viewer's brain receives each image and retains it long enough to seamlessly link it to the next picture due to a process called *persistence of vision*.

INTERLACE SCANNING

The scanning beam skips every other line as it traverses the screen back and forth from the top down, which is called **interlace scanning.** Engineers working for the **National Television System Committee (NTSC)** decided in 1941 that interlacing was the best way to fit the TV picture into 6 megaHertz of bandwidth. The process divides the frame in half and sends out two fields of odd-and-even numbered lines.[8]

How fast is interlace scanning? It scans 30 flames per second (fps), and there are two fields per frame—one for even-numbered lines and the other for odd-numbered lines—which means that 60 fields, or 30 frames, are scanned each second. Figure 4.7 illustrates the process by showing how electronic guns vary the currents to each of the dots in the pixels so that they glow at different intensities. In TV receivers, the picture information is fed to electronic guns that paint each scene one dot (or phosphor) at a time on a *kinescope,* a type of cathode ray tube (CRT). The color phosphors on the screen of the CRT illuminate the pixels in shades of red, green, blue, white, or black at the same intensity as the scene captured by the camera lens.

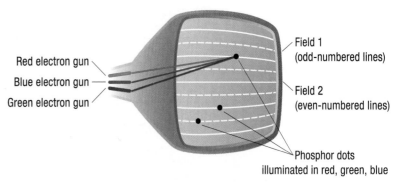

(a) Analog Interlaced Scanning

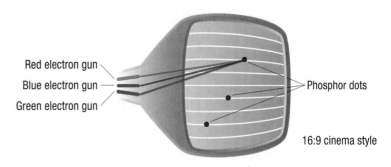

(b) Digital Progressive Scanning

FIGURE 4.7 *Scanning Systems: Interlaced and Progressive*
The National Television System Committee (NTSC) uses interlaced scanning (odd- and-even numbered lines) in order to economize on the bandwidth by sending out half a picture (one field) at a time. Inside the kinescope tube, the picture process that takes place in the camera is reversed. Beams of red, blue, and green are scanned by electron guns across the screen one line at a time—first even-numbered and then odd-numbered lines. In progressive scanning, the video frame is scanned from top to bottom without skipping any lines.

Video Passengers

Television pictures are held in check by a **sync generator.** Not only does it generate horizontal and vertical pulses to keep pictures from fluttering or rolling but it also keeps the camera's output in step with the TV screen. In addition, the sync generator adds a black bar at the bottom of the frame, known as the **vertical blanking interval (VBI).** The bar appears when the scanning beam reaches the bottom of the frame. In the moment before it returns to the top, information is sent on the black bar, holding the picture in place with other data, such as closed-circuit captions.[9]

Stereo and foreign language on television broadcasts come over the channel's *second audio program (SAP)* transmissions. Technically, the SAP channel in television is much like radio's subsidiary communication authorizations (SCAs), noted earlier, only the sound is of a higher quality.

PROGRESSIVE SCANNING

An important distinction in the scanning process separates screens for TV and computers. Computers use **progressive scanning,** which paints the lines across, one after another in sequence to form a whole frame, unlike the even- and odd-

Interlace scanning Alternates odd- and even-numbered lines.

National Television System Committee (NTSC) Standard definition of 525 lines, 30 frames per second.

Sync generators Hold TV signals in order during transmission.

Vertical blanking interval (VBI) Forms a black bar when the electron gun reaches the bottom of the screen.

Progressive scanning Visual lines of output in ascending order to create one frame at a time.

protalk
Fred Lass
CHIEF ENGINEER, WRGB-TV

The transition to DTV has been an awkward one for several reasons. The 8-VSB standard was incompatible with COFDM technologies preferred by Europeans, and TV station owners wondered where they would find the money to finance replacing old analog equipment with expensive, new digital facilities. Stations would need to buy everything from antenna towers to digital tapes and all of the hardware and software in between.

Chief Engineer Fred Lass of WRGB-TV in Schenectady, New York, oversees the transfer of technology for a pioneer station in broadcasting, but he sees obstacles in the road ahead. "The FCC put the deadline of 2006 out there as a carrot so everyone would work toward it." That deadline, however, is based on the number of digital TV sets that have been bought by American consumers. "There's absolutely no chance whatsoever that there is going to be 85 percent penetration of digital TV by 2006," Lass says, "Zero chance." However, the FCC's Media Bureau plan proposed to include local digital televi-

> **"There's absolutely no chance whatsoever that there is going to be 85 percent penetration of digital TV by 2006."**

sion signals that cable companies "down-convert," or translate back, into analog format at the set-top box. The NAB protested, saying that counting local TV signals cable companies down converted into analog would turn the Congressional plan on its head. Millions of Americans would potentially lose their television service if analog television sets were rendered obsolete by the move to DTV.

career focus

TELEVISION CAMERA OPERATOR

An estimated 93,000 broadcast, audio, and video equipment technicians work in the United States. Television stations employ, on average, many more technicians than radio stations do. Broadcast and sound engineering technicians install, operate, maintain, and repair the equipment used to record and transmit radio, television, and cable programs. They work with TV cameras, microphones, lights, audio- and videotape recorders, transmitters, antennas, and other electronic apparatus.

In the control room of a radio or television broadcasting studio, technicians monitor the audio and video signals for strength, clarity, and stability before and during the recording and transmission of programs. In smaller stations, titles such as *operator, engineer,* or *technician* are used almost interchangeably to describe various jobs. At larger stations and networks, the jobs are more specialized.

Audio and video equipment operators are responsible for recording productions, either live television programs or studio recordings. They also edit tapes for compact discs (CDs) and cassettes, or transmit programs for radio and television broad-

casting. **Transmitter operators** monitor and log television signals and operate transmitters. **Master control room operators** regulate the station's signal strength, clarity, and range of sounds and colors for TV broadcasts. They also monitor and log outgoing signals at the transmitter. **Maintenance technicians** set up, fine-tune, check, and repair electronic broadcasting equipment.

Recording engineers operate and maintain video and sound recording equipment. **Field technicians** set up and operate portable field equipment for audio and video recording outside the studio. **Chief engineers** supervise the technicians who operate and maintain broadcasting equipment. **Television and video camera operators** operate TV studio cameras or double as electronic newsgathering camera operators.

Master control engineers make sure that a television station's scheduled program elements—such as satellite and microwave feeds, prerecorded segments, and commercials—are recorded. They also are responsible for making sure these elements are inserted into the scheduled programming. **Technical directors** are in charge during the production of a program and direct the studio and control room staff.

Television station engineers generally hold college degrees in electrical engineering or associate's degrees from technical schools. Some are also members of IBEW (International Brotherhood of Electrical Workers). The FCC licenses engineers to maintain the transmitter and tower facilities. The Federal Communications Commission no longer requires the licensing of broadcast technicians; this requirement was eliminated by the Telecommunications Act of 1996. Certification by the Society of Broadcast Engineers has become the accepted standard of professional competence with experience.

numbered rows scanned in NTSC television. The system adopted in Europe, uses progressive scanning, and in the United States, the ATSC authorizes specifications for both interlace and progressive scanning digital formats.

ASPECT RATIO

The size of the screen is defined by its **aspect ratio,** which is the relationship of the horizontal to vertical dimensions. Under NTSC standards, the ratio is about 4 units horizontal by 3 units vertical (4 × 3). That remains consistent regardless of whether the screen is large or small, 36 inches or 12 inches in diameter. Television's aspect ratio soon will resemble the rectangular dimensions of a cinema screen, which is about 16 × 9, 16 units across and 9 units down.

Aspect ratio The width-to-height proportion of a television picture.

DIGITAL VIEWING

Chances are that your television set at home uses a kinescope picture tube, which is shaped like a narrow glass cylinder and widens to form a viewing screen. Russian inventor Vladimir Zworykin invented this glass screen, also known as a *cathode ray tube (CRT)*, in 1929. Recent research, however, has developed new technologies for producing flat screens that can be hung on a wall. The flat-screen devices in your computer are called liquid *crystal displays (LCDs)* because stored between the sheets of glass are thin films of liquid crystals with transistors etched on the surface.

Liquid crystal displays are either *passive matrix*, which uses a continuous stream of power to the pixels, or *active matrix*, which is turned on and off. Active LCDs are superior in image quality, color, brightness, and stability. The LCD, however, offers limited peripheral vision to the audience. As with a computer, viewers need to be seated directly in front of the screen.

Another screen for future TV viewing is the plasma-display panel (PDP), which uses a grid of tiny pixels filled with a gas mixture. The grid illuminates red, green, and blue phosphors with ultraviolet light when charged with electricity. The PDP offers a better peripheral view and is relatively thin and light in weight, but its price is prohibitive for some consumers.

Summary

The technical business of broadcasting requires both competition and collaboration. It is only through the collision of timely ideas born in a competitive environment that progress has been made in the development of electronic media. Each creative burst of technology has served to show what ideas will work and those that will not serve to better harness signal waves so they can be used to present messages through speakers and screens of radio, television, and computers.

In the encoding phase of the technical process of broadcasting, audio waves are reflected by microphones and recorded on digital or analog tape or discs. In television, scenes of light are encoded through a camera lens and eventually converted to picture elements on a TV screen and edited on digital tapes and discs. Storage formats have seen a progression from discs of vinyl to oxide tape to metallic discs engraved with beams containing digital information. In each stage of development, when quality, cost, and convenience support the new technology, the marketplace has proven receptive. Digital television and radio must now compete with new channels of telecommunications that will offer video and audio via satellite and the Internet.

Broadcasting delivery is basically the business of harnessing invisible waves of electromagnetic energy and encoding them with information—audio, video, graphics, and text. These waves behave according to certain principles of physics—the higher the frequency, the shorter the wavelength.

Radio and television are wireless communications that broadcast over the electromagnetic spectrum, which can be envisioned as lanes or channels. Broadcast stations are assigned frequencies referring to the number of times an antenna sends out electromagnetic cycles within one second. Another principle regarded by those licensed to engineer AM radio signals is that lower frequencies are more desirable because their longer waves travel farther than the waves transmitted by high-frequency channels.

The International Telecommunications Union (ITU) is the global manager of spectrum, allocating bandwidth and settling broadcasting disputes between neighboring countries. Channel assignments are left to governments to license; in the United States, the Federal Communications Commission handles that duty. Choice spectrum space is constantly in demand for a variety of purposes. Commercial demand is high, and special interests seek to broadcast, as well, so the FCC has created low-power radio and television broadcasting.

Delivery and reception of digital television have raised a number of interesting questions because they require new thinking in the broadcast industry about what the future of television will look like. Television stations are looking for new revenue streams to make the digital transition possible and must choose how to broadcast digitally in high definition, or over multiple channels, or both—once the digital conversion is complete.

Food for Thought

1. The five stages of technical production—encoding, editing, storage, delivery, and reception—suggest the necessity of compatible digital technologies for each stage of the process. Should the U.S. government impose standardization or leave it up to market forces? Why or why not?

2. The international community speaks in different formats when it comes to the technology of telecommunications. Should the United States push for a universal digital standard for radio and television?

3. In certain channels of electronic media, foreign countries lead the United States. For example, Great Britain completed its transition to digital radio while American broadcasters lag behind. Should the U.S. government treat telecommunications as it did the race for space and aim to be first? Why or why not?

4. Revealing breakthroughs in new media technology can be controversial, especially when patents and profits are involved. Still, some of the most successful innovations owe their status to sharing the "recipe" in order to increase its distribution and visibility. What do you see as the plusses and minuses of sharing such secrets?

5. If you were stranded on a desert island and could have only one electronic media device with you, what would it be and why?

Digital Domains

Chapter 5

Political pundits called it his "I Have a Scream Speech." Howard Dean, the pugnacious governor from Vermont, led a wave of web-based discontent with President Bush's "War on Terror" when he sought the Democratic Party's presidential nomination. After finishing second in Iowa, Dean gave an impassioned speech to rally his supporters, which began to sound more like taps than a bugle charge. "Not only are we going to New Hampshire. We're going to South Carolina and Arizona and North Dakota. . . . And then we're going to Washington, DC, to take back the White House. . . . Yeaaaarrrrhhh!" Dean's televised exuberance gave critics all that was needed to draw a bull's eye on his back.

Like a missle, blogs and e-mailed messages from candidates, such as Dean, Senator John McCain, and even President George W. Bush can reignite campaigns. As Dean described it, "Along comes this campaign to take back the country for ordinary human beings, and the best way you can do that is through the Net. We listen. We pay attention. If I give a speech and the blog people don't like it, next time I change the speech." *Blogging* is derived from two words, "web log," and it usually means the art of keeping track of the day's events with commentary and welcomed responses from others. Dean's bloggers liked the "trickle up" theory of political action; their ideas got responses once expressed on-line.[1]

After Sen. John Kerry's campaign surged ahead of Dean's, he expressed his full faith in the power of the Internet. If campaign contributions were any indication, Kerry's commitment paid off. His campaign collected $44 million over the Internet, eclipsing his direct mail and telephone contributions.

Republican campaigners for President George W. Bush also encouraged the party faithful on their website, GOP Team Leader, which offered credit points to redeem for partisan merchandise for Republicans who showed up at rallies with their friends. Unlike Gov. Dean and Sen. Kerry's blog, GOP Team Leader defined *blogging* as a place to post press releases and suggest talking points, but not to invite comments from outsiders.

The age of digital convergence is more than a political phenomenon. Many different types of information—audio, video, text, and graphics—are transmitted over new channels. Digital media put more control in everyone's hands, shrinking the world to a virtual reality that is both immediate and mobile. It offers new choices that drive consumers to use even more media. Ultimately, digital convergence speaks the language of computers, the Internet, radio, television, and the future.

The Digital Challenge

PRINCIPLE 1

Digitization allows video, audio, and text information to be compressed for more efficient transmission.

Depending on how you look at it, **digital convergence** is either here or has yet to arrive. Computer chips are faster and have room to store huge files for information and entertainment. Americans think nothing now of downloading, storing, and playing music and video on computers. DVD players have essentially replaced VCRs. Broadband and Wi-Fi (wireless fidelity) connections are commonplace. (Wi-Fi uses radio frequencies rather than wires to allow computer users to connect to the Internet.)

In terms of media production—audio and video recording and editing—the transition to point-and-click tools is nothing new. Yet, in terms of radio and television reception, loose ends now are being tied.

Investors and owners in radio, television, telephone, and cable enterprises have embraced digital technology for the **compression** necessary to send pictures, sounds, text, and graphics over the Internet, but their transition to digital broadcasting is a work in progress. Plug-and-play digital television sets are on sale in stores, while radio stations look forward to the rollout of HD (high-definition digital) radio. Copyright protection issues top the list of problems to be solved. Nonetheless, the future is digital, and to understand it, we need to know the theory behind the practice.

Alphanumeric Character		Binary Code for Bytes
5	00110101	00110100
8	00111000	00110111
G	01100111	01000110
R	01010010	01010001
S	01010011	01010110
e	01100101	01100010
l	01101100	01101011

Group of 8 Bits
ASCII Code

FIGURE 5.1 *Bytes*

Compression Reduces the bandwidth needed for electromagnetic energy transmissions in video, audio, or text.

Digital convergence The trend toward merging what were separate media (radio, television, telephone, and computer) into one medium for purposes of communication and commerce.

Bit Abbreviates *binary* and *digit* to represent a unit of data with two options—one or zero.

Byte Groups bits together according to units of eight in the American Standard Code for Information Interchange (ASCII).

DIGITS DEFINED

Simply put, digital recording and signal transmission are more efficient than analog. They are faster, easier, and, like a department store sale, give you more for less—that is more signal with less noise. When scientists and technicians began harnessing the electromagnetic spectrum, they relied on the dots and dashes of telegraphy to communicate. It's actually not that different with digital television. Binary language is the computer's way of translating text, graphics, pictures, and sounds.

The binary digit is the fundamental building block. A **bit** is simply the consolidation of two words, *binary* and *digit.* One bit is just another way of saying you have turned a switch on or off, to one or to zero. The point is that digital has become the universal language of technology expressed in bits and in larger collections of bytes. A **byte** is typically eight bits of information; in Figure 5.1, the code is broken down to show byte-sized morsels of data.

BEGINNING WITH THE BITSTREAM

Speeding along the digital lanes, radio and television broadcasters seize bitstreams to transport audio and video signals at volumes and speeds expressed in *bits per second,* or *bps.* Such numbers grow larger than radio frequencies, and are likewise described in *kilobits, megabits,* and *gigabits.* When engineers speak of *terabits,* they mean trillions of bits flowing in data streams. Figure 5.2 shows the speedometer for the Internet and spells out the bit-rate abbreviations. But for a shorthand description of how this electronic language developed, drag the cursor back in time.

NYQUIST'S NUMBERS

The theory behind the engine of the digital age belongs to a Swedish-born scientist and inventor, Harry Nyquist. He immigrated to the United States as a teenager and earned degrees in electrical engineering and physics from the University of North Dakota and Yale schools of engineering. Nyquist's interests covered a wide range of electrical phenomena, particularly telegraphy. After taking a job at Bell Labs, he penciled in the formulas for measuring the pulse of electromagnetic energy.[2]

Binary Is Born

Nyquist's idea was to rethink analogous waves of energy in terms of dots and dashes, a type of Morse code to modulate electromagnetic signals in mathematical arrays. He envisioned breaking down the waves into numbers, binary codes of ones and zeros. The key piece to the Nyquist puzzle was **sampling.** In a 1924 article, "Certain Factors Affecting Telegraph Speed," he defined sampling as the means for capturing the voltage of the original wave at fixed intervals, and coding it into pulses to represent the original signal. If the sampling rate is fast enough, the gaps between measures of sound and light waves will not be heard or seen. Today, engineers talk about this process as **pulse code modulation (PCM).** It means not capturing the whole wave but drawing a sample of it.

Sampling Solves the Equation

Sampling is just the first step in converting an analog signal to digital. The waveform is broken into a series of narrow pulses by **quantization,** in which the amplitude is measured and converted into digits. Nyquist's theory would work so long as sampling rates measured enough of the original wave. This system set the stage for compressing video, audio, and other forms of information so it could be easily shipped in bundles. That, in a nutshell, is a sketch of digital compression. So what happened after Nyquist? Just about everything. Digital machines were invented, replaced, and reinvented—all due to the wonders of bits and bytes.

20 kb	20 kbps (20,000 bits per second)	Kilobits
20 kB	20 kBs (20,000 bytes per second; 160,000 bits per second)	Kilobytes
20 Mb	20 Mbps (20,000,000 bits per second)	Megabits
20 MB	20 MBps (20,000,000 bytes per second; 160,000,000 bits per second)	Megabytes
20 Gb	20 Gbps (20,000,000,000 bits per second)	Gigabits
20 GB	20 GBps (20,000,000,000 bytes per second; 160,000,000,000 bits per second)	Gigabytes
20 Tb	20 Tbps (20,000,000,000,000 bits per second)	Terabits
20 TB	20 TBps (20,000,000,000,000 bytes per second; 160,000,000,000,000 bits per second)	Terabytes

FIGURE 5.2 *Bit Rate*
The flow of information on the Internet is clocked according to the flow of bits or bit rate. From Tassel, Joan Van. *Digital TV over Broadband—Harvesting Bandwidth.* Woburn, MA: Focal Press. 2001, 50.

Computer Evolution

As the infancy of radio and television was marked by rancorous disputes over patents and court cases, the computer had something of a troubled adolescence. The engineers who created its source codes, the key to the machine's system of logic, either shared or guarded them as closely held secrets. There were also those who converted dot-com discoveries into wealth and fame, or disappeared during the downturn of 2000–2001, only to re-emerge years later with new ideas and inventions.

Sampling Digitizes analog signals through periodic measurements of continuous waves.

Pulse code modulation (PCM) The sampling process where the amplitude of an analog signal wave is converted into its numerical equivalent for digital processing.

Quantization Assigns a numerical quantity to a sample of a signal.

MAINFRAMES TO PCS

The early days of the computer's invention were spurred by military necessities. Computer prototypes sprouted up before World War II when the U.S. Navy sought to solve ballistic errors in its antiaircraft guns using precise computations. Bell Lab technicians in 1939 hooked up electronic relays and switches to typewriters to help aim the weapons. With keyboard controls, the results resembled today's personal computers. Harvard mathematician Howard Aiken connected his machine's electrical components to rolls of perforated paper tape and called it the Mark I. Backed by International Business Machines, Aiken's analog machine was computing numbers in 1944. The British also made their contribution. The Colossus was invented to crack the Nazi's secret codes during World War II, and it did just that by using binary code.

After the war, the University of Pennsylvania set aside an auditorium to demonstrate a tube-powered calculator that was 80 feet long. Its named was almost as long: the Electronic Numerical Integrator and Computer (ENIAC). Designed by John Mauchly and J. Presper Eckert of the Moore School of Electrical Engineering, the 30-ton monstrosity was built with a half-million-dollar grant from the Army Ordinance Department. Like the Navy, the Army hoped for a machine that could correct the problems of artillery fire. After ENIAC, Mauchly and Eckert entered private business and designed another computer for the government in 1951. The UNIVAC I (Universal Automatic Computer) processed population data for the U.S. Census Bureau.

ENIAC—ELECTRONIC NUMERICAL INTEGRATOR AND COMPUTER. It was built in 1946 by the University of Pennsylvania at government expense. The ENIAC was inspired by the war effort, and performed computations for the hydrogen bomb in the early 1950s.

UNIVAC I UNIVERSAL AUTOMATIC COMPUTER. Scientists J. Presper Eckert and John Mauchly operate their invention for the Remington-Rand Corporation. The UNIVAC was used by the U.S. Census Bureau to tally the population, and later projected the 1952 presidential race.

The Punch-Card Era

If you know the phrase "Do not fold, spindle, or mutilate," then you may recall something of the computer era between the arrival of the big UNIVAC machines and the advent of personal computers. The mainframe age was marked by punch cards that tabulated numbers and reflected IBM's dominance in computing centers around the world. The firm held important patents on the punch-card process, which its competitors eventually circumvented by changing the number of rows and the shapes of the holes to devise their own "electromechanical tabulators."

The Smithsonian Institute's Steven Lubar observed how these punch-cards were binary—their holes were either punched or not, and their constellations figured into a larger scheme of numbers. Punch-cards also symbolized an era of depersonalization in the national consciousness. The U.S. Census Bureau started using punch-cards to identify people by the numerical dimensions of personal data. The fear of computerized "Big Brotherism" was visualized in Stanley Kubrick's *2001: A Space Odyssey*. The menace of a malfunctioning computer HAL, whose name were the letters preceding IBM in the alphabet, had to be defeated.[3] The next decade would see punch-cards give way to solid-state computing and a new way of life.

Advent of the Personal Computer

The first personal computers appeared in the United States in the 1970s. Technicians and hobbyists at Osborne, Kaypro, and Commodore companies built the machines small enough to fit on a desktop. Soon, other firms drew on the same concept in the labs of General Electric, NCR (National Cash Register), and Xerox.

Planting the Apple Seed

The story of the personal computer cannot be told without placing two California teenagers in the script. Steve Jobs and Steve Wozniak formed a computer club and, at Jobs's insistence, began to create a homemade prototype. The Apple I was built in Jobs's garage; it had a keyboard and processing unit tightly fitted into a briefcase. When a marketing specialist joined their team in 1977, Apple Computer, Inc., was incorporated. Within six years, Apple had created its prototype personal computer, the Lisa, using a mouse to execute commands by pointing and clicking on icons. This system was described as "gooey" or **GUI** for **graphic user interface.**

The Apple Macintosh was released in 1984, with a Super Bowl spot depicting an Olympian runner hurling a hammer at a screen where an Orwellian dictator appeared to be ordering the masses to march together in lockstep. The dictator's face, representing IBM, was shattered by the Apple Runner's hammer.

COMPUTER DEVELOPERS STEVE JOBS, JOHN SCULLY, AND STEVE WOZNIAK. This team introduced personal computing with the Apple II in 1977. In 1984, they reconfigured it and sold the Apple IIc. Jobs holds the keyboard, while Scully and Wozniak stand behind the early monitor.

1984 SUPER BOWL AD. The year 1984 was symbolic of a nation under the control of "Big Brother," based on a novel by George Orwell. Apple's new computer defeats the menace in this spot.

BIG BLUE VERSUS APPLE

The nation's largest computer manufacturer, IBM, famous for its big blue logo, took notice of Apple's success. In 1980, IBM executives called for a personal computer to compete in this growing market. The director of IBM's development lab in Florida recommended contracting with hardware and software suppliers from outside the company. Instead of designing its own microprocessor, IBM adapted one from the Intel model designed in Albuquerque, New Mexico.

To program the computer's operating system (OS), IBM contacted a software company near Seattle represented by a Harvard dropout. Bill Gates of Microsoft agreed to provide the operating system—on one condition: His company would retain ownership of the license. Gates bought an OS program written by Tim Patterson of Seattle Computer Products. Patterson had dubbed it QDOS (Quick and Dirty Operating System). Gates simply changed its name to MS-DOS (Microsoft-Disk Operating System) and reaped a fortune from his $50,000 investment.

After marketing its first personal computer, the PC, IBM came up with a smaller version called the PC junior, and then dove deeper into the portable personal-computer market with its first laptop model. Other firms began chasing their

Graphic user interface (GUI) Added to the computer's operating system a set of icons and visual links rather than text commands to handle data.

share of the PC market, including Commodore, Atari, Texas Instruments, and Radio Shack—all building machines that shared the same basic component parts.

PC Components

The main elements of the PC are its hardware, operating system, software applications, and peripherals. The most important piece of hardware is the central processing unit, or CPU. This is the brain of the computer. These microprocessors house one or more silicon chips to perform a variety of functions, all using some form of binary language. They respond to the software applications through the machine's operating system.

The **operating system (OS)** provides the means by which the software applications enable the computer to display graphics and text, perform calculations, and print data. Usually, an operating system is tied directly to one particular model of computer, and the OS along with that compatible model eventually become obsolete. The reason is chip efficiency. Gordon Moore, one of Intel's founders, said in 1965 that chip memory and power would double every one or two years. As a result, computer speeds would increase along with storage capacity. His observations proved correct over the years, and became widely accepted as Moore's Law.

PC Versus Macintosh

Because of its decision to share the source code of its original OS with software manufacturers, IBM soon saw clones of its original PCs marketed in competition. Apple, on the other hand, retained all of its rights to the system technology and effectively kept clones off the market until the 1990s. Personal computers dominated the PC market due to the price competition between the Windows OS systems and its imitators.

By 1995, Apple had only about 7 percent of the global market for personal computers when it decided to embark on its own cloning program, which relied on a licensing system for manufacturers that would agree to pay a royalty for its software. For two years, buyers would be allowed to purchase the PowerPC clone and run a Mac OS. However, when Apple's founder, Steve Jobs, returned to his old company, he tried to renegotiate the cloning agreements and failed. Eventually, the PowerPC cloning program was cancelled. However, Apple did continue to keep its video-streaming program, QuickTime, as an open source code. These changes have put more information in the hands of consumers and allowed them to handle data in ways that older, analog machines could not.

PRINCIPLE 2

Digital media put more control in the hands of consumers than older, analog media do.

The Internet Age

The Internet was the offspring of the Defense Department's Advanced Research Projects Agency, or ARPA. This agency was a branch of the Pentagon formed in response to the Soviet Union's launch of *Sputnik I* on October 4, 1957. The leap in space by a hostile superpower galvanized American resolve to enact a new system of defense.[4] Under the watchful eye of its military leaders, ARPA began meeting with the Research and Development (RAND) Corporation of California to perfect a fail-safe system of communication in case of nuclear attack. The project was called **ARPANET,** a system of communication links that would function even if enemy missiles disabled parts of the overall infrastructure.

Operating system (OS) Determines how a computer's central processing unit (CPU) will read, process, and store data.

ARPANET Pioneered the effort to network host computers to withstand the threat of nuclear attack. It is considered the genesis of today's Internet.

Apple's share of the global market lags behind leading PC vendors such as Dell and Gateway, but there is something that sets Apple apart. Founder Steven Jobs is not shy to talk about it either. He calls it simply "great design," but he does not mean how Apple computers and notebooks look and feel—he means how they work.

Perhaps the best example of Apple design is the popular iPod. In 2001, Apple released this new digital music player weighing less than 7 ounces but holding more than 1,000 songs in the MP3 format. In three years, millions of customers found iPod to their liking, and it soon replaced earlier competitors. For at least one quarter in 2004, Apple sold more iPods than it did Macintosh computers. In fact, when Michael Dell began promoting his version of the MP3 player, DJ (Digital Jukebox), he took direct aim at the iPod by offering a $100 rebate to customers who sent in an

iPod for recycling. What was it that made the iPod so special?

Apple's answer was simple. In an age when hand-held devices light up with confusing arrays of buttons that confound the buyer by their sheer complexity, the iPod stood apart as defiantly simple. One main control, the scroll wheel, was easy to spin with the thumb and delightfully navigable for any music lover. When users press a button to choose a tune, they only have to adjust the volume with the wheel and sit back to enjoy the music. As digital technology grows more complex, designers will think about the iPod's simplicitiy and the maxim about less being more.

MILITARY-TO-COLLEGE TRANSITION

In design, at least, ARPANET was revolutionary because it replaced top-down hier-archies with an egalitarian model. Terminals were created equal—if any one computer went down, others would bypass it until all links were restored. By 1966, the government was ready to establish such a network and initiated SAGE—the semi-automatic ground environment project—exploring interactive links between computers using telephone lines and video display terminals.

In 1969, computers at Stanford University and UCLA were linked, and the following year, UC Santa Barbara and the University of Utah were added to the network, soon followed by MIT, Harvard, and Carnegie-Mellon. The Internet, which began as a military secret, swiftly migrated to university campuses. Today, it covers the globe with instantly exchangeable information.

NETWORKS ON TOP OF NETWORKS

The Internet has become our most familiar network, but it's actually thousands of networks cooperating with each other to direct information to a final destination. At the top level, the **backbone** networks with high-message capacity to travel wide distances to link smaller networks. Like an interstate highway, backbones cover the longest distance at the highest speeds carrying the biggest loads. Wide area networks (WANs) serve regions of the country; local area networks (LANs) interconnect computers within a building or campus area (see Figure 5.3).

Packets and Protocols

Internet cooperation means that networks of computers abide by the same rules, or protocols. **Protocols** enable computers to speak with each other by sharing small bundles of information known as **packets** (see Figure 5.4). Packets move along different routes to reach their destinations, where they are reassembled by the receiving computer that gives meaning to the message. The idea of **packet-switching** was the inspiration of a Welsh physicist, Donald Davies, who observed that a brain overcomes neural damage by sending messages along alternate routes. If there is an impasse at any one juncture, the brain finds a detour. The national system of computers was designed to function in the same manner.[5]

Backbone Forms the overarching route linking, by wire, smaller networks of computer terminals.

Protocols Determine how computers are going to make sense of each other's data in terms of transmission and reception.

Packets Carry the information on the Internet by breaking it down into independent segments.

Packet-switching Allows data to be separated into bundles and reassembled at their destination.

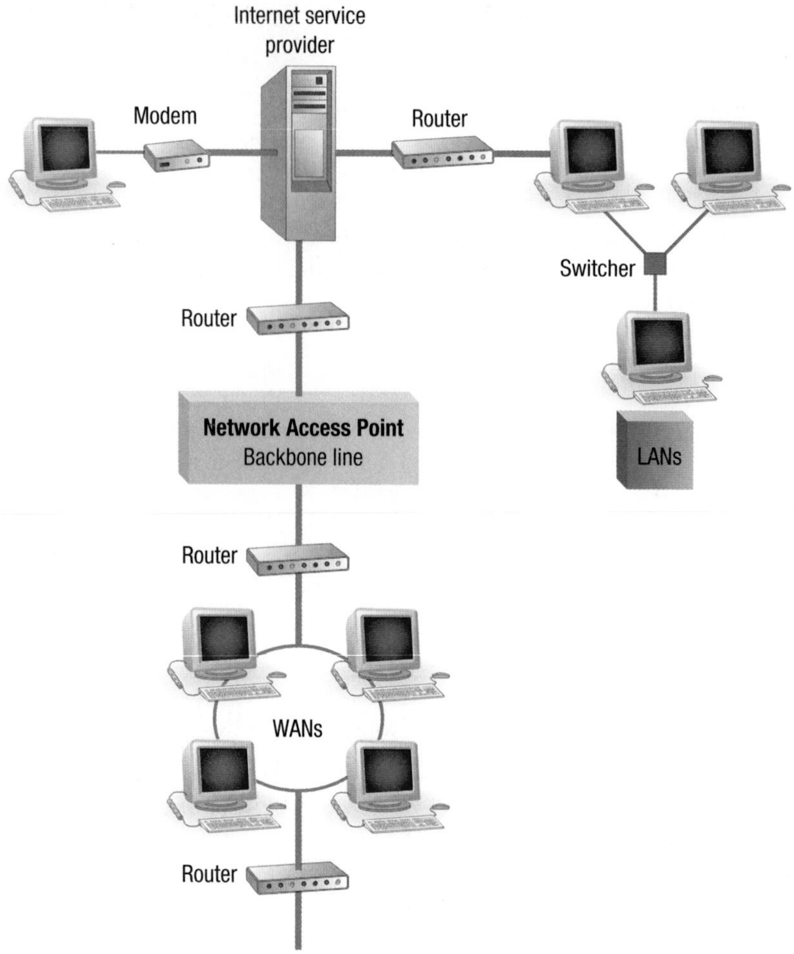

FIGURE 5.3 *Local Area Networks (LANs) and Wide Area Networks (WANs)*
Individual computers feed information into the Internet by Internet service providers or by local area networks at schools, businesses, or other organizations. The providers and networks transmit signals through routers to wide area networks at the regional level and enter the backbone line at a network access point.

Internet service provider

Modem

Router

Switcher

Router

Network Access Point
Backbone line

LANs

Router

WANs

Router

Uniform resource locator (URL) Finds a home page or website on the Internet.

Top-level domain (TLD) Identifies websites with terms describing their address by general application (edu = education, mil = military, org = organization, com = communication, etc.) and specific location (uoregon.edu, ablongman.com, pentagon.mil).

Detours in Cyberspace

On today's Internet, computers speak to each other in cyberspace using two sets of rules, or protocols, that make shipping data easy. Transmission control protocol (TCP) is what separates data into bundles of about 1,500 characters for shipping as packets to the destination computer. The TCP also labels the packet with a "check-

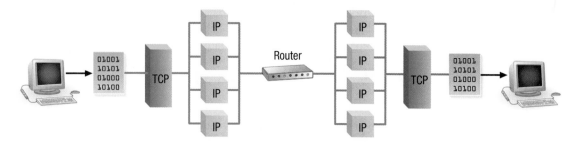

FIGURE 5.4 *Packet-Switched Network*
The sender creates a message, which is translated into binary code by his or her computer, and assigns a destination address to receive the message. The message is transferred to the TCP (transmission control protocol) computer, which breaks the message into smaller packets of information that can travel easily through the Internet. The IP (Internet protocol) computer routes the data through a router to the receiving computer, where the information is received by the destination's computer, recombined into the original message, and translated from binary code to a readable format.

sum" if any errors contaminated its information cargo during the journey. If flaws are detected, TCP quickly discards it and asks the host computer to resend fresh packets.

Internet protocol (IP) is how information finds its way along the quickest path. The IP is like an address on an envelope, so that as the packets travel, routers examine their destinations and direct them to the quickest route. On the Web, these addresses are called **uniform resource locators (URLs)**. Thus, mutually accepted protocols allow data to travel between computers in the most efficient way possible.

Domain Names at the On Ramp

Every IP address is a series of four groups of numbers separated by dots, such as 587.34.903.32. Letters replace numbers in the addressing stage. The last half of the address is called the **top-level domain name (TLD)**, and the first half is the user's identification. To the right of the @ sign is the designation for the groups of computers where the residence is located. There are 15 TLDs assigned by an international corporation called the Internet Corporation for Assigned Names and Numbers

> PRINCIPLE 3
>
> The Internet works due to the common language through which computers exchange information over a web of networks.

E-mail (electronic mail) Sends text messages digitally between two or more computers through wired or wireless networks.

Several people have taken credit for the ground-work that facilitated the invention of the Internet. Paul Baran of the RAND Corporation, for example, designed the programming code whereby computers could talk to each other. Dr. Vinton G. Cerf worked on the program from 1976 to 1982, and, in collaboration with Bob Kahn, developed the packet-switching protocols to link radio and satellite communications to the Internet. Kahn and Cerf did much to promote its development, and came up with the name: *Internet,* a compression of *interconnection of networks.*

Another candidate for the Internet's founding father is a scientist who shunned publicity and was known to his friends as "Lick." Joseph Carl Robnett Licklider took over the Information Processing Techniques Office (IPTO) in 1962, and began working with the Pentagon's Defense Advanced Research Projects Agency (ARPA). While directing the IPTO, he shifted contracts away from independent corporations to

sidebar
WHO IS THE TRUE INVENTOR OF THE INTERNET?

academic computer centers that shared his vision of interactive computing. Mainframes were preferred by big business, but Licklider thought of computers in terms of communication, not numbers, and explored concepts such as human-computer symbiosis.

Licklider prophetically nicknamed his team of specialists the "Intergalactic Network," and coauthored a paper in 1968 titled, "The Computer as a Communication Device," which was revolutionary for its day. Licklider predicted that by the year 2000, millions of people would be on-line, connected by a global network. While the initial ARPANET messages were exchanged in 1969, it was not until the 1980s that the National Science Foundation became involved and began beefing up the Internet's capacity with more bandwidth. Shortly thereafter, Internet service providers began to encourage consumers to use the Internet and experience their first **e-mail.**

As with all inventions, several scientists contributed time, talents, and creative energy to make the Internet a reality. In addition to Baran, Cerf, and Kahn, Leonard Kleinrock at MIT wrote the first paper and book on packet-switching. Lawrence Roberts and Thomas Merrill connected the MIT computers with the California terminals. And the list goes on. If you are looking for one individual who deserves the title of "the inventor of the Internet," J. C. R. Licklider is the name.[6]

OXFORD-TRAINED ENGINEER TIM BERNERS-LEE. He invented the World Wide Web in 1989 while trying to better organize his notes. At a Swiss think-tank, Berners-Lee designed software enabling computers to tap into each other's hard drives.

PRINCIPLE 4

The variety, speed, and access of information on the Internet have made it a major world medium.

Internet service provider (ISP) Offers Internet access to consumers by linking their home or office computer terminals to the Internet.

Hypertext transfer protocol (HTTP) Indicates documents written in hypertext markup language (HTML) that can be linked to other sites on the Web.

E-commerce (electronic commerce) Facilitates transactions of goods or services through the exchange of information or currency between computers.

(ICANN). These include ".com" for commercial, ".edu" for educational, ".org" for organization, ".net" for network, ".mil" for military, and ".gov" for government sites, among others. Right before the TLD is the specific location, sometimes called the organizational identifier, and to the left of the symbol is the specific user's identification.

In order to call upon these addresses, users gain access through **internet service providers (ISPs).** These ISPs put millions of people on-line every day, including America Online (AOL) Earthlink, SBCglobal, Netzero, and hundreds of others. The most frequent activity for ISPs is carrying e-mail, which produces trillions of on-line exchanges each year as compared to only billions of postal items.

WEAVING THE WORLD WIDE WEB

The World Wide Web Consortium (W3C) runs the most dynamic network on the Internet. The WK is headquartered at the Massachusetts Institute of Technology (MIT). It is an industry consortium made up of many private companies. A British software engineer, who now heads the W3C, made this his contribution to the Internet revolution in 1991.

Tim Berners-Lee, known to his friends as "TBL," developed the system for transferring documents at the CERN laboratory for particle physics in Geneva, Switzerland.[7] He named this innovation in computer programming **hypertext transfer protocol,** or **http.** He based it on the idea that linking documents by clicking on key words and phrases is a natural activity because it follows the user's intuitive curiosity. Websites greet users with a homepage that introduces its contents and directs readers by hyperlinks to other headings or websites. The ease of navigation and the rich content available made it clear that the Web would become the world's first truly interactive system of multimedia communication.

ELECTRONIC COMMERCE

The gold rush of the Internet's boom days during the 1990s was inspired by the lure of big profits through electronic commerce. **E-commerce** actually predates the Web; it began with a network known as the *electronic data interchange (EDI).* The creation of EDI came about because manufacturers wanted to swap with retailers' product data on prices, inventory, and shipments via their company's computers. It grew through the Internet's expansion into two main systems of commerce: business-to-business (B2B) and business-to-consumer (B2C).

B2B Commerce

Cisco Systems first established an e-commerce site in 1996, and Microsoft began moving fast on its heels. Within a month, Microsoft created its "merchant system software" for making sales on the Web. Then IBM elected to find its niche in e-commerce through electronic banking and financial advice. The eIntegrion Financial Network of IBM marketed "e-business strategies." Microsoft and Netscape contributed to the growth of business-to-business on the Web by offering data exchange services. There were so many firms struggling to survive in B2B e-commerce that it was inevitable that some would fail; many did during the dot-com downturn of 2000–2001.

B2C Commerce

The success of e-commerce for consumers is easy to understand in view of the pioneering work of Amazon.com's Jeffrey Bezos. The founder of on-line book sales ignored his critics, including those who said his "1-click" system for buying books was just too easy. Bezos sued companies that imitated his ideas by selling merchandise through a seamlessly convenient interface, and Amazon.com became one of the hero stories of e-commerce.

Bezos's dream of selling books on the Internet was not inspired by any personal retail experience. What this Princeton graduate in computer science actually knew how to do was organize books by computerized lists of other sellers and publishing houses. His on-line store that began in 1994 gave customers a huge selection of titles from which to choose. Bezos added his personal touch by keeping track of trends and customer preferences, and recommending to the buyers new titles they would like. Amazon.com grew well beyond its book-selling roots, offering video, music, and other merchandise. In so doing, it proved just how much e-commerce could allow Americans to enjoy shopping without ever leaving home.

E-COMMERCE IN THE FUTURE

More than $63 billion in retail revenue was made through on-line sales in 2003, and that figure is expected to double in three years. Credit for such spectacular growth belongs to companies such as eBay and Amazon.com to be sure, and also

AMAZON.COM FOUNDER JEFF BEZOS.

sidebar

EBAY'S CEO

The silicon-valley stories of computer-class dropouts who turned fantastic ideas into huge fortunes before riding off into the Internet sunset have now faded, but there is one worth recalling. It is not a story about getting rich from B2B or B2C, but from C2C commerce. Pierre Omidyar left his major in computer science at Tufts University after interning in software programming in northern California. And yes, he did create the world's largest on-line auction house, eBay, in 1995. But then he did something different before exiting stage left—he turned it over to someone with a new perspective but who kept Omidyar's formula intact. Now, what makes this an unusual tale is that Omidyar's concept of e-commerce is a far cry from business textbooks.

Rather than acting like a profit-driven Web merchant eager to grab shoppers and squeeze them for every last dime, Omidyar insisted that eBay live

by the golden rule. There would be no Web pitches with advertising causing "banner blindness," he said. Ebay's interactions with web customers were to be on their terms. If you had something to buy or sell at eBay, he wanted to work with you, and not the other way around. It was just that simple.

Before he moved to France in search of his family roots, Omidyar turned over the company's leadership to Margaret "Meg" Whitman. Whitman was a Princeton graduate with a Harvard MBA, who formerly held management positions at companies such as Hasbro, Disney, and Procter & Gamble. So, she was understandably shy about grabbing eBay's reins.

Whitman took over as CEO of eBay in February 1998, and eight months later the on-line auction house went public. And she has done it without compromising any of Omidyar's original vision. Success continued to follow the web company under Whitman's leadership, with market share prices rising to astronomical levels, including a secondary offering of $1.1 billion, one of the largest ever for an Internet firm. Perhaps it's not easy keeping eBay's two million-plus members happy while managing the company's brand image and employees, but Whitman made it look that way.[8]

smaller ones—Bizrate.com, Mypoints.com, Columbia House sites, travel sites, and greeting card sales, to name just a few. The difference between successful and unsuccessful e-commerce is founded on the same principles of the interpersonal world of retailing: *quality control, customer trust, product satisfaction,* and *brand loyalty.* The main difference is it all must be achieved in a media-converged environment.

▌ Broadband Future

Convergence comes in many colors, shapes, and sizes. There is the convergence through mergers of media enterprises forming huge conglomerates; there is the convergence of professional roles as journalists become skilled in the rich media tools of print, audio, video, and graphics; there is the convergence of channels available via satellite, cable, antenna, and the Internet. All are converging in the broadband world of electronic media.

Broadband refers to the higher speeds and larger **bandwidth** it takes to deliver multimedia communication and commerce on-line. The Pew Center discovered in 2004 that almost one-third of adult Americans (68 million) have broadband access to the Internet at home or work, and more than 75 percent of the nation has at least some access to the Internet.

For broadcasters, digital convergence is both a means of production and a means for streaming content on-line to new audiences. That point was underscored at the National Association of Broadcasters Convention in 2004 when then chairwoman and CEO of Hewlett-Packard, Carleton (Carly) Fiorina, complimented broadcasters for recognizing how important the convergence of computers with radio and television networks had become to their mutual survival. Fiorina, however, did not survive as Hewlett-Packard's chief executive due to sagging profits.

STREAMING MEDIA

Television stations have been **streaming** programming over the Web to encourage audiences to convert their computers into TV sets. Streaming media requires a server to send audio and video content from one computer to another. It follows two protocols for streaming: real-time streaming protocol (RTSP) and real-time transport protocol (RTP).

Streaming was introduced in the mid-1990s by Xing Technology and Progressive Networks (both now part of Real Networks). At first, the aim was to listen to audio through the computer. By 1997, RealVideo had been developed and interest in streaming grew. Today, three companies compete for streaming content: Real-Networks, Microsoft Windows Media Players, and Apple's QuickTime. QuickTime, in contrast to other Apple innovations, uses an open source code. Streaming media faces the same challenges regardless of the player, which include bandwidth, accessibility, and downloading. There also are copyright issues, which discourages radio stations from streaming on-line.

ABC News' Digital Media Group serves its broadband audiences, ABC News Live, a 24-hour news channel via the Internet. ABC calls this service its iPod for video because it gives viewers the chance to watch the news anytime, anywhere—so long as the viewers' laptop computers and Wi-Fi (wireless fidelity) connections are handy.

INTERACTIVE TELEVISION

Interactive television (ITV) has always held great promise but has never quite realized its full potential. Cable first experimented with it in 1977. The "Qube" allowed viewers to talk back to their TV sets in Columbus, Ohio. The experiment consisted

Bandwidth Measures the information capacity of an electromagnetic conduit.

Streaming Carries audio and video information over the Internet to the computer by downloading and buffering data while playing it back.

Interactive television (ITV) Converts television from a one-way medium to one responding to viewer requests for information and entertainment.

of questions directed to the audience, which they could answer with a handy touch pad. Viewers were nonplussed, and Qube folded within three years.

European models of interactive television (ITV) were more successful, using videotext and teletext to interact with the audience. British Broadcast Corporation's Ceefax service and France's Minitel system offered banking from home, in addition to train and airplane scheduling, and even directory assistance. In Great Britain, nearly four million viewers used ITV for e-mail and other information services through a Sky Digital box. Advertising charges and the commissions on sales secured profits for British ITV.

Other ITV experiments conducted in the United States included GTE's Viewdata and Knight-Ridder's Viewtron. Both gave customers newspaper text, weather, and agricultural data but failed to reap the profits necessary to sustain their expenses. Virtually all American videotext ventures suffered financial losses despite the fact that the FCC encouraged them by setting aside frequencies for ITV.

DIGITAL CABLE

Digital Cable has not only produced tiers with hundreds of channels and premium services that are multiplexed for viewers but it is also gearing up for more high-speed Internet access through interactive program guides. By 2004, there were over 22.2 million digital cable subscribers, and the number of bundled services and channels continues to grow.

PRINCIPLE 5

Greater media choices and easy access for consumers is the driving force of digital convergence.

▜ Satellites in the Digital Age

For years, America's satellite television industry was focused on serving homes unreachable by broadcasters or cable systems. That was when several firms were competing for satellite television business in the United States. Five companies once competed for **direct-broadcast satellite (DBS)** customers; currently there are two offering digital television and data services, DirecTV and EchoStar's DISH, but their reach is growing faster and further than ever before.

SATELLITES AND MPEG VIDEO

Satellite television has led the way in adapting to digital technology, reflecting a decade of digital services produced by a group of engineers that made it all possible. The **Motion Pictures Expert Group (MPEG)** was the brainchild of an international "dream team" of engineers who first met in Ottawa, Canada, in 1988. Their mission came from the International Standards Organization (ISO), which was to design the most practical means for digitally recording and transmitting pictures and audio for a variety of electronic media.

Toward that objective, the group met over the years since 1988 and authorized compression standards for still pictures, audio, and video. In November 1992, the ISO announced its design for digital video compression, and the Motion Pictures Expert Group responded with its first standard Level 1, or simply MPEG-1. MPEG-1 travels at a bit rate of up to 1.5 megabits (Mbs) and is the standard for video games, compact discs, and multimedia. MPEG-2 became the standard for digital video discs (DVDs). MPEG-3 became famous as the most popular means for compressing audio files and transfering them on the Internet. A fourth standard to handle the requirements of multimedia is MPEG-4, which works with media players by compressing text, animations, and graphics for computer presentation.

Direct-broadcast satellites (DBS) Beams television signals in the Ku-band directly to viewers at homes, at hotels, or in businesses using dishes to receive them. Also labeled as *direct to home (DTH) satelite service* by the FCC.

Motion Pictures Expert Group (MPEG) Determines appropriate standards for compressing audio and moving video images.

protalk

Mike Luftman

TIME WARNER CABLE

Michael Luftman has been in corporate communications with Time Warner Cable for years. He found that his company's merger with AOL in 2001 produced new digital products based on a "lot of very complimentary skill sets" necessary to develop interesting media. Luftman says their research showed that TV viewers wanted full video on demand (VOD) in order for them to be able to call up the TV show or movies of their choice simply by pointing and clicking on the screen. Luftman's company did not have that technology, but through the equipment of Scientific Atlanta and Silicon Graphics, Time Warner Cable was able to upgrade their network and create the necessary mechanics to meet the demand for VOD. The VOD rollout began in the markets of Boston, Honolulu, and Tampa, then spread to other Time Warner Cable franchise cities. Another interesting venture is the high-speed Internet access available through the Roadrunner system, which connects Time Warner customers to the Web.

Luftman recommends a shift of attitude for future professionals hoping to work in the multimedia world that Time Warner is building. "People need to know coming in that the industry is changing dramatically from a one-product industry over a one-way pipeline with limited competition into a multi-service, multiple-product business." Even

> **"College graduates have to be comfortable operating in an environment that requires a lot more technical expertise."**

though Time Warner is regarded as something of a behemoth, it competes with a wide range of companies, from local phone service to direct-broadcast satellite. "So, I think college graduates have to be comfortable operating in an environment that requires a lot more technical expertise." Luftman says the future digital universe will be infinitely more competitive and more complex.

DVDS AND VCRS

In 2001, the Consumer Electronics Association (CEA) charted how fast DVD player sales were taking off, especially in comparison with videocassette recorders (VCRs). From 2000 to 2001, VCR sales dropped 35 percent while DVD player purchases increased rapidly. According to the electronic manufacturers' group, DVD players achieved 25 percent penetration in U.S. households faster than any electronic appliance in history. Not surprisingly, DVD player prices fell dramatically by more than two-thirds in just one year—from about $1,500 in 2001 to $500 in 2002. By 2004, prices were $100 per set or less, and 71 million-plus DVD players had been sold.

Concurrently, home DVD recording sales hit an all-time high, passing $25 billion, which began to change the way Hollywood did business. Certain motion pictures, especially action films, could reliably make as much money or more through DVD sales than at the box office. Two such action flicks, *Dark Blue* and *XXX*, surpassed their theater revenues in DVD sales.

DIGITAL VIDEO RECORDERS

The digital medium with the most potential for changing TV viewing habits is the digital video recorder (DVR). It is a streamlined set-top box that represents a cross between a computer and a VCR. It uses a hard drive and modem like a computer, but it stores and plays back video programs like a VCR. Rather than using videotapes for storage, it records programs digitally, using hours of storage space on the hard drive.

Here's how it works technically: The DVR takes an incoming television signal and compresses it into MPEG-2 format on its hard drive. Doing this enables the machine to continuously record the TV signal and store hours of video.

The DVR also "learns" what types of shows the viewer prefers and records them automatically, making it unnecessary to program the device's memory each time the impulse strikes. DVRs allow the viewer to pause a live TV show, jump past the spots, create instant replays, and record more than one channel at a time. Perhaps most important is that DVRs enable viewers to pause or even skip commercials during a show. One marketing survey showed that about two-thirds of DVR users enjoyed skipping through TV commercials.

At first, several major brands engineered DVR machines, including Microsoft's Ultimate TV and America Online. However, only two survived: TiVo, created by Sony and Phillips Electronics, became the leader, and ReplayTV, which was sold in a bankruptcy auction by SONICblue but continues to merchandise its products.

The concept also caught the attention of satellite and cable television operators. The DISH Network produced the first DVR alternative in 2002, and began offering it virtually free to its subscribers. Comcast cable promoted a Samsung DVR cable box using existing cable lines to distribute video to other cable-connected televisions around the house. Other DVR makers include Motorola, Scientific-Atlanta, and Pioneer.

DVR owners gush about the machine's capabilities. Former Federal Communications Commission Chairman Michael Powell once called it "God's machine." Late-night talk-show hosts rave about "TiVo-ing" their favorite shows, while consumer research indicates levels of consumer

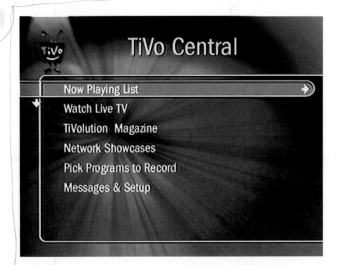

PERSONAL DIGITAL VIDEO RECORDER. TiVo uses a hard-drive disc instead of videotape to electronically store television programs. It allows viewers to skip commercials by pausing the TV program in progress then resuming it after the commercial break.

satisfaction exceed 80 percent. TiVo and ReplayTV have licensed their technologies to other consumer electronics makers in order to boost sales of digital video recorders. TiVo also added home networking components to its software, enabling machines to share digital photos and play digital music. It also began offering TiVo Basic, a service included in DVD players.

Combined DVR/DVD machines add mobility to the DVR's convenience. The DVR story indicates how digital convergence is not only driven by adding more choices and increasing control but also by the convenience of mobility.

Wireless Networks

The vision for the future is to make the Internet as portable and mobile as possible, and to remove the wires that connect users to the Web. In order to transfer data along telephone lines, modems link through the phone, the cable television line, or a special phone wire dedicated just for Internet access, the **digital subscriber lines,** or **DSL.**

The next generation of telecommunications media—third generation, or 3G— deploy wireless protocols with even greater bandwidth and faster speeds. Colleges and universities were among the first to adopt the wireless Internet in order to give students freedom to access the Web in their dorms and classes without using a cable and socket. The future of wireless Internet use is not just in the classroom—it is where people travel, shop, or enjoy a cup of coffee.[9]

WIRELESS HORIZONS

It all began in 1971 when the Federal Communications Commission opened the way for cellular phone traffic by allocating spectrum between 800 and 900 MHz. Another decade passed before all cellular trials had been held and technical issues resolved. The first commercial cellular phone service began in 1983, and it grew until 62 percent of the American public used a mobile phone 20 years later.

Six wireless *telcos* (abbreviation for telephone companies) emerged as dominant players in the mobile phone market: AT&T, Cingular, Nextel, Sprint, T-Mobile, and Verizon. They used second-generation systems that were based on a global system for mobile communications, or GSM. This was the name given the digital cellular communications, which has been the most popular system in the United States.

THIRD-GENERATION TECHNOLOGY

Digital subscriber line (DSL) Forms data loop, sending digital information over copper wires by bypassing the circuit-switched lines.

Wireless application protocol (WAP) Transmits telephony or Internet signals using laptop computers.

The focus shifted to third-generation (3G) technology for wireless data services. Wireless fidelity, or Wi-Fi, uses networks that carry audio and video data. It is sometimes confused with **wireless application protocol (WAP),** which is another standard for gaining wireless access to the Internet. WAP technology is used for mobile phones, including models sold by Nokia. By 2004, the number of WAP users in Europe was estimated to be more than 200 million, and its growth was driven by the introduction of Bluetooth. WAP is not controlled by any single firm, but has more than 100 members who say it affords the best means for linking to other computers or personal digital assistants (PDAs).

SMART PHONES

Smart phones are wireless instruments capable of sending and receiving both voice and data messages. Some people use built-in PDAs, such as Blackberry or Treo. These handheld machines may come with jam-packed keyboards, a touch pad, and a small black-and-white liquid crystal screen. The idea is that not only can you make phone calls but you can also keep track of your appointments, send faxes, and take notes. Smart phone users compare competing qualities, such as their software aspects, games available, screen color, browsers, and speed of e-mail delivery.

Smart phones using WAP have less memory and lower connection speeds than a PC, but they do allow network transmission to computer terminals. The Intel Corporation, IBM, AT&T Wireless, Cingular, and Verizon Communication demonstrated their preference for the competing 802.11 standard, which caters to Wi-Fi instruments. Verizon launched the nation's first 3G wireless network in 2002, and soon produced another media choice for consumers, wireless video games. In 2003, cellular-phone–based games amounted to a $500 million business in the United States and is expected to grow into a multibillion-dollar enterprise.

VIDEO GAMES

The emergence of video games in the American culture has been the subject of both celebration and controversy. On the plus side, video games are a tool for teaching and recruiting. The U.S. military, for example, introduced a game called "America's Army" on its website designed to bring in new recruits. After they sign up, Microsoft's Xbox console teaches them how to organize company missions by playing "Full Spectrum Command," or how to maneuver in a battle zone by playing "Full Spectrum Warrior."

On the minus side, criminal sociologists have noted a disturbing denominator that young, serial shooters seem to have in common: a dark penchant for violent video games. Jurors in the murder trial of the Washington, DC, sniper, Lee Boyd Malvo, saw excerpts from five video games that taught him how to kill as a sniper, including one titled "Ghost Recon," in which the shooter is invulnerable. The perpetrators of the Columbine massacre in 1999 obsessively played a video game called "Doom" before taking the lives of 12 students and 1 teacher.

University of Wisconsin at Madison Professor James Paul Gee dismisses video game critics by arguing that even excessively violent ones teach "strategic modes of thinking that fit better with today's high-tech, global world than the learning they are taught in school."[10]

Video games are rated for sex and violence so that parents can make distinctions and monitor the differences between such games as "Grand Theft Auto," which rewards players points for killing prostitutes, and "The Sims," which invites players to participate in the municipal zoning and transit planning of a make-believe city. The good news is that Will Wright's "Sims" video games have sold a combined seven million copies and are usually at the top of sales charts. On the other hand, a Gallup poll showed teenagers who preferred "Grand Theft Auto" where more likely to have been involved in fist fights.

Many adults who grew up with Nintendo and Atari games in the 1980s still play them, and on a variety of platforms from cell phones to laptops. In fact, the average video gamer is an American woman in her late 20s, who may average 6½ hours a week interacting with video board games or designing her own virtual city. Men in their late 20s lean toward action games, sporting contests, or first-person shooter simulations. A Pew Center study charted video gaming for younger ages and found

The Industry

protalk
Robin Sloan
INDTV PRODUCER

Robin Sloan is one of those people on the journalism edge of digital convergence. Sloan has been obsessed with the Internet since he posted his first comments about an on-line magazine.

While enrolled at Michigan State, the lure of high-speed access introduced Sloan to the world of on-line news, and for four years he became a "total nytimes.com junkie." It was when he began studying in Bangladesh that Sloan discovered how much fun posting web dispatches for his friends and professors could be, and the idea struck him: Why not do web journalism professionally? So, after graduation, he packed his bags and headed for St. Petersburg, Florida, where he put his web skills to work writing and producing content for Poynter.org, and then moved to California to continue experimenting in digital formats. There, by the Golden Gate, he signed on to become a producer for INdTV, former vice president Al Gore's new cable network.

Sloan sees the digital convergence as inevitable for reasons familiar to his fellow bloggers. First, there is the abundance of choice and convenience. "People want to be able to choose the chunks of media that are most interesting to them," he says, "and then read or view them on their own schedule." Second, there is all that creativity spreading around. "More and more people are going to become media producers as

"More and more people are going to become media producers as well as consumers."

well as consumers." Finally, there is the inspiration of collective intelligence. "Say good-bye to the lone editor. Instead, we'll get news based on what big groups of people—people like us, people not like us," are thinking about.

For those who cannot (or will not) make the leap, Sloan has a dire prediction. "In the years to come, clumsy mass-media technologies (I'm looking at you, newspapers and broadcast TV) are going to suffer as personal technologies—stuff that's directly connected to your life and choices, from IM [instant messaging] to TiVo [digital video recording]—gain more and more traction." The good news for young professionals is their career paths will be easier to follow. "As long as they follow that natural intuitive path—and don't revert to old methods and outmoded media because that's what they think they 'should' do—they'll be fine."

PROGRAM DATABASE

The jobs in on-line media cover areas of content, system management, and business affairs. Depending on the size and structure of the media organization, the jobs may be discretely different or may include elements of all three—maintenance of business accounts, creative design, and maintenance of the system.

Content jobs offer the most creative challenge and include graphic design, animation, and audio and video production. Others create and maintain databases for a media outlet. They are also called information providers.

System management people are more involved in the programming and software applications for websites. Those positions require technical knowledge and expertise beyond simple application skills. Then there are those who work in the business end by promoting and selling the services of the staff. Following are a few of the job descriptions in these areas:

Web managers update and maintain websites. They set the agenda for news flow and Internet publication times in coordination with either the station news director or the newspaper editor. They develop and create content for the cable or satellite system, or the broadcast station or network, and forge links to relevant sites. Their role is to find ways to complement and enhance repurposed content on websites. A web manager may or may not have a support staff, depending on the size of the organization.

Web/graphic designers are creative types who manipulate content by designing a website and choosing lines, images, and graphics to engage the audience. They become familiar with software applications and they translate message objectives into logos and layout designs on the computer.

In addition to knowing how to use the coding and mark-up languages, such as JavaScript, HTML, and XML, system management people, or **web authors,** decide what content will be used. They work in cooperation with **web designers.**

Web developers are familiar with software and server applications. They know how to build networks and how to place security locks into the systems. They are familiar with databases and they know how to operate PC- or MacIntosh-based computer systems, HTML/web languages (usually Java or Java Script), and a variety of software applications. This is a job for people who enjoy working at computers most of the day.

Multimedia specialists use two or more media (audio, video, graphics, text) to create content for distribution over a variety of channels, including compact discs, the Web, television, and so on. The job usually involves developing Internet-based instructional materials and websites, as well as training others in the use of instructional technology.

that about two-thirds of American college students were actively involved. Not only did they report playing video games each week, but also about half of them said they did so when procrastinating doing their school work.

Whether student, soldier, or businesswoman, video game players look for a multimedia experience that feels real but transports them into a fantasy realm far removed from their present situation. That's the good news for industry leaders such as Sony's Playstation, Microsoft's Xbox, and Nintendo's GameCube, as well as competing media that offer an escape into virtual adventures.

INSTANT MESSAGING

Instant messaging (IM) was created in 1996 by a group of avid computer users in Israel, all in their 20s. They formed a company called Mirabilis to promote a new way to talk over the Internet. These young Israelis realized that millions of people were connecting to the Internet but actually were not interconnected. They created a technology that would enable e-mailers to locate each other on-line and chat. They called their technology *ICQ*, (I seek you), which Mirabilis released in November 1996. Within six months, 850,000 users had been registered by "word of mouse." In less than a year, Mirabilis was able to handle 100,000 concurrent users.

Major Internet companies noticed the success of Mirabilis, and began making their own plans. America Online bought out the firm for $287 million in 1998, and created its own Instant Messenger system. Microsoft called its IM network MSN Messenger, and Yahoo! created a third major IM service. These IM services were incompatible, forcing customers to limit their "IM-ing" to fellow subscribers.

Instant messaging abuses were uncovered by a British research firm that found about 40 percent of business employees in United Kingdom used instant messaging in the office, and many did so in order to escape e-mail detection. About half of the workers answered yes when questioned whether "IM-ing" decreases their productivity when instant messages pop up on their screens.

Summary

Digital theory traces its origins to a Swedish immigrant who worked in Bell Labs on the nature of telegraphy, but the military's demand for a rapid transfer of information over long distances drove the development from the digital roots in telegraphy. The wartime push to improve ballistics and break the German code gave rise to computer mechanics. Similarly, the specter of nuclear war gave rise to the Internet's aim to communicate via multiple pathways should the national telecommunications infrastructure be attacked.

The computer developed as the means for collecting numbers and conducting business; it moved from military purposes to serve economic and academic needs. The language of digits in the computer not only could compress information, but also eventually afford a more efficient means of carrying pictures, sounds, and words around the world once the Internet digitally linked computers to share information over its networks.

In a similar way, the World Wide Web became the medium where purposes as well as content converged. In voice, data, television, and computers, Americans gained greater control of their media for information and entertainment. Digital compression schemes allowed the migration from a wired world to a wireless one. Electronic media have converged, with broadband access the common denominator. Merged media corporations will wage battles to see who can become the dominant carriers of e-content and e-commerce. Broadcast, cable, radio, television, and telephone networks will serve customers by selling bundles of media. As one technician put it, the digital age is "pipe agnostic." Whether the binary information is transmitted by air, cable, satellites, or microwave towers, it gives greater media choices and easy access to consumers.

Food for Thought

1. Critics of video games say that people have failed to recognize what powerful teachers these games can be, especially the violent ones, and cite as examples mass murderers' preference for such games. Do you think any pressure should be applied to stop video game manufacturers from rewarding violent or murderous actions in that context?

2. Instant messaging has become a popular activity that one British study shows is distracting office workers from their jobs. Do you believe that the tempting variety of media activities on office computers threatens worker productivity? If no, why not? If yes, what should be done about it?

3. The resurgence of e-commerce has come amid new concerns about security issues, such as identity theft. What factors influence your decision to buy on-line?

4. The sales record of digital video recorders (DVRs), such as TiVo, has not reflected the anticipated degree of customer satisfaction with the machines. Why do you think people have not bought DVRs yet, and do you think that will change?

5. What types of content do you like to view that is streamed on-line, and what would make you want to see more programming over your laptop or desktop computers?

The Industry

Chapter 6

Take a cheap pocket camera to a riot, a carnival, or a political convention. Point it in any direction and "click." Check the results, and you probably will find that your random photo-making has captured something interesting, maybe even mysterious. So it goes with the loosely knit multibillion-dollar industry built on electronic media. Every angle presents a noteworthy view, often one that makes us thirst to know more.

Consider mid-2004. That late spring and summer brought blue-chip drama to some of the largest players in corporate media. An appeals court threw out much of the Federal Communications Commission's 2003 decision to relax rules on media ownership. That ruling affected some of the biggest, richest corporations in the United States, and almost certainly failed to end their long drive to control more of what we see, hear—and buy. At the same time in the election year of 2004, the standards and values of broadcasting drew attention from the government: Radio giant Clear Channel paid a record $1.75 million fine for a series of on-air incidents that offended listeners. Two weeks later, the Senate passed a steep hike in broadcast indecency fines.

The United States was still a free and diverse country, of course. MTV announced plans for the first gay-focused network. Fox launched *Trading Spouses,* stealing the idea from NBC, which had licensed the British reality show *Wife Swap* and was planning to air other swap-fests (ah, values). CBS focused its crown jewel program, *60 Minutes,* solely on former President Bill Clinton one Sunday, giving him heavy air time to plug his heavy new memoir (957 pages). Ex-anchor Walter Cronkite, at 87, did political analysis for MTV, the pioneering youth channel, which said it might invite him back for more. There was also room for sports in the mid-2004 media panorama. Preparing to broadcast the Athens Olympics were NBC and its new corporate siblings, USA Networks, Bravo, and Telemundo, an unprecedented mix of outlets. And on public (noncommercial) television, the stately, issue-focused *NewsHour* steamed proudly toward its thirtieth birthday—while debate raged over whether the Public Broadcasting Service (PBS) exerted too much power over local stations' programming.

While all this was going on, cybermedia kept growing in stability and appeal. In early 2004, revenue from advertising on the Internet surpassed its previous quarterly peak. A media research group reported at mid-year that about 10 percent of Americans had viewed harsh images of the Iraq war and occupation that the television networks declined to air.[1] Of those Americans who *go* on-line, another study showed, 70 percent get news while doing so.

Glimpses, flashes, quick looks: Many snapshots here. Like photos from family albums, they only hinted at the complex drives and relationships behind them—in this case, the endless race of a huge media industry developing and satisfying changing demands of humankind.

Fundamentals

This chapter will examine commercial electronic media's ways of operating and the people who keep them operating. We'll discuss their functions within an industry, the varied purposes of their products and services, some principles that can help you understand these media's potential and limitations, and how industry insiders feel about their often fascinating work.

Let's start with a preview of the *principles* to keep in mind while reading this chapter; in a way, they define the boundaries within which the electronic media function. The overriding objective of commercial electronic media is to make money. Every dollar they spend—even charitable donations—must in some way support that objective. Hence, the chapter's first principle: *In commercial broadcasting, most public-service activity must support business objectives by promoting or otherwise benefiting the company.* Most businesses exchange their products or services for money, but that's not true of broadcasting. Its revenues come mainly from advertisers who hawk their wares to the audiences that radio or TV programs attract. That leads us to a second cornerstone: *The industry rests on a three-legged stool of economic factors: broadcaster, audience, and advertiser.*

The ceaseless push for profits leads commercial electronic media to chafe under the government regulation that goes with their critical role in society. They prefer to operate as freely as possible, which suggests another principle: *Commercial electronic media will oppose—and may actively resist—efforts to regulate their business.*

Businesses in the United States generally are free to concentrate ownership in order to cut costs, offer more goods, and reach more consumers efficiently. Public concern about big-media control of information has not counteracted this powerful fact: *The urge to consolidate ownership is fundamental to commercial electronic media, as it is to many other industries.*

Individual stations and Internet media sites seldom initiate public stock offerings, but big corporations owning many layers of media issue stock shares that attract major investors. They provide financial fuel, but can demand higher and higher profits and thereby influence daily decisions that can affect what people see and hear on the air or on-line. Clearly, *consolidation increases electronic media dependence on approval from Wall Street.*

It costs a great deal of money to produce technologically sophisticated programming and maintain an "instant news" capability in our digital age. At the same time, advanced tools make it possible to operate with smaller staffs. Therefore one can say that, *because of its technologies, broadcasting generally is more capital intensive than labor intensive.*

A DUAL ROLE

Note this basic fact: In the U.S. system, *local stations are the only true broadcasters.* To win a government license to transmit programming "over the air," a station must have a local address and a specific audience to serve. This audience usually is considered to be everyone within a clearly defined geographic area that the station's radio waves can reach.

#1 →

#3 →

Because these invisible waves are considered to belong to the American people, all of us are, at least in spirit, consumers with a special stake in the broadcast industry. Laws and regulations have long required that local broadcasters perform public service. The Federal Communications Commission once required that stations air substantial programming intended mainly as public service, not as advertising vehicles. But government pressure to maintain this practice slacked off in recent decades. Unprofitable programs, such as public-affairs shows digging into important public issues, have mostly withered away from commercial programming.

Now, the primary form of public service undertaken by most U.S. commercial broadcasters is exemplified by what happened in two middle-sized towns a few years ago: A January tornado slammed into Owensboro, Kentucky, destroying or damaging hundreds of homes. This was big news 32 miles away, across the Ohio River, in Evansville, Indiana, where WFIE-TV general manager Lucy Himstedt realized it was more than news—it was a human crisis. She decided to see what her station could do for the tornado victims. "We knew right away they needed help," Himstedt recalls, "and we'd been doing the Muscular Dystrophy Association **telethon** for 30 years. Now, in just two full work days, we pulled together a telethon that generated $500,000 in cash for the Red Cross."[2] That feat resulted not just from the generosity of Himstedt and the viewers within WFIE's reach but also from the station's readiness for action. The many complex tasks within a television station tend to produce a high degree of interdependence among workers. That spirit paid off in a rapid and broad-based scramble to come to Owensboro's aid.

"We got the people in news involved, as well as our [commercial-time] sales manager," Himstedt recalls. "She went to local businesses . . . and said, 'We know you'll donate; we want to put you on the air to present us with a check.' We gave everybody something to do. Our business office knew we would have to have guards around when we put money in the pot, and they took care of that. We ran phone banks too."

A single television station—with its unique ability to rally and organize other businesses and citizens—thus helped a nearby city recover from disaster. That clearly was public service, and of a sort that, besides helping people in need, also polished the image and enhanced the civic status of the station and its advertisers. That's the type of payoff that broadcasters' public service *must* provide, now that most compa-

Telethon (from "marathon") Continuous TV broadcast for hours or days, usually soliciting donations to a charity or cause.

protalk
Lucy Himstedt
GENERAL MANAGER, WFIE-TV

When she was hired to run WFIE-TV, Lucy Himstedt had to move quickly. First came her need to study and understand every cranny of her station to a degree of detail she had never needed as a news director.

"You only know your niche," says Himstedt, "and now, suddenly, you have to learn the big picture. What are the problems that are facing the traffic department? What are the problems Operations is dealing with? And how do we function as a *group* of departments?"

Himstedt had to choose what to discard, what to change, and what to leave alone: "I had one-on-one interviews with everybody on the staff—not about work, but 'getting to know you.' That helped me learn a lot." Himstedt realized that some major tinkering was needed, but that it would take time.

Meanwhile, as the new "GM," Himstedt would have to supervise the station's overall budget and keep subordinate managers focused on the bottom line. She would need to spur ad sales, programming, and promotion staffers toward maximum performance. And there was a people-to-people

> "The sales department and the business office are going to fight sometimes, but between those times we pull together for the good of all."

dimension to the job that brought her special satisfaction. She might have a staff member help monitor community factors on which WFIE's broadcast license depended—but she wanted the buck to stop at her desk.

"Part of my job is to be on boards and to be in the community," she says, adding that she also stays personally close to the station's employees—as a broadcast manager needs to do in a business of rapid change and psychological ups and downs.

"The sales department and the business office are going to fight sometimes," Himstedt acknowledges, "but between those times we pull together for the good of all."

nies work almost solely to increase profits and hold down costs. Another example of pubic service: In the 2004 election campaigns, some stations and station groups gave candidates free air time, possibly in civic altruism but also hoping to stave off criticism that broadcasting had forgotten to serve the public.

PRINCIPLE 1

In commercial broadcasting, most public-service activity must support business objectives by promoting or otherwise benefiting the company.

FREE TO DO BUSINESS

The federal government today grants great freedom to radio and TV stations to do pretty much what they want. They're expected to keep activity files open for public inspection and to show social responsibility by informing the community about important issues from time to time. However, the FCC doesn't spell out how much information is enough. Nor does the government slap broadcasters' hands unless they stray into forbidden zones—"indecency" during most daytime hours, obscenity anytime, rigged contests, payola, and so on—in which Congress has authorized the FCC to act.

Most important, perhaps, the system does *not* interfere with profit making. Unlike broadcasters in most countries, U.S. broadcasters were permitted to take a sharp turn toward private enterprise from the beginning.

THE BUSINESS DYNAMIC

Commercial broadcasting as a business is unusual in that it doesn't receive money directly from its consumers. Instead, stations deliver programs to listeners or viewers in exchange for their *attention.* That creates audiences, which are then sold to advertisers who seek to horn in on that attention and profit from the resulting consumer interest in their products or services. Advertisers simply buy the rights to time periods for their commercials, which surround (and often seem to interrupt) the programming.

If this dynamic works as intended, everybody—station, advertiser, and audience—is satisfied. Those revenues result from programming that comes from a variety of sources, which make up some of the major segments of the industry. They begin at the local level, but if they ended there, the content of broadcasting would be narrow and dull indeed.

Evansville's WFIE, like most TV stations today, does produce some local programs, but high costs and small audiences have eliminated most original community shows. Newscasts still attract many "eyeballs" daily and help businesses sell their goods, so they receive heavy investment. Stations also buy shows from **syndicators**—companies set up to acquire programs and sell them to stations, cable systems, and networks. Most of all, WFIE and other stations need the heavily promoted, nationally known entertainment that their affiliated networks provide day by day. Network dramas and comedies draw many viewers to the "affiliates," and the viewers often hang around, sitting through commercials, when local news or other offerings come on.

Networks retain control of most of the advertising slots in the programs they "feed" to stations. However, they also release some of those commercial minutes and seconds to the affiliated stations so that they can insert some local ads to bring in more revenue. It's a barter arrangement in which both sides benefit (see Figure 6.1).

PRINCIPLE 2

The industry rests on a three-legged stool of economic factors: broadcaster, audience, and advertiser.

AN EXPLOSION OF CHOICES

Broadcasting once held the deed to a "national hearth"—a metaphor for the way most Americans typically gathered to watch a single hit television show at the

Syndicators Agents or companies that sell programs to radio and television; most successful network programs are later sold in syndication.

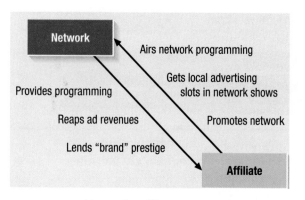

FIGURE 6.1 *Network-Affiliate Ties: Relationship between a Typical Network and Its Affiliates*

#5

same hour of the same night in the mid-twentieth century. The medium was new. Channel choices were few. The hits were on at the same time every week. Everybody talked about them, so everybody watched them.

Now, not even the best television programming draws a vast and reliable horde to the same cozy metaphorical fire on a given evening. Radio programming, which galvanized families every night during and after World War II, has split up and moved into **niches** where select clusters of listeners dwell. Not many shows in either medium draw massive audiences. Most of the time, each individual watches or listens to something different from the show being enjoyed by the folks next door—even in the next *room*, since American homes average about 2½ TV sets apiece. What's more, neither medium dominates people's leisure moments as automatically as it once did. Too many activities are packed into the normal day, too many channel choices, too many other electronic toys—and a serious challenge from the Internet along with other new media.

As noted earlier, broadcast-style programming still begins at, or is relayed to us by, nearby broadcast stations. Many changes have come along, though, since radio and television each had its "golden age" through the mid-twentieth century:

▶ Since the 1970s, many people have been watching broadcast television through local and regional cable systems. These systems reach more of us geographically than over-the-air broadcasting can, and they provide clearer and more stable signals for many areas.

▶ The 1980s brought the start of a cable revolution in which movie companies and others created programming just for cable. This original content now competes directly with shows produced by old-line broadcast stations and networks.

▶ Yet another delivery technology—satellites—has captured millions of loyal fans. Companies can send us programming directly, with no stations or cables in the loop, by bouncing it off human-made objects in space to antenna dishes at our homes. Satellites also play a huge role in the delivery of programs to local cable systems, which transmit them to subscribers and place commercials in them.

▶ Although we still turn to either broadcast, cable, or satellite channels to swim in the pop-culture mainstream, that may be changing. Many people now use their computers to hear radio-style audio or watch **streaming** versions of TV video.

▶ Possibly more important, the Internet has spawned scores of news, information, and entertainment enterprises of its own. Some operate without reference to the "old" media—broadcasting, newspapers—using tools and techniques to draw young audiences, while others draw on more familiar media names.

These other pipelines haven't destroyed broadcasting, however; its access to virtually everyone and the superior value of its air time to advertisers continue to make it the nation's media powerhouse. Many sources say local stations still reap profits of 30 to 50 percent a year, so they remain among the nation's most successful businesses. Still, newer options for consumers have forced the broadcast industry to review its budgets, tighten its belt, look for new revenue streams, reorganize its stations, and generally become more innovative.

Niche Submarket in which radio station or TV channel can find consumers interested specifically in its unique programming (e.g., outdoor sports, foreign-language shows, etc.).

Streaming Transmission of sounds or images as continuous stream of data over Internet.

Levels of Operation

When early founders and their investors launched the first radio networks, most of the local stations around the country had local owners. Rarely were stations as homespun and intimate as a mom and pop general store, but their managers knew many of their audience members by name, primarily because they were neighbors.

Was such utterly local business any better for society than the widely dispersed but rather centrally controlled modern broadcast industry? Opinions about that are all over the map. So, probably, are the people who control most of the stations in *your* town. Today, companies known as **station groups** have acquired many local outlets and oversee their operations from whole time zones away. The older networks and some of the newer ones are members of much larger corporations, some of them **conglomerates**—that is, having interests in many businesses besides broadcasting. The policies and practices found at all of these levels ultimately must satisfy stockholders who pour their investment dollars into the companies.

We'll start our examination at the top level, among the largest media corporations and conglomerates. By the time we reach the local-station level, a number of important realities should be clear. One is that however WFIE-TV improves the measured size of its audience **(ratings)** in Indiana, Lucy Himstedt doesn't hold all the strings to her station's future. Far from it.

CORPORATIONS

There always has been some benevolence at the roots of the electronic media; few of their leaders, as good citizens entrusted with the public airwaves, could forget their public obligations. By the late twentieth century, however, the potential of these media as *businesses* clearly came first in the minds of the mighty. Many moved toward **consolidation** of ownership and/or control—a move to run or direct whole masses of media companies from central hubs of power.

Objectives

Media owners pursued consolidation for sound economic reasons. One was to achieve "economies of scale," the savings and earnings that only a large business could expect. It's possible that some owners also wanted to acquire *monopolies* in some media **markets** (targeted commercial areas, usually cities), hoping to control all or most of their supply of products, despite the intent of U.S. antitrust laws. There were cases, too, when acquiring a few highly successful middle-sized companies simply would improve a corporation's overall "bottom line" and make its business more lucrative and also more predictable.

That phenomenon is called **synergy,** which is when the sum of the parts is greater than the whole. For example, even a movie studio's lower-quality movies might pay off if a sister network or station group gave the movies new life on television. Or perhaps cable companies could extend the geographic reach (and advertising power) of their sibling radio and TV stations. Over recent years, FCC regulation has come closer to making most kinds of combinations permissible.

In one drive toward synergy, Time Inc., a print and broadcast concern, and Warner Communications, the cable branch of a movie/media company, merged in 1989. The major purpose was to make more money by combining movie, cable TV, and magazine units of the merged companies in creative ways. At first, it proved to be a winner. Time Warner invented programs and product lines, bought CNN and other properties from flashy media mogul Ted Turner, and went on to consummate the biggest media merger ever, joining with AOL in 2000.

The fusion of America Online, an Internet giant, with Time Warner, a titan of print, broadcast, movies, and more, produced a case study in what can happen when media conglomerates tie the knot, and fail to show a profit. This marriage brought 11 subsidiary companies under one roof (see Figure 6.2), the total value of which was estimated at $350 billion—enough money to fund the U.S defense program.

#6

Station group Cluster of broadcast stations owned by single company; such groups now own most U.S. radio stations.

Conglomerate Corporation made up of subsidiary companies or holdings in a number of different industries or industry sectors.

Ratings Numbers showing size of audience for particular show or time period.

Consolidation Centralization of ownership or control of a number of entities; many radio and TV stations, as well as cable channels, have been consolidated recently.

Market In broadcasting, specific population or area to which programs are targeted, as in "the Louisville market"; an industry term is *DMA*, for designated market area.

Synergy Process by which combining two or more units results in an outcome greater than the sum of the parts; consolidation usually is aimed at achieving synergy.

FIGURE 6.2 *Time Warner Companies (Partial List), 2004*

Few questioned the wisdom of the merger in business terms. AOL's experience and resources in the on-line communications business (it had 22 million subscribers) would appear to complement Time Warner's old-media strengths and its capacity to produce content of all sorts. The result, some analysts speculated, eventually could be dominance of twenty-first-century media by a single supercompany.

Then blue skies turned grey. AOL Time Warner's stock prices fell 66 percent in 12 months and the company assumed billions in debt. Some blamed overexuberant investments in new technologies and ventures such as AOLTV interactive television (which later was cancelled). Others said too much power was consolidated in New York, robbing division heads of autonomy and the flexibility to compete.

In any case, the giant corporation's new CEO, Richard Parsons, was faced with saving one of the world's largest conglomerates from failing. After losing many tens of billions of dollars, the company snipped "AOL" from its name—but not from its operations—and by late 2003 was making profits again, primarily from its movie and cable businesses. Much of commercial broadcasting now is part of a few huge corporations. One is General Electric, a "diversified conglomerate" that is the world's highest-valued company, worth several hundred *billion* dollars.

This sort of financial weight tends to alarm critics who contend that placing great media power in just a few hands is certain to reduce the variety of voices in today's culture. Others argue that powerful companies actually

PRINCIPLE   3

Commercial electronic media will oppose—and may actively resist—efforts to regulate their business.

produce a wider variety of programming than smaller companies can. In any case, the concentration of electronic-media control continues to this day.

Government Limits

With backing from Congress, the Federal Communications Commission might be able to slow this process of media concentration. However, the FCC, most notably since the years of Republican President Ronald Reagan, has worked to reduce or dismantle controls on group ownership of broadcast outlets. Many citizens have supported or acquiesced in this, believing it is in keeping with the American legacy of freedom for all. Business people with the courage, vision, and capital to build media corporations would have increasing freedom to do so. Without their enterprise and risk taking, many have said, the nation would never reap the full benefits of what technology and big ideas could bring about.

Post-1980 FCC rules (and loopholes) encouraged many people—some of them with no known prior interest in broadcasting—to start buying stations as *investments.* Looser federal controls created a sort of national broadcast-license bazaar, in which stations that once might have been sold one at a time, with considerable gravity, were peddled in batches. Some that were sold were almost immediately resold at a profit.

By the late 1990s, under the 1996 Telecommunications Act, one person could buy eight radio stations *in a single town* and could own as many stations nationally as that person could afford. This attracted not only committed broadcasters, but also financial speculators and absentee owners. Neither group seemed bent on contributing much to the stability of local communities or to fulfilling broadcasting's public obligations. Ownership restrictions are intended to prevent monopolies and to encourage diversity. However, but some corporations that own electronic media (an example is in Figure 6.3) insist that if they can't expand their control, competitors will engulf them and deny their services to consumers.

PRINCIPLE 4

The urge to consolidate ownership is fundamental to broadcasting, as it is to many other industries.

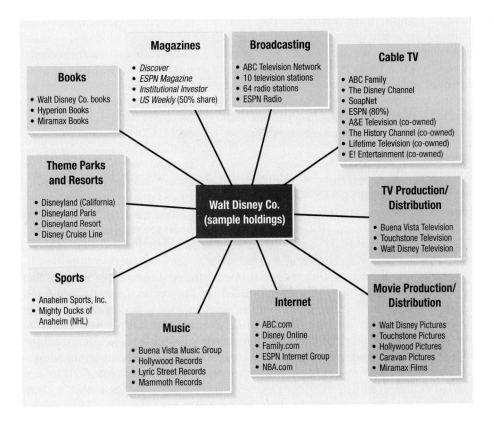

FIGURE 6.3 *Walt Disney Company Businesses (Partial List)*

Surprising things can happen when twenty-first century electronic-media owners sit down together to make deals. Sometimes they lead to disaster.

In electronic media, **vertical integration** means owning everything in the system from creation of the product to its delivery to consumers' homes, cars, offices, or neighborhood theaters. Here's how one series of deals unfolded early in the twenty-first century:

Deal: Vivendi, a French-owned global media giant, purchased Universal Studios for $30 billion in 2000.

Motive: Vivendi wanted a bigger presence in the U.S. marketplace. It already owned 43 percent of USA Networks, a cable-TV and movie company.

Deal: In December 2001, the renamed Vivendi Universal bought 10 percent of the U.S.-owned satellite broadcaster EchoStar Communications Corporation, for $1.5 billion.

Motive: Vivendi Universal, which now had plenty of production capacity, wanted more U.S. distribution for its products. Since EchoStar owned the direct-broadcast DISH Network and had just merged with DirecTV, it would have a monopoly on much noncable home service. That, in turn, would give Vivendi Universal films and shows a great U.S. outlet.

Deal: A few days later, Vivendi Universal struck a deal with the head of USA Networks, Barry Diller, to acquire his company, owning 93 percent of its stock.

Motive: More distribution. This deal would give Vivendi Universal control of USA cable, Sci-Fi, Home Shopping Network, and other channels—with access to more.

Aftermath: In 2004, NBC vought Vivendi Universal's entertainment assets, creating a multibillion-dollar global media company: NBC Universal.

Whether services to consumers are always a top priority was questioned during one "clash of the titans" in spring of 2000. Time Warner Cable, locked in a contract dispute with Disney over **carriage** of its ABC and other networks, disconnected ABC-owned stations from 96 cable systems for 40 hours. This denied programming to 3.5 million viewers in New York and other major cities. Worse, it happened during May, a ratings **sweeps** month in which audience size helps to determine advertising rates. (See Chapter 9, The Audience, for an update on the controversial sweeps practice.) The FCC, evidently reluctant to punish Time Warner Cable severely, accepted a "voluntary" $72,000 payment to the U.S. Treasury as a settlement gesture from the company. Time Warner Cable also gave some free services to the affected viewers. The incident indicated vividly that media companies caught in business disputes might not always put consumers first.

Shareholder Influence

Except when someone starts and runs a small company (such as an Internet venture) using his or her own savings, media businesses need massive and continuous investment. Advertising revenue alone won't cover all corporate obligations and keep a business growing. Instead, companies sell stock to investors, use the proceeds to create and market programs, make profits (usually), pay dividends to the investors, and thus keep attracting new investors.

People have been investing money in media for centuries, often in ways that benefit society. Benjamin Franklin cofounded the *Pennsylvania Gazette* with funding from his partner's father, who is said to have chipped in on the condition that his booze-loving son keep Franklin around to provide stability.[3] Much later, in the early 1900s, Guglielmo Marconi found little support for wireless telegraphy in his native Italy. However, he was able to win over British and American backers who helped make him—in some people's eyes—"the father of radio." (If he was the father, however, it took other pioneers, such as Reginald Fessenden, to turn Marconi's breakthroughs into the mass medium of radio.)

Vertical Integration Buying companies to support the buyer's business (e.g., a Hollywood studio purchasing cable systems to air the studio's films).

Carriage One medium's delivery of another medium's content, as when a cable system "carries" (delivers) a TV station's programs.

Sweeps Periods four times yearly in which measurement firms "sweep" (intensively sample ratings in) broadcast markets to measure audiences.

Electronic media are based on fairly simple scientific principles but require expensive facilities and constant upgrading, mainly for competitive reasons. Most important media companies today have "gone public"; that is, they issue stock on public exchanges where anyone can buy it. People willing to buy shares of stock, putting capital into a growing corporation, are crucial allies. In return, they receive voting power over corporate policy. Those with the most stock can become company directors, helping to make the biggest strategic decisions.

Throughout broadcasting, managers say that the mere knowledge that shareholders *exist* exerts tremendous influence today. Much of it is positive, through investments that lead to upgrading, but there's a long-simmering argument over the influence of investors and the profit motive on the *nature and quality* of media products. Some critics (and broadcasters) complain that shareholders, concerned about the value of their stocks, focus too much on efficiency and profits. By doing so, goes this critique, they pressure companies to lay off too many employees, slash production spending, cancel challenging programs, and otherwise compromise media quality.

The other side of this argument holds that commercial media always have been businesses first and cannot be expected to settle for inefficiency. Companies must be lean and productive to bring the public what it wants and needs. Everyone has a right to seek profits. Anyway, runs this line of defense, most shareholders simply hover in the background, hoping for financial gain as their dollars stimulate growth.

Sometimes there are apparent conflicts between private profit and public service. The Disney Company has been a target of criticism along this line. Disney's holdings in 2000 included—by chief executive Michael Eisner's accounting—"seven theme parks (with four more in the works), 27 hotels with 36,888 rooms, two cruise ships, 728 Disney Stores, one television broadcast network, 10 television stations, 9 international Disney Channels, 42 radio stations, an Internet portal, 5 major Internet websites, interests in 9 U.S. cable networks," and more.[4]

In many respects, those widely varied leisure and media businesses help one another do business in ways that are both profitable to Disney and pleasing to society. However, among those on the corporation's megapayroll are the journalists of ABC News. They've been employees of the corporation since it acquired ABC in 1996. Some observers (including journalists) had misgivings about Mickey Mouse's entertainment empire buying a major news operation; they were concerned that corporate profit goals might influence journalistic performance. Soon, there were reports that, indeed, the network had suppressed some ABC stories that portrayed Disney negatively.

ABC denied misconduct, and the controversy died away. Today, advocates of public service over profit can find little comfort in a declaration at the corporate website: "Disney's overriding objective is to create *shareholder value* (italics added) by continuing to be the world's premier entertainment company from a creative, strategic, and financial standpoint."[5]

WALL STREET. Trading in media stocks on the nation's financial exchanges affects the media's behavior. Shareholders want value and profits, placing demands that can lead to controversial steps to win audiences.

PRINCIPLE 5

Consolidation increases electronic media dependence on approval from Wall Street.

NETWORKS

NBC, CBS, and, later, ABC built much of American broadcasting. They have made a business of producing or buying highly polished programming and broadcasting it to every comer of the country through local stations that are affiliated with them.

THE ANNUAL MEETING. A media corporation's yearly gathering of shareholders emphasizes their power to wield quiet influence over how broadcasters make decisions.

KATIE COURIC. This one-time local-TV reporter became a national sweetheart to morning TV audiences as co-host of NBC's *Today* show, signing a $15-million-a-year contract in 2002.

In return, the stations grant the networks advertising time or revenue. For decades, simply gathering audiences to hear and watch satisfying shows drew enough advertising to bring ample revenue to the "Big Three."

Since the 1990s, however, the trio of older networks has acquired rivals: Fox, WB, and UPN all operate as networks, owning some stations and in affiliation with others. But changes in the stature of the Big Three—in fact, *all* of the networks, since they're now owned by even larger companies—didn't begin near the close of the twentieth century. The 1970s brought the start of some profound and permanent changes in the landscape of electronic media.

Regulation's Toll

In those days, two FCC actions—known as the financial interest and syndication ("**fin-syn**") rules—were squeezing the networks hard. Both were intended to reduce the power of the networks to dictate programming and to profit unduly from it at the expense of its producers. The first rule prohibited networks from acquiring financial stakes in programs produced by others. The second rule barred networks from selling programs to local stations. This left many programs available for sale only in syndication, where others could reap profits from what the networks had created.

Another tether on the networks was the prime-time access rule (**PTAR**), which limited network programming to three hours a night during the four-hour prime-time period. Like the syndication rule, this encouraged stations to create shows or buy them elsewhere. They did, and soon at least one hour every night in most cities featured inexpensive game or talk shows or other products from independent production companies.

These government limits substantially broke the big networks' choke-hold on what Americans could watch, helping to diversify TV content. The limits also badly damaged network profitability. Soon, new networks and cable TV grabbed viewers from the Big Three.

#8   #9

MONDAY NIGHT FOOTBALL. The savvy, often wise-cracking announcers of *Monday Night Football* were a long-term success on ABC—until the games moved to ESPN in 2005.

DAVID LETTERMAN. Late-night personality Letterman has brought amusement to viewers of more than one TV network.

Within a couple of years in the mid-1980s, ABC, CBS, and NBC were sold to or merged with richer companies. Sapped of strength, each one-time media giant had become little more than "an office building," in the assessment of media writer Ken Auletta.[6]

Auletta exaggerated: Each network still owned lucrative local stations and was affiliated with many more that would broadcast its programming. The Big Three maintained large business and production complexes in Los Angeles and New York. On the other hand, fin-syn had limited their power to make money, which would not satisfy the bottom-line–oriented corporations that bought them. So "downsizing" arrived: many rounds of cut-backs and layoffs in the interest of overall corporate efficiency.

Adapting to a New Age

Meanwhile, the networks were having increasing difficulty with their affiliated stations. In the 1990s, the two sectors squabbled over how revenues would be shared. However, 1995 brought the abolishment of the fin-syn rules, restoring to the networks some of their old leverage over programming. This came too late to prevent many new competitors from horning in on the marketplace.

The networks had not lost all of their financial strength or their resilience, however—not even close, as is evident from today's financial statistics. For instance, NBC, which had become NBC Universal (see "Vivendi Goes Vertical," page 124), made $2 billion in *profit* in 2003. On the other hand, the old networks know they no longer rule alone and have branched out widely. They've acquired Internet outlets, formed alliances, and generally reinvented themselves to compete with new rivals.

PRINCIPLE 6

Because of its technologies, broadcasting generally is more capital intensive than labor intensive.

Fin-Syn Financial interest and syndication rules, which kept networks from profiting from production and sale of programs; abolished in 1995.

PTAR The FCC's prime-time access rule, which limited network TV programming during high-viewership evening hours; abolished in 1996.

Still the networks faced not just new competition but also older fights. In 2001, more than 600 affiliates complained to the FCC that the Big Three had been illegally bullying stations over programming and business decisions. FCC chair Michael Powell wanted to stay out of arguments stemming from network-station contracts, but by 2004, he was under new pressure to act because of a public flap over broadcast *indecency*. Local affiliates complained that they were being forced to run network shows that violated or might violate their communities' moral standards. The head of the rebellious Network Affiliated Stations Association charged that the networks were "changing the [TV] system to be nationalized and homogenized" in defiance of viewers' sensibilities. The urge to protect viewers from broadcast programs was perennial; for years, many citizens and some politicians had been pushing the FCC to enforce its own indecency rule and penalize offending broadcasters. The rule defines indecent content as "language or material that, in context, depicts or describes, in terms patently offensive as measured by contemporary community broadcast standards for the broadcast medium, sexual or excretory organs or activities.")

The indecency issue drew fuel from incidents including singer Janet Jackson's breast-baring "wardrobe malfunction" during halftime of the 2004 Super Bowl game and repeated offenses by a radio personality called Bubba the Love Sponge. Lawmakers warned broadcasters, held well-publicized hearings, and introduced legislation proposing six-figure fines for broadcast indecency. All of this added force to the network affiliates' pleas for more freedom to choose safe programs.

STATION GROUPS

Changing regulatory and economic realities as far back as the 1980s created huge demand for broadcast stations as investments. TV stations in particular had been "cash cows," and it sometimes seemed that everybody wanted to own a station—or 20 of them. Belo, a Texas-based media company with a solid civic image and news reputation, is an example of the station-group explosion that followed the Reagan-era FCC's moves to loosen ownership rules. Belo was born as a newspaper company but launched Texas's first network-affiliated radio station in 1922. Not much else in the way of broadcasting entered the Belo fold for six decades. Then, in 1983, the company brought four medium/large-market TV stations in what for that period was a mammoth transaction. The company had to sell off two other stations to meet an FCC ownership "cap," but wasn't about to stop expanding.

Selling stock in its own business to raise capital, Belo bought two big-city network affiliates in the early 1990s. It then acquired the Providence Journal Company in 1997, and with it eight more TV outlets. As of 2004, Belo owned 19 TV stations—6 of them in high-revenue top-20 markets—plus 6 cable news channels. Careful acquisition had created a powerful media corporation that could, if it chose to, heavily influence the programming being pumped into millions of U.S. households every day.

One industry insider decided to measure the effects of consolidation of station ownership in the mid-1990s, and came up with a startling result: Between 1994 and 1998, the number of owners of full-power commercial TV stations spiraled from 658 down to 425—a drop of more than one-third.[7] Radio stations have become especially ripe targets for consolidation of ownership. After many years when no one was permitted to own more than 40 stations, the 1996 Telecommunications Act repealed national ownership limits. Buyers were allowed to hold as many as eight stations in a local market. Within the next year, the total number of owners of U.S. radio stations dropped almost 12 percent as station groups new and old bought up all they could. For example, Citadel Communications Corporation, which owned 36 stations, in 1997 sold a million shares of its stock and bought 61 more.[8]

Motives for Growth

Big groups got bigger, and even small groups strained to buy more stations, for a number of reasons, including these:

▶ The larger a station group becomes, the more it's able to reduce the average costs of doing business, a basic efficiency move.

▶ Acquiring many stations at different market levels, or in different geographic regions, allows a station to offer an advertiser a single buy that exposes her or his products or services to many audiences at once.

▶ When a recession hits or other events curtail advertising—the September 11th terrorist attacks had such an effect—the resulting financial stress can be spread over many stations or allocated to the most cash-rich family members.

▶ The more stations a group owns, the better able it is to negotiate the costs of programming it acquires for them. An owner of many stations has much greater bargaining leverage against networks and syndication firms than the owner of only one or two stations has.

There can be other bonuses to bigness. Lucy Himstedt's TV station in Indiana is owned by Liberty Corporation, based in Greenville, South Carolina (see Figure 6.4). The company also owns 14 other network affiliates and also owns or controls a cable-advertising sales company, a video production house and a firm that sells broadcast equipment, as well as financial stakes in Internet-related companies. These companies can be resources for one another—for example, a Liberty station can buy gear from the Liberty equipment firm and the money stays in the family.

Despite their growth, some station groups have had to struggle to keep profits as high as shareholders want them. One reason is that, under FCC orders to convert gradually to high-definition TV (**HDTV**) technology, TV broadcasters have been forced to spend millions of dollars *per sta-*

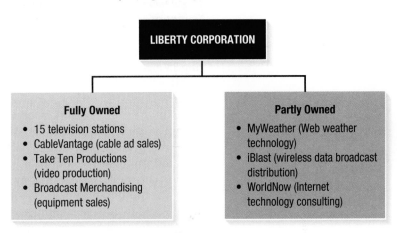

FIGURE 6.4 *Liberty Corporation*

tion. Sinclair Broadcasting Group, owner of 62 TV stations, has had to refinance more than a billion dollars in loans to avoid defaulting on them.[9]

Concern for Localism

Some critics accuse broadcasters of placing faraway owners' financial motives above the needs of local audiences. Certainly "localism" was a founding tenet of broadcasting, and some station groups maintain that they give their local managers freedom to serve their communities. However, business pressures on commercial stations appear to have eliminated most local programming except for newscasts—and have torpedoed some of those as well. Station groups also have provided the same programming to all of their members, which tends to homogenize on-air content from city to city.

One major objection to electronic-media consolidation is that it reduces the *diversity* of programming and thus denies viewers and listeners access to many points of view. Many industry leaders and others say more media mergers are likely to increase, not reduce, the array of voices. "The reason is simple," wrote James L. Gattuso, a research fellow at the conservative Heritage Foundation. "While owners with only one station each may all compete for a lowest-common-denominator

#10

High-definition television (HDTV)
New digital format offering exceptionally sharp pictures.

market, owners with several stations each are able to target niche markets with different programming on each station."[10]

However, opponents of consolidation insist that content diversity will wither if too much power falls into too few hands. In a hearing on media mergers, North Dakota Sen. Byron Dorgan quipped: "When you talk about more voices, are you talking about more voices by one ventriloquist?" He went on to ask, "Is this not a case where when people talk about more diversity there is in fact less diversity?"[11]

PRODUCTION COMPANIES

This is a good point at which to consider production companies, an unsung but important sector of the electronic media that none of the other sectors can do without. Production companies that most of us rarely hear of (not the big-name networks) create most of the programming seen on television today. They literally build programming from the ground up, conceiving, shooting, editing, producing, and polishing the shows.

Pioneering TV

Although megacorporations own some production studios today, the traditional mode, responsible for most TV programming over the past half century, is to contract with an *independent* production firm. Starting as TV first entered U.S. homes in the 1940s and 1950s, independent companies—often paid by **sponsors** (advertisers who financed entire programs)—created many important shows for the fledgling TV networks. One series, *Dragnet*, began at a company owned by its star, Jack Webb, who sold the program in 1949 to NBC Radio and later to television.[12] The classic comedy *I Love Lucy* premiered in 1951, starring Lucille Ball and her husband Desi Arnaz; again, the stars owned the production company (Desilu).

Independent producers eventually came to play an even larger role in our entertainment media, largely because of the FCC's fin-syn rules, imposed in 1971. These restraints on network control of programming created big opportunities for production companies, and soon they were mushrooming, selling completed shows to networks and to cable television. By the time the fin-syn rules were lifted in 1995, many companies were healthy and, thanks to the proliferation of cable and other non-network programmers, had developed a broad customer base. Today, some production firms are so diversified that they lack only broadcast stations to get their products on the air—and, if they've become part of a huge media conglomerate, they may even have stations.

Some successful production companies are massive, like the combined operation created when NBC and Vivendi Universal Entertainment (VUE) merged in 2003. That merger united NBC Studios and Universal Pictures, which together controlled 32,000 episodes from decades of network television—many still in circulation on cable TV, making money—plus more shows in current production. Most companies, however, are much smaller; some want to stay that way and others hope to create breakthrough hits, as Scout Productions did. After starting in the mid-1990s as a team of three people, Scout struggled toward fame for years before reaching it with *Queer Eye for the Straight Guy*. This offbeat show rode a "makeover" craze to huge success, including spin-off programs overseas and "Queer Eye" merchandise.

New Opportunity

Some companies offer not just conventional production expertise but **multimedia** design work—digitally combining audio, video, and other media—that may show up on the World Wide Web or perhaps on hybrid computer/television screens. Besides creating major TV programs, even some of the largest firms rent out their skills and facilities to nonbroadcast businesses for special projects. Disney's production

Sponsor Traditionally, advertiser who bears full cost of a program, gaining exclusive rights to commercial time, sometimes including advertiser's name in show title.

Multimedia Computer-based presentation of integrated media; can combine text, audio, video, animation, and graphics.

arm offers moviemakers their choice of seven big stages—a synthetic business street or Western town, for example—and dozens of services.

To the aspiring electronic-media professional, it's not the huge, Hollywood-focused production companies that offer the most personal opportunity; rather, it's the medium-to-small companies scattered not only throughout Los Angeles and New York but also across the continent. The smallest (consult your town's telephone directory) may do good business by serving short lists of customers—from area broadcast stations to nearby manufacturing plants or even used-car sales managers who want commercials taped.

In the middle range of size, companies can take on more ambitious productions. Their staffing may start with managers who prowl for business and work with creative specialists who come up with sharp program concepts and write scripts to bring them to life. There may be producers to oversee field and studio work, videographers, music specialists, graphics experts, sound designers, animation wizards, and other critical support players. Some survive as freelancers, available to companies that can't afford to keep these specialists on their payrolls full time.

Production companies hire the most talented and ambitious (though not necessarily in that order), including new graduates. Some offer internships to students. The experience to be gained can be transferred easily to bigger production arenas or translated to other electronic-media fields.

The Local Picture

Little of what local broadcasters do every day is directly related to the organization of the industry. With whatever resources they command, local managers generally try to "fly below the radar" of large corporate concerns and to keep their audiences and advertisers satisfied.

Like Lucy Himstedt of WFIE-TV, most local broadcasters organize their forces and facilities tightly, spending only what it takes to do the job. This generally means fairly small payrolls. Even in flush times, broadcasting never has had a large "rank and file" by the standards of some other major industries. Government researchers found that about one-third of a million broadcast jobs existed nationwide in 2002.[13] Perhaps more interesting is that about 70 percent of those jobs were in broadcast operations with 100 employees or more. This reflects the fact that, besides the networks, it's big-city TV stations with hours of newscasts every day that need the most workers (see Table 6.1).

RADIO

In 2004, there were more than 13,400 radio stations in the United States.[14] Most are small, often with fewer than a dozen employees, but their potential influence in their communities is hard to overstate. These stations can steer listeners by providing traffic alerts, weather bulletins, air-pollution warnings, and disaster reporting. They can wage editorial campaigns or provide air time to interest groups. Talk-radio formats dominate AM radio today, and these help energize national policy debates.

Thanks largely to the Telecommunications Act of 1996, however, thousands of radio stations are controlled by a handful of media corporations. Critics say this has resulted in homogenized radio formats that emphasize music and conservative viewpoints to the exclusion of broader local discussion and news. Still, radio is a powerful advertising platform and an invaluable public communication resource.

TABLE 6.1	U.S. Broadcast Jobs, 2002–2012 (in thousands)		
	2002	**% OF TOTAL**	**EST. CHANGE, 2002–2012**
Announcers	40	12.0	–17.9
On-air newsworkers	18	5.3	+6.8
Producers/directors	19	5.6	+0.6
Camera operators	10	2.9	+1.1
Editors	4	1.3	–4.7
Sales-related workers	45	13.3	–9.4
Computer specialists	7	2.1	+32.8
Audio/video equipment techs	5	1.5	+12.1
Art and design workers	4	1.3	+11.6
Top executives	10	3.1	+10.3

Source: Bureau of Labor Statistics, 2004.

Managerial Functions

Almost any radio station must have a leader, a *general manager* or *station manager* who supervises the operation. This person plans the broadcast day (often a 24-hour "day"), lays out an overall budget based on available or projected revenue, oversees all personnel, and finds ways to establish or enhance a public image for the station. Whether running a large-market organization of 50 workers or a tiny-market team of 5 (or even fewer), the top manager bears ultimate responsibility and sometimes fills several key roles.

The larger the station, the wider and deeper the array of management tasks and the greater the need to hire "middle" managers. Each must take on one or more key functions (see Figure 6.5):

▶ *Programming:* A schedule of programs is carefully fashioned to attract the largest and most desirable audiences. The programming manager not only designs the lineup but must oversee the production of local shows and the acquisition of other shows from outside producers. Some general managers, keenly aware of their responsibility to make the business profitable, like to do their own programming.

▶ *Advertising Sales:* This function involves pursuing clients from among local or regional businesses that need (but don't always know they need) radio advertising to recruit customers. Once, the objective was to reach the largest audience; now, numbers remain important but demographic factors (age, gender, etc.) have grown more important as clients go after specific audience groups. Sales work is more complex as a result; even so, sales managers in smaller stations often have skeleton staffs and must do much of the time-selling themselves.

▶ *Business Affairs:* Monitoring and guiding where the money goes is the responsibility of the business manager. He or she controls or heavily influences the investment of precious revenue in station equipment and personnel—a fact that may put the manager at odds with department heads, but that also prepares him or her to move up to general management.

▶ *Engineering:* The engineering manager maintains, upgrades, and operates all of a station's critical electronic systems. This is the heart of station operations. The engineering manager (or chief engineer) monitors all transmissions to be sure they meet the standards required by federal regulators, and thus safeguards the station's license to operate. This manager knows the equipment and facilities intimately and always is brought into discussions of station design and expansion.

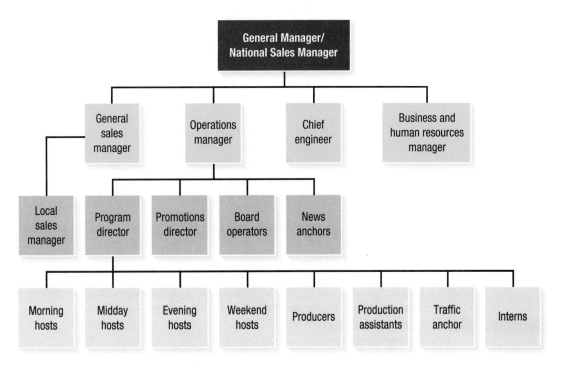

FIGURE 6.5 *Organizing a Radio Station*

▶ *Human Resources:* This function involves recruiting and screening new employees as well as administering pay, benefits, company discipline, and the orderly departures of employees who leave. However, "HR" managers generally are excluded from hiring and firing key on-air "talent"; general managers and news directors handle that.

▶ *News:* The news director is responsible for the origination (or at least relay) of news reports and other important information for local listeners. News is the primary and sometimes only station department that generates content locally.

RADIO STAFFS. Economic pressures in the 1990s and early 2000s led radio stations to slash their workforce and increase their automation in efforts to cut losses.

protalk

Joe Smelser

**BROADCAST OPERATIONS MANAGER,
CHAMBERS COMMUNICATIONS**

Joe Smelser is a man who plants and cultivates. He produced a successful 13-week series of shows on practical gardening, and went to work on another series to air over 26 weeks in 2005. A lifelong devotee of television production, he now does such work in the hothouse of high-definition television. It's the much anticipated wide-screen TV format that is expected to dominate the medium, and Smelser is experiencing it as one more change of seasons.

"I've seen television change in major ways three different times," Smelser recalls. "It changed from black and white to color, from film to video—with all that portability in the equipment—and now it's changing from analog to digital."

Smelser first breathed the air inside a TV station as a 5-year-old visitor back in Iowa. He entered the business in 1966 as projectionist at a TV station—a specialty that vanished once TV stopped using film. He went on to other technical and production jobs before moving to Oregon.

As broadcast operations manager at Chambers Communications, in Eugene, Smelser helped develop a system for shooting University of Oregon football games

> "It's been my life, and if I can take a fraction of the technology and the creativity I've learned and pass it on, I want to do that."

by switching remotely among four stadium cameras over a fiber-optic cable system. He's an innovator with a long memory and a constantly fresh outlook, who spreads the roots of his knowledge to university students whenever he finds time to teach production classes.

"It's been my life," says Smelser, "and if I can take a fraction of the technology and the creativity I've learned and pass it on, I want to do that."

Besides routine news reports, the news director arranges for critical traffic and weather alerts that can ease local emergencies. In many smaller stations, this manager is a news staff of one and takes on non-news duties as well.

▶ *Production:* This area usually involves the creation of local audio material that is not produced by the news department, which can include commercials, station promotion spots, and coverage of special (but non-news) events such as religious ceremonies or telethons. In addition, the production unit may be a valuable resource pool for general station use.

The Rank and File

Personnel calculations were simpler when radio began: Working *alone*, an engineer in 1919 transmitted recorded music from his Pennsylvania garage. The subsequent rise of the medium nationally created demand for thousands of writers, performers, announcers, and technicians that lasted, largely undisturbed, well past mid-century. However, the 1980s brought industry reorganization and shrinkage of all sorts. Local broadcasters found ways to import much more content and automate many functions. This allowed them to reduce their staffs. Actions by the FCC virtually eliminated the need to do unprofitable public-service programs. Local radio in the United States fairly rapidly became what it is today: a giant web of stations, most of which have absentee owners and drum-tight payrolls.

TELEVISION

The picture, so to speak, is somewhat different for television. Because its technologies are more numerous and complex and its projects are visual, TV almost always needs larger staffs than radio ever did. Still, its basic organization is roughly analogous to that of radio.

Managerial Functions

Today, almost every TV station has a *general manager* to deal with owners, monitor the station's profitability, and make decisions about long-term projects, alliances, and business strategies. Many a station today, however, gets along without a traditional number-two person, the *station manager*, charged with shoring up community support of the license and providing day-to-day oversight of operations.

The station still needs its *business manager*, who, sometimes with a small staff, supervises the flow of spending and revenue. This person's targets often include major equipment purchases and upgrades, which may become bones of contention as technologies change and departments wrestle with tight budgets. There's a *sales manager* to keep advertising revenue flowing in. Sometimes, in fact, there are two managers: One for the sale of local or regional commercial time and the other for the sale of commercials to run in national programming, although that is often handled by a national account representative. In some cases, a *general sales manager* will oversee both areas. Sales, being the station's primary revenue source, usually gets the personnel it requires.

The management corps also might enlist a *community affairs director* to polish the station's local image. It almost invariably includes a *news director* to steer the coverage of daily local events. A *program director* usually is in place to locate and acquire programming from many sources, although some general managers do much of this work. The *production manager* guides technical operations with skilled assistance from the *chief engineer*, all the way to the final step in television work—transmission.

MINORITY WORKERS. Electronic media employ relatively few members of ethnic minorities; some critics say that limits not only career opportunities but also the diversity of broadcast programming.

The promotion track for these managers varies widely. Sales managers, being the hunter-gatherers of station revenues and thus essential to the business, often seek to become general managers; many current "GMs" began their careers in sales. News directors generally have backgrounds as journalists but increasingly have had broader executive training, preparing them for general management. A chief engineer primarily must have talent and experience in designing, running, and trouble-shooting highly complex electronic systems. Without business or management training or experience, such technical experts find it difficult to ascend to the top station jobs.

The Rank and File

Most managers not only have important tasks but they also have other people to help execute those tasks. In television, that means a wide assortment of professionals. It's mainly among their ranks that the action—and opportunities—can be found, and mainly in news. That's because, in most cities today, news is virtually the only locally generated programming; other forms proved too costly for the ratings they received.

From a business standpoint, as outlined earlier in this chapter, there are lots of reasons *not* to hire. One is pressure for profits. By the mid-1990s, fully three-fourths of the TV stations in the top 100 U.S. markets were owned by station groups.[15] Most general managers today know that corporate officers are, or might be, constantly looking over their shoulders. This tends to hold down all but truly necessary hiring.

Payrolls could shrink even further as advances in digital technology enable smaller teams to produce TV content. Already reporters at some stations edit tape at their desktops rather than handing it off to specialized editors. The industry can be counted on to keep developing ways of operating that will require fewer paychecks.

ACCESS

Members of ethnic minorities and, to a lesser extent, women still face the greatest challenge getting full value from broadcast careers. Studies consistently show their hiring, pay, and promotion opportunities are significantly lower than those afforded to white men. According to the Radio-Television News Directors Association, the average percentage of minority employees in a broadcast newsroom in 2003 was hovering just above 21 percent. That represented little or no significant improvement over the past decade—a decade of rapid increases in the diversity of the U.S. population. Similar complaints come frequently from the National Association for the Advancement of Colored People.

Broadcasters routinely declare their commitment to diversity, but few have managed to bring about substantial change in the diversity of their staffs. When it comes to *owning* broadcast or cable outlets, the situation is at its grimmest: In 2004, of about 14,000 radio stations across the United States, 4.2 percent reportedly were minority owned. Worse, of about 1,700 full-power TV stations, only 1.4 percent were minority owned.[16]

protalk
Don Browne
**CHIEF OPERATING OFFICER,
TELEMUNDO NETWORK**

D on Browne spent years trying to sell NBC on the idea that getting into Spanish-language TV programming would be smart. He won the debate and today helps to run the result—as chief operating officer of the Telemundo network, now a subsidiary of NBC.

Telemundo claims to reach 91 percent of the U.S. Hispanic audience through stations it owns or helps to program in nearly 120 markets across the country. Browne, who once directed news coverage of Central and South America, became an NBC executive and helped engineer the purchase of Telemundo. He sees his current position at its helm as an extension of long-ago news events.

"I was, in the 1960s, totally affected by how NBC and CBS covered civil rights," he says. "They covered it with courage and they covered it with attitude. They used television to do something very positive in this country. That's why I got into the business.

"Having joy and passion and relentless commitment to [broadcasting] is an absolute prerequisite if you're going to enjoy the profession."

"Having joy and passion and relentless commitment to [broadcasting] is an absolute prerequisite if you're going to enjoy the profession or make it in it. People at every turn will tell you there's something you can't do. You've gotta defy that. You've gotta have the inner strength and determination to prevail."

His own commitment to civil rights and the role that broadcasting can play in social progress have won Browne high praise. He won the 2004 Ida B. Wells Award for his work "to ensure newsrooms more accurately reflect the diversity of the communities they serve," said the National Association of Black Journalists and the National Conference of Editorial Writers in making the award.

▊ Into the Future

[handwritten: probably meant 2012]

[handwritten: #16 →]

Federal forecasters projected in 2002 that employment in broadcasting would probably rise by about 8.5 percent by 2112. Further industry consolidation, coupled with
still more technological change, could cut that increase in jobs as some of today's
occupations grow obsolete.

THE CONSUMERS

Much of what's ahead depends, as always, on what people will buy, reject, or long
for as the term *media* stretches and changes before our eyes. Traditionally, it has
been applied to our means of communication—newspapers, magazines, television,
radio, and more recently the Internet. Increasingly, however, younger consumers
(and some older ones) are turning to devices that stretch or step beyond the well-
worn *media* monicker. Some offer interactive experiences and put more and more
technology in the palm of the user's hand. These include portable video games that
can receive advertising messages and personal digital assistants (PDAs) that can
send or receive most of the information or entertainment that traditional media
offer.

The booming popularity of games, especially among people in their teens
through mid-thirties, already has led to production of some big-ticket Hollywood
movies in the visual style and tone of video games. Game websites have prolifer-
ated, national leagues have been formed, and some advertisers are rushing to place
their products and services before the players. So if these electronic diversions aren't
exactly media, they stand a strong chance of capturing some advertising revenues
that commercial media traditionally have enjoyed. In fact, Viacom and Time Warner
are looking for ways to buy into the video-game business. Viacom, Inc. chairman
and CEO Sumner Redstone said, "It's a great business. No major media company
can afford to ignore it."

Although some video games may not affect real media other than to distract
their audiences, PDAs and versatile cellular telephones are media *delivery systems*
that could influence how users view the world—and want to view it. Similarly, the
relative absence of local news on radio stations run by corporations that "central-
cast" the same content to everyone may become a plus, not a minus. It's unclear
whether young listeners miss the local newscasts now supplanted by syndicated talk
hosts and prepackaged music shows from far away. Of course, local radio employees replaced by automation are likely to miss their jobs—jobs that are unlikely to
return.

Some market analysts are now predicting continued bullish growth for the new
service called *satellite radio.* By one account, it had 4.6 million users by mid-2005
and was winning praise as a Wall Street investment. If central casting by station
groups is preparing people to sever ties with traditional radio altogether, today's
electronic media will have another reason to keep changing.

Bottom line: As consumer habits shift, so will the generally resilient industry
of electronic media. That's why the rate at which it may hire college graduates is
notoriously difficult to project.

MEDIA PROSPECTS

The electronic-media options (and necessities) of the near future will be closely
tied to technological advances and organizational moves within the commercial
industry. This by no means should suggest that *noncommercial* media-makers will
play no role in big future developments (see Chapter 14, Public Broadcasting), but

TELEVISION REPORTER

More and more electronic-media job ads are calling for digital skills, and the changeover from analog to digital equipment may be complete within a few years. For the job hunter, this puts a premium on researching any appealing sector of the industry to find out exactly what skills it demands. The *versatile* worker has the greatest chance of both getting hired and riding out future technological changes.

Many college broadcast programs now offer training on digital video and audio equipment. For that matter, many of today's college students have become adept with the newer technologies through their personal computers or their home-entertainment gear. Here are some job categories that may provide openings at most television stations and, to a far lesser extent, at radio stations. Some of these, however, already are on the way out.

Engineers (radio/TV) currently are licensed technical wizards who put broadcasts on the air or into other, newer distribution pipelines. Some also handle field transmission of news video back to their stations. Because unions have been weakened and digital tools will replace older electronic ones, the number of engineers is expected to shrink in the coming years.

TV directors at stations and networks absorb streams of information and turn them into commands to their crew members. These **control-room personnel** are trained in use of audio, video, and graphics control equipment—and in reacting quickly. **Studio personnel** include floor directors and camera operators who spruce up anchors and hosts, line up "shots" with seasoned precision, and maintain calm in the studio once live broadcasts begin. **Advertising sales representatives** seek out clients, persuade them to "buy time" for commercials, and work to ensure that the commercials will be seen or heard when their target audiences are on hand.

Among the most digitally sophisticated workers in many TV stations are the **videographers**—photographers and video editors who work on news, special programs, and commercials. They often are filling requests from **producers** who conceive, organize, and write newscasts and other programming. **Reporters**—increasingly rare birds in radio—are the broadcast journalists who dig up news, write stories and edit them or supervise their editing, and present them on air. **Newsroom personnel** include assignment editors, news planners, writers, production assistants, and sometimes helpful (and unpaid) interns from nearby colleges.

Administrative/clerical workers range from secretaries to accountants, from traffic (ad scheduling) clerks to maintenance employees, from news librarians to security guards to drivers.

it's the deep pockets of companies nourished by advertising revenues that have brought most major innovations into general use. Are any coming soon? At this writing, several appear:

▶ *HDTV:* Although early sales of high-definition television sets were slow, adoption is rising (and programs proliferating) rapidly enough to convince one consulting firm that 60 million U.S. homes will own the new sets by 2008.[17] Their stretched-horizontal screen shape (16:9 "aspect ratio") reportedly will show scenes much as we view reality with the human eye—a first.

▶ *Other digital:* Owners of home computers and PDAs can now use them to download music legally, make professional-quality CDs, view movies, edit videos, tap into obscure entertainment sources, create multimedia performances of their own—and (by the time you read this) much more. Any of these may either enrich or compete with the industry.

▶ *Cable:* More surprises are due from cable companies, many now owned by rich corporations and positioned to extend their roles in home media. Cable already brings most broadcast TV into homes, has spawned hundreds of channels, produces innovative programs, is upgrading old lines to fiber optics, and could launch an interactivity decade without much help.

▶ *Satellite:* Once dependent on cable as a way into the home, satellites now transmit directly to home antennas, bringing rural, cable-free viewers newscasts from the nearest city and leading all electronic media in cutting-edge entertainment programming. But HDTV's progress and consumers' use of digital video recorders (DVRs) is expected to influence satellite TV's ability to compete in the future.

Summary

Like most U.S. industries, broadcasting for many decades had a stable structure within which predictable sets of employees did their work. Today, competitive pressures, new technologies, and financial strains are prompting massive industry change. Within it, broadcast employees do much of what they've always done, but often more rapidly and under heavier corporate pressure to perform efficiently and to adapt their products to new media. Shareholders gradually raise their profit demands, forcing networks and station groups to look for efficiencies such as shared programming that can be purchased once and used by many stations in a system. Some of these shifts will help divert money and energy into quality broadcasting.

As federal deregulation proceeds, making it easier for companies to corner the market on entire categories of content, the trend toward concentration of ownership continues. Broadcasters still hire most of the same kinds of specialists they have always hired, but in smaller numbers and with a premium on versatility and digital know-how. The most successful job seekers will be those with college degrees or advanced training. The sophistication of today's radio or television station demands knowledge and flexibility—one reason that internships are a virtual necessity for some specialties. Another reason is that, amid media turbulence, no media worker—except perhaps the creator of content who can adapt to new forms—has a lock on lasting employment.

Industry growth for the next few years is projected to be slow, partly because consolidation is rampant and will swallow some current companies. However, new consumer crazes and breakthrough technologies could produce sharp bursts of growth. Video streaming has already brought television to the Internet; much smoother systems for integrating old media into new are being developed.

This general process is called **convergence.** Already a cliché, the term probably represents much of the media's near future—the interweaving of broadcast, Internet-based, and even print media in ways that make more formats available to more people.

> **Convergence** Joining or blending of older media forms with new, as when TV or radio pass through the Internet or a computer is used to edit animations digitally.

Food for Thought

1. Is it reasonable to expect broadcasters to balance the profit motive against their public-service obligations? Evaluate arguments on both sides of the issue.
2. The FCC has worked to deregulate electronic media for the past two decades. Should it take a different approach? Explain your answer.
3. If you could buy a large share of stock in one electronic-media company, which one would you choose, and why?
4. Does it matter that most of us get our daily news from electronic-media units of big media corporations? Argue your case.

Programming *and* Distribution

Chapter 7

Here's NBC's (poorly written) description of a stunt on its TV show *Fear Factor:* "Child is placed in a box filled with cockroaches, the parent must free the child by transferring roaches with their mouth onto a scale which will reveal a key that will unlock the roach box."

Disgusting? Loathsome? Halfway through the first decade of the twenty-first century, "reality" was perhaps the most powerful programming force in American television.

So much for the forecasts of some media critics and scholars. One had predicted soon after the terrorism nightmare of 9/11 that younger viewers would abandon reality shows: "They want something that takes them away from stress, that takes them away from danger—where reality TV puts it right in their face."[1] Reality ratings did drop for a while. However, only months after 9/11, Fox's *Joe Millionaire* was openly demeaning women—and scoring big viewership numbers—defying citizen groups that attacked what a TV historian called "humiliation television."

By 2004, it was clear that not even terrorism would keep some viewers away from reality shows for long. Scores of programs—no, hundreds—had been tried in the United States and elsewhere. Many

had failed, sometimes under attack: Southern protest prompted both CBS and NBC to drop plans to revive *The Beverly Hillbillies* by putting impoverished rural people into Hollywood mansions. But Fox broke taboos and drew strong summer ratings from *Trading Spouses,* just as ABC was planning *Wife Swap.* UPN braved Congressional criticism to air *Amish in the City,* in which young people left their provincial religious lives for a few weeks of moral–cultural education from hip Hollywood youths.

None of this necessarily meant that tawdry "reality" would prompt television to turn its back on older program forms. King World Productions signed talk host Oprah Winfrey to a contract through 2010–11, which would give her an amazing run of 25 years in syndication. And in a presidential election year, NBC struggled hard to revive the ratings of one of its most honored shows, *The West Wing,* which had brought a fictional White House team to vivid life on-screen.

If all these developments presented a single clear message about electronic-media programming, it was that battles for the attention and loyalty of audiences rage across many fronts day after day, year after year. Some battles can never be clearly "won."

▌What Programs Are

All programs that reach the broadcast schedule serve the purposes of media enterprises, most of which are out for financial profits. However, to accomplish this, programs first must *attract* audiences, then must *meet their needs or desires*. In the course of a day—even a single hour—radio, television, and the Internet can be undemanding roommates, intrusive lecturers, enlightening town criers, amusing guests, sympathetic friends, and much more.

Not even the most disheartening content can drive every listener and viewer away. Far from it, in fact: Studies have shown that families of air-crash victims find solace in watching TV coverage of the very events that bereaved them. September 11, 2001, brought millions (possibly billions worldwide) to their electronic media for live programming of the most compelling sort. Broadcast executives quickly pulled violent fictional dramas off the air to avoid upsetting viewers, but by December, as reality shows began to bloom again, movies with terrorism plots started to reappear. Recognition of television's potential effects on us had changed programming to inform and console us; awareness that we were ready to resume our normal patterns changed it back.

The art of programming does not often require such keen sensitivity, and the production of successful shows is not always costly or complex. Just as news dispatches can form a short radio program, the spontaneous horseplay of teenagers "caught on camera" can occupy half an hour of TV. Practically anything interesting can be program material. What will "work" with an audience at any given time is the elusive target of programmers' continual quest.

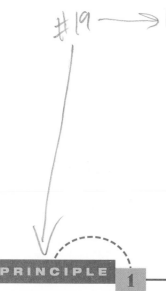

#19 →

PRINCIPLE 1

Anything that attracts people and holds their interest can be successful as programming.

RADIO

Most radio today draws from a sparse menu of programs to fill the broadcast schedule. It was not always this way: Between the 1920s and 1950s, every sort of public spectacle, industrial announcement, musical performance, vaudeville act, moral/religious lecture, drama, suspense or adventure series, newscast, documentary, esoteric chat, or government information project seemed to make its way onto the radio airwaves. As networks signed contracts with stage and movie stars, some of the biggest names in entertainment became far bigger, "visiting" homes across the nation night after night. Others—new to show business—gained their first popularity from radio. News, sports, politics, comedy, celebrity, and the intimacy of familiar voices emanating from local stations made the medium popular across every age group.

Later, when television rushed onto the electronic landscape a few years after World War II ended, commercial radio faced daunting competition. A lot of advertising dollars moved to TV. In response, radio began to narrow its range of offerings to what it could do best and with the greatest commercial results. Today, a search of the airwaves can turn up great variety, but most stations emphasize only a few program types.

Music

Recorded music has reigned for decades as radio's top program content. The people who keep it going include musicians, their producers, their agents in getting tunes onto the air—never a sure thing in a crowded music marketplace—and audio engineers at the stations. Just as radio broadcasting depends heavily on the recording industry to supply content, the record-makers and distributors rely on radio to popularize their songs.

Radio continues to be the cheapest, most ubiquitous medium for reaching listeners with music, but change is in the air. First, not just broadcast but *direct satellite* transmission of music and other radio-style content is growing in popularity. The more people pay satellite services to send them tunes around the clock—in their car, where most radio still is consumed, or in other locations—the smaller the audience for conventional radio. That would mean reduced ad revenue for radio, and so would continued growth in the downloading of records from the Internet. The "early adopters" of this new music medium have been computer-savvy college students and other youths, many of whom already were less devoted to old-style radio than are their elders.

Music always has constituted most FM programming, and it's still found on AM stations as well. Any sense of randomness in record selection by disc jockeys (when they are not "automated" or absent altogether) has been virtually eliminated from commercial radio. Stations, consultants, and the record industry have cooperated over recent years to divide music into genres and subgenres so that listeners with specific tastes can be better targeted by advertisers. The typical station sticks primarily to one music format on which most of its promotion can be based (see Figure 7.1). This strengthens its ability to win a possibly narrow but loyal group of listeners.

Talk

Apart from music, it's talk that commands most airtime on today's radio stations. Many are specialized to the "news/talk/information" format (see Figure 7.1). It's a subgenre in which news or other information appears around the start of each hour on the clock and discussion of social or political issues fills the balance of the time.

Some of the best-known names in broadcasting have emerged in the talk arena. Many of them—Rush Limbaugh, G. Gordon Liddy, Michael Reagan, Sean Hannity—lean to the conservative side of political and cultural issues and have found large, sympathetic audiences. Limbaugh, a talk veteran, gained special notoriety for goading and irritating Democratic politicians; he even inveighed against a 1990s movement to revive the Fairness Doctrine by branding it the "Hush Rush Bill."

Limbaugh's prominence and the arrival of other right-wing talkers created an image of talk radio as invariably conservative. However, 2004 brought the shaky debut of Air America Radio, a Democrat-backed talk network with a liberal bent, spearheaded at the microphone by comedian/critic Al Franken. Air America, short

Active Rock	80s Hits	Hit Radio
Adult Contemporary (AC)	Educational	Rhythmic Oldies
Adult Standards/MOR	Ethnic	Soft AC
Album Adult Alternative (AAA)	Gospel	Southern Gospel
Album Oriented Rock (AOR)	Hot AC	Spanish Contemporary
All News	Jazz	Spanish News/Talk
All Sports	Mexican Regional	Spanish Oldies
Alternative	Modern AC	Spanish Religious
Children's Radio	New AC (NAC)/Smooth Jazz	Spanish Tropical
Classical	New Country	Spanish Variety
Classic Country	News/Talk/Information	Talk/Personality
Classic Hits	Nostalgia	Tejano
Classic Rock	Oldies	Urban AC
Contemporary Christian	Other	Urban Contemporary
Contemporary Inspirational	Pop Contemporary Hit Radio	Urban Oldies
Country	Religious	Variety
Easy Listening	Rhythmic Contemporary	

FIGURE 7.1 *Radio Station Formats (as measured by Arbitron)*

of money, stumbled early through delays, layoffs, and missed payrolls. After a year, it was claiming affiliation with only 53 U.S. stations plus the XM satellite network. But its audiences, at least in New York and Chicago, contained an impressive number of young adults—the potential customers that advertisers most want to reach—a promising sign.

Still other types of talk programming that have succeeded on U.S. radio address audience concerns including finances, health, "parenting," and the complications of romance. Rather than on commentary and opinion, these formats rest on their hosts' asserting the authority to *advise* listeners. While a single station may pepper its schedule with an assortment of such syndicated counselors, the station usually maintains its clear talk focus—the basis of its branding, emphasized in ads on buses and billboards.

News/Information

Complementing and fueling talk radio, news/information formats can trace their roots to the origins of U.S. radio; some of its earliest broadcasts were reports on the 1920 elections. Radio always has been a vital link from the scenes of historic events. Its well-defined news programming typically is free of overt opinion (other than what *newsmakers* say), restricted instead to what journalists regard as neutral presentation of current matters. This holds down its entertainment value—by contrast, many people find opinionated talk-show hosts highly entertaining—but secures its status as public-service programming to which some listeners turn for regular reports and in times of crisis.

In that sense, news/information programming can be considered indispensable. However, commercial radio executives in recent years—spurred by the consolidation of stations under a few big profit-driven owners—have largely ceded journalism to public radio. In their own stations they have reduced news staffs, to zero in some cases and to a handful in most. Today, in all but the largest cities, it's difficult to find radio news on the air more than two or three minutes an hour, and then usually from far-off central newsrooms. Has this process cost listeners knowledge of the world around them—especially their own communities? Probably, since for a variety of reasons, relatively few people habitually listen to public radio. That's where in-depth reporting, documentaries, and offbeat information or essay programs such as *This American Life* still thrive.

Sports

Network and local radio, between them, carry all major professional and college sports, and "sports talk" has helped to fill many hours in radio since the 1980s. All-sports ESPN Radio, for example, has hosts who hour after hour gab with athletes or take calls from listeners who are passionate about games. In today's highly commercialized sports environment, broadcasting brings in big ad revenues and—through contracts between stations or networks and the teams they put on the air—returns some profits to amateur and professional franchises.

Money that remains with broadcasters can be invested in future coverage, which often is promoted heavily within the sports-news/talk shows. This cooperative apparatus keeps the sports radio genre thriving. It has spawned some powerhouse all-sports stations, such as Detroit's Sports Radio 1130 AM (The Fan) and Los Angeles's Fox Sports 1150 AM.

TELEVISION

At first, television didn't have to spread a smorgasbord of programs before its viewers. Some people were so excited just to have the new medium available that they would sit and watch the "test patterns" left on the screen between shows. Within a few years, though, TV programs ranged from civic conversations to soap operas to

CSI: MIAMI. The gritty work of crime scene investigation made this CBS police drama a big TV hit. A suite of *CSI* shows set in several cities fills various time slots, multiplying ad revenue. This formula also turned *Law and Order* into a tradition on NBC.

prime-time comedies to music to boxing matches to broadcasts of challenging plays. Today, purely in terms of programs available to watch, television is more diverse than in any previous time. New networks and hundreds of cable channels provide hundreds of shows a day, many of which never would be scheduled by over-the-air stations. At the same time, there's a constant countertrend: Competition has created a "copycat" environment. A TV network or station that sees a hit emerging on another channel is tempted to imitate the newcomer, and may drop one or more of its existing programs to make room. The effect of that, of course, is to *reduce* diversity.

PRINCIPLE 2

Program diversity works until one channel rushes to imitate another's hit show—thus reducing diversity.

What once was simply the movie industry now is the leading maker of prime-time TV content. Its studios and production houses crank out thousands of hours a year of network shows and cable "originals." These producers also license programming—often former network hits—in an open market called **syndication** that helps fill many program slots every week across the United States.

Dramatic Series

Some of the most popular programs in *prime time*—four viewer-heavy hours each weeknight—are tautly scripted and elegantly produced stories. They depict ordinary people, families, public-safety officers, or care-giving professionals facing life's challenges. Television revenues often influence this enterprise early. By the time a production is clearly conceived and at least partially planned, producers are meeting with electronic-media companies, setting up deals. Those involving a new high-budget show usually call for a "pilot," a test episode to help broadcasters gauge a new program's potential appeal. Even if enough appeal is established, the show still can do fabulously well, adequately, or poorly.

Behind the original network and cable series are inventive writer-producers. Many have their own companies and work under contract to the networks. All have developed ways of striking chords in television viewers. Chris Carter, creator of the *The X-Files, Millennium,* and other spooky-realistic series, taps into apocalyptic lore from The Holy Bible. "There's lots of good stuff in there," Carter told an interviewer.[2] David E. Kelley (*Ally McBeal, The Practice*) said he portrays the legal system by noting what is absent, not what is present: "We take the law and we say, 'OK, as a gen-

Syndication Process or marketplace in which programs are "sold" (actually rented) to broadcast or cable components for original use or reuse.

DAVID E. KELLEY. A former attorney, this highly successful TV writer–producer began with legal series (*L.A. Law, The Practice*) and branched out to other themes—but had two shows quickly cancelled. "We're all dealing with the fact networks can make (reality) shows at a cheaper cost and yet get big ratings," Kelley told *Television Week*.[4] In 2004, he announced plans for an extended break from Hollywood, but not before creating *Boston Legal* for ABC.

eral rule it works, but how about those individual equations where it doesn't?'"[3] By playing off real life, creators of prime-time television programs hold us in thrall.

Situation Comedies

Situation comedies get their name from the reassuring continuity of the situations in which their characters live, coping comically with problems and one another. Almost any situation can be harvested for laughs. Dating back to 1940s–1950s series such as *The Goldbergs, Life with Luigi,* and *The Life of Riley,* **"sitcoms"** have been a staple of television's popularity and growth. Almost all are at a compact half-hour length and thus useful in many schedule slots. Some plots and styles become dated, yet many hold up well over time—perhaps due to their reliance on family roles (traditional and nontraditional) as well as their nostalgic power over viewers—an important factor in today's broadcast marketplace. Scores of sitcoms circulate through rerun seasons on cable channels and local stations for years after their network runs. They have far greater staying power than do dramas.

Movies

Hollywood creates feature-length productions specifically for TV, and also earns billions from the sale (or **licensing,** a sort of rental) of its theatrical movies to television. Movies still emerge from studios that have survived since film's golden age, as well as from newer production companies. They have found popularity across the moving-picture media, from TV to home video to continued theater runs. All of these forms seem to feed Americans' continuing love for long-form, fully resolved stories on the screen—especially the ones that end happily. Better movies not only fill two hours of key programming time but also attract big-spending advertisers consistently. Christmas 2001 brought huge audience response to warm family movies, three months after the September attacks on New York and Washington, DC.

However, after years of fat returns from airing films, networks after 2000 were scaling back their movie schedules. Lifetime and other cable channels had been using original movies (not ex-theatricals) to establish their identities and further divide TV audiences. Networks were making more room in their schedule for other attractions, including "reality" shows.

"Reality" Shows

As the first section of this chapter indicated, reality programs feature non-TV-professionals in scenarios that generate drama, suspense, or hilarity—as in an early show that placed young married couples on a desert island seeded with marauding (and attractive) singles. These programs are *unscripted*—although insiders insist that little is left to chance by the hidden producers—and their "cast" members are camera-hungry outsiders who often are paid less than $100 to participate. These factors account for the shows' seeming spontaneity as well as their relatively low production costs.

The average bill for a reality-show episode had long totaled about half of what a scripted show would cost to make, but by 2005 that was changing: The search for winning ideas was leading to higher costs, sometimes topping a million dollars or more per episode. Reality, in other words, was beginning to lose some of its edge of seductive economic appeal to broadcast and cable programmers. Still, to advertisers, this TV genre was the clearest path to large groups of young adults, so it seems likely to continue in some volume for years.

Sitcom Situation comedy (i.e., a comedy series with continuing characters, settings, and/or themes).

Licensing Sale of broadcast rights to a program, usually for a fixed period or number of showings.

News

Journalism gives TV an inexhaustible source of programming. This is demonstrated by the success[5] of cable channels now covering stories 24 hours a day. Fresh reports on real-world happenings also are the red meat of conventional newscasts airing morning, noon, and night on most stations and broadcast networks.

Americans now have access to a greater volume of TV news than anyone else ever has. A fair amount of diversity in anchor "casting" and content is available on cable and satellite channels that bring us international and special-topic news. On the other hand, the once-mighty networks have refocused their newscasts to produce reliable ratings, which has narrowed the news menu. They obtain video worldwide but send reporters to far-flung news sites less frequently than in the past. Close-to-home issues, such as health care and social conflicts, as well as high–human-interest crimes, such as kidnappings and celebrity murders, can command as much network air time as more broadly "important" stories. In the current dynamic of broadcasting, news must compete with entertainment programs for viewers and advertising revenues.

One way networks help news shows succeed in this environment is by turning them into part-time entertainers. Morning shows, which are both newsy and frothy as well as draw big ratings and major advertisers, are a good example. Still atop the heap at this writing is NBC's *Today*, which generally leads its rivals in ratings. The show is run by the network's news division but at times is heavy with celebrity chat, plugs for other NBC shows or movies, and stunt features with popular anchors Katie Couric and Matt Lauer. News or not, however, *Today* fulfills the paramount requirement of making money: The show reportedly accounts for fully half of the network's $500 million in annual profits.[6]

Newsmagazines

The newsmagazine genre is another answer to the competition problem that news faces across the major viewing hours. Slickly made newsmagazine programs now occupy plum spots in the prime-time schedule. They emphasize emotional story-telling, celebrity profiles, and consumer investigations. Their segments, like most modern TV news stories, tend to focus on simple themes and events—but, unlike the brief reports in today's newscasts, a newsmagazine story can run on for as long as its capacity for drama holds out.

Although not always meeting standard definitions of "news," newsmagazines' raw materials are the same as those of traditional news and the wildly popular reality shows—which means they're cheap compared with more heavily produced, expensively cast programs. Moreover, except when they spin off a day's top news story, newsmagazines can be moved around in the broadcast schedule to do the most competitive good.

The newsmagazines' usually dramatic presentation of real-world events can have unforeseen consequences, however. In 1993, *Dateline NBC* showed a General Motors vehicle bursting into flame in an impact test—without telling viewers that a small rocket had been attached to the car to make sure it caught fire on camera. Fatal wrecks and owner complaints had set off the network investigation, and some of its findings of possibly dangerous truck design were at least arguably accurate. But GM found the rocket's remains and threatened publicly to sue; producers were fired, anchors apologized on national television, and NBC News learned how such dramatic occurrences in prime time can backfire. Its president soon left the company.

Daytime Dramas

Television started trying out the famously slow-moving genre of daytime dramas—often called *soap operas* because of their early household-product sponsors—as early as the 1940s. By the early 1950s, they had arrived to stay. Rich in vivid characters, unresolved family crises, unrequited love, and often a streak of treachery, they continue to rule many midday television hours. One daytime drama, *The Guiding Light*, enjoyed its fiftieth anniversary on the air in 2002.

Strong new elements of sexuality and violence entered the daytime arena in the 1980s and added to the "soaps'" popularity. This also set the stage for change in the twenty-first century as broadcasters chased ever-shrinking slices of the media revenue pie. As they became too edgy for some family audiences, soap operas opened the daytime door to a style of drama prevalent in most countries: the *telenovela*. These are stories with faster-moving plots and shorter lives than the season-long U.S. perennials that return year after year. Because they focus more on romantic couples than on families, and attract viewers including some young adults, *telenovelas* may well be adapted by Hollywood for U.S. viewers—including a burgeoning Hispanic American population.

Talk

Many daytime and early-evening hours on TV are filled with conversations between hosts (usually sympathetic but sometimes testy) and show-biz stars, newsmakers, or ordinary people with problems. Some shows simply expose people's sad or tawdry stories; others try to provide solutions and support.

Talkers can be controversial, usually not a bad thing for television personalities. Jerry Springer, a host known for tastelessness and riotous studio audiences, provoked an anchor walkout when he showed up to do political commentary at a Chicago station. Geraldo Rivera, once a noted mainstream TV reporter, became a successful talk-show host who made unsuccessful forays back into news after 9/11. Undoubtedly, the most respected talk host—not to mention one of the country's richest and most influential women—is Oprah Winfrey. Less celebrated but addictive to many young viewers was a vulgar radio call-in show host, Howard Stern, who cost stations millions of dollars in FCC fines—until he signed with Satellite Radio for $100 million a year.

Sports

Networks continue to broadcast a handful of major sports for thousands of hours a year. Until Fox bumped CBS out of its National Football League partnership in the 1990s, the "old" networks dominated coverage of the most nationally popular sports. Much has changed, however. Cable television, rich in available air time, paints from a wider palette that includes the most esoteric athletic specialties. Many of them make their way onto the schedule of ESPN, one of the most successful channels in television history.

PRINCIPLE 3

With fewer restrictions, cable television can range over a wider array of topics than broadcasting can.

Launched in 1979, ESPN was purchased by ABC five years later. Cable channels were still new and small in number. But, thanks in large part to CNN pioneer Ted Turner, the cable revolution was gaining steam. Since then, ESPN's seemingly endless stream of game coverage, instant scores, and lively sports-related talk shows has transformed the way Americans consume sports. It outperformed other sports-news TV outlets so thoroughly that local stations gradually shortened their own sportscasts. ESPN marked its twenty-fifth anniversary in 2004, and ABC soon moved *Monday Night Football* to the cable network.

One final sports note: Although few fans probably give it any thought, the annual Super Bowl professional football game always provides a creative challenge for programmers. Competing networks "stunt," trying to woo women or other groups from the Super Bowl audience with specially targeted shows. And the network that "owns" the year's bowl spectacular encourages lavish pageantry before, during, and after the game to hold as many viewers in place as possible. It usually works: Nearly 90 million viewers watched the 2004 game.

What viewers saw, by the way, included not only a late win by the New England Patriots over the Carolina Panthers but also a remarkable event at half time: the assertedly accidental display of one of Janet Jackson's breasts in what later was termed a "wardrobe malfunction." The revealing moment triggered weeks of bad jokes and touched off a national flap about broadcast "indecency" that brought new and unwanted scrutiny to television. (The 2005 Super Bowl drew 86 million viewers, off slightly from 2004.)

A RATINGS MOMENT

If every program delighted audiences and never lost its appeal, radio and television would be sleepy, unchanging media. That's far from true, however. Failure is at least as deeply embedded in the creation and presentation of programs as success is. We can see that in the fortunes of TV talk shows. Stephanie Drachkovitch, who develops new programming ideas for television, pointed to the survival rate of such shows over the 1990s: Only 3 of 55 stayed on the air. "There's a high failure rate in daytime—more channels, more choices—and it's hard to get a program noticed, and to break existing [viewer] habits."[7]

SPORTS VIEWERS. Audiences for TV's sports coverage aren't always huge, but tend to consist of fiercely dedicated viewers—who respond well to targeted advertising.

In commercial broadcasting—and sometimes even in the noncommercial realm—nearly every program's fate rests largely on audience-measurement statistics, usually called *ratings*. They're so important that this book devotes much of another chapter to the subject. However, since ratings' most direct and powerful impact is on programming, let's look briefly now at how that system works.

Objectives

What *networks, stations, and cable channels* primarily need to know is how many people tuned in to each of their programs. Knowing this helps executives decide whether to stick with a program or give up on it and take it off the air. Reminder: The U.S. broadcasting system depends almost totally on the willingness of advertisers to pay money to broadcasters for commercial time. The broadcasters use that money to make or acquire shows to attract audiences, who then will see or hear the commercials. That's why broadcasters strive to track the popularity of each program as precisely as possible.

Measurement

The leading audience-measurement firm in radio is Arbitron; the primary ratings company in television is Nielsen Media Research. Each calculates the size of audiences by using a mix of two methods: diaries and meters. Listeners or viewers fill out *diaries* by noting which programs they were listening to or watching at specific times in the broadcast day. Critics of this method say it puts too much responsibility in the hands of media users, who may not remember their programming choices or forget to mail in their diaries promptly. Such critics usually prefer *meters*, electronic devices that record listening or viewing as it occurs; they can transmit data to the researchers' home offices, reducing chances of human error.

Broadcasters through the years pushed for technology that would tell them not just which programs drew an audience but who was in it. One answer was Nielsen's Peoplemeter, which could be assigned individually to each member of a household. However, the meter depended on the user to identify himself or herself by entering a personal code. The new millennium brought tests of a portable Peoplemeter about the size of a pager that each TV viewer carries around. This promised much more thorough measurement of audiences; Canada was first to adopt it.

Next came a *local* version of the Peoplemeter; among other improvements, it was expected to increase the number of minority viewers counted—a perennial topic of complaints about Nielsen. The company continues to wrestle with advertiser-broadcaster demands, public concerns, and the rapid changes in technology available for what theater people call "counting the crowd."

Results

When a measurement firm has meter or diary information in hand, it can calculate the ratings of a network or station program or **daypart** (segment of a broadcast day). In television, which mostly is metered, the first value to be established is called **HUT (homes using television)**—or just "the HUT level." It's the total number of households in the measurement area that had their sets turned on during a specific period. Once that's known, a program's **share** can be calculated—that is, the percentage of HUT that were tuned in to that specific show. The third value of importance is the rating itself. That's the percentage of the area's total number of homes that were tuned in to the program. So, programmers who learn that a show received "an 8 rating and a 19 share" know that it was seen in 8 percent of all the homes in the area, which was 19 percent of all the homes using television.

PRINCIPLE 4

The power of ratings lies in how programmers interpret them and act on them.

#23

Most critical, of course, is how programmers *interpret* and *act on* those numbers. Projected ratings come into play even before a show is born, and a record of ratings follows it until its last cancellation. In between, it can pull many millions of ad dollars into a network's coffers. It also can suffer the life of a vagrant, drifting from time slot to time slot, from cable channel to backwater station, at the whim of programmers carefully tracking the ratings.

How Programs Move

#22

Broadcast programs move from their origins through a complex distribution system. A bit of history—what Hollywood writers might call a *backstory*—underlies how this works. Americans who now are middle-aged or older grew up with three broadcast networks, the big-city stations they owned, and many network-affiliated and independent stations. It was a simpler world in which most programming seemed to flow from the networks. Then, in 1970, the federal government barred networks from holding financial stakes in the production of programs and their later resale. This quickly reduced the networks' revenue and stripped away much of their power to shape and control programming.

These so-called fin-syn (financial interest and syndication) rules lasted more than two decades. During that period, program producers, networks, and other buyers set up new relationships and supply lines that in a sense replaced the old distribution system. Today, networks again are investing in programs from their inception, and still routinely supply many shows to the stations they own and to hundreds of affiliates. Yet, these once-cozy relationships have been strained and often weakened by battles over cost sharing and, especially, by the invasion of more and more powerful players in the trading game. There are so many stations and cable channels, so many new owners in radio and TV, and so many programs available from so many distributors that the industry seemingly has become a sprawling web of distribution paths among vendors and customers.

Daypart Multihour period often devoted to a single type of program (e.g., soap operas or news/information shows).

HUT (homes using television) A measure of available viewers at a particular time.

Share The percentage of homes using television (HUT) that tuned in to a specific program.

WHO DELIVERS, WHO RECEIVES

Let's clarify: Much of what happens as programs leave their creators' hands and move through the marketplace is not really *buying* and *selling*; it's a variety of less permanent (and sometimes less secure) transactions. Whatever content somebody produces in a form fit for TV broadcast—fiction, sports, news, or special events—can be "sold" or syndicated. As in most industries, everything's negotiable. Figure 7.2 shows some of the ways in which distribution works.

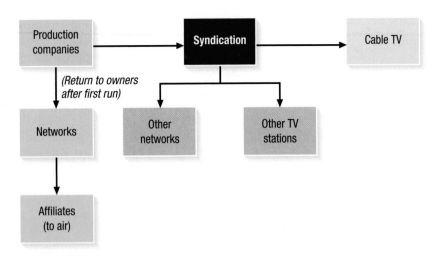

FIGURE 7.2 *Paths to Profit*

Production companies often create programs for TV networks (or sell movies to them), which pass the shows to their affiliated local stations to air. When air rights expire, the programs return to their owners—often the producers—who then can make more money by syndicating them to cable TV, other stations, or other networks.

THE RADIO MARKETPLACE

Some of the biggest radio names are in syndication. One is Paul Harvey, a household name across the United States and beyond. His homespun news–commentary at this writing is heard on more than 1,200 U.S. stations, plus another 400 Armed Forces Network stations around the world.

Even one person with some audio equipment and a place to transmit can put a show into syndication. Companies have arisen to manage the work of performers and producers, promote their ad-revenue potential to stations, and broker the resulting deals. Cox Radio Syndication's offerings to stations include talker Clark Howard, billed as a can-do source of solutions to financial problems. Another syndication company, Murray-Walsh Radio Programming, distributes *Science Update*, a 90-second daily report that can be plugged into many different station formats.

A syndication contract between a supplier and a station (or station group) specifies very narrowly just what is being "sold." A station typically acquires the rights to broadcast a program for a certain length of time or through a specified number of repeated uses. It's possible to acquire not just individual features such as the Rush Limbaugh or Don Imus shows but programming that fills *entire dayparts*. There are suppliers standing ready to provide everything stations need to fill the broadcast day—all on a *turnkey* (self-contained or "canned") basis. Such programming can fit seamlessly into a local format that once was produced at a station with local hosts. Fed by satellite, the imported programming makes possible near-total automation of everything listeners hear.

Major radio companies moving toward higher return per dollar tend to reduce their on-air lineups to the most marketable personalities and music programs. Many are nationally syndicated, such as Premiere Radio's *American Top 40 with Ryan Seacrest (AT40)*, a show founded in 1970 (with a different, long-time host, Casey Kasem) as a traditional record-spinner. It died after a quarter-century but was revived in 1998 in modern form; fans can vote on the Internet to get their favorite tunes played.

The resurrection of *AT40* demonstrates that, even if Clear Channel and other radio giants reduce program variety, syndication can help restore the medium to its feast-of-plenty status. Stations around the country now offer comedy shows, trips through history, pet-talk forums,

PRINCIPLE 5

The distribution of programs through *syndication* is a cornerstone of content variety in broadcasting.

religious hours, jazz sessions, computing seminars, fishing shows, and much more. Live shows such as ball games, news specials, and music-awards ceremonies might seem to be transitory, but even they can be sold for replay as "historical" events or as time-fillers for Sunday mornings.

THE TELEVISION MARKETPLACE

Television programs have especially large drawing power, production costs, ad-revenue value, and economic and social impact. They move through an especially high-powered, high-stress distribution system. Everyone bargains and competes for an edge. Some of the most experienced negotiators are at the fountainhead of dramas, comedies, and movies: Hollywood.

Production Companies

In one sense, Hollywood and other program makers provide their products to broadcasters in exchange for money. Often, however, something closer to a joint venture is involved. For example, as mentioned earlier, a network often enters production in the early stages and helps to pay for a pilot. Soon enough, the Hollywood studio or production firm asks the network to commit to buying a certain minimum number of episodes. The license fee that goes with that commitment is in real money and pays for more production. Producers still may go into the hole—working on a deficit basis with borrowed money—until the new show goes on the air. That's a major reason that many programs are aired, then sold, then resold again and again: to pay off old debts.

A studio that retains ownership of a show will regain control of it after a network or station group's license to the show expires. Producers typically will make a show for a non-network customer only when stations or cable systems in 70 percent of the nation's local markets agree to clear air time for it. That helps guarantee its exposure to a broad enough audience to pay all the bills. In both network and non-network cases, it often takes a few years of licensing to foreign markets as well to bring in enough money to pay off all production costs.

Once production debt is erased and a network owns a program free and clear, it can go into "off-network" syndication and bring the owner big money. A single episode of a hit TV series often fetches more than a million dollars on the broadcast-and-cable market. Besides programs created expressly for television, Hollywood studios have huge libraries of movies to license for television use. In 2000, MGM licensed 13 of its always-popular James Bond movies to ABC for a period of two years, during which the network could use the films to attract millions of dollars in ad revenues. Studios also license batches of their lower-profile or older films to pay-TV and cable channels, where some seem likely to survive until television's last day on earth.

The rapid consolidation of media companies in the 1990s brought many syndication deals "in-house" and may have reduced competition. A case in point is the popular *Cops* series, in which cameras follow real police officers through their tense work days (and nights). The show's creator, John Langley, sued Fox, claiming it wrongly cut him out of his fair share of millions of dollars in profits by syndicating *Cops* between Fox's production arm and its own TV stations. That put the Fox company "on both sides of the bargaining table" in syndication, the lawsuit claimed. Fox reportedly had settled other suits involving *M*A*S*H* and *The X-Files*, while Disney—which produced *Home Improvement* and then aired it on its own ABC network—dealt with a similar legal assault.[8]

It's easy to infer from all this squabbling that the distribution of programs is a massive money-making enterprise, and that major producers have worked to maximize their gains from it. That's "good" business in pure economic terms. Whether it always is *fair* or even *legal* is for competitors and the courts to decide.

Networks and Affiliates

The world's greatest television program is worthless unless it gets on the air, and local stations continue to perform that function for producers. However, anyone who believes that networks supply their high-cost programs to stations for *free*, just to get them (along with commercials) on the air, is wrong. Here's how the system really works.

A network sets up contractual agreements with its affiliates, promising to make the same programs available at the same time, day after day, week after week, to all of them. Along with each program comes the network's cachet—its glittery brand-name power—as an aid to each station's local audience building. In return, the station promises to hold open ("clear") regular times in its schedule for the network show. Viewers in every market thus learn to schedule a regular daily or weekly time to get to a TV set and see the show, or to set their recorders to capture and keep it for them to watch later. Thus, the network creates national viewing habits, making much of its money by exposing—in many cities at once—its highly lucrative national commercials for cars, cereals, home-cleaning products, and, increasingly, prescription drugs.

These slick national commercials have the earmarks of the big-money network shows, but stations get selling opportunities, too. That's because the network leaves an agreed-upon amount of blank time within a program to allow a station to insert its own commercials for car dealerships or regional store chains. This is called **barter** time, since it's a noncash consideration. The amount of local barter time granted in a contract often is a point of disagreement between networks and affiliates and a lingering sore spot even after a deal is signed. A local station is not without any leverage, though. It often insists on the right to shift a network program from one time slot to another to make way for a sports event or other popular show purchased in syndication. A station also usually retains the right to reject an episode that it finds potentially offensive to audiences. Such issues are fluid and a constant source of disharmony. Besides the practical and financial impact, this tug-of-war between networks and stations amounts to little more than a power struggle.

Something usually called **"comp,"** compensation money that networks pay to stations to carry network shows, has become especially divisive in recent years. Cost and revenue pressures have led networks to cut, or threaten to cut, these payments. For example, the nine Scripps TV stations in 2001 were earning a total of about $10 million annually from network comp, down substantially from $13 million in the late 1990s.[9] No amount of compensation would pacify local broadcasters who accused the networks of pressuring them to carry all network programs without fail. During the 2000 national election campaign, some leading stations refused to do this, spurning network sports coverage to carry presidential debates.

These and related trends helped to forge an alliance among more than 600 affiliates that, in 2001, asked the Federal Communications Commission to curb such network pressures on local stations. After that, NBC agreed to pay comp to affiliates in Detroit and Houston—with much of the money deferred over the next decade—and some broadcasters entertained hopes that the network iceberg was finally melting.

As the industry changes, so does the power balance in the distribution of programming. Networks no longer depend on their affiliates alone to carry network shows. They now are licensing programs for use by cable channels immediately after airing them for the first time, and sometimes *during* their debut runs, a practice called *repurposing*. An example appeared when *The Conan O'Brien Show,* a popular late-night comedy-talk program on NBC, was picked up by Comedy Central. The cable channel began re-running episodes almost as soon as they had aired for the first time on NBC.

Barter Commercial time in a program, "bartered" back to its distributor by a station programmer as partial payment for rights to broadcast the show; this practice gives both parties access to advertising revenues.

Comp Compensation that commercial networks traditionally have paid to stations to carry network shows.

HOME IMPROVEMENT. The "everyman" comic Tim Allen used do-it-yourself situations and family antics to make this series a hit. Launched in 1991, it still was going strong in syndication in 2004.

Syndication: TV's Afterlife

Syndication brings an endless stream of content to broadcasters. Stations have done away with so much of their own production capacity that many couldn't begin to fill their schedules without importing shows. One technology analyst, noting syndication in print media as well, asserts that "without syndication, the American mass media as we know it would not exist." A station general manager in the Northwest explained some of the variables facing program buyers:

> You can see programs in this market running anywhere from no cost—"Hey, would you just run my show?"—up to $1½ million a year for an *Oprah,* or even more for something in **access,** maybe $4 to $4½ million for a *Friends* or a *Home Improvement.* . . . You 're charged based on, first, market size—households, basically—then from there on, it's just a competitive marketplace and what it will bid.[10]

Increasingly, program owners insist on being paid not only in cash license fees but also in barter time that the stations otherwise could sell to their own advertisers.

In a hot syndication market that could tie him up in 25 "pitch" meetings a week, that station chief shopped among programs priced at up to $3 million a year (e.g., *Oprah* for a daytime slot). Some command as much as $4 or even $5 million (e.g., *Friends, Home Improvement* for pre–prime-time hours).[11] Can a station afford such a show? Is it the right show for the job? To answer, the station must project the *cost per point*—in effect, how much money it would be paying for each audience-ratings percentage point the show brought in. Ratings over the two or three years of the "buy" would be predicted as accurately as possible, yielding an estimate of expected ad revenue.

"You take all of that gross revenue," says that Northwest station executive, "and you deduct things like sales commissions and promotion spots that you're gonna put against that program to make it successful. And then you say, 'Well, what do we want our [profit] margin to be on this program?' Some stations are willing to accept a 5 percent margin if they really want the program, because it's going to help put them on the map. Others will accept only a 50 percent margin." *Only* a 50 percent

Access A short term for *prime-time access,* the early-evening period established under "fin-syn" rules to give local stations access to evening air time.

profit? "Right. And then that dictates what license fee you're willing to bid for the program."[12]

The rise of station groups, with their superior buying power, raised the bidding on many programs. Off-network comedies, in particular, seem to earn huge revenues in syndication. By 2000, the famed situation comedy *Seinfeld* reportedly had earned $1.5 billion (yes, that's a *b*) in U.S. syndication alone. A few years later, the NBC reality show *Fear Factor* got off to a good start in syndication when the cable channel FX paid around $300 million to run the show for four years. The entertainment publication *Variety* noted that the dollar figure paled in comparison to what FX already had spent to air off-network dramas such as *NYPD Blue* and *The X-Files*.[13] With programs so costly and sure-fire hits pretty scarce, some station groups had decided by 2004 to start creating and syndicating their own shows.

Programming Radio

What matters most to those of us watching television or listening to radio is what our local outlets *do* with the programs they've produced or acquired. We want to see and hear electronic-media presentations that meet our needs, at convenient times, and in predictable rhythms. Arranging that for us is the task of *programmers*, the managers who specialize in placing certain shows into certain time slots around the clock.

British television viewers in the mid-1960s enjoyed—but, judging from the shortness of its run, did not enjoy *greatly*—a satirical BBC series called *Not So Much a Programme, More a Way of Life.* That title, whether with satiric intent or not, captured the mission of programming in a market economy: to forge a way of life that will attract audiences. Once they're tuned in, a programmer's next objective is to keep people from reaching for their remote controls.

BASICS

Radio is our oldest electronic medium, and its programming has helped to solidify and accelerate national change. In the 1930s, Irna Phillips and other writer-producers developed the daytime radio soap opera into a virtual addiction for housewives—one that greatly spurred U.S. consumerism. This arose in a time when advertisers were gaining greater control over broadcasting, partly by agreeing to **sponsor** entire programs. Researcher Marilyn Lavin has written that Phillips "adjusted story lines to meet the selling needs of her sponsors; she used soap opera characters as effective product spokespersons; and she designed program promotions to stimulate product sales."[14]

Today, with advertiser influence as ubiquitous as radio itself, the manipulation of shows and their scheduling to attract and retain listeners as *consumers* is universal practice. Once consumer attention is assured, even the old practice of full sponsorship, rather than of purchase time slots for ads, becomes feasible. For example, WCLV-FM, "Cleveland's only locally owned commercial FM station," has broadcast classical music for four decades and has originated Cleveland Orchestra concerts since 1965, building loyal listenership. The station in 1998 began broadcasting *complete* classical works every Monday. Who would sponsor a weekly classical marathon? Many an advertiser, as it turned out, since the charge for gaining five hours of access to a devoted audience would be just $2,750, the program's production cost.[15] Programming, like advertising, is a matter of dollars and sense.

One calculation—which programming fits best on *AM* (amplitude modulated) radio and which on *FM* (frequency modulated)—is over. Because of its superior sound quality, FM long ago became the radio band of choice for music lovers. There

Sponsor A person or company that pays for all ad slots in an entire broadcast.

Radio dayparts are weekday periods of several hours each in which established lifestyles put certain groups of consumers near their radios. This is how they line up (in Eastern and Pacific time zones; one hour earlier in Mountain and Central):

▶ *Morning Drive:* Running from 6:00 to 10:00 A.M., this period is when most people are on their way to work; their car radios make this the first daypart that draws large audiences, often to news or to music-variety programs with joke-cracking hosts.

▶ *Midday:* Running from 10:00 A.M. to 3:00 P.M., this is a time when people in offices may listen to light rock or light classical music, while homemakers and retirees are taking in radio talk shows.

▶ *Afternoon Drive:* The period from 3:00 to 7:00 P.M. is when people head home from work, tired and leaning toward sprightly pop music, talk, and news.

▶ *Evening:* The time span between 7:00 P.M. and midnight is when stations schedule jock/talker/hosts to chat or play music in a variety of formats.

▶ *Overnight:* These are the hours when night workers, insomniacs, and many college students search their radios for, among other things, edgy talkers and offbeat music formats. Like evening, the overnight daypart brings far fewer people to radio than do morning and afternoon drive dayparts.

#27

are some talk shows on FM, largely through public (noncommercial) radio, but AM commands most of the news/talk programming and sports on the air today.

RESEARCH

Early in 2002, the Arbitron audience-measurement firm released survey findings that formed a partial profile of U.S. radio listenership. It showed that, although virtually every home has radio, people with college degrees and household incomes of over $50,000 a year listen to more radio than do less affluent or less educated people. It also confirmed that adults do most of their radio listening in their cars, and mostly in morning or evening "drive time." That's good national information, but only a starting point for the local programmer.

#20

A radio station's first step toward building an audience is to select a format, a particular category of programs that will become the station's signature "sound" and provide a strong platform for promotion and ad sales.

Formats in Use

Formats in use refers to what the competition already has established in the market. If one station has done well with, say, an adult contemporary format, choosing the same format could touch off a wasteful struggle for listeners that a newcomer is likely to lose. A major Hispanic broadcaster reflected this when he told *Billboard* magazine that, in selecting a format for its new station in a fast-growing California market, his company would follow its standard blueprint: "We'll go in with no assumptions and find what's missing."[16] Still, a large city often has a number of stations using the same program format and jostling constantly for listeners. No initial format is set in stone; managers occasionally see some weakness in a rival station's programming and abruptly adopt the same format, believing they can move in and lure away part of the audience. In general, however, a station that finds, attracts, and maintains a sizable and loyal audience segment for its advertisers becomes a successful business.

#28

Most stations (or their group owners) continuously or repeatedly use audience or marketing **consultants,** surveys, focus groups, and other aids to acquire knowledge of the market. At many radio stations, recent downsizing has eliminated the program director and put the general manager directly in charge of programming. This links programming even more closely to overall business concerns and keeps

Consultants Professional advisors to broadcast stations; sometimes controversial for perceived negative influence on program quality.

the general manager's nose in trade publications that bring word of successful formats and strategies.

Television hits have helped to show that certain programming plays are almost certain to grab radio listeners, too. One such type is the "confessional" or highly personal show in which the stars or their guests expose to public view some of their most private moments or thoughts. Katie Couric of NBC's *Today*, whose husband had died of colon cancer, had her own 1999 colon exam televised on the show in an effort to spread health awareness; she also won ratings and a prestigious Peabody Award. In Los Angeles, KBIG-FM put the birth of disc jockey Leigh Ann Adams's baby on the air live and posted photos of the delivery on its website. Although the main purpose may have been promotional, the program director told of hearing from a listener who "had to pull her car off the road, so she could wipe away the tears" (women are the station's core audience).[17]

SPECIAL TOOLS

Once its format is in place, a radio station must set up a program schedule that will reach listeners and hold them for as long as they're listening in the car or at home. Whether a station has a music, news/talk, or other format, its minute-by-minute and hour-by-hour programming strategy is important. Not everyone is able to listen closely at all hours of day or night; music or even news or talk often acts as background to work or study activities, reducing the power of commercials. Sometimes, listeners' involvement in having dinner or managing children prompts them to tune out radio altogether. For these reasons, a programmer must understand and work with *time*, a fact that places something called the *hot clock* prominent in the arsenal of weapons to win the war for ratings.

A "Hot Clock"

Figure 7.3 displays the programming within any important segment of broadcast time, arranged around what looks like the face of a clock. There are many different ways to set up such a timepiece, and it can serve several purposes. One is to make visible the rotation of various types of programming around an entire 24-hour broadcast day. Another is to help a programmer see how his or her ideas might work to support the station's format within a given hour or period. Still another purpose of the hot clock is to show how well the station competes with its rivals in delivering certain features—blocks of music, news breaks, weather reports—and in showcasing commercials. Simply put, the hot clock shows how a station is doing its job, airing programs to suit appropriate consumers at times when they naturally gravitate to radio. Versions of the hot clock are useful in music and nonmusic formats.

Playlists

The *playlist* is a rundown of all the pieces of music featured in a music station's schedule (see Figure 7.4). These lists are useful to programmers evaluating their formats and comparing their "sound" with that of other stations. Playlists also help record companies discover how much air time their products are getting, and, increasingly, help corporate overseers fine-tune each station to reach its audience surgically.

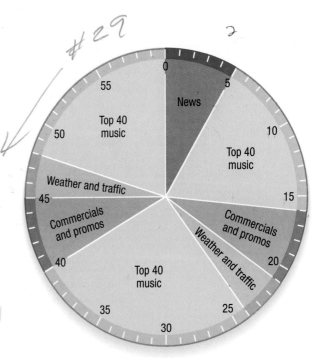

FIGURE 7.3 *A Hot Clock*
A "hot clock" is used to give a quick look at a radio station's programming plans. The clock can be designed to show an hour (shown here), day, or other time period.

Artist	Title
Big & Rich	Save a Horse
Jimmy Buffet	Hey Good Lookin'
Kenny Chesney	I Go Back
Billy Currington	I Got a Feelin'
Sara Evans	Suds in the Bucket
Josh Gracin	I Want to Live
Toby Keith	Whiskey Girl
Tim McGraw	Live Like You Were Dying
Brad Paisley	Whiskey Lullaby
Keith Urban	Days Go By
Gretchen Wilson	Here for the Party
Trace Adkins	Rough and Ready
Gary Allan	Nothing On but the Radio
Dierks Bentley	How Am I Doin'
Terri Clark	Girls Lie Too
Amy Dailey	Men Don't Change
Andy Gdggs	She Thinks She Needs Me
Alan Jackson	Too Much of a Good Thing
Joe Nichols	If Nobody Believed in You
Rachel Proctor	Me and Emily

FIGURE 7.4 *What Six Stations Play*
Above is the "hot rotation" playlist used by a small Midwestern station group in the summer of 2004.

ISSUES

Reducing playlists obviously reduces variety, as well; it already has taken much of the surprise out of tuning in stations at random. Now, some promoters of "satellite radio," which offers extensive music variety, are using the popularity of conventional-radio playlists as a sales point in their own favor. They provide hours of uninterrupted music from almost any genre.

Clear Channel Communications owns more than 1,200 outlets nationwide and, as of recently, a concert promotion company that helps the stations organize music events, plus an outdoor-advertising firm to give them wide exposure. Critics claim that Clear Channel's **vertical integration** of music-related businesses is sure to discourage dealings with outsiders, thus suppressing competition and reducing listener choices. Already the company pays a relative few disc jockeys to record programs for many of its outlets, homogenizing content across those markets. That's a sensible financial move but no help to radio's content diversity or to the "localism" of which the industry often boasts.

When a young radio show is cancelled, it's usually a clear sign that a programming decision went sour or a hotter prospect showed up. In Spokane, Washington, radio station KXLY-AM dropped the syndicated *Rick Emerson Show* barely six months after putting it on the air. According to a local journalist, "Few people listened, according to the station, and those who did only called to complain."[18] However, another factor also came into play: A more popular nationally syndicated show had become available as a quick replacement in Spokane. For radio personality Emerson, the incident held some irony. *His* show had been the replacement a few months earlier when another talk-based program was yanked from the air in Albuquerque, New Mexico, and other cities because of offensive language.[19] So it goes; such are the hazards of radio programming in a time of noisy, "edgy" competition.

▌ Programming Television

#24.

Unlike most radio stations, which must find content wherever they can, most television stations today are linked to networks. ABC, CBS, and NBC send programs to their affiliates that can fill all or most of a long broadcast day. Four newer networks—Fox, UPN, WB, and PAX—provide less than a full schedule of shows to their stations. Regardless of the amount, whatever the networks supply to local broadcasters is almost sure to have greater audience appeal than they could manage on their own. That's why a network's programming philosophy and practices are of vital concern to all.

BASICS

One evolutionary note may be revealing: Under the eyes of their cost-conscious owners—and because the business has changed—many TV stations that once had full-fledged program directors and staffs have retrenched. Many a general manager does all the programming personally. It's relatively simple these days: Most larger TV stations produce lots of news and a sprinkling of other local programs, then plug

Vertical integration Ownership by one company of all steps from production to distribution.

a few syndicated shows into key schedule slots. For the rest of the long broadcast day, they run the networks' main programs just as the networks scheduled them to run. That's how the "nets" want it; if their nationally promoted programs come into viewers' homes in every city with rigid regularity, huge audiences congregate simultaneously. With that, the confidence of major advertisers is maintained and the ad revenues keep flowing in. Stations benefit from this as well, since they sell advertising within network shows.

A Network's Burden

This dynamic puts great responsibility on the shoulders of a network's programmers, who usually report to a vice president or even a division president. He or she controls many millions of dollars in program expenditures—and must choose well from among program ideas. Networks don't just buy finished shows; they pay to have important shows developed by outside producers. Any big mistake can waste production money and cause a dip in ratings that might take months or longer to correct. Two or three flops in a season can taint a whole schedule, depress the rates a network can charge for commercial minutes, and damage its potential to attract future advertisers.

Thus, TV networks take few real programming risks unless someone else has taken them first. Nothing better illustrates this caution than the late 1990s arrival of network "reality" shows.[20] The Big Three networks kept a lofty distance from the genre until it broke into CBS through *Survivor.* By then, even once-maligned Fox had loudly criticized reality shows, but almost no one was backing away from them. The size and loyalty of reality audiences were too powerful to resist.

NBC reportedly paid $13 million per episode for its long-running hospital hit *ER.*[21] That kind of spending required programmers to make sure that an already good show *stayed* good at bringing audiences to advertisers. Bullet-quick feedback on gross ratings and—more important—on viewer demographics tell a programming executive whether to order more episodes of a show as it is, have it retooled to increase its appeal, or put the program out of its expensively unwatched misery.

Targeting Viewers

Programming obviously plays the lead role in establishing and maintaining a network or station's image among its desired viewers. This can be a tricky undertaking. Sometimes, when trends already have turned negative, it involves "damage control." CBS, for example, has a somewhat older audience than the other major networks—not a happy fact to advertisers who want to reach young adults and their families. How to remedy this?

The most obvious approach is to schedule shows filled with youthful actors in comic settings or dramatic predicaments. However, that could alienate the stable, loyal older audience on which many of the network's advertisers depend. Indeed, CBS aimed a 1995 show called *Central Park West* at younger viewers, but it failed to reach them and practically everyone else. In 2000, the network's fortunes began to change with the launch of *Survivor,* a reality show. Younger viewers flocked to it. The average age of CBS viewers dropped, while that of *other* networks' audiences rose. What once was regarded as the staid but top-flight "Tiffany Network" came out of the shadows to beat its rivals in several ratings periods. Older viewers mostly stuck around. CBS suddenly had become the one to watch.

On another level, networks that reigned alone over broadcasting for many decades have yielded ground to spunky new rivals. A company that starts with a modest slate of shows and works systematically to appeal to a key audience can find success. In the late 1990s, the WB (Warner Brothers) network trained its sights on women ages 18 to 34. It launched youth-slanted *Dawson's Creek* and *Buffy the Vampire Slayer;* then, when each show had secured a solid following of young women, the WB added *Felicity,* a series about a college student's troubles and triumphs.

protalk

Helen Little

CO-CHAIR, DANGEROUS ENTERTAINMENT

H elen Little embodies close relations among different segments of the modern music industry. She went to college hoping to do research on sleep and dreams—an offbeat ambition for a person who someday would be charged with keeping people awake, not snoozing. Perhaps fortunately, she dropped her psychology major, switched to broadcasting, and today looks back on a two-decade record of radio programming that lifted her to the top of her field.

After helping to run some big-market radio stations—in Dallas, New York, and Philadelphia—Little became brand manager for Clear Channel Communications, the country's largest radio-station owner. She worked with programmers in San Francisco, Chicago, and Detroit.

Moving into music production, Little served as president of U.S. operations for the hip-hop label RuffNation Records in a joint venture with Warner Music Group. Next, she became co-chair of a management and consulting group called Dangerous Entertainment.

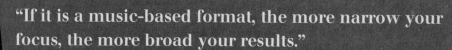

"If it is a music-based format, the more narrow your focus, the more broad your results."

She has played a large part in *producing* "urban" music, often working directly with R&B and rap groups. One executive credited her with having "an all-encompassing, across-the-board knowledge of the ever-changing urban landscape."[22] Back in her days as a radio programmer, Little had shown her savvy and experience by adapting the long-lived "Top 40" format to urban music, keeping favorites constantly in the mix.

"If it's a music-based format, the more narrow your focus, the more broad your results," she says, noting a simple truth of popular music: "There are not but a few songs that people want to hear."

Soon, a 30-second commercial during *Felicity* was costing the advertiser more than $100,000 (based on what's called CPM—the cost per thousand viewers to whom an ad is exposed). This was vital revenue for the WB, since each episode of *Felicity* cost about $1.3 million to produce—only one-tenth as much as an *ER* installment for NBC, but a great deal of money to a fledgling network.[23]

SPECIAL TOOLS

Television programmers have developed a toolkit of tactics that's a good deal larger than radio's bag of tricks. Some of them involve the best placement of a single show in a schedule in order to reach certain viewers. Others focus on arranging several hours of programming to boost audiences for multiple shows, which, in popular parlance, will "lift all boats." Still others are aimed simply and directly at knocking the competition off a single half-hour perch.

Head-to-Head Programming

If a rival is getting high ratings from a particular program, a station may be tempted to schedule a similar show directly against it, at the same time in the schedule. This is not subtle; it's often called *blunting.* For example, CNN personality Greta Van Susteren jumped to Fox in 2002 to begin anchoring a program opposite CNN's main anchor, literally a face-off. Stations routinely schedule head-to-head local newscasts, which often leaves them scheduling their networks' newscasts head-to-head, as well.

FUN WITH NEWS. Comedy Central's *The Daily Show with Jon Stewart* attracts a desirable demographic. Stewart's witty takes on the news have earned him serious respect and journalism honors.

Like radio, television organizes its broadcast day into dayparts. However, TV's clock is more complex than that of radio. Here's how programmers usually divide a 24-hour day (times stated in Eastern and Pacific zones):

► *Early morning* runs from 7:00 to 10:00 A.M. and encompasses network morning news-interview shows, often augmented by similar local shows.

► *Daytime* is a long stretch of hours, from 10:00 A.M. to 5:00 P.M. Its programming includes network (and syndicated) talk shows and soap operas aimed primarily at homemakers, as well as some children's programs in late afternoon.

► *Early fringe* lasts two hours, from 5:00 to 7:00 P.M., during which networks and stations typically air their early-evening newscasts while cable channels offer alternatives.

► *Prime access* runs from 7:00 to 8:00 P.M. and is usually filled with syndicated light entertainment or local programming. The term *access* is a legacy of the FCC's 1971–1996 prime-time access rule (PTAR), which pulled networks out of the 7:00 to 8:00 P.M. hour and granted stations the right to program it.

TELEVISION DAYPARTS

► *Prime time* is TV's crucial evening block from 8:00 to 11:00 P.M. This is when the networks trot out their strongest comedies and dramas (family-oriented first, then more adult), as well as newsmagazines.

► *Late fringe* is half an hour, from 11:00 to 11:30 P.M., and is almost always occupied by local newscasts except on non-network stations and cable.

► *Late night* is for viewers still up between 11:30 P.M. and 12:30 A.M. It offers mostly talk shows, comedy, and fiction.

► *Overnight* on TV stations and cable channels of all sorts is an electronic Lazy Susan in which the viewer who's awake between 12:30 and 7:00 A.M. can find almost anything. Networks have attempted, with mixed results, to succeed with night-owl newscasts during this period.

FRIENDS. This show about a closely knit group of young adults was such a huge hit that it was used to boost neighboring shows through "hammocking" and other techniques. Its six stars were earning $1 million per episode by the time *Friends* left the air in 2004, ending a 10-year run.

Counterprogramming

#30

When a network or station sees that its competitor will be airing a blockbuster show—such as the Super Bowl—the only answer may be to go after a different audience. Ratings rivals for years have run female-oriented shows against *Monday Night Football*. That's called *counterprogramming*. Movies or old episodes of a strong drama series might draw people who care little about sports.

Hammocking

Sometimes a show that has yet to find its audience can draw strength from the programs that surround it. Putting a weak program between two stronger ones, hoping they'll pull it up in the ratings, is called **hammocking**. NBC tried helping a new show called *Jesse* by scheduling it between the hits *Friends* and *Frasier*. Early ratings indicated that most *Friends* viewers were hanging around for the new show, but later they drifted away and the hammock collapsed in the middle; *Jesse* was cancelled. Some observers are skeptical of hammocking, believing that impatient viewers with remote-control zappers simply won't linger through a so-so program.

> **PRINCIPLE 6**
>
> Programming success depends not only on selecting shows but also on using them in the best ways possible.

Tentpoling

Doing the reverse of hammocking—that is, placing one strong show between two weak ones, hoping to pull them up—is known as **tentpoling**. Sometimes audiences tune in well before a highly popular show begins and will hang around after it's over. The young-adult series *The O. C.*, on Fox, showed so much strength among prized young-adult viewers that it was used to help surrounding shows. The search for programs strong enough to hold up an evening's prime-time programming tent is continuous.

Hammocking The placement of a weak program between two strong ones.

Bridging

Tentpoling The placement of a strong program between two weak ones.

When a program routinely ends a minute or more past the half-hour or hour mark on the clock, chances are the programmer behind it is engaging in *bridging*. This

#31

technique aims to hold viewers in place long enough to make them miss the start of a program on a competing channel. Local news departments sometimes schedule a strong feature story or personality to run from, say, 6:28 to 6:32 P.M. That can freeze some of the remote-control zappers that otherwise might have jumped to another channel at 6:30. A zap postponed is often a zap forgotten.

Block Programming

When a broadcaster strings together similar programs, they sometimes hold particularly interested viewers in place for hours. Prime examples of *block programming* show up on autumn weekends, when football fans may watch pregame shows, games, postgame shows, and then newscasts featuring game video. Soap operas march through daytime in similar fashion and for the same reason: to mesmerize loyal fans so that advertisers can reach them easily over an extended period.

Stunting

Networks and stations *stunt* by breaking the mold. This can mean showing unusual attractions, or routine attractions at unusual times, to draw viewers away from competitors' favored programming. Any heavily promoted twist can have that effect. When the script of a drama series suddenly makes room for a quick visit from a hot movie actor, that's stunting. When a popular show suddenly moves one hour later to challenge a rival during a ratings "sweeps" period, that probably is short-lived, and it's stunting. NBC in 2001 briefly considered putting a celebrity-studded reality show against the sweeps-month premiere of a heavily promoted Fox series—a stunting idea. It died when NBC realized it had too little time to do its own promotion.

Stripping

Most often used with syndicated shows, *stripping* entails scheduling an episode of the same series—comedy, drama, game show, other—at the same time five days a week. Stations typically strip daytime talk shows and adventure programs such as *Xena: Warrior Princess*. Network programmers may decide to strip especially popular programs, such as *Who Wants to Be a Millionaire* (before it faded, partly from overexposure). Cable channels endlessly strip old "off-network" shows; some specialists like the Sci-Fi Channel had little choice at first if they were to fill their schedules.

THE ROLE OF CONSULTANTS

Programmers look for help wherever they can find it, and sometimes that means outside experts. A company called Frank N. Magid Associates, founded by a social psychologist in 1957, launched a subindustry of *TV consulting* that has advised stations and networks about their programming ever since. The central objective almost always is to increase the size and/or demographic desirability of audiences. Consultants have injected viewer-based thinking, derived from surveys and focus groups, into decision making about TV programs as historic as NBC's morning show, *Today*. National operations continue to use consulting; for example, Fox News in 2001 signed a contract with a veteran audience researcher to bring his expertise to the Fox News Channel and other offerings. However, since networks are inclined to employ their own research directors, most of the consulting firms' work is focused on local stations—especially their news programming.

After a New York news director named Al Primo launched his *Eyewitness News* concept in the late 1960s—putting street reporters on camera speaking directly to audiences—consultants helped to propagate Primo's seminal approach. They also spread the use of light "happy talk" between anchors and other attempts to keep news entertaining. In recent decades, TV consultants' techniques have raised ratings—but also eyebrows and, sometimes, hackles—in hundreds of towns and cities across the United States. Under names like Audience Research and Development (AR&D) and Broadcast Image Group (BIG), researcher-consultants have made norms and routines out of what once were experiments in audience appeal. This keeps the focus on satisfying viewers, a plus to them even if it often is a minus to broadcast journalists who find consultants likely to thin and even distort news content.

An important role that consultants can play—helping station executives decide exactly what programs to acquire—has been greatly broadened by changes in the network-affiliate relationship. As noted in Chapter 1, local stations are the only true broadcasters, and as such have been protected from total network dominance partly by the interventions of regulators. The result: Stations that once had to survive on their own newscasts and whatever their networks fed them for other time slots are now free to buy shows from almost anyone. With a growing marketplace of independent producers and syndicators spread before them, local broadcasters can thus provide their viewers with much more diverse content. Another effect, of course, is the continued weakening of networks' power and primacy in an age of multichannel competition.

PRINCIPLE 7

In an open society, program content that attracts some listeners or viewers is vulnerable to desertion by others.

ISSUES

Television draws complaints on many grounds—most having to do with what is or is not visible in its programming, which is the medium's public face. Some objections or demands come from people who believe their lives and concerns are distorted or ignored on television, and are aimed at the writers and producers of network shows. Critics also may attack the local managers who decide whether and when a controversial show goes on the air in their communities.

The environment in which television operates has changed greatly since its earliest years of national exposure, when viewers mostly seemed grateful that at last it was available to them. Often ugly social clashes in the 1960s over civil rights for African Americans, the Vietnam War, and the demands of women for expanded opportunities drew new scrutiny to programming. Today, everything people see on "the tube" is vulnerable to criticism that can blossom into national campaigns against specific program content. Issues include those mentioned below (some of them also addressed in Chapter 9, The Audience).

Children's Viewing

Many people, particularly parents, are concerned about images and messages that enter children's minds through the television they watch. Of greatest concern has been rampant violence in all sorts of programs—even cartoons—shown during hours when children are likely to be watching. More than 1,000 studies, according to the American Academy of Pediatrics, have linked TV violence to subsequent aggressive behavior in children.

Television already offers guidance to parents through program-content ratings; efforts to *impose* controls on TV violence have foundered on First Amendment guarantees of free expression. Still, goaded by powerful lawmakers, the FCC in mid-2004 called for public (and expert) comment on the idea of a government-imposed standard on TV violence. The commission asked "how such a standard could be implemented in a manner that is both clear to the industry and practical to administer." Free-expression advocates immediately objected, and a new battle was joined.[24]

On another front, a New Zealand study indicated that two hours a day of TV-watching by people between 5 and 15 years old makes them fatter, less fit, and likelier to take up cigarette smoking.[25] Organizations including the Parents Television Council and Action for Children's Television have led consumer battles on such issues, which also include the display of junk-food commercials during kids' shows.

Diversity

Some critiques of TV programming point to the obvious and perennial shortage of ethnic-minority members in prominent on-air roles. This complaint spans genres from fictional shows to newscasts. It leads to questions about hiring and promotion practices of networks and stations, on the theory that more minority power within the industry would mean more diverse TV programs and more diverse roles for actors of color. The National Association for the Advancement of Colored People (NAACP) puts constant pressure on broadcasters to diversify both their payrolls and their programs.

Success has been mixed. On one hand, African American actor Dennis Haysbert won wide attention as the President of the United States in an FX suspense series, *24*. A few standout minority comedians have broken out with their own shows; some have gone on to good roles in Hollywood movies. But most TV comedies and dramas continue to feature African American, Asian American, and Hispanic American players only in secondary roles.

Gays and lesbians campaign vigorously for parity on (and within) television. Since comic actress Ellen DeGeneres made history in the 1990s with her own situation comedy, others have made progress in her wake. A light-hearted but groundbreaking "makeover" show on the cable channel Bravo, *Queer Eye for the Straight Guy*, had trouble winning major advertisers—until its ratings began to soar and it won a big, diverse audience. The media giant Viacom planned a channel named Logo for a 2005 launch; among its early offerings would be a series, *My Fabulous Gay Wedding*.

In the no-show category was *Seriously, Dude, I'm Gay*, a projected two-hour Fox "reality" special in which two straight men would have competed for $50,000 by trying to pass themselves off as homosexuals. Among advance critics was the Gay and Lesbian Alliance Against Defamation, which called the special "an exercise in systematic humiliation." Fox dropped the idea. (Working positively, as well, GLAAD annually honors TV programs that combat homophobia.)

Indecency

February 2004 was a tough month for television efforts to push the limits of public and political acceptance. A scene in the soap opera *The Guiding Light* that showed a woman pulling down her boyfriend's underpants, baring his bottom, appeared to be the reason for a producer's abrupt firing, although CBS denied it. That same month brought the assertedly accidental exposure of singer Janet Jackson's left breast during halftime of the Super Bowl game. Among many millions of viewers at home watching, with their children, were members of the Federal Communications Commission.

Soon, outrage over Janet's "wardrobe malfunction" started an FCC crackdown and a national debate over programming. Both houses of Congress passed legislation sharply increasing fines for on-air indecency. Much questionable content appeared to be constitutionally protected but was offensive to at least some TV viewers. That included explicit sexual movements by actors in episodes of *Will and Grace* and *Buffy the Vampire Slayer*, both of which faced complaints before the FCC and survived them.

Lawyers said they were confused by the FCC's "flip-flopping" in indecency cases. The rules prohibited airing sexual or obscene material between 6 A.M. and 10 P.M., but there was leeway as to what was indecent. So the issue was left to be resolved case by case by the FCC, even as parents, politicians, and broadcasters made their own decisions about programming. Some radio stations cut their ties to offensive "shock jocks," and a popular commentator was fired for using the "F word" on a Los Angeles station. The court of public opinion can be stern.

protalk
Jim Lutton

GENERAL MANAGER, WSTM-TV, SYRACUSE, NEW YORK

Jim Lutton knows programming. He's been doing it for more than 30 years at TV stations in large and middle-sized cities. He's even called the tunes for large station groups. One thing he knows is that many people don't like to see much change in television. They have habits, they like them, and broadcasters should be grateful for it. "That's why the most successful syndicated shows in history have been *Wheel of Fortune* and *Jeopardy,*" Lutton says. "They're straightforward. They don't need a lot of promotion."

Promotion—including those loud and flashy little blurbs between commercials and programs—often goes unacknowledged, but for some shows, it's a necessary and powerful factor. As examples, Lutton cites two highly popular syndicated programs dealing with entertainment. "If we run *ET* [Entertainment Tonight] or *Access Hollywood,* we have to promote it topically," he says. "That's because every day brings a new audience. If Brad Pitt is taking his shirt off at the Cannes Film Festival, we have to get the word out.

"Prime time, on the other hand, changes a great deal. New shows come in, and soon people are watching in a very active way. . . . They figure out what the good programs are on the networks; they find what they like. People recommend a show—

> "People recommend a show . . . and if you haven't watched it, at the water cooler the next morning you feel deprived."

maybe *Survivor* or one of the other cult shows—and if you haven't watched it, at the water cooler the next morning you feel deprived."

When it comes to television news, adds Lutton, many viewers like to stay put: "People don't 'sample' news shows too much. They've usually been in the market too long. They follow an anchor, or they've just decided that somebody in town has the better news. News audiences move a lot less than some other audiences." So do older people in general: "Court shows were a big staple a few years ago—*Judge Judy* still is—and they got fairly big audience numbers in large markets. But the demos [demographics] were almost all [age] 50-plus individuals.

"In this PC environment, we don't talk about that much, but advertisers want viewers [age] 18 to 49."

Some of the most serious challenges to broadcasting's status quo come from advocates for more ethnic diversity in programming. Most prominently, the NAACP has pushed for more opportunity for minorities to own and develop programs, changing the face of television from the inside. In January 2002, a year after joining agreements for change with the four biggest networks, NAACP president Kweisi Mfume complained of slow progress and threatened economic sanctions against television.

Programming is the public surface of radio and TV, and as such draws the sharpest barbs aimed at them. One of the earliest and most penetrating came back in 1961 from an FCC chair, Newton Minow, when he called nightly television a "vast wasteland." Although TV occasionally has reached great heights since then—the landmark 1977 mini-series called *Roots* and the early coverage of the 2001 terrorist attacks—much of its programming still seems expendable, and millions have stopped watching. With more and more new competitors for Americans' leisure time, the giants of commercial electronic media may have to become more responsive to critics. Whether these companies will respond fully to anyone other than audiences is the big question.

Into the (Digital) Future

Most of this chapter has focused on traditional broadcasting, but there's a far larger story now being written. Digital technology is revolutionizing what broadcasters can do with their programming hours. It also is encouraging people who hope to find information and entertainment on-line that blends all the virtues of broadcasting and the Internet.

Since 1997, pushed by Congress, broadcast and cable TV have been labeling every program (except news) according to the potency of its sexual and violent content. This rating shows up in the upper left corner of the screen during a show's opening moments, giving concerned parents a chance to change channels or divert their children from the set.

Besides giving a visual cue, the rating signal on the screen interacts with the V-chip device that the FCC ordered included in virtually all new TV sets starting in 2000. The V-chip allows parents to set a rating level, and then all programs that have received that rating will be blocked out of their home. Here's how the rating scale looks:

> #32

Rating	Appropriate Audience
TV-Y	All children (content is for the very young)
TV-Y7	Older children (content may be mildly violent or frightening)
TV-Y7-FV	Older children (fantasy violence may be more intense or combative than in other TV-Y7 programs)
TV-G	Anyone (nothing objectionable)
TV-PG	Parental guidance suggested (some foul words, mild sexuality)

sidebar

A DIGITAL TOOL FOR PARENTS

TV-14	Parents strongly cautioned (too bloody or racy for many kids)
TV-M	Mature audiences only (too much of everything for those under age 17)

In addition to this scale, the TV-screen rating box displays one or more letters of the alphabet to tell viewers and parents a little more precisely what types of risky material may be present in a show. **D** is for dialogue that may have sexy undertones; **S** is for sex; **L** is for language that is probably offensive; **V** is for violence, and **FV** is for fantasy or cartoon violence. These symbols, too, can be programmed into the V-chip.

However, while the guideline codes may be helping parents who spot them on the TV screen, the V-chip has failed to catch on with many. Studies have shown that most viewers don't even know whether their sets contain V-chips, and that only a small percentage of those who do know about the devices take the trouble to use them. Having to program the chips confounds many adults; as a researcher put it after one test, "It was just maddening for the parents."[26]

TELEVISION APPLICATIONS

Visual images and audio in digital form resist interference, distortion, and deterioration as they pass from point to point. This is a great advantage to satellite and digital-cable companies as they bring new channels and new types of programming into millions of U.S. homes. Digital recordings of radio and TV content will not literally fall apart as they age, a great weakness of traditional video- and audiotape. Television in the digital age will more effectively expand and preserve our popular culture—and already has begun.

High-definition television, although still spreading rather slowly, is providing some consumers with the clearest video in history. It also promises to transform how we view and experience TV programming. One big factor is the shape of the screen; rather than a traditional 4:3 "aspect ratio," in which every four inches of screen width brings three inches of height, HDTV sets offer a 16:9 picture. The screen is much *wider* than it is high, meaning it will cry out for new content that's framed in that expansive way. It also means producers and directors will have room to give fantasy and reality greater spatial dimension and perspective than we see on television now.

Besides expanding the range of programming content, the digital age has begun to challenge and sometimes disrupt some of the arrangements undergirding the broadcast industry. Digital video-disk recorders now are in many homes, playing movies recorded in DVD format. Viewers can use digital tools to watch TV programs while fast-forwarding commercials (hardly a happy development for the retailers who keep broadcasting afloat). The new recorders also enable viewers to "time-shift" easily—that is, to watch programs at any time and in any sequence they wish. So much for the carefully crafted sequences of shows on which programmers build their reputations and sales departments sell advertising time! Digital technology makes viewers into programmers *themselves.*

RADIO APPLICATIONS

Digital technology also will influence radio programming, although perhaps not as dramatically as it affects TV. Stations that for years have played music from digital compact disks (CDs), but had to broadcast it along nondigital paths to the listener, have begun to improve on that. The industry and the FCC promise that digital radio systems will send out sound just as clearly as it was recorded—with the quality of CDs or, says the commission, the acoustics of a fine concert hall.

DIGITAL VIDEO RECORDERS. TiVo, the best-known DVR, gained converts at the 2005 Consumer Electronics Show—and in millions of homes. Viewers use DVRs to time-shift shows and fast-forward commercials; this disrupts the sequencing of programs and makes ads ineffective, worrying the TV industry.

This could prompt some already daring stations to experiment with programs in which delivery of near-perfect sound is crucial. This might include educational performances of unfamiliar genres of music, with unfamiliar instruments; field adventures by documentarians and by naturalists using their (digital) microphones to capture wildlife sounds; and the coverage of spontaneous news events such as protest marches in which faint voices might carry important information.

All of that is an optimistic reading of digital radio's programming potential. In recent years, most experimentation has been dampened by cost-cutting across commercial radio. Still, even a pessimist will recognize that live music performances in particular will become more "real"-sounding through digital delivery. (Other currently popular programming is less certain to benefit; indeed, some talk radio might suddenly be heard a little *too* clearly.)

Clear Channel Communications, often accused of homogenizing radio, announced in 2004 that it was joining with iBiquity Digital Corporation to speed development of high-definition broadcasts at 1,000 (of its more than 1,200) stations. This could help Clear Channel in two ways: Its stations could begin sending ads and other messages to listeners along digital paths even while music was playing. And a digital leap would help radio stations compete with the digital sound already being transmitted by the space-relay audio providers XM Satellite and Sirius.

PROGRAMMING THE INTERNET

A time when the Internet might bring entertainment to computers much as TV delivers it to viewers is edging—no, walking—ever closer. One reason is that many people are rushing to acquire high-speed broadband connections, showing they want no barriers to whatever content becomes available. Broadband's size as a "pipeline" will allow computer users to move past the current "streaming" video technology, with its jerkiness and glitches, to Internet images about as fluid and clear as what TV now delivers.

Some study groups reported that between 2003 and 2004, the number of people using broadband had pulled ahead of the number still using "dial-up" connections. Besides facilitating better video delivery, this transition could speed the development of interactive programming on the Internet—that is, shows that viewers could change or at least influence *while they're "on the air."* In addition, depending on how entertainment is funded and organized, consumers might escape the tyranny of broadcasting's traditional time schedule, which measures shows by the half-hour or hour.

So far, most of the messages traveling the Internet continue to originate from individuals, which supports the notion that it can have a democratizing influence on society. However, big media companies are moving into position to attract people away from the randomness of the present toward better-produced material not too different from radio and TV's packaged shows. As we wait to see if high-quality entertainment is headed for a committed relationship with the home computer, both industry and education seem to be playing Cupid:

▶ Nickelodeon, the child-oriented cable channel, announced that some of its top TV attractions were about to show up, interactive, on two new websites. (Kid-friendly quote from the announcement: "Fans of the hit shows can play undersea shuffleboard with SpongeBob SquarePants and surfer-squirrel Sandy Cheeks.") This approach to young minds could guarantee future on-line audiences.

▶ Some companies had already been taking rental-DVD orders over computer lines when the Starz Encore satellite-movie firm began letting customers *download* movies at their computers. A user could choose from more than 100 titles at any one time and then, using broadband, download a film in as little as 20 minutes.

career focus

PROGRAMMER

Programming jobs in radio or television provide fascinating opportunities to influence the daily menu for listeners and viewers. However, careers in programming don't come easily. It takes years of experience to acquire the skills and insights to do the work consistently well. Most professional programmers have worked in other broadcasting jobs on their way up, learning how TV and radio affect audiences and what it takes to attract them.

Some universities include programming courses in their broadcasting sequences. Students should take a very broad view of the industry and learn a variety of skills before hitting the job market. Here are excerpts of one "help wanted" ad for an entry-level position in television:

PROGRAMMING ASSISTANT: *Work under the direction of Director of Programming as assistant, help to find, screen and evaluate appropriate programs for broadcast. . . . Enter & update program offers to (tracking software.) . . . Research distributors for new program daily using Internet. . . . Maintain an organized and complete file for every broadcast program. . . . Screen & edit programs for language & content. . . . Research new programs with direction from Director of Programming via Internet and other program sources . . . .*

Qualifications: *Must possess excellent word processing, spreadsheet and database management skills. . . . creative thinker. Able to juggle multiple assignments. . . . Pleasant and professional phone manner. . . . 4-year college degree, or 2 years with equivalent experience in professional office setting. Access to own transportation . . . . You are on call in the evening, weekend (and many other times—ED.)*

Our excerpts of that job ad mention only intensive program-related work, but the ad also specifies that tape/disk management and production duties are part of the job. That's good: The more varied the experience to be gained, the better. That's why certain other jobs would provide good background for a programming career.

The ambitious worker who rises through the system may someday encounter a much higher-level opportunity, in an ad worded like the one below, which gets right to the point. This radio station wants a bigger share of the local audience:

PROGRAM DIRECTOR: *How do you turn a potential 6 share into a 12? If you understand that a radio station must be about MUCH more than just the music it plays, then you're the person we're looking for . . . the next program director for (XXXX)-FM. . . . You should be exceptional in these areas: imaging, street presence, team building, & morning show evolution . . . . Understand the importance of following a strategic plan, but be insightful enough to bring needed changes to the table when necessary. Prior Country experience would be helpful, but not necessary.*

That ad says nothing about years of experience being needed, but clearly it's not aimed at radio rookies. A program director's job, in radio or television, can put a college graduate on track to become a top executive of a station or station group. But again: People who shun new, unfamiliar challenges are unlikely to rise to the upper levels of the industry.

▶ Reversing the flow, the cable music channel VHI scored strong ratings in the key 18–49 age group—strongest of all among young males—with a show adapted from the sardonic website *CampChaos.com*. VH1 called it "the first time a successful Internet entertainment brand has migrated to television." Programs traveling so flexibly were likely to continue blurring the lines between older media and the digital world.

▶ Top universities made it easier for students to obtain music—probably their most consistent form of entertainment—free over the Internet. Yale University planned to let students download 700,000 songs legally for about $2 per month. Whether ending up at portable iPod devices or at desktop computers, such downloads make everyone a programmer—self-producing music shows and competing with radio.

Summary

Programming is the face and voice of broadcasting, and has begun to play similar roles on the Internet thanks to digital advances. The basic challenge is to create and arrange content that will attract and hold audiences. Backed by research and graded by audience ratings, programmers use a variety of techniques to reach listeners or viewers. Their work is exhibited not only through their affiliates and their vertically integrated corporations but also through a wide-ranging distribution market known as *syndication*.

Programs' success determines the economic health of their producers and contributes much to people's decisions about how to use all electronic media.

Food for Thought

1. If you download music or spend hours daily surfing the 'net, is it because of the *technology* more than the content? What kind of programs might increase your time spent with traditional radio and TV?
2. How much does it matter that only a tiny percentage of all electronic-media content is generated *locally*, where you live? Do you think U.S. consumers care much anymore about local programming? Support your answer.
3. Based on the job ads shown in Career Focus, and other factors, would broadcast programming be a good career for you? Why or why not?
4. If you watch 10 hours or more of television a week, to what shows or program types do you keep returning? Why are they important to you?
5. Do you believe parents who ignore program content ratings (TV-Y, TV-PG, etc.) are failing their children? To what degree do you think television influences personal behavior or social conditions?

Broadcast News

> *He who laughs last has not yet heard the bad news.*
>
> —BERTOLT BRECHT, *German playwright*

> *Radio news is bearable. This is due to the fact that while the news is being broadcast the disc jockey is not allowed to talk.*
>
> —FRAN LEBOWITZ, *humorist*

Chapter 8

It was the end of April 2004, and Ted Koppel couldn't buy a break.

Did he deserve one?

The celebrated TV journalist was about to air a *Nightline* show called "The Fallen," reading aloud the names of U.S. war dead in Iraq and showing their pictures. The idea had old roots: In 1969, *Life* magazine had run pictures of the 200 troops killed in just one week in Vietnam. Now, 35 years later, what was the Koppel program's central message?

"Just look at these people," he replied to an interviewer from the Poynter Institute for Media Studies. "Look at their names. And look at their images. Consider what they've done for you. Honor them."[1]

During 41 years at ABC News, Koppel had aroused sporadic criticism with controversial reports. Now—promising to show Americans their entire inventory of human losses in the Iraq war and its aftermath, in one sitting—he drew vehement criticism on political grounds. Some suspected him of aiming "The Fallen" at President George W. Bush's Iraq policy: "There's only one goal in mind: It's to turn public opinion against the war," said the president of the conservative Media Research Center. The Sinclair station group agreed and said its stations would preempt the show.

Others wondered about Koppel's timing: The "Fallen" would air in an important ratings "sweeps" period. After *Nightline* executives denied even having realized this,[2] a columnist wrote: "Imagine, nobody at ABC News stopped to think that telecasting this thing on the second night of the May sweeps might appear like an unseemly sweeps ratings grab."[3]

To suggest this was to hint that one of the nation's most prominent broadcast journalists might be looking cynically for gain from simply counting off Iraq casualties. "I'm somewhat surprised that anybody would object," insisted Koppel's executive producer. But might the show undermine support for the war by focusing on U.S. losses? Perhaps, but the executive director of the American Society of Newspaper Editors said: "It's a very important part of the coverage. And I don't think its political. It's reality."

The program aired as planned, and as this *Nightline* episode passed into history, a loud and caustic political environment—greatly amplified by the news media—kept the national pulse thrumming fast. Koppel had used his late-night hour to stimulate an already vigorous national war debate. One year later, its ratings down, *Nighline* faced imminent revamping or cancellation.

Now let's consider how and why broadcast news is what it is to millions—and the threats to this modern electronic institution.

▜ How News Matters

Americans first snapped awake to the miracle of news entering their living rooms over radio waves in the long-ago year of 1916. That's when a talented inventor named Lee De Forest transmitted election results from his primitive New York station to perhaps several thousand listeners. By 1920, Pittsburgh station KDKA was cutting a deal with the *Pittsburgh Post* to broadcast its news of election returns. These developments themselves were news because they were history. They showed that technology could transmit far and rapidly the kinds of information we had passed around slowly, from person to person, for thousands of years.

AT ITS ROOTS

Broadcasting basically is a process by which other humans give us the news *by word of mouth;* speech still seems essential to most communication. The modern electronic media thus have roots running back to ancient **oral cultures;** we can look to the "town criers" who shouted the news through England and colonial America. Some early news broadcasts were not much more sophisticated—just readings by radio actors using quite formal-sounding English. Soon enough, however, journalists with more of a man-in-the-street style were found or developed. Also, broadcasters learned to "write for the ear," meaning to break complex news stories into short, simple sentences that seemed almost *conversational.* Radio reports could therefore help the listener's brain create vivid images of events. All of this made news accessible and appealing to a wide range of listeners, turning radio into an all-American medium.

The correspondents assembled by CBS before and during World War II won particular notice for the clarity and directness of their reporting. Edward R. Murrow, broadcasting from the bombing blitz of London, set the standard. In waves of staccato phrases that sounded almost poetic, Murrow painted word pictures of war and suffering that carried the power of the European catastrophe through the listeners' ears and into their hearts.

PRINCIPLE **1**

The appeal of broadcast news rests on humans' ancient need to *tell* one another of the day's events.

The subsequent growth and branching of broadcast journalism turned it into a vibrant subindustry of the electronic media. It always has been subordinate to broadcasting's main product—entertainment—and has swollen or contracted depending on ad revenue. However, as a practice and institution, it spans the nation, and today, U.S. radio and TV news operations provide livelihoods for about 20,000 on-air employees and thousands more behind the scenes.[4]

IN MATURITY

Not all news employees today have to take the sort of scary public risk that anchors face when they tackle touchy issues. All journalists, however, share in journalism's public *responsibility* to ask questions and report answers. Many news broadcasters theorize that viewers or listeners tune in to answer, first, the most urgent of questions: Is my world safe tonight? The fortunes of broadcast news ride heavily on the trust that people place in anchors and reporters who are paid to answer that question and others.

Audience Support

The *intimacy* of broadcast media tends to create visceral relationships with listeners and viewers. Research indicates that the public responds not only to news but also, in complex psychological ways, to the people who present it (helping to explain the

Oral culture Social environment in which person-to-person speech, unaided by writing or other technologies, was the primary form of "news" communication.

elevated salaries of on-air "talent").[5] In general, listeners and viewers seem to approve of and consume broadcast news above other forms—but by no means unconditionally.

Radio and TV journalists usually assert that they convey only facts and the opinions of others, not their own, and that they do so as accurately and neutrally as possible—at least trying to be fair. Audience surveys show, however, that viewers and listeners perceive bias in broadcasters. In our contentious "argument culture," as one social scholar dubbed it, efforts at balanced reporting often fail to satisfy critics.[6] Fox News Channel flaunts its slogan, "Fair and Balanced," while openly working to showcase conservative perspectives. On the other hand, even though most journalists admittedly lean Democratic, the liberal media-watch group Fairness and Accuracy in Reporting (FAIR) charges that journalists transmit mostly *right*-of-center political messages.

Broadcast news cannot please everyone; anyway, politics is not at issue in all journalism and fairness is not the only challenge. Credibility is another—but here too the public response is mixed: In a study following one of the Clinton-era political scandals, two-thirds of respondents said reporters were failing to check facts carefully before reporting them. However, three-quarters of respondents said they still supported network journalism.[7]

Yet, TV-news viewership has been declining overall for many years. This is visible primarily in **ratings,** which are estimates of the number of households tuned into a program. Ratings for the nightly newscasts on ABC, CBS, and NBC reportedly dropped 34 percent between 1994 and 2004—falling almost 60 percent below their peak in the late 1960s. Local TV news declined too, but less steeply, and in surveys dominated most other media in popularity. Cable news operations, dating only from 1980 (the birth year of Ted Turner's CNN), had always survived with relatively tiny audiences, and still did. Radio news had become a weaker segment of electronic journalism, with much narrower coverage and a shrinking payroll.[8]

Bottom line: By the first decade of the new millennium, traditional broadcast news had lost much of its hold on the U.S. public. Without strong ratings, money coming from advertisers diminishes and in some cases vanishes with the disappearing audiences. This means that electronic-media journalism's struggle to survive, dating at least from the 1980s, would have to continue—and, indeed, intensify.

Technology's Role

Broadcast journalists have employed techniques that would send an old-fashioned town crier into a swoon on the cobblestones. During the 1990–1991 Persian Gulf War, correspondents in hotel rooms and on rooftops in Saudi Arabia could go on camera and report *live* to U.S. audiences as missiles flew overhead. A national boom in use of cellular phones enabled radio journalists to outrace television, report instantly from news scenes and even reach traveling newsmakers.

Satellites in space, which have been around since the Soviet Union's *Sputnik* startled the world in 1957, have been relaying broadcast signals since Telstar's launch in 1962. Using **portable earth stations** that can be packed in their cars, TV crews bounce news reports off satellites more than 22,300 miles from earth and back to their stations or networks. Add field-to-base **microwave** technology—made familiar to us by those high-masted vans parked at news scenes—and there is little that

EDWARD R. MURROW. The CBS correspondent's dramatic wartime radio reports and trailblazing TV documentaries turned his face and name into icons of modern journalism.

Ratings Tally of the size of broadcast audiences, usually in number of households.

Satellite Space vehicle containing communication relay devices; broadcasting uses satellites in "geostationary" orbits—that is, in fixed positions relative to earth.

Portable earth station Bundle of electronic equipment that fits in a truck and can "bounce" signals off a satellite in space and back to a distant receiving station on earth.

Microwave Electromagnetic wave of extremely high frequency; can be generated by a portable device to carry audio or video signals from one location to another.

NIGHT VISION. Despite some murky scenes, news video shot nocturnally with high-tech cameras conveys the turmoil of war. In this footage, British soldiers patrol an unidentified location in Iraq.

television cannot show quickly. Stir in **digital** tools such as advanced cameras and editing systems, and both radio and TV reach the cutting edge of realism, precision, and speed.

As the United States went to war with terrorists in Afghanistan in 2001, network *videophones* went along, sending back live reports from remote places that a few years earlier would be seen only in "file" video or on maps and slides. By the mid-1990s, some TV stations were acquiring night video of police operations by using helicopters equipped with darkness-piercing **infrared** cameras. Live aerial coverage of developing crime or traffic dramas is now routine. Reporters with cameras as small as campaign buttons have probed commercial and government practices, sometimes exposing matters of real importance. Broadcasters tend to acquire new technical devices partly for journalistic reasons and partly to give their news programs a cutting-edge look, sound, and image.

◼ Traditional Paths

RADIO

The twenty-first-century citizen can avoid television but so far has had more difficulty sidestepping radio; it reaches into almost all U.S. households, often through multiple sets (the average home has more than five radios). Radio thus permeates our environment and keeps most of us from falling utterly out of touch with important events. However, its news role has diminished and it rarely is heard above the other electronic voices in today's crowded universe of choices; this is ironic, in that radio started the whole process.

At Its Roots.

One of the most memorable early demonstrations of radio's ability to present news dramatically wasn't live, but sounded like it and turned out to be very big news. In 1937, announcer Herbert Morrison broke into sobs as he watched and described the dirigible *Hindenburg* bursting into flames, killing 36 people. What was intended

Digital Technology that records and transmits sound and images through the use of numerical values or "digits."

Infrared Electronic technology that uses part of the light spectrum to create visible images from the heat given off by humans, cars, trees, animals, and so on.

to be just an off-the-air event recorded as an experiment with disc technology for Chicago's WLS became a prolonged and heart-rending news report. Its subsequent broadcast on network radio is said to have been the first use of recorded sound in radio history. Recording technology good enough for routine field use wouldn't arrive until the 1940s.[9]

With radio spreading and World War II approaching, the networks organized strong teams of journalists and support staff. Their broadcasts emanated from New York stations, passing through telephone lines to the networks' affiliated stations—called **affiliates**—across the United States. This mimicked a pattern set earlier by **wire services** that fed stories to member or client newspapers (and to radio newsrooms). The networks even hired journalists who had honed their skills on "the wires." These fast, broadly informed writers could meet radio's demand for crisp news that, heard only once, could be clearly grasped. Their reports became familiar to Americans as Europe suffered through a German firestorm.

After the war, networks extended their news capacity through more U.S. and foreign bureaus and a mushrooming web of local stations. By the 1950s, however, radio's primacy as a medium was overtaken by the raging growth of television, for which radio had paved the way by assembling audiences in the living room.

HINDENBURG DISASTER. The spectacular and lethal incineration of a German dirigible was narrated by a sobbing announcer; his recorded account made radio history.

In Maturity

Today, ABC, CBS, and NBC continue to be brand names in radio news. However, where those brands fit in the industry picture has changed, because that picture has changed. Commercial broadcasting today is a marketplace not just for the sale and delivery of news and entertainment but for a brisk trade in radio *ownership*. A company called Westwood One now owns, partly owns, or distributes all radio newscasts under the CBS, NBC, Fox, and CNN brand names. Westwood One controls much of the "network" radio news we hear, and has many foreign clients as well. Its main rival is ABC Radio, which still distributes news, talk, sports, and entertainment programs to stations from within the ABC network.

Consolidation has affected local radio profoundly as well. Station groups' acquisition of hundreds of stations has stilled many hometown news voices in favor of syndicated news. News departments' airtime and staffing have been reduced even as opinion-rich talk shows proliferate. Disc jockeys frequently stand in for journalists. Some newscasts have shrunk to little more than copy from the wire services. New corporate owners have found ways to cut newsgathering costs even more severely. A station group can tap into digital technology that permits one staffer to feed newscasts to several stations without ever going near their towns.

Clear Channel Communications, the nation's largest owner of local radio with more than 1,200 stations, practices this form of networking. In Ohio, for example, Clear Channel stations in smaller cities feed stories to a hub station in Columbus, which feeds them back as part of a regional newscast. A 2003 survey by the Radio-Television News Directors Association and Ball State University found that money-saving consolidation was enabling radio owners to spend more on digital equipment and use it more often for producing news.[10]

Although it's hard to fund much truly local radio news in some small towns, there are big-city stations that maintain robust news operations. For example, in Seattle,

Affiliate (noun) Local radio or TV station affiliated with—but not owned by—a network.

Wire service Journalistic company or collective that contracts with broadcast stations and other media to provide news material, sometimes including video.

PRINCIPLE 2

The quality of proximity makes local broadcast news appealing to most people.

#17

at KIRO-AM, owned by the Entercom station group, there is a balance between local news and network newscasts. The station claims to have the largest radio news staff in the Pacific Northwest, with 10 reporters—a staggering workforce by today's industry standards. On the telephone and on foot, using digital audiotape recorders and other modern tools, these journalists dig the raw ore of stories out of their bustling region, then write, narrate, edit, and present the news.

Today's radio journalist usually is an all-around worker; she or he has to be. Not all of this is due to recent business changes; most radio stations never have been heavily staffed. The majority have no more than three news employees; many have only two; most probably employ only one, perhaps aided by a part-timer. A reporter typically covers at least two or three stories a day—often many more in a small market—and uses basic audio production skills in the field and the studio. The biggest rewards probably are psychological, not financial; salaries in radio are substantially lower than those in TV. Some radio journalists' greatest source of satisfaction is in "reach," which means that their newscasts are exposed to thousands of rush-hour commuters (in towns large enough to have rush hours).

NETWORK TELEVISION

A **network** really is just a relationship among stations and a central production core that feeds them programming. In return for this bounty of content, stations agree to clear air time for *prime-time* (evening) shows and the commercials that accompany them—the networks' primary revenue source. Recent economic pressures have squeezed TV-network news, permitting rival news sources to compete. Still, the Big Three networks (ABC, CBS, and NBC) continue to dominate national news viewing through their local affiliates and cable TV, and have moved onto the Internet, as well.

News Structure

Let's step back again, but only to the 1970s. By then, radio was drawing less than one-fourth of the broadcast-news audience nationally. CBS's Walter Cronkite and other big-name anchors pulled millions of people to their TV sets nightly. Network managers instituted news structures, routines, and specialties that, for the most part, persist today.

Each of the Big Three networks places news in a division with its own leadership, work culture, and budget. Until the 1970s, a news division had been an acceptable *loss leader* for networks; it could lose money but contribute prestige, and that was all right so long as entertainment programs were paying off. However, attitudes toward "loss" were already changing when CBS's *60 Minutes* turned that model on its head by gaining high ratings, those all-important viewer headcounts. Soon, the show was raking in a million dollars a week in advertising revenue versus only $150,000 in expenses, according to its founder-producer Don Hewitt.[11]

The network news divisions soon came under pressure to follow in Hewitt's footsteps toward profit. Producing more television **newsmagazines** was one approach. ABC's *20-20* made a (disastrous) debut in 1977 and, after major retooling, became a staple. Other efforts did not fare as well; NBC News lost millions of dollars annually—perhaps more than $100 million, by some accounts[12]—until General Electric took over the network in the mid-1980s and cracked down on the budget. This was part of an unprecedented series of corporate acquisitions that triggered change in network economics. Takeovers of ABC (by Capital Cities Communications) and CBS (by Loews Corporation) pressed their news divisions to toe a new bottom line. They would have to cut costs as well as find more money-making ways to practice journalism.

Networks Large media companies that own or are affiliated with smaller broadcast companies, supplying content to them in exchange for advertising access.

Newsmagazine Radio or television program format that combines news and features in style analogous to print newsmagazines.

protalk
Heather Bosch
KIRO-AM REPORTER, SEATTLE

Seattle radio reporter Heather Bosch describes her often rigorous work day:
"I hit the ground running . . . I come in at 4:30 A.M. unless I'm called in earlier. . . . I usually have a couple of ideas I think will be good stories. The producer in charge also will have good ideas, and the producer will usually win. . . . I go on live at 5:00 A.M.—that is, the network news is at the top of the hour, then the local. Today, I covered the trial of an alleged terrorist, then followed up with the death of a Des Moines [Washington] police officer. I was on live at 20 [minutes after the hour] and at 50.

"When the cop was shot, I was live at 4:00 A.M. My goal in radio is to get as much information as possible quickly and update continuously. By 4:00 that morning I had information on the officer. I talked to people who were caught in roadblocks. I went to a coffee shop and interviewed local folks. I was on, live, 16 times that morning. You've gotta get different audio, angles, updates. Otherwise, it's the same old stuff, and why does anyone want to listen to that?

"Also, it's more interesting to me as a reporter to get more information. I was able to watch a huge public presence build up as they tried to find the cop killer. There are days like that, when I'm going from the get-go. There are other days when I'm just

"When there's something like an earthquake, you're running on adrenaline for three days. This is what a lot of reporters live for."

working the basics. Today, I did a phone interview with a terrorism expert early on. All of the early stuff goes on the air during 'morning drive,' 5:00 to 9:00 A.M. . . .

"When there's something like an earthquake, you're running on adrenaline for three days. This is what a lot of reporters live for. You get out there and get the pain of it, the visual picture, and you let people know what's going on. It must be similar to being an athlete in the playoff games. . . . Radio is so wonderful. Getting things on the air immediately, that's part of it. . . . News radio is baptism by fire; there's nobody to fall back on.

"The pieces I do for afternoon, I write and prepare. I do series pieces about once a month, maybe a five-day series, with each piece a minute and a half or so. That allows for more in-depth interviews. It's a wonderful privilege, as a journalist, that right to ask all kinds of questions, to sit down with all sorts of people."

The resulting drive for ratings generated newsmagazines rich in emotional stories with dramatic structure—similar to their fictional siblings in network entertainment—and very light on complex political and government-policy stories. The new programs fit nicely into prime time. Elsewhere in the schedule, the networks' morning shows moved increasingly into **soft news** and talk, emphasizing stories of *personal* interest—not relatively abstract big-picture news.

By the 1990s, NBC's morning flagship, *Today*, was moving away from **hard news** and into consumer-oriented discussions, crime and tragedy features, chats with Hollywood stars, and more live music. Health topics were dramatized in 2000 by cohost Katie Couric, who had lost her husband to colon cancer; she soon had doctors perform her own colonoscopy on the air live, prompting a 20 percent national increase in such screening exams.

News in any form remains important to the Big Three networks, for its prestige as well as its role in helping network affiliates meet their generally unofficial **public-service** obligations. Moreover, startling or sweeping news events (like 9/11 or the war in Iraq) and protracted human-interest stories (like 2004's Kobe Bryant and Scott Peterson prosecutions) can help raise a network's profile and its advertising revenues.

News Process

All across broadcasting, news systems and practices have evolved similarly. Managers and workers moving from company to company continuously reinforce this homogenization. Following are a few key functions that are common among networks and, in roughly this form, across most of television news.

Management. Top executives are paid to keep a news division journalistically competitive, strengthen the "brand" of its programs, and watch the bottom line. A division chief (often with the title of news president) has subordinates who oversee areas such as daily news, political coverage, morning programs, and special events. Besides steering the ship, executives sometimes have to jump into topical issues affecting the image and integrity of the network. After the vote-projection errors of Election Night 2000, news divisions assigned vice presidents to investigate.

News Selection. The process of choice among possible news stories competing for airtime begins many hours before a network newscast. Some stories have been hatching for days or longer. Assignment editors arrange for key bits of video, do research by phone and Internet, and set up interviews. In a morning story conference, editors who focus on different types of news (general, political, foreign, health, legal, consumer) pool and debate their ideas. The lead anchor at networks and some stations may act as a managing editor, advocating some stories and discounting others. Then, hour by hour, early decisions are refined, revised, or discarded. To the degree that a single newscast provides all the news many people absorb in a day, its final shape is crucial, summarizing the most compelling or important situations and events.

Reporting and Production. New technology, mostly of a digital nature, keeps reducing the number of people required to make a network newscast happen. Fewer employees perform more and more tasks. However, some general activities remain distinct: "Line" producers rank stories in importance, interrogate reporters, arrange supporting visuals, and turn the results into a newscast. **Photojournalists** shoot the pictures. Correspondents and field producers get the facts, write the scripts, catch the planes, and meet the deadlines. Many of these jobs may be blended together and probably will be over time.

Presentation. Graphic artists create computer-generated charts and visual effects to enhance the storytelling on the TV screen. A director and his or her control-room crew plan the timing of special sound elements and the route of each rolling studio

Soft news Human interest or non-urgent news stories.

Hard news Serious news of importance to a broad audience—on such topics as politics or foreign affairs—as distinguished from routine news items or features.

Public service Basic responsibility of U.S. broadcasters to create socially important programming without regard to ratings, in exchange for free use of publicly "owned" airwaves.

Photojournalist Professional news photographer who, in television, also edits material for on-air use; often called *videographer.*

camera (a function sometimes handed to *robot* cameras these days). They also must consider the aesthetic values of the program; again, TV's entertainment shows keep the production-quality standard high, and news cannot escape it. The anchor—the only team member most viewers know—is the *last* link in the presentation chain. He or she must tell the day's stories while projecting conviction that all the stories are useful or important to the viewers.

Local TV Stations

The largest sector of American broadcast journalism is local television news. Newscasts flow from more than 800 TV stations. Some are in small cities and remain small themselves; others are in metropolitan centers and have grown with them. News operations are local TV's primary profit engines, but the number that are profitable slipped for years—to just 54 percent of all local newsrooms in 2002, bouncing up to 55 the following year.[13]

Structure and Process

The local-news director must design, staff, and support a special kind of team. Its members maintain the flow of news programming. The size of the team varies from the smallest towns to the largest, but in general certain key players are in place everywhere, reporting to the news director. Briefly, an executive producer oversees all elements of newscast organization and assembly; line producers are in charge of one or two newscasts each; one or more writers merge visual and verbal skills to produce voice-over scripts for anchors to read live, among other material; an assignment editor deals with quick story checking and arrangements for interviews and other "shoots"; reporters and photojournalists stalk the landscape for story elements, and sometimes specialized videotape editors apply another sensibility to the storytelling.

Even many stations in small markets broadcast several hours of news every day. To the journalist, this means deadlines can arrive repeatedly from before dawn until late evening or beyond. At most stations, the news day falls into roughly the same functions or phases as the network news process explained earlier. However, there are variations, including these:

► The morning meeting to consider potential stories embraces all ideas, national or local, that might interest large groups of viewers. This includes national stories with transcendent appeal or local "angles." It also includes situations that staffers discover in their neighborhoods or in the morning newspaper.

► Local anchors tend to work in pairs. By contrast, networks have had little luck pairing anchors on the evening newscasts since the glory days of NBC's Chet Huntley and David Brinkley decades ago.

► When a newscast ends, phones ring; viewers are calling. Many skip over substantive issues to complain that an on-screen name was misspelled or that the anchor's hairstyle was unflattering. However random or trivial, though, viewer calls are a window on community opinion and a link to grass-roots viewpoints that network journalists rarely hear.

► The end of a TV newscast doesn't mean the staff is in for a long snooze. Increasingly—thanks to consolidation and technological change—local journalists convert all or part of their daily stories to content for other media, including their corporate siblings. One study found that in 2003 about 70 percent of all newsrooms were supplying news content to their stations' websites. About 43 percent serviced radio news, and more than 16 percent of local TV-news operations sent stories to other stations or to cable.[14]

Rewards—and Shortfalls

Many TV journalists find the stresses they face daily to be a small price for all that they're allowed to do, and the chances they have to make a difference. A news

protalk
Patrice Goya
LOCAL TV-NEWS PRODUCER

Patrice Goya produces the 6:00 P.M. newscast at KGMB-TV, the CBS affiliate in Honolulu. It's her second producing job at a medium-market station since her graduation from the University of Oregon. Now in her mid-twenties, she's a tough-minded young news pro but, like many journalists, feels empathy for the people whose stories fill her work.

While producing in Tucson, Arizona, Goya spent September 11, 2002, helping build homes for low-income people to commemorate the human losses from the terrorist attacks of 9/11, exactly one year earlier. Her compassion had emerged back in journalism school. One morning in 1998, a disturbed teenager who had murdered his parents shot up his high school in Springfield, Oregon, killing 2 students and injuring more than 20. Goya and other broadcast students rushed to the school, and she signed up to help a network news team cover the incident.

When the producer asked her to check out a home where the young gunman's sister was rumored to be staying, Goya recalls: "I went to the house, but couldn't get out of the car to go knock on the door. It just didn't seem right" (to pursue a woman whose family had just been devastated). She sat in the car for a while, struggling with her principles. They won.

> **"It's slower here in Hawaii than it was in Tucson; we have to work a lot harder to find news."**

Later, in Tucson, Goya plunged into a hard-news environment that often seemed to test principles at every turn. The city averaged almost two murders a week, tragic stories were common, and the news director even issued a "viewers' bill of rights" to assure the audience that the station stood for ethical reporting. There was open debate among staffers when dilemmas arose. "This is a very open newsroom," she said back then. "We can just stand up and say 'I completely disagree with you.'"

On slow news days in Tucson, producers could always reach out to Phoenix for strong enough stories to put in their shows. Moving to Honolulu has meant having to work with whatever news Honolulu generates, since there are no larger cities in the state from which to draw bigger news. "We have to work a lot harder to find news," she complained, but it was a gentle complaint; she was married in spring 2005—for Patrice Goya, the biggest news yet.

job can provide some of the most intense and rewarding experiences available anywhere, which is why some choose to make it a career. Among the dividends, though not necessarily paramount, is money. Pay is highest at the networks, where anchors make millions annually while correspondents and top producers can earn in the hundreds of thousands. Most budding journalists, however, begin in local news, where pay is good—eventually. On average, it starts at little more than $20,000 annually, the lowest in all of journalism, but salary growth outstrips cost-of-living increases and improves sharply as the TV reporter climbs to larger markets.[15]

LOCAL TV REPORTER. Entry-level pay usually is low, but today's local-station TV reporter enjoys richly varied work and, in a large city, can earn a six-figure salary.

At the top of local news, pay can be called lavish: A survey in 2000 by the American Federation of Television and Radio Artists (AFTRA) showed TV reporters in Los Angeles averaging $100,000 to $200,000 a year. Anchors fared even better, averaging more than $300,000 annually.[16] A few in big cities have multiyear, multimillion-dollar contracts. Besides Los Angeles, other top-10 markets such as New York and Chicago pay their "talent" highly. Even a few news workers in radio, which creates fewer celebrities, crack the $100,000 line annually.

Detracting sharply from this on-air salary harvest is the fact that *women* do not share in it equally with men. The Los Angeles survey showed women earning, on average, 28 percent less than the men with whom they typically shared the **news sets** where anchors sit (or stand). Nor, in radio, did a woman *solo*-anchor any newscast; lone men anchored most of them. As for the racial balance among TV anchors and reporters, the landscape in many cities is even worse. A 1999 Gallup Poll showed that although 60 percent of whites believed the number of black anchors on TV was "about right," 57 percent of blacks said there were "too few." Progress has been slow; in many communities, the on-air racial mix does not reflect the local population, leading to some concerns about bias in news coverage, as well.

Fairly or unfairly, legally or illegally—and some cases do get to court—stations juggle race and gender among their on-air personnel. Producers, however, are less vulnerable to such manipulation and are in greater demand. TV economics in recent years heightened this: With competition intense and budgets tight, many stations now emphasize news *packaging* while holding the line on expensive newsgathering. To do this, they need producers with broad knowledge—especially of current events—and keen verbal and visual skills. They need them badly enough to pay some of them $50,000 a year and more in large markets.

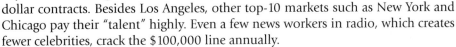

Newer Channels

CABLE NEWS NETWORK

Cable News Network (CNN) was launched in 1980 and seemed to trigger an invasion of competition that would loosen the old networks' grip on the news market. At first, this 24-hour news channel seemed little more than a gamble by a rich, flamboyant Atlanta entrepreneur named Ted Turner. As CNN emerged in scores of cities through cable-TV systems, the insomniacs who wanted news at 4:00 A.M. could get it with *pictures*—as could the midday viewer otherwise trapped among the soap operas. CNN kept journalists—some of them citizens of newsy countries, not "parachuting" U.S. correspondents—working around the world even as ABC, CBS, and NBC cut back their foreign bureaus. From a starting strength of eight U.S.

News set Table or arrangement of furniture where anchors sit or stand during newcasts; sometimes called a *news desk*.

bureaus in 1980, CNN grew to nearly 40 bureaus worldwide over the next 20 years. Hiring local reporters and old-line–network journalists whose contracts had not been renewed, CNN defied the skepticism of veteran broadcasters.

Turner's brash network made a spectacular thrust during the Persian Gulf War (1990–1991) by airing live reports from correspondent Peter Arnett behind enemy lines in Iraq. Other big stories gave rise to what would be labeled the **CNN effect;** this term is applied sometimes to live TV reporting's tendency to speed up diplomacy, and at other times to live reporting's ability to draw wide attention and responses to a previously unnoticed human disaster. CNN was seen to have *both* effects by injecting far-flung world news into living rooms, offices, sports bars, and airline terminals night and day. It also covered slowly unfolding major events—such as the aftermath of the 9/11 terrorism—continuously for many days. When networks went back to their regular programming, CNN stayed with news (*its* regular programming), sometimes to the point of overkill.

The network's reach and stamina have not brought it big ratings by conventional standards. This may be attributed partly to the fragmented attention of twenty-first-century viewers, many of whom flit among channels while others leak out of the television audience altogether. CNN also is vulnerable to the aggressiveness of the rival Fox News Channel. Recent changes in management and on-air style in Atlanta show that, if CNN loses, it won't lose quietly.

PRINCIPLE 3

Live reporting abroad can draw mass attention to neglected disasters—part of the "CNN effect."

OTHER TV ROUTES

Fox News Channel (FNC) was brought to cable TV by conservative Australian press magnate Rupert Murdoch, owner of the Fox entertainment empire. He said he intended it as a counterbalance to "liberal" network news. Its ratings went up sharply during and following the 2000 election, when Florida's unresolved ballot count triggered weeks of news, commentary, and debate. Surprising industry observers, FNC managed to keep its ratings high after the election dust settled: February 2001 ratings showed the newschannel outpacing all of its rivals. Its *Special Report* was the top-rated weekday political show on TV.[17] FNC further heightened its profile (as did most channels) from the immense viewership generated by Osama Bin Laden's attacks on New York and Washington, DC. Its news-with-an-attitude approach also struck a chord with many viewers.

MSNBC—the Microsoft/NBC channel, a TV-Internet partnership—is an offshoot of NBC, using the network's anchors, repackaging stories from the network's newscasts, and also doing some reporting of its own. Like other cable-news channels, MSNBC made audience gains in the post-2000 election period, as did its Internet site; the political confusion moved millions of Americans to try new information sources. Also like other channels, MSNBC put a dramatic running title—a strategic "brand"—on its coverage of the aftermath of the 2001 terrorist hijack-bombings. Three months after the attacks, its title was "America at War."

CNBC (Consumer News Business Channel) was formed by NBC and Dow-Jones, owner of the *Wall Street Journal.* It hasn't gained high ratings—but, as analysts have noted, *household* ratings don't capture viewing by people watching cable news at the office, club, or gym. One study reportedly found about 700,000 *Wall Street Journal* subscribers watching CNBC from such outposts.[18]

Bloomberg TV, part of a large media/financial enterprise owned by Michael Bloomberg, who was elected New York mayor in 2001, sells data services directly to businesses but also reaches broader audiences through cable TV. News about personal investing and business draws high-income viewers, readers, and computer users. Knowing this, Bloomberg, NBC, and others have worked to turn financial news into a powerful specialty.

CNN Effect Viewer compassion and charity after human suffering is shown on TV.

Health, hobbies, sports, and other interests also have given rise to cable-TV channels catering to those enthusiasts. ABC-owned *ESPN (Entertainment and Sports Programming Network),* a 24-hour service, has spawned successful imitators, including Fox Sports, which runs a number of regional networks. ESPN rushes scores and video to sports fans so quickly that it has prompted some local-TV sports anchors to favor features and follow-ups over quick-action reporting.

On some cable systems, *regional* news channels covering multicounty and even multistate areas are finding their audiences. In 2001, the National Cable Television Association listed 18 regional news channels; by mid-2004, the number had grown to more than 70. Marketers were using regional channels to reach prime audiences through cable television. In Michigan,

COMBINATION NEWS. TV/Internet enterprises such as MSNBC benefit from growing use of computers to find news, especially in complex situations like the 2000 election aftermath.

Comcast was aiming at men from 18 to 45 years old by devoting a new channel to sports—specifically, as one writer put it, "high school and college games that don't get as much airtime as Big Ten games."[19]

Major League Baseball moved toward a mid-decade launch of its own national baseball channel, to compete with regional sports networks around the country—Fox owns 12 of them—for advertising sales. The lure of ad revenue continued to complicate the TV landscape, already beyond the old "you-can't-tell-the-players-without-a-program" status.

One unique offshoot of interest in television and government is a nonprofit enterprise called *C-SPAN (Cable-Satellite Public Affairs Network).* Funded by license

TV MEETS INTERNET.
Faced with changing audiences and DVR use, CNN has moved in recent years to restructure its operations and integrate its TV and Internet efforts. The aim is to draw news fans by any path available.

fees paid by cable systems that put it on the air, C-SPAN serves up unfiltered politics and policy by televising public appearances, live, unedited, and with little or no journalistic intervention. This appeals to Americans who've grown suspicious of the mainstream press—increasingly accused of drawing conclusions so quickly that it makes careless mistakes or even of reporting while under the influence of political bias. C-SPAN's audiences tend to be small but devoted, fond of watching uncut presentations on serious subjects. Interest in its coverage spreads wider, however, whenever developments drive people to seek more firsthand information than standard network newscasts provide.

INTERNET NEWS

Since the 1970s, broadcasting in its original sense—the spread of messages over airwaves to an anonymous mass audience—has been crowded and weakened by new competitors. The newest are Internet based. Broadcasters have set up websites to be sure they're on the playing field of the future; but the imagination of new players seems likely to outstrip that of the older industry (which could lead it to buy them out).

Demand

A government study found that "Americans' use of information technologies grew at phenomenal rates in 2001."[20] This certainly occurred partly because of the frenzied fact-finding that started with the September 11 terrorist attacks. At any rate, the study found that one-third of Internet users were looking for news and information while on-line, a promising omen for broadcasters who could adapt.[21] Not much later, in March 2004, Nielsen reported that three of every four Americans had access to the Internet from home. Moreover, fully half of this on-line population was using *broadband*, the "bigger pipeline" that permits users to receive complex transmissions—potentially including TV's rapid-fire images—at a high-quality level.

There also were reports that millions of Americans were ready to move from their traditional passive media activities (such as watching TV) to more active ways of gathering information and enjoying entertainment. Soon, newcomers were following the old Big Three TV networks in establishing websites, some of them with comprehensive news coverage plus links to other news sites. Today, smaller networks and hundreds of local stations have such sites; some run their own, while others "outsource" the management and technical chores. Some TV-station sites already have won prestigious Edward R. Murrow Awards from the Radio-Television News Directors Association.

Structure and Work

The front of an Internet news site often resembles a newspaper's front page; in fact, many sites are the Internet arms of major newspapers, supplying not only headlines and news briefs but also **hypertext** links to full stories. The viewer/reader who activates such a link is tapping the World Wide Web's capacity to open long, deep corridors of information. By combining broadcasting's brevity and visual allure with its ability to provide detailed background on a news topic, an Internet site can appeal to a large cross-section of news consumers. At least, that's the idea.

Many sites depend on advertising rather than charging subscribers directly; unfortunately, many potential advertisers have not yet developed confidence in the Internet's ability to generate customers. Even a brief economic downturn can cost a news dot-com lots of money, especially when it relies on advertising for its revenue and when advertisers suddenly cut back on spending. Such reversals have led even big-name news sites—run by media giants such as the *New York Times* and Knight-Ridder—to lay off workers. Some promising sites went out of business early. Others

Hypertext Collection of World Wide Web documents (or "nodes") containing cross-references or "links" that allow the reader to move easily among documents.

began charging subscriber fees but later withdrew them; a few subscription sites survive, and more are promised.

CNN claimed in 2001 that its website, CNN.com, had reached that "wonderful and vital" stage and it began charging monthly fees to access its news video from the site. The company said it led the pack according to the number of *page impressions* and *gross usage minutes (GUM)*, both indicators (something like TV's **Nielsen** ratings) of just how and how much a site is used.

Internet news publishers have recruited workers of all sorts. They include young, "computer-hip" journalists never initiated into traditional news media; mature ex-print journalists; a sprinkling of former broadcasters; and even some computer whizzes who aren't particular about content. Anyone with quick reflexes and information skills might find a place in this *avant-garde* realm. News sites employ reporters, editors, writers, graphic designers, and other print or broadcast specialists who can adapt quickly to packaging news for cyberspace. (Irresistible language from a help-wanted ad for a World Wide Website "anchor": "Ideal candidate is a female anchor in her twenties whose style is a mix of MTV and Fox News.")[22] Because some sites offer video clips of news stories, they also hire producers. Some come from fast-paced broadcast newsrooms, and yet they learn quickly how to organize and refresh multimedia sites.

News and Society

There are some special influences on electronic journalism and its relationship with society, and there are some controversial issues involving news management and practice. Let's consider a few of these factors.

RATINGS AND RESEARCH

Because prices for commercial air time are pegged directly to ratings and to **demographic** targeting of listeners or viewers, these statistics have been a powerful force for decades. They apply to news just as to other kinds of programming. Much emphasis still is placed on sheer numbers of households watching a newscast or one of its segments. Ratings firms can estimate viewership in time periods of five minutes or less; this can be used to gauge viewer responses to an *individual story* in a newscast. One of the authors once heard a manager order an Emmy-winning investigative reporter to avoid a certain topic because her brief report on it the previous evening had drawn mediocre ratings. Question: If viewers knew of this, would they object to the notion that audience measurement can sway the judgment of news organizations?

Beyond just counting, the use of social-science analyses of audiences has accelerated the trend toward "news you can use" (or at least "want"). It's now possible to learn, for instance, how much an audience likes "live shots," which then can be added to or subtracted from the news menu accordingly. This can be harmless in regard to the live shot, which often is just a dramatic device that can be dropped without altering the substance of news. However, what if viewers or listeners indicate little interest in foreign news or political stories? Should those stories be dropped from the agenda, too? Journalists often claim an obligation under the First Amendment to tell even unpopular stories. It's a point they sometimes discuss with their own managers, who often note that good journalism is useless if people aren't watching or listening and that a station has to maintain ratings and revenue. This issue is far more passionately debated within the broadcast community than

PRINCIPLE 4

Broadcasting's urge to serve changing audience desires always clashes with some journalistic values.

Nielsen A. C. Nielsen Co., a pioneering audience-measurement company.

Demographic Any audience segment as defined by its members' age, ethnicity, income, or other characteristics that may be targeted by advertisers.

among the general public—another ominous sign to journalists who fear that audiences' commitment to news is diminishing.

TECHNOLOGY

Technology has been a boon but it can also be double-edged. It has brought great speed to the transmission of news, but that speed has encouraged some journalists to skirt their profession's best practices, such as careful corroboration of tips from anonymous sources. TV's resort to high-tech devices has helped attract audiences and inform citizens but has also put journalists under ethical scrutiny. They are criticized for invading privacy when hidden cameras or microphones expose unsuspecting characters. They are reviled for endangering hostages and police by looking over the shoulders of SWAT teams and showing their maneuvers on live television, which a hostage-taker could be watching.

With budget "bean counters" more influential than ever before, managers who purchase expensive tools feel daily pressure to *use* them, whether news justifies it or not. The tension seems especially high when a helicopter is available. In Baltimore one evening, viewers of WBAL-TV were treated to live aerial views of a car fleeing police; the drivers were suspected only of car theft—one of thousands a year in that city, thus hardly a big news story. However, said the news director, "It was compelling to watch."[23] That thirst for spectacular drama, rather than importance or context, drives many news decisions today.

Behind the scenes, economic pressures are affecting news content. Merged media conglomerates are trimming payrolls in the newsroom, and are adopting new technology for greater efficiency. A Florida company, ParkerVision, has sold to some TV stations a computerized system that enables *a single employee* to perform most of the functions that make a newscast happen.

TRUST AND THE CONSUMER

No challenge holds more long-term potential for disaster than the pressures to build ratings and reduce expenses, which go to the heart of electronic news media's purpose. The long-term usefulness of news to society hinges on its quality—that is, how well and truly it informs citizens—which depends, in turn, on its accuracy, its independence, and its credibility. In the end, the whole journalist-audience relationship hangs on a measure of trust.

Credibility

All important social values matter, but *credibility* comes first in news, and has suffered some serious blows. Americans often base their general opinions of media on what they read or hear about a few specific incidents, not on the steady performance they witness daily on the air. Unfortunately, not long before word of unethical deeds by the *New York Times'* Jayson Blair and *USA Today'*s Jack Kelley harmed the credibility of all journalists, radio and TV news provided some disturbing cases.

CNN had to retract a false report that U.S. troops used nerve gas in Vietnam. A survey found CNN "most believable" among TV news outlets, but with only a 39 percent credibility rating. Even lower were the ratings of the Big Three network news teams. High-profile anchors ate their words after reporting errant 2000 election-night projections. Fox sent the discredited reporter Geraldo Rivera to cover war in Afghanistan, where he was accused of new on-air distortions. Those are a few of the more glaring signs of the unreliability of some broadcast information Uncounted lapses at the local level may have gone undetected or little noticed

The ephemeral nature of the spoken word—the cornerstone of oral cultures—and the nuances of language have added to the credibility problem. This is also true of the hectic routines of broadcast reporting, always rushing people, video,

protalk
David Marash
NIGHTLINE CORRESPONDENT

David Marash is that rare American journalist who sounds unpretentious finishing a sentence in Latin. Having once taught college (at Rutgers in New Jersey), Marash sees journalism as, in large part, education. He has worked for more than 15 years as a correspondent for ABC _Nightline_.

"_Nightline_ is unique—alas—in American television and probably in the world," Marash says. "It's not just that we have an unlimited editorial ambition but that we have a _network budget_—which, even in the days when network budgets are shrinking, is still better than any other alternative."

The money pouring into the show's accounts has long come from advertisers who appreciate reaching four to seven million viewers every night. Marash claims a large blue-collar audience but also notes the nightly ads for expensive cars, "a tip-off that we have educated consumers. . . . If you want to address the policy-making elite in America, there's no single more efficient vehicle than _Nightline._"

Marash travels widely and often, bringing back eye-opening stories from Kosovo, Zimbabwe, and other points foreign and domestic.

"_Nightline_ is unique—alas—in American television and probably in the world."

As a cum laude graduate of Williams College doing literature studies, Marash did not at first seem a likely broadcast journalist. However, he did small-town radio while at Rutgers and then held some New York radio jobs, a stint with ABC's _20-20_ and other slots as a TV reporter and anchor. In 1989, he joined _Nightline,_ where he does original reporting constantly and investigative projects 10 to 15 percent of the time.

"I am free to range from sports to profiles of people . . . ," Marash says. "I know that I will never have less than five minutes [of air time] and may have as much as 17½ minutes to tell my story."

In early 2002, Marash's lavish air time—indeed, _Nightline_ itself—faced a challenge from ABC's own leaders. Coveting comic David Letterman's ratings, they tried to lure Letterman away from CBS by offering him _Nightline's_ time slot. Letterman turned down the offer, and the crisis soon passed, but for Dave Marash and the news program he treasures, the corporate earth seemed to have shifted.

The programs's mature audience, however prestigious, was not drawing in money as it once had. Media business analyst Tom Wolzien said, "An advertiser will probably pay at least double, if not triple, to reach young people, 18 to 34, than reach people over 55." And in 2004, _Nightline_'s ratings dropped below four million.

Dan Rather was famous as a hard-driving journalist, willing to defy authority. He had *been* CBS News to most viewers since inheriting the hallowed Walter Cronkite's anchor chair more than two decades earlier. So in 2004, when Rather insisted that he and producer Mary Mapes had a lock on a super-hot story—that old documents indicated a young George W. Bush had been allowed to shirk his duties in the Texas Air National Guard—network news executives didn't make him prove it to them.

Evidently they should have. Rather broadcast the Bush report on *60 Minutes Wednesday* on September 8. Only after that fateful act did the really sharp questioning begin. As critics built a case that Rather and Mapes had depended on an unreliable source whom they did not sufficiently challenge, Rather stubbornly defended the story for days—while CBS sank into a credibility meltdown. Finally, 12 days after broadcasting it, the anchor went on the air to apologize for the Bush story, piling responsibility on his main source, a retired Texas National Guard officer. A former U.S. attorney general and a retired Associated Press chief were brought in to investigate the report. The upshot: They blamed CBS.

Conservatives complained not only of error but of political bias against President Bush. Some people called for the dismissal of Rather even before his planned March 2005 retirement. Mapes and three others in the upper ranks of news producers were fired, after which Mapes offered to share blame for the debacle with executives all the way up to CBS News President Andrew Heyward.

The episode was the most violent blow in years to the credibility of CBS News. What many had once called "the Tiffany Network" had dealt not in journalistic diamonds this time but in far less glittery stones.

and information into little boxes of time and space, seemingly incapable of conveying certain ideas, and even facts, with total accuracy. Beyond factual errors, blown projections, bad sources, and verbal missteps lies perhaps the cruelest problem of all: Broadcasters very rarely are willing to *correct* their misstatements on the air, especially with any prominence. Internet journalists have much more opportunity to do so, but it's not clear that they are any more inclined to come clean.

Commercialism

One survey found that close to half of all consumers believe radio news is improperly influenced by "marketplace factors such as ratings, profits and advertisers. One-third of TV news directors said they'd felt pressure to skew news coverage in favor of advertisers.[24] That's the audience and the management agreeing that commercialism is distorting broadcast news. Certainly stations and networks are expected to seek profits, but whether that effort should intrude into the crucial act of reporting the news is anything but certain.

If *no* intrusion at all were equivalent to a touchdown, many media companies would appear to be running toward the wrong end zone. KOLO-TV in Reno, Nevada, offered any business elaborate and positive news coverage in return for a $5,000 sponsorship. That followed a similar incident in Knoxville, Tennessee. Some local TV (and radio) managers may simply grow so desperate for profits to report to their bosses and stockholders that they cross the "line" between advertising and news. A station in Pittsburgh, Pennsylvania, was caught using new technology to condense an NFL game unnoticeably—enough to squeeze in an extra 30-second commercial.[25]

Further complicating the rise of "stealth" commercialism in newsrooms has been the ascendancy of the **video news release (VNR).** This is a press release about a product, service, or institution, delivered to a station in video form and usually scripted much like a news story. Many small stations have been known to run these releases as if they are real news stories, without acknowledging their commercial origins. Even some large-market producers routinely assign staff reporters to "re-track" VNRs with their own voices—another kind of deception. To

Video News Release (VNR) Press release tailored to the needs of television, providing video related to a product or a service.

LOCAL RADIO STATION NEWSCASTER

Local stations are the main entry points to a broad-cast-news career. Radio journalists so often are "one-man bands" that they need prior or on-the-job news and production training. The makeup of a local TV-news team varies from small towns to large, but certain key players are found almost everywhere, reporting to the news director:

▶ An **executive producer** oversees the organization of newscasts.

▶ **Line producers** are assigned to construct each newscast.

▶ One or more **writers** (if available) write stories and lead-ins to reporter **packages.**

▶ An **assignment editor** pursues news leads and angles throughout the news cycle.

▶ **Reporters** follow up leads, generate new stories, interview sources, and write the self-contained narratives called *packages.*

▶ **Photojournalists** and, in larger stations, special-ized **videotape editors** (or the two jobs combined in **videographers**) shoot and/or edit news foot-age, working alone or with reporters or producers.

Reporters and sometimes photojournalists may be assigned to **beats**—special areas of news inter-est, such as police, health, or city hall—or to general subjects in which they have expertise. Even if they do have beats, however—given the viewer appeal of live breaking news—reporters often are kept on a general assignment (GA) footing so that they can reach crime and accident scenes quickly. Similarly, in most stations, photojournalists move from story to story without regard to content (but must *know* content to do a good job).

The jobs mentioned here can be realistic targets for almost anyone who can complete college broad-casting courses. Most news directors want to see at least one *internship* on a job applicant's record; internships often are a deciding factor in hiring.

Anyone aspiring to an electronic-news career is well advised to read trade publications *(Broad-casting and Cable, Columbia Journalism Review, American Journalism Review).* These keep the reader abreast of ethical issues, economic realities, and hiring trends.

use a VNR in that way, even if identifying its source, may soften resistance to more serious commercial assaults on news content.

Broadcast news does important work every day. Thousands of people strive to make it excellent. However, they too often encounter opposition, subtle or blunt, from commercial forces within and without. To con-sider a career in this field is to face some degree of ethi-cal responsibility to make public service, not private gain, paramount.

PRINCIPLE 5

Doing broadcast journalism entails a responsibility to put public service fast.

Into the Future

There is no guarantee of eternal success for broadcast news. A major question of the Internet age is how audiences will respond as newer delivery forms redefine the electronic-media world. As communication increasingly "goes digital," pressure to shift resources will grow—but will not move far ahead of consumers.

Don Hewitt, originator of *60 Minutes,* has urged the traditional networks to merge their three flagship evening newscasts—to pool them into just one "block-buster," with the networks' top anchors rotating. That hasn't happened, but it highlights the predicament of television's managers as the economic and technological ground keeps shifting.

The issues that rage between, and within, electronic media and society should not be seen as deterrents to aspiring journalists. Wonderful opportunities await today's college graduates, but no one can gauge the scope of their future. There are journalists who already may find themselves reporting or editing for radio, TV, print, and on-line news in the course of a single work shift.

If the work of electronic journalism rarely has been dull, hold on: We may soon enter an era in which there will never, ever be another "slow news day."

Summary

Broadcast news is a demanding, sometimes scary line of work but a vital public service. Network television anchors such as NBC's Brian Williams suffer in the public eye when their "facts" turn sour. This signifies the importance of radio and television journalism careers as well as the pressures they bring. News has been a major component of broadcasting in the United States since early in the twentieth century. The value of radio information was compounded by its oral delivery, which drew on society's roots in the spoken word.

Modern broadcast news—still "writing for the ear" but also employing new electronic tools—can transmit human experience through both sound and pictures, sometimes enhanced by special graphic elements. Audiences bond to broadcast journalists as people telling them important things; and they deliver harsh judgments when they believe those people have failed to check facts and purge biases. Although local stations provide most people with most of their broadcast information, it is major networks that remain the most nationally prominent and often most controversial of news sources. In times of national crisis or simply elevated national interest—9/11, the war in Iraq, a literally earth-shaking tsunami—networks use their resources to showcase the news.

News no longer is permitted merely to break even or lose money; its profits must compete with those of the networks' entertainment programming. This has given birth to sensation-seeking "newsmagazines" in prime time, a narrowing of news agendas, and cuts in newsgathering budgets. Critics say such moves have weakened broadcasting's commitment to providing public service, as it is required to do in exchange for free use of public airwaves.

A major question is how much and how rapidly the growth of Internet news will affect broadcast news. Networks and many stations have established websites and use them to offer news and to promote programming. The developing audiences for news sites such as MSNBC.com and CNN.com reflect broadcasters' convictions that the Internet is an important news medium, which means new job opportunities for journalists—and pressure for them to learn new skills.

Food for Thought

1. If broadcasting is an extension of our ancient oral culture—and reading among youth is in decline—then can it be that we still live in an oral culture? Argue your points.

2. Some critics and journalists argue that broadcast news would hold audiences and grow if it stuck to news, avoiding "infotainment." Would that work? Explain.

3. What can our government leaders (or anyone) do to influence the news that broadcasters cover and how they cover it? Is that a good public-policy goal? Explain.

4. Is there real potential for Internet news sites to take the place of either broadcast news or newspapers? What role can these sites best play? Explain your analysis.

The Audience

Chapter 9

The Bernie Mac Show hit the Fox airwaves in 2001, and if it wasn't all things to all people, it was mighty good to most. This sitcom about a married man who takes in his drug-rehabbing sister's three children—drawn from an experience in the lead actor's own life—appealed to many types of viewers. For starters, one critic observed that, compared with other TV comedies, the show brought "a little less fighting and a lot more loving." (This could include tough love, as when Bernie decided one child's school was not tough *enough*.)

With veteran stand-up comic Bernard Jeffrey McCullough just emerging into TV in his early forties, and with a corps of sharp writers, the series—like most sitcoms—ranged over the personal and family effects of commonplace situations: Bernie and his buddies traveled to Las Vegas. His "children" encountered thorny school and social issues. His wife became jealous when a woman joined the men's circle.

To Fox executives, of course, ratings—a sizable audience for advertisers to reach—were the primary reason for scheduling and promoting The Bernie Mac Show. It did well enough in ratings to be moved next to weaker shows to give them a boost. Once,

Fox placed it directly opposite another black-oriented show, threatening to kill off both, but that crisis passed. Bernie Mac was tops among African American viewers early in its life. In the 2003 NAACP Image Awards, the show's positive reflections of African American life earned it best-comedy honors and Mac was named outstanding lead actor in a comedy series.

On a broader level, its often warm and upbeat approach to domestic strains won the show the 2004 Humanitas Prize. Perhaps most remarkably, the Parents Television Council—always alert for clean-minded, family-friendly programs—put Bernie Mac on its "Top 10" list. The show flourished for three years, was renewed for a fourth, and went into syndication as well. "For too long, networks have been coasting on all-black comedies with stereotypical characters and cheap, low-brow jokes," noted one critic. "Mac has raised the bar."[1]

The hanging question: How long, in an era when five years is considered a good run, can The Bernie Mac Show continue to *clear* the bar not just in quality but in ratings? It has tough adversaries: Competing comedies, rival black-oriented shows, the trend toward unscripted "reality," and television's history of dropping programs with African American themes and casts as soon as they even begin to lose steam.

The NAACP and black opinion-makers fought Nielsen Media Research over a new system that they said would aggravate the "undercounting" of minority viewers. If that happens—or is happening—it could sabotage even good shows. Perhaps, it seems, Bernie Mac would do well to savor his good fortune one day at a time.

The Mysterious Crowd

The audience factor will resist understanding if we look at it only in terms of "what's on." In commercial broadcasting (and increasingly in Internet programming), what's on that's most important is not programs but *advertising*—the economic basis of the business. Media companies don't pay millions for audience measurement, research, and analysis just to make programs that thrill or amuse us or that serve high-minded public ideals. They invest in shows in ways that will help them attract advertising dollars. Programs, in effect, are not the fish but the bait.

Will an expensively researched TV show bring in the ad revenue needed to pay its costs? Is a drive-time radio personality holding enough commuters to make key advertisers happy? The answers to such questions directly affect year-end profits, and are the primary objective in measurement and studies of the audience.

A NECESSARY EVIL?

Almost no one defends ratings as wholly reliable head counts of the audience. Indeed, several years into the twenty-first century, interest groups forced the Nielsen ratings company into a pitched battle in the courts and public forums. The hot issue here was whether ethnic minority members were being counted fairly and correctly in ratings samples. And this was not the first such brouhaha; after three-quarters of a century, the art and science of tallying and analyzing a broadcast audience represent works in progress. Even measurement professionals sometimes acknowledge that their techniques are imperfect, or worse. Ratings have made a lot of enemies, especially among media executives to whom a single rating point can mean hundreds of thousands of dollars' difference in advertising revenues.

There also are industry managers who fret about what *viewers* may be losing. "Why do you have ratings?" asks one news director, disgust in his voice. "To establish

Nielsen Dominant TV-ratings firm.

Sweeps Period of intensive audience measurement three times a year.

ad rates. There's no other reason for them. That's the bottom line. We have no better system."[2] He believes ratings have distorted news coverage. So do many others. When CNN anchor Bernard Shaw announced his retirement, he huffed to the *Hollywood Reporter*, "Don't talk to me about ratings. I'm a journalist."[3] To critics such as Shaw, ratings might as well be hostile space aliens taking over the planet.

CALCULATIONS

Ratings come into play at the conceptual phase of a new program, when it's still being "pitched" to broadcast or cable by a writer, producer, agent, or other bearer of new ideas. Before production can go forward—often with big money at stake—broadcasters assess how large and desirable an audience the show can attract and hold. This projection of "eyeballs" that will be wide open for TV commercials must be presented to advertisers as they shop for so-called *upfront* commercial time before a new season begins.

PRINCIPLE 1

Projecting future ratings is an early step in the development of programming.

In summer of 2004, advertisers put an upfront hold on more than $15 billion worth of commercial time on broadcast and cable television. The money would buy ad slots during the 2004–2005 season. Those advertisers were agreeing to pay ad time rates that were calculated mainly on the expected size and makeup of fall-season audiences. However, those agreements to advertise were not yet final. Two months later, giant corporation AT&T, having a bad financial year, flip-flopped and canceled its spending plan—subtracting $110 million from the industry's upfront advertising total.[4] (Many millions would be added as ad buys occurred during the season.)

In electronic media, as in farming—which long ago "broadcast" seed across fields, gambling on germination—no returns are guaranteed. That's why advertisers and television companies today place great emphasis not only on raw numbers but also on *who is in the audience*. It's in combination with such information that ratings, which basically count households, have become an immensely powerful force.

Companies such as Nielsen use scientific methods to estimate from samples the number of people watching TV, tapping the Internet, or listening to radio at any given time. However, the results of this work often do not seem sufficiently scientific (or financially promising) to broadcasters and their clients, the advertisers. And even if the number of households tuning in is not in question, it may be a weak indicator of ad value without *demographic* data projecting the age, income, and other characteristics of the audience. The technical and tactical power to acquire such information has been growing for the past half-century.

THE EARLY LISTENER

In radio's infancy, people sometimes would gather in a public hall to listen together over a single receiver—and it was pretty easy to "count the house." As home radios became cheaper and use of them spread nationwide, audience measurement became a broadcast priority. In today's ratings-driven environment, it's hard to imagine being stuck in a studio having no idea whether your work was reaching a dozen people or thousands.

Measuring even local audiences was a challenge to measurement pioneers, and the U.S. population in 1930—about 123 million—loomed large enough to be daunting. Radio's spread had accelerated in the 1920s, but had not yet reached into every cranny of the country. The fact that one of every three Americans was still living on farms put complete tallies of a network's audience beyond researchers' grasp. They simply would have to work with the listeners they could find.

protalk
Stephanie Drachkovitch
PROGRAM DEVELOPER-PRODUCER

Seasoned TV-program developer Stephanie Drachkovitch finds *daytime* broadcasting—the long stretch between 9 A.M. and 4 P.M. weekdays—to be a challenging zone. Audience concerns make it so. She creates programs for networks and stations, often as "strip" shows to run at the same time every weekday. Their fate, says Drachkovitch, hangs on *ratings* (measurements of audience size) and *demographics* (measurements of audience composition):

"Whenever you're in the 'daily' business, you have to be careful about how you respond to ratings. 'What did we do yesterday? How do we do it again?' You have to take in all the factors: What was your lead-in yesterday? What competition were you up against? Were your HUT [homes using television] levels higher?

"There are certain topics and approaches that show a 'spike' [jump in ratings]. Think about it: Why are there so many episodes of the talk shows that deal with DNA testing for paternity, or troubled teens—'My teen dresses like a slut,' or 'My teen's combative' or 'My teen's in trouble'? Because those themes do rate well consistently over *time*. . . .

"In daytime, we've lost a big share of audience to cable. Now it's hard to get above a 2 rating in daytime. A couple of years ago, if you got a 2, you were off the

> ## "The demo you're looking for is upscale. It's the $75,000-plus household with an educated female."

air. Now, if you're delivering a 'pure' demo [the targeted audience] in that rating, you're fine. . . .

"The demo you're looking for is upscale. It's the $75,000-plus household with an educated female. They make buying decisions and lifestyle decisions in families. They're the ones buying the new car, buying the PlayStation, buying the Cokes. That's an attractive consumer, that woman who makes buying decisions for families."

Ways of Counting

One of the first ratings services, **Cooperative Analysis of Broadcasting (CAB)**—created by and for radio advertisers—began trying in 1930 to measure the national audience. No special tools for this task had been invented yet. Most of those that would be invented later would only build on and refine the solid, basic approaches with which CAB started.

SCIENCE OF THE SAMPLE

Archibald Crossley, CAB's lead researcher, directed squads of *telephone interviewers* who worked to identify radio listeners around the country and their favorite programs. A call to a home in a certain city, asking what people had listened to the night before, left the audience in the driver's seat: An interviewer had no choice but to trust the memory and candor of whoever was on the line. Still, the information was better than nothing. Armed with the phone-call findings, Crossley used a method called **quota sampling** to extrapolate national numbers from what local viewers had revealed.

PRINCIPLE 2

As an inexact science, audience measurement is open to challenge by its stakeholders.

Scientists understood that this was not the most statistically sound method. For one thing, it required the researcher to organize groups of listeners according to the proportion of their individual characteristics in the U.S. population as a whole. This might or might not yield results that accurately represented the whole radio audience. So quota sampling in radio soon gave way to **random sampling,** where everyone was supposed to have a chance at selection, which was intended to produce reliable cross-sections of the population. It reduced the likelihood that some of the results of this expensive research might be simply a matter of chance.

Whatever the sampling formula, the primary survey method in those days was the same: telephone calling. Even this basic information-gathering task could be approached in different ways, however. Within a few years, another media research firm, Clark-Hooper (later simply **Hooper**), was commissioned by broadcasters to do **coincidental** calling of radio owners. That involved calling a home while a certain program was on the air to see who, if anyone, was listening. This method obviously helped people report their program choices more accurately, without depending on what sometimes were faulty memories. Hooper used it as well to help determine the *available* audience for any program by eliminating from the total of radio owners all those who weren't at home when the program was on the air.

The company gradually implemented other techniques to increase the credibility of radio ratings. As scholar Karen Buzzard has pointed out, Hooper's ratings—typically much smaller totals than Crossley's, but more precise—tended to displease the broadcasters who were selling commercial time while pleasing the advertisers who were buying it.[5] In other words, Hooper provided better data per dollar on the real audience for a show, counterbalancing some of the broadcasters' grand claims.

Virtually from the start, a key to determining ratings was that it was neither possible nor necessary to count *every* listener. Instead, ratings consultants came to use *households* as the basic unit of numerical measurement—and didn't even count all of those, instead using samples.

WHAT NIELSEN CHANGED

In 1923, a company arrived that would come to be more closely associated with broadcast ratings than any other. Engineer Arthur C. Nielsen at first tracked retail food and drug purchases, but in the 1930s, he entered broadcast audience research.

Cooperative Analysis of Broadcasting Early audience-measurement cooperative among radio advertisers.

Quota sampling Study of a specified number of subjects of one type—say, 10 teenage girls or 20 male retirees; because subjects are not *randomly* recruited, results are open to challenge as not valid across a population.

Random sampling Study of subjects drawn from a larger group, with each individual chosen by chance; this gives all members of a population a chance of being chosen, making results potentially valid all across that population. (In *simple* random sampling, all members of a population have an *equal* chance of being chosen.)

Hooper Early ratings firm that pioneered use of coincidental telephone calling to determine radio listening patterns.

Coincidental Technique by which researchers call households *during* a program to see if they're watching or listening to it.

TELEPHONE INTERVIEWERS.
Using an early communication device, broadcast researchers since 1930 have probed audiences to learn what makes them listen or watch.

A. C. NIELSEN. Well before his name became synonymous with broadcast audience measurement, this researcher was tracking purchases of food and drugs to help marketers plot strategy.

Diary Booklet in which a listener/viewer's channel choices are recorded for analysis by a measurement firm.

Arbitron Audience measurement firm, working primarily in radio.

Two professors sold Nielsen rights to a mechanical "black box" that would become known as the *audimeter*. Placed inside someone's radio, it would record what station (frequency) was being tuned in and for how long. After several years of development, Nielsen put his audimeter into 800 homes in 1942—and the use of meters as a ratings method was born.[6]

Broadcasters who paid for the newly available data would know how many homes (at least homes with audimeters) had tuned into which radio programs. Such numbers could convince advertisers that their investments in commercial time would be sound. Soon, Hooper went further, using listener **diaries** to supplement its telephone interviews. The company was acquired in 1950 by Nielsen, which then stepped into a new medium through its Nielsen Television Index and its local Nielsen Station Index. The company had come to dominate audience measurement of a mushrooming sector of American cultural life.

Most of the emphasis in the rush for ratings had been on national broadcasting—that is, programming and advertising at the network level—but that soon would change. The Federal Communications Commission in 1948 had imposed a four-year freeze on the expansion of TV, as a means of sorting out conflicting frequencies and letting the industry improve its technology. When the freeze ended in 1952, television began a growth spurt that carried with it new pressure to measure both national and local audiences.

The result was a mad scramble among ratings companies to establish beachheads across the landscape. A new company called *ARB*, for *Audience Research Bureau* (today's **Arbitron**), used viewer diaries and also pioneered the famous *sweeps*—regular samplings within TV markets in certain four-week periods of every year. Nielsen, meanwhile, worked to refine and extend the reach of its meters. The logistical demands of the interview technique tended to confine it to major cities. Meters had an edge in reaching far-flung audience sectors. Nielsen's electronic monitoring approach had gained a foothold that helped it lead the pack in measuring network TV.

THE NUMBERS NOW

Broadcasting's boom through the latter half of the twentieth century placed increasing demands on ratings services. They responded by developing better ways to collect audience members that could be assessed in combination with demographic and other data about whom those audiences included (see "Analyzing the Audience," page 205). The measurement field today still is dominated by two large operations: Arbitron in radio and Nielsen Media Research in television. Each traditionally has used a mix of methods in tallying audiences, relying primarily on diaries and ever-evolving meters.

Diaries

Unlike what most of us mean by "diary" (a private record of personal experiences), the ratings diary is a structured log to be filled out by a viewer or listener after he or she has watched or listened to a program. The log is then returned to the measurement firm for analysis. People are supposed to use diaries in a steady, regular way, filling in their program choices throughout each day and evening, usually for a week at a time. Four weeks in a row have made up each of the traditional "sweeps" periods, normally occurring in February, May, July, and November.

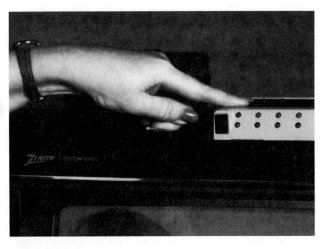

AUDIMETER. As early TV viewers enjoyed programs, Nielsen's monitor lurked inside their sets, recording channel choices and paving the way for audience meters in wide use today.

The diary generally indicates when the TV or radio is turned on or off, which is helpful information for the programmer or advertiser who's trying to pin down audience routines. Ratings firms compile that set-use information into totals called **HUT (households using television), PUT (persons using television),** or **PUR (persons using radio).**

After statisticians determine how many people or homes were tuned to each program, they project that snapshot across the total *potential* audience to come up with a total viewing estimate. Stations receive this information as two numbers: the **rating,** which is the percentage of all homes or people owning radio/TV who tuned in to a show, and the **share,** which is the percentage of all homes or people *currently using* radio/TV who tuned in.

Nielsen Media Research said it was contacting a million homes a year in all 210 TV markets, asking viewers to keep diaries during the sweeps periods. Arbitron was using a similar process, sending each household up to nine diaries so that every radio listener could fill one out—and paying a cash premium for each completed diary.

Participation in a diary system is voluntary, of course, and audience response never has risen particularly high. In the late 1990s. Nielsen's diary response rate reportedly dipped below 20 percent in some markets, and it took cash payments to bring the level above 30 percent.[7] Even that level of cooperation was inadequate, however. CBS's veteran research chief David Poltrack was quoted to the effect that a participation rate under 50 percent probably was "statistically invalid."[8]

As measurement technology advanced, the relative credibility of ratings drawn from diaries slipped. Arbitron, which abandoned its TV work in 1993, still claimed that diaries were the most reliable source of ratings for radio. While using diaries for sweeps, Nielsen also kept meters in 5,000 homes nationally and drew data separately from 49 metered markets.

Meters

In the six decades since A. C. Nielsen installed his first audimeters in radios, electronic monitoring that records program choices automatically—not depending on viewers to remember their choices—has grown tremendously. In fact, ACNielsen Corporation, now owned by a Dutch-based conglomerate, has become the world's largest *market* research firm. Among other business lines, it can draw floods of

HUT Figure indicating number of homes using television at a particular time.

PUT Persons using television.

PUR Persons using radio.

Rating Percentage of all television-owning homes that are tuned in to a particular station or channel.

Share Percentage of homes using television (HUT) that are tuned in to a particular station or channel.

information from electronic scanners in stores to help the food industry work more efficiently.

The company's TV-ratings unit, Nielsen Media Research, heavily influences the purchasing of tens of billions of dollars in advertising time every year in the United States and Canada. The meters it uses are intended to be not only accurate but also convenient for TV households—quite a change from the 1940s, when the audimeter arrived in radio and required almost as much of listeners as diaries do. Because the device picked up signals and recorded them—rather than transmitting them electronically to Nielsen headquarters—each household had to mail the tapes back to the company. This introduced human factors that left the metering totals incomplete, imprecise, and thus of limited usefulness to broadcasters.

Nielsen threw millions of dollars into developing technology, and in 1987, the company switched to *people meters,* automatic devices that required viewers simply to punch in personal codes whenever they started and stopped watching. This gave broadcasters a useful reading on exactly who in the vicinity of a television set tuned to a program actually was watching it. With that information came some of the viewer's demographic details. However, all of this still left open the possibility of human error or neglect in entering codes, especially when children were the viewers doing so.

Active/Passive Meters. Nielsen Media Research still hoped for an entirely "passive" technology, requiring no viewer action at all, and in the late 1990s began developing and testing an *active/passive* meter. It looks not merely at channels but at specific *programs* coming into the home, each carrying a distinctive electronic code. This can facilitate ratings based not just on television but also on whatever people watch through their digital video recorders (DVRs), devices that are spreading rapidly in the early twenty-first century. (Without this capability, Nielsen employees installing meters have skipped DVR homes altogether!) If the company reaches its design goals, it will perfect a meter that can separate TV, video recordings, video on demand (VOD) movies or programs, and other sources. Thus, without their involvement, what people choose from a growing menu of viewing choices can be sorted out more reliably and in greater detail than ever.

Some broadcasters and advertisers have complained that, even when meters do sort out DVR viewing from TV, they may not report it rapidly enough to allow quick changes in the placement of commercials. That's where ratings development is headed: toward audience-measurement that yields information on a *daily, year-round* basis. Although daily numbers have been reported under a sweeps system, when massive attention is focused on gauging audiences only every three or four months, returns had not so far been available each day, with overnight speed, all year long. However, Nielsen reported that active/passive meters with this capability would go into service by mid-2005.

Local People Meters (LPMs). For years, Nielsen worked to develop the *local people meter,* designed to supplant the sweeps-and-diaries approach to acquiring local TV ratings. Whenever a viewer prepares to watch television, he or she needs to enter a code. Then, when the show registers automatically as having been watched, the LPM shows *which viewer* did the watching. It also provides pre-loaded demographic data about that person to the broadcasters and advertisers waiting anxiously to see what sort of audience their show (and commercials) drew in terms of size and makeup.

Problem: Early trials of the LPM in Boston and New York returned ratings far lower than usual for many programs, including some prized by minorities as displaying good values and images—or as just being fun to watch. Many numbers were surprises, others were just hard to explain: New York African Americans spent 54.2 percent of their March 2004 viewing time watching cable TV, about 15 percent more time than diaries had recorded. Latino American viewership of Telemundo was up 22 percent above diary levels. Blacks increased their viewing of 90 channels and decreased their viewing of 17.

The National Association for the Advancement of Colored People objected to the LPM, which was scheduled to be installed in other cities later in 2004 and activated by 2005. Other groups joined in a coalition that sued for relief from the new technology. Congress held a hearing. Black and Hispanic leaders called for action. Nielsen joined some of them in forming a task force to iron out the problems. An improved LPM system seemed likely to provide the rapid, detailed information the television industry had been seeking for many years.

Portable People Meter. This device is the product of a Nielsen-Arbitron partnership aimed at capturing data from people listening to radio or viewing television outside the home. Such media consumption was known to be a hallmark of the young, especially young males who, as it happened, were the primary target of many advertisers. The *portable people meter (PPM)* hears or senses whatever show the user is watching or hearing and identifies it through codes. The downside: The human element is large. The user must pick up the device in the morning, carry it around all day, return it to its "hub" at home, and transmit its data to the measurement firm's remote computers. These data include not only programs watched but also known demographic descriptions of the user.

The PPM is not expected to be nationally deployed until later in the decade, but early trials reportedly have shown that it can be an effective adjunct to the diaries that radio still uses, and to the LPM headed soon for general television use.

Analyzing the Audience

In a hypercompetitive age, profiles of audience members have become far more important than mere head counts. Assuming that we know or believe that Viewer X is watching a particular show, many questions remain:

▶ What do we know about *X*'s age, gender, race, income, and education?
▶ How about his or her preferences in personal-care products, cars, appliances, credit cards, and clothing?
▶ How much income would he or she be likely to spend on such a product or service?
▶ Is Viewer *X* retired? Disabled? A teenager? A retiree? Sedentary? Athletic?

Broadcasters know the mass audiences who built broadcasting have been shattered into fragments by cultural changes and new choices of entertainment and information. They work to cultivate those fragments as successfully as they once farmed the mass. Particular slices of the old mass audience are especially lucrative. A top NBC programming executive could have been speaking for most networks when he said, "We need shows that draw a large audience that's desirable to advertisers—the eighteen-to-forty-nine demographic with high income and Internet usage."[9]

With an extra minute's thought, the executive could have listed many more qualifications for his target audience, further narrowing it. That would make every dollar his network spent still more efficient in reaching the people likeliest to heed the commercials. That, in turn, would improve his own odds of succeeding, show by show, day after nervous day.

PRINCIPLE 3

The characteristics of an audience usually are more important than its size.

DEMOGRAPHICS

Years ago, it was believed that TV and radio would be forces for unity on the U.S. landscape. Theorist Marshall McLuhan spoke of a symbolic "national hearth" around which television would gather us all. For a while—through the 1950s and later, at least—this notion seemed valid. A small number of stations and networks designed the small number of programs that attracted most of the population.

Today, an army of networks, hundreds of cable-TV channels, thousands of broadcasters, and a growing but uncounted fleet of Internet sites are *dividing* the population. Statisticians chop it into smaller and smaller units based on personal characteristics. Whether or not that's beneficial in civic and political terms, it has helped networks design programs that attract selected viewers or listeners, which helps advertisers find ready buyers for their goods. A few key markers sought by audience researchers are especially important:

Age

In today's culture, fairly or not, age often determines how people are viewed by others; that's why a commercial in which a teenaged "slacker" tries to chat with his elderly aunt hits our laugh buttons.

Targeting age groups can perpetuate stereotypes. Still, people with similar (or just similarly short or long) track records in life do tend to band together and share discoveries about products and services. Some especially tight-knit groups may place high value on a single commercial item.[10] Age turns out to be a good indicator of a person's disposable income, interests, and other realities. It's hardly surprising, then, that advertisers and broadcasters do try to gather age-defined audiences and sell shows to them.

Like the NBC executive mentioned earlier, most broadcasters are out to find viewers in one particular bracket: ages 18 to 49. People in that group tend to have autonomy, consume media (including *new* media) avidly, and spend money on their own. From a chronological standpoint, they're prime targets. Unfortunately for broadcasters and advertisers, heavy media competition has made that "demographic" harder and harder to attract consistently. As a result, the TV networks are pressing advertisers to broaden their prime target area to encompass people over age 50—baby boomers.

Gender

Women buy more products of more types than men do. Men do shop passionately for certain items that reinforce the male image (as *Home Improvement* often noted), but, to many TV advertisers, women remain the most desirable demographic in most age groups. That's why female-oriented programming is manifest through much of the television day. Success has come to advertisers who pitch their goods to women during morning programs rich in domestic-issue stories, daytime soap operas, afternoon talk shows (where *Oprah* has reigned supreme) and prime-time newsmagazines that emphasize human dramas. Many more women today work outside the home than just a decade ago. That means they're not at home as much to watch television, and it means that audience researchers must keep unearthing information to help programmers and advertisers appeal to this moving target.

Race and National Origin

The unsteady march of racial integration in the United States has put more disposable income in the hands of many African Americans. What's more, the growing number and prospective success of immigrants from many lands will create stronger markets for goods and services. Now, for broadcasters—remember, that term suggests a *broad* arena—race, ethnicity, and national origin are not always crystal-clear indicators of who will respond well to advertising. Buying within families falls into changing patterns as the nation grows more multicultural.

The 2000 U.S. census brought in a few relevant facts: More black households than white households contain children; the traditional two-adult family grouping is more common among Asian Americans than among whites, and the number of black and Hispanic families with both parents present is relatively low but rising.[11] These sorts of data are sure to influence programming strategy if they show up as part of audience research. They also will cause advertisers to shift their aim—and their spending.

Income, Education, and More

Several key factors or variables round out the profile of a household and help to predict the behavior of viewers or listeners within it. Researchers find that income and education track rather closely together; the size of a home, its geographic location, the number of occupants, the number of TV sets, and so on, add to the picture. If interviews and questionnaires determine that generally small, well-educated families live on good incomes in comfortable homes in a middle-class neighborhood, certain advertisers will conclude that their messages will find a friendly reception there. When such attributes appear across larger populations, then research firms may create strategic maps for the advertisers to follow.

PSYCHOGRAPHICS

Once broadcasters determine what viewers and listeners look like demographically, they can move toward discerning what viewers *feel, think,* and *do.* This information will bring advertising and programming much more precisely in tune with the audience.

So-called **psychographics** has been around since roughly the 1970s. Surveys and other methods already had made strides in linking the behavior of individuals to their demographic groups. Now, researchers began bunching broadcast consumers into clusters—who shared values and lifestyles and thus could be targeted by advertisers. It was advertisers who paid for much of this deep description, as a way of ensuring their future investments in commercial time. The prices for time, set by broadcasters, ultimately translate into **CPM,** or **cost per thousand**—what it costs an advertiser to reach each 1,000 viewers.

Pressure to make every dollar count has increased rapidly with the increased competition for select viewers and listeners. It makes good business sense to narrow the audience reach for some shows to people who are favorably inclined toward related products. This focus on how all consumers view the world has launched an almost surgically precise differentiation of some programs from others. For example, it was clear early that college graduates generally are less mesmerized by TV than are less-educated people. Still, both groups watch TV along rather similar time patterns after work, so the same commercials theoretically could reach both. What if an advertiser, trying to spend money wisely, doesn't want to pay to reach both types, but wants only the more culturally upscale audience? To be able to sift through viewers' passions and prejudices and display them "psychographically" and in other dimensions, comparing one show's audience and time slot with those of another, can help achieve that goal.

FINDING THE FACTS

No station, network, or advertiser can obtain all the audience data needed. Nor have ratings giants Nielsen and Arbitron provided a complete scan of each potential broadcast audience. Other companies with a complex array of experience have stepped into the field. Just like academics whenever they must furnish tables or charts, audience researchers tend to choose one or more proven methods before going about their work. That's because at some point, they must account for their

Psychographic Term for behavior profile of electronic-media users; psychographic research builds on ratings and demographics.

Cost per thousand (viewers or listeners) (CPM) Refers to cost of a commercial based on its expected audience reach.

methods' *validity* and *reliability*—two trusted yardsticks of how accurate and meaningful the data really are.

A central question to ask of any research report is: How did the observers extract data from their **sample**—all the people who cooperated in the study—through the specific questions they asked? One company, Mediamark Research Inc. (MRI), tries to allay concerns among broadcasters by stressing its systematic objectivity and "lack of interview bias," saying it reaches populations "difficult to survey through traditional means." Systematic and reliable methods applied to the same problem will yield the same results repeatedly—like a true set of scales or a good thermometer.

SRI Consulting Business Intelligence

SRI Consulting Business Intelligence is a broad-based research company that homes in on media audiences through a system called VALS.™ (VALS comes from an accepted abbreviation for *values, attitudes, and lifestyles*.) It claims to uncover the "underlying psychological makeup" of consumers—especially those traits that motivate buying—which strikes to the heart of what broadcasters and advertisers want to know. The VALS system approaches this by posing statements that encourage each person to think hard about his or her deeper impulses. Agreement or disagreement with such statements clearly can provide clues to what people might or might not buy, as well as to their tastes in programming. The VALS approach not only characterizes viewers and listeners through psychographics but also identifies where they live and seeks out related patterns. This can be called **geodemographics** and is practiced by other research firms, as well.

Symmetrical Resources

Symmetrical Resources is the parent company of Simmons Market Research Bureau, long noted for reports on how different demographic groups see broadcasting and the advertising it carries. In 2001, Symmetrical introduced a new way of "segmenting" television viewers according to their strongest preferences in programming—categories of shows with helpful labels such as Evening Soap Drama, Real Crime TV, Ethnic Sitcom, and Late Night Junkie. Through advanced statistical work, the system produced 23 types of viewer clusters that advertisers can locate within the great mass of viewers, using Nielsen ratings. Then the advertisers are able to target these groups in their commercials and choose shows in which to run the commercials—obviously a matter of critical interest to broadcasters.

Marketing Evaluations, Inc.

For many years, this company's TVQ rating has helped local stations decide whom to hire and fire, by canvassing audience members to see which performers—usually news anchors—are most "appealing." Critics often grouse that good anchors are fired because of imperfect "Q ratings" or because managers can use these ratings as a cover for the real reasons. In addition, TVQ tests the popular appeal of specific products and brands. Each May and November, the company interviews

Sample Group of subjects chosen for study in audience research.

Geodemographic Correlation of a household's location with demographics of its members; helps in determining broad patterns.

---

sidebar

DIGGING IN OUR BRAINS

A VALS questionnaire asks: Do you agree or disagree with these statements?

▶ I am often interested in theories.

▶ I follow the latest trends and fashions.

▶ Just as the Bible says, the world literally was created in six days.

▶ I consider myself an intellectual.

▶ I am very interested in how mechanical things, such as engines, work.

▶ I am always looking for a thrill.

▶ I like my life to be pretty much the same from week to week.[12]

1,800 adults to learn what different brands mean to them and how loyal they are to certain products.

Issues and Concerns

As noted at the opening of this chapter, nobody seems to be neutral toward ratings; for decades, ratings have seemed to hold almost absolute power over the content of the most popular mass medium in the world. Often forgotten is that ratings are simply numbers, buttressed by cold demographic facts—and by personal information that some of us give directly to media researchers. These bits of knowledge alone have no inherent power, but they acquire great power, indeed, when someone puts them to use.

That someone usually is a radio or TV executive weighing options for the near future, a situation often riddled with uncertainty and peril. Options include the launching of expensive programs that nobody may care about and the cancellation of ongoing shows that millions love dearly. Thus, while hoping conscientiously to serve his or her stockholders, superiors, advertisers, and possibly even the audience, the broadcast manager clenching a ratings sheet can probably smell its metaphorical fuse burning. If so, one of two important constituencies—the executive's own fellow broadcasters or the audiences they serve—may have lit it.

CONSUMER INTERESTS

Consumers of radio and TV programming—and, at one time or another, that's almost *all* of us—have a stake in the system that gives such influence to ratings. However, we have little clear control over it. All commercial systems set up ways of measuring product effectiveness, and that's what ratings are. Consumers always can try to outshout the ratings—force a hearing of their arguments—by focusing public attention on them. In a few noteworthy cases, they have effectively done just that.

"Quality" over Ratings

More than three decades ago, the now-classic *Star Trek* was cancelled, then was kept on the air a couple of seasons longer after a reported 100,000 letters flooded NBC. In 1984, CBS put the buddy-cop show *Cagney and Lacey* on "hiatus," an often permanent status, but a group called **Viewers for Quality Television (VQT)** quickly rallied letter writers. Thousands wrote to the network, and the show was spared for another four seasons.

Another popular but ratings-challenged program, *Designing Women*, also won a reprieve; it lasted another six years. If those were gentler times, when the industry seemed kinder—reading and sometimes heeding viewer mail—they soon ended. With competitive and budget pressures mounting, only ratings could sustain the prices the networks charged for commercial time. By the 1990s, television might as well have said to viewers, "Okay, from now on, it's no more Mr. Nice Network!" Shows would be cancelled and then stay cancelled. That happened in 1993–1994 when NBC removed a highly praised but lightly watched family drama, *Against the Grain*. Through the news media, the show's co-creator appealed directly to Nielsen households to rescue it from pending oblivion. He told a reporter, "No offense to all other Americans, but right now the only people we really care about who watch our show are Nielsen families, because that's the only thing that can keep us on."[13] It didn't happen: NBC stuck to its guns, replacing *Against the Grain* with the action show *Viper* (one newspaper dubbing that the year's "worst programming move").

Viewers for Quality Television, often in concert with independent viewers, mounted more letter-writing campaigns, but to little avail. Their founder, Dorothy

Viewers for Quality Television Citizens' group launched in 1984 to pressure networks to improve television; disbanded in 2000 when founder Dorothy Swanson became ill and retired.

DESIGNING WOMEN. The actors—Delta Burke, Annie Potts, Meshach Taylor, Jean Smart, and Dixie Carter—brought this popular comedy series to life. Eventually its ratings dropped to the cancellation point, but passionate viewer appeals won the program another six years on the air.

Swanson, saw that viewers were most effective in strengthening shows early, not in trying to save them later. Still, grassroots action did keep a popular young-pal series called *Party of Five* on the air *somewhere* for awhile; launched on Fox in 1994 and cancelled six seasons later, it moved to the Lifetime cable channel but then faded from the lineup.

When Swanson became ill and disbanded VQT, it was succeeded by another national audience group, Wisconsin-based Viewers Voice, which had been pressuring broadcasters on its own. It, too, faced long odds.

PRINCIPLE 4

Broadcasters do not give all audience members equal weight in programming decisions.

Diversity Campaigns

Besides campaigns to promote "quality" broadcasting, there are crusades to bring more prominence to various minorities in the program content and management of television in particular. In 2001, NAACP demanded that the networks give nonwhites more authority over programming. The move came nearly two years after the NAACP held heavily publicized meetings with top TV executives, and even though overall casting of nonwhite actors improved, some movement leaders held out for more lead roles and greater influence behind the Hollywood scenes. Pressure also has come from advocates for Hispanic Americans, an especially fast-growing population group. Cable and satellite channels—looking for niches and seeing diversity as less of a ratings risk than the networks do—have created some notable minority-based shows and offer extensive Spanish-language programming.

Concerned viewers and career activists can work more surgically toward their goals, as demonstrated by the Gay and Lesbian Alliance Against Defamation (GLAAD). With support from other homosexual groups, GLAAD campaigned against Dr. Laura Schlessinger, who had been controversial on radio and was now launching a TV advice show. The campaign worked: Complaints that Schlessinger was antigay cost her advertisers, then low ratings helped push her off the screen before she had completed a season. Conversely, gay and lesbian organizations have been fervent backers of shows such as ABC's mid-1990s *Ellen*, featuring a gay comedian-lead-character, and the Showtime cable hit *Queer as Folk*, adapted from a popular British program.

In Public Broadcasting

It may seem incongruous to think of ratings in connection with public broadcasting—that special web of stations and program suppliers that depends on individual donors, foundations, and government for most of its financial support. The official ethos of public broadcasting holds that it can take on topics and treatments that commercial broadcasters wouldn't dare touch. However, audience research has influenced decisions in public radio and television for decades. That's a source of simmering controversy as public stations creep toward a commercial model under which they would run true commercials—not just brief "underwriting" spots—to bring in revenue.

Since the mid-1990s, the Corporation for Public Broadcasting has used Arbitron ratings as a basis for granting funding to community and public radio stations. Critics of this practice say it tends to knock the corners off programming, pressuring stations to take fewer risks in their choice of issues to tackle and the content they air. The role of ratings was a factor in turmoil at the nation's first community radio station, KPFA, in Berkeley, California. The social activists who made the station a fixture in community life wanted less emphasis on audience size, as measured by Arbitron, than KPFA's leadership preferred.

The Radio Research Consortium publishes ratings and listener demographics for many major public-radio programs, to help stations find and nurture shows to hold local audiences—prospective donors—in place. Consultants now advise public TV programmers to keep viewers engaged in the weeks before on-air "pledge drives," through which most membership donations flow in. In general, audience size and loyalty have not kept pace with programming costs, leaving a yawning and persistent financial gap. Nor has Congress been consistently supportive. In this environment of need, the use of ratings and audience analysis seems likely to grow in public TV and radio.

BROADCASTER INTERESTS

At this writing, a combination of ratings and audience-analysis data governs the daily content of TV and radio programming across the United States. That's hardly news; what may seem more surprising is that pressure to follow ratings has intensified in recent years. At the same time, resistance to the old ways of tallying audiences has grown within the industry. Broadcasters accept the need to "count the house," but they want Nielsen, Arbitron, and other researchers to do their jobs better. Behind that drive lies a need to deal with issues that range from annoying to threatening.

Drifting Attention

The added pressure to generate high ratings was produced largely by two related trends: the upsurge in competition from other media and the rapid reorganization of the broadcast industry. First, the proliferation of broadcast, cable, and Internet competition has split the attention of millions of media users across the United States. Although a relative few channels dominate most viewing, the scores of TV options today obviously make it difficult for any one network or station to rule the field. Moreover, for viewers armed with today's rapid-fire remote-control devices, "surfing" can turn a widened spectrum of choices into a colorful blur; it's the despised enemy of something broadcasters love: viewer loyalty. When forces like that drive ratings down, they're hard to win back.

The migration to cyberspace for entertainment and information poses an even greater challenge for broadcasters. For example, the viewer who buys her or his first Internet-empowered computer may well become infatuated with the new world of bits and bytes and watch less TV.[14] As his or her enthusiasm spreads through the population, it has diverted one viewer after another from the usual viewing diet—certainly not all viewers, but many. Some may return to the old shows, but

protalk
Pat Cashman

DISC JOCKEY, KOMO-FM, SEATTLE, WASHINGTON

Audience power can rise up against broadcasters at the drop of a pink slip. Angry calls and e-mails poured in after Seattle radio station KIRO-FM cut a comic "personality" named Pat Cashman from its lineup in favor of syndicated programming. One listener told a newspaper that when he heard of the change, "I felt as though a member of my family had died." Two major advertisers cancelled their commercials.

The station's program director was left to explain that he took Cashman off the air because he clearly appealed to two audience segments: female listeners and families. Why was this a bad thing? Because it detracted from the sex-talk of the station's raunchy afternoon performer, whose show commanded a huge young, male audience. KIRO didn't want to blur its macho image.[15]

Cashman recalls, with a kind of sardonic wonder, the day he lost his job: "It was April of 1999, my day of reckoning. I was called into the general manager's office. They had the human-services guy there, and the PD (program director), and the GM. I said, 'Gee, I don't believe I've seen you all in the same room together before.' After that meeting, doing the final show was a little tough. . . .

"But the most unusual thing about the firing was what happened next. . . . There was a group that formed as a result of my dismissal. They were big fans of the show,

> **"I was called into the general manager's office. . . . After that meeting, doing the final show was a little tough."**

and formed the 'Pat Pack.' These people found a collective gathering place on a website. They'd use it to hold rallies. Even the PD who fired me said he had never seen anything like this. People just deluged that station with cards, and even picketed outside the building."

Before long, Cashman—highly popular on Seattle radio for years—found himself behind a microphone at another major station, KOMO-FM. Now, having moved to yet another Seattle station, he looks at the radio industry and finds fault with the relationship between its decision making and its analysis of audiences. "Consolidation bears much of the blame," says Cashman. "Local stations at their best are part of the fabric of a town, so there can be quirky types that populate your shows. You're part of the local scene, the local uniqueness. But when the people making the decisions are back in Dubuque, they don't have an appreciation of that. All they look at is what the rating book says." Cashman left radio in 2005 to make other use of his speaking and performing skills.

ratings still slip in the meantime. This, of course, keeps many executives struggling to give advertisers a large enough and eager enough audience to justify their spending on ads. The result can be either me-too imitation of successful programs or risky innovation in search of demographic appeal.

An Economic Jam

The reorganization of the industry largely involves wealthy media (and nonmedia) corporations that buy up weaker companies. These include networks, station groups, and local stations. This power shift upward tends to harness small-market broadcasters to Wall Street players who own the parent corporations' stock. The effect is to place greater pressure on stations to hit ratings home runs at every opportunity.

PRINCIPLE 5

Industry consolidation increases pressure for high ratings.

If a station can't raise ratings, and thus ad revenue, high enough to meet profit demands, it must cut costs—which can involve laying off workers. It also can drive stations toward look-and-sound-alike syndicated programs, though a few stand out from the crowd. One of these, Rush Limbaugh's mostly political talk show, has been a particularly productive choice for radio managers eager to gain ratings. Since 1989, Limbaugh's success reportedly quadrupled the number of AM talk-radio stations, and was capable of single-handedly turning a ratings-losing station into a winner.[16] One former station-group chief is quoted as calling the impact of ratings on local radio "fabulous for shareholders, but terrible for listeners and employees."[17] Many industry voices have been raised against this web of financial causes and consequences, but almost always behind a mask of anonymity and to little lasting effect.

Doubts about Numbers

Broadcasters often blame the ratings methods for costly shortcomings that keep "the numbers" too low. An example occurred when the head of CBS Sports complained publicly that Nielsen was "undercounting" TV audiences for sports events. He said this was because many fans watched not just from homes—including "Nielsen homes"—but also from other venues, ranging from bars to college dorms. Technology may remedy some ills, but unfortunately, over time, ratings advances have failed to stave off broadcasters' concerns.

Discontent with the quality of Nielsen's numbers boiled over in 1996. A Wisconsin broadcaster confided starkly to the *Washington Post:* "It's about time we called the baby ugly."[18] The Fox network threatened legal action against Nielsen; ABC, CBS, and NBC put money into an effort to develop a ratings alternative called **SMART (System for Measuring and Reporting Television).** The networks' main complaint was that Nielsen's sampling process underreported the young viewers. If true, this was bad for all broadcasters. Another beef was that Nielsen *overcounted* cable-TV homes, giving cable higher ratings than it deserved and putting over-the-air TV at a disadvantage.

In 1999, assured by Nielsen that its in-house tinkering would improve ratings, the networks withdrew their money from the SMART project, and it folded. This did not discourage two other companies—the cable-system operator Comcast and the home-shopping network QVC—from pushing another alternative. Called **TargetTV,** this would use set-top boxes to record program choices. It would not answer demands for a mobile device like Nielsen's Portable People Meter, and had other limitations; but it would record channel use *every five seconds* to keep up with channel surfing. By spring 2001, TargetTV was in 60,000 homes and its developers were planning to install it in many more, but development of newer meters was overtaking them.

Diaries still are a widely used tool for obtaining demographic data. Stations have complained—evidently with some validity—that very low response rates from diary households were distorting results. In some cities, those responses reportedly dropped below 20 percent. Nielsen answered this by giving viewers $5 apiece for sending in their diaries during sweeps, and it worked: The response rate began to

System for Measuring and Reporting Television (SMART) Unsuccessful 1990s effort by TV networks to design alternative ratings system to Nielsen's.

TargetTV Effort by Comcast Cable and QVC (shopping channel) to develop alternative to Nielsen techniques, using set-top boxes to track TV-channel choices.

TALKING TOM. As a widely syndicated personality and a ratings success for urban stations, Tom Joyner is one of few African Americans to gain strong popularity in talk radio.

rise. Still, fewer than one in three viewers provided Nielsen with data, which seemed to support charges that the all-powerful ratings system was badly flawed.

The audience-counters strive to improve, but in the super-heated cauldron of broadcasting today, measuring audiences may be a no-win proposition. When the National Association of Broadcasters asked TV stations nationwide to indicate their overall level of satisfaction with Nielsen, 23 percent said "low," 40 percent said "neutral," and about 30 percent said "fair."[19] This reinforced impressions that broadcasters feel deflated by the system. Worse, many emphasize—they're paying for it.

A Collision of Values

The deepest concern about ratings-driven broadcasting is aimed not at measurement firms but at the industry's own managers. It is they who convert ratings into decisions that sometimes confound the values of their own programming. It also is the top managers who, critics say, have ceded all control over broadcasting to its advertisers—as in the early days when sponsors dictated content.

Nowhere is the impact of audience numbers more onerous than in the newsroom. Journalists who complain that the ratings obsession is diluting news run the risk of being branded mavericks; still they complain. News executives often privately criticize (or curse) the influence of ratings over their upper management. Too often, some say, a single low **book** (sweeps period) leads the boss to order staff cuts; since the news department usually is the station's largest, it's the most inviting target for layoffs.

Carping about the power of ratings is a familiar refrain in the news business, and has been so at least since Edward R. Murrow and his producer, Fred Friendly, made CBS a temple of journalism in the mid-twentieth century. Murrow the journalist chafed at business pressures on landmark news programs such as *See It Now*—then gave way to Murrow the entertainer, the chain-smoking host of a celebrity show, *Person to Person*. Friendly, on the other hand, became president of CBS News, only to quit in 1966 when the network dumped live coverage of Senate hearings on Vietnam, cutting to the ever-popular *I Love Lucy*—in reruns. Ratings realities could overpower even the icons of journalism.

In the 1970s, as ABC News sought stability and respect, stars Ted Koppel and Sam Donaldson beamed their pride in the network's news coverage and paid little heed to its huge audience deficit, according to one account.[20] As the seventies ended, veteran news executive Richard Wald could still say, "The question always is, will you put something on that may not get ratings, but may be important? So far, television news, like other news organizations, has always said yes."[21] Today, Wald could not honestly say "always." Unable to escape the ratings trap, news managers struggle in its grip. Prime-time newsmagazines cater to curiosity about family violence and Hollywood, dispensing gossip amid the "news-you-can-use" format. Not just individual stories, not just entire newscasts, but news *departments* may be thrown out if the ratings go sharply against them. In 2001, years of low numbers led Indiana's WEVV-TV to cancel all of its newscasts, as some other stations have done since.

THE GRIM SWEEPER

Like salmon in the Pacific Northwest, local TV news directors are a perennially threatened lot. Their departments produce what in most stations is the only locally generated programming. Sometimes news directors lose their jobs after developing "creative differences" with their superiors. More often than not, it's ratings that usher them to the exit. Their average job tenure is not much over two years; nobody is immune. John Mussoni is a case in point. He had spent almost 15 years running

Book Another word for *sweeps*, a period of intensive audience measurement several times a year.

the news operation at the Fox station in Philadelphia when the general manager fired Mussoni, and told a reporter why: "It's ratings, pure and simple."[22]

Such dismissals usually occur in either December or June, soon after the end of one of the two most important sweeps periods of the year. During sweeps, local managers quickly obtain minute-to-minute feedback on any newscast's ratings; this enables them to tweak a show in time to get better results the following day. Because tweaking is possible, it has become routine.

Sweeps periods are noted for the loud hyperpromotion of news and for the prevalence of sensational, even salacious, material. There are promotion firms that specialize in fashioning sweeps promos, and TV consultants who track stations' ratings and pick up their highest-scoring sweeps ideas to peddle to other clients. Many viewers do respond to sweeps hype, which includes contests and even cash inducements to viewers, a kind of reverse pay-per view. When these intense periods are over, ratings typically tail off, making it as hard as ever for stations—and their advertisers, who have big money at stake—to determine a newscast's true drawing power.

Some executives say they yearn for a better, calmer, longer-lasting way to raise their ratings and revenues. They disdain the necessity to do garish, low-common-denominator programming four times a year, "but we have to put on our party clothes, because that's when they take the pictures," said one general manager.[23]

The Digital Future—Now

At the outset of this chapter, we posed several questions: Should the rating system be changed? How? Should people change the way they *use* ratings?

Beyond any conclusions you may have reached thus far, other possible answers are stirring in the minds, offices, and laboratories of ratings specialists. Some broadcasters believe new approaches to measuring and evaluating their audiences must be found as media competition increases. Otherwise, they warn, radio and television will not be sufficiently self-aware and nimble to anticipate consumer desires and hold their own against future alternatives.

The wild card in these projections is that new players with new game plans are contributing to rapid technological change across all electronic media. This already has begun to restructure broadcasting, and it could necessitate entirely new approaches to every aspect of all media—including the assessment of their audiences.

PRINCIPLE 6

No modern media—even "new" media—can escape the imperative to count and evaluate audiences.

THE INTERNET AS A MEDIUM

The success of the Internet in reaching Americans focuses more audience research on this medium every year. Several veteran ratings companies, as well as some new players, are mapping the growing audience for on-line news, information, and entertainment. At this writing, the most popular news sources on the Internet are established newspapers and magazines. When the World Trade Center was attacked in September 2001, researchers measured the audience's news quest to see if websites belonging to traditional news organizations such as CNN and BBC were preferred (they were), and if network TV news coverage exceeded the Web's popularity (it did).

As expansion comes—and it's coming rapidly—audience measurement and analysis will grow with it. Already, among others, ACNielsen Company, its now-separate offshoot Nielsen Media Research, and an Internet-focused branch called Nielsen NetRatings have collaborated to survey the fan base across this widening landscape.

protalk
Matthew Zelkind

STATION MANAGER, WRIC-TV,
RICHMOND, VIRGINIA

H e has a good track record, a respected staff of journalists, and a growing audience. In several ways, however—all related to ratings—Matthew Zelkind was trapped. "We are at the mercy of a flawed system," he said, "a monopoly on how the numbers are determined. Our viewer base is diminishing, and that creates almost a situation of desperation to attract viewers."

As long-time news director of a Nashville, Tennessee, TV station, Zelkind lived under the tyranny of audience ratings. Nielsen Media Research had reported an overall decline in local-news viewing in the Nashville market. It probably was tied to cable and satellite-TV competition. Whatever the cause, it was damaging to Zelkind's enterprise, and he believed that ratings companies make such trends worse by only *sampling* newscasts, exaggerating the importance of short time periods.

"It unfortunately has turned into a situation where sound journalism is secondary to attracting reviewers in a metered market for 300 seconds at a time," Zelkind said. "You

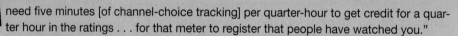

> **"Our viewer base is diminishing, and that creates almost a situation of desperation to attract viewers."**

need five minutes [of channel-choice tracking] per quarter-hour to get credit for a quarter hour in the ratings . . . for that meter to register that people have watched you."

Zelkind saved his heaviest barrage for rival stations that "stunt" during ratings sweeps, not only airing sensational material but also running contests to draw viewers. "They literally bribe viewers to watch their newscasts four times a year," he said of one station. "They give away cash, trips, houses—$500 every half hour."

For years, Zelkind kept his newscasts competitive. In 2004, he moved up to a job as station manager of a Richmond, Virginia, station. There, ratings would continue to be an ongoing challenge.

Nielsen/NetRatings says it's charting "actual click-by-click Internet user behavior measured through a comprehensive, real-time meter installed on the computers of over 225,000 individuals in 26 countries worldwide both at home and at work."[24]

Besides the established audience-measurement companies, there are new-media research firms that already take the pulse of a variety of on-line communication channels, for a variety of purposes. For example, Jupiter Media Metrix analyzes the impact of the Internet and e-marketing on consumers.[25]

LINES OF CONVERGENCE

Unlike broadcasting's viewers and listeners, computer users tend to operate interactively through individual links to the Internet. This makes them easier to count and identify. The difficulty of obtaining demographic profiles of them could depend on the users' willingness to answer questions—an update of the decades-old diary-response issue—and on researchers' ability to analyze these consumers through surveillance methods if interviews and surveys fail.

Interactive television is becoming available to consumers, offering them more control than ever before over the media they use. Many of these media will draw support from paid subscriptions, from per-use charges such as movies-on-demand, and from the purchase of time and screen space by advertisers pushing their wares.

Summary

Ratings show managers of electronic media how many households are tuned in to their channels. Audience research extends that knowledge to encompass many of the personal characteristics of viewers, listeners, and Internet users. Armed with this information, managers reach some of their biggest decisions about what we see and hear over radio, television, and now the Internet.

It's an imperfect system of estimates and extrapolations, since no one (so far) has discovered how to sample *every* user of an electronic medium at once with reliability. Measurement and research firms have experimented for decades with methods to improve their audience snapshots. Systems have come far—from simple telephone calls to 1930s radio listeners to see if they remembered their recent listening patterns, to today's sophisticated monitoring of TV viewers' behavior through electronic sensors. In demographic and psychographic research, greater texture and detail define media users. Meanwhile, the digital age has given birth to new media and is spawning new kinds of data-gathering techniques to identify their audience and its desires.

Food for Thought

1. ABC dropped the nighttime *Who Wants to Be a Millionaire?* after several successful seasons. Why did the show lose its ratings (or demographics) power? Are audiences really so prone to impatience or boredom, or do they just give way to *new* audiences?

2. Broadcasters who change their programming in response to ratings or research usually say they're simply giving audiences what they want. Is this good, bad, or both? Why?

3. What are the practical limits of meter-based audience measurement? Experimental meters can see or sense who's in the room and who's watching or not watching TV. Is that invasive or not? How and why?

4. When radio or television presents programs targeted to narrow slices of society instead of the mass audience, is it threatening our social fabric? Why? Does broadcasting have enough power to do harm to social cohesiveness? Explain.

Advertising *and* Promotions

Chapter 10

Something has changed about going to the movies, besides the surround-sound speakers and the stadium-style seating. Ads spliced in the previews of coming attractions and the products placed in the movie itself are now playing at a theater near you. Advertising at the movies is a fast-growing business—at a rate of around 38 percent per year from 1999 to 2004. Compared to network television or newspapers, it is still a small enterprise, but it is growing.

When a brand-name product appears on the screen as part of the action, advertisers call it *product integration* in film and television. Some agencies now specialize in this not-so-subtle form of advertising, promising that the company's name and image can become an integral part of the action conveying implied endorsements.

Internet advertisers are just as ingenious in using the techniques necessary to create a memorable brand image. TV-style commercials neatly tucked behind web links that read "view our ads" appear with a couple of mouse clicks. A viewer clicks on a web page that begins downloading, and with one more click, a TV-styled spot starts running.

Is this all a form of subterfuge, savvy business practice, or both? Contemporary advertisers know that in order to reach fickle audiences, they must cut through the clutter. Cinema ads, product integration, web commercials, and other nontraditional media are the latest weapons of choice used to strike the target audience.

Business Basics

Advertising is based on planning a media strategy designed to reach target audiences, and expose them to a message more than once. It is all a part of marketing, which is the science of converting product knowledge into consumer information in order to persuade potential customers.

Commercials are created and priced according to the audience's size, the number of impressions they make, and the media used, among other factors. Regardless of the medium, government regulation in advertising *differs* markedly from other information sources. When truth in advertising is at stake, it is not just a case *of buyer beware* but also *advertiser beware*. Legal and ethical issues about the nature of advertising were among the reasons why broadcasters were reluctant to accept commercial sponsorship as a source of revenue in the first place.

A Brief History of Broadcast Advertising

Advertising trade publications, such as *Printers' Ink*, were wary of broadcasting as it entered the commercial realm in the early days. Radio was a privileged medium for the sanctity of the family circle, and "advertising has no business intruding there unless it is invited."[1] Once radio's ability to inspire the imagination and create loyalty among listeners was realized, advertising grew to become the lifeblood of America's electronic media.

Radio ads were introduced to listeners in 1922 (as noted in Chapter 2) when a station in New York City helped to lease apartments on Long Island. Almost two-decades later, American television entered the field of advertising when the first TV commercial for Bulova watches aired in 1941 during a baseball game in New York City. Today, television's advertising volume is second only to newspapers, and although radio's advertising fortunes have tapered off, it remains a robust commercial medium.

RADIO'S FIRST ADS

The first company to experiment with radio advertising was the phone company. American Telephone & Telegraph's radio station in New York sold what was called "toll time" to a Long Island realtor, the Queensboro Corporation. WEAF broadcast five "talks" of about 10 minutes each to promote apartments for lease. These early "infomercials" seemed to work, and big-name sponsors followed Queensboro, including Colgate, Macy's, Metropolitan Life, among others.

The demand for decorum inspired self-imposed limits on early radio commercials. Until 1927, most shows were *sustaining*, which meant they aired without advertising. Those that did include ads had to play by the rules. They limited the pitch to one—that's right, only *one*—mention of the company name or product per show. No word of prices was allowed on radio until 1932.

THE JACK BENNY SHOW. In 1934, General Foods signed on to *The Jack Benny Show* as its sponsor. Benny began greeting audiences with "Jell-O again" for every Sunday evening over a 10-year run. Radio audiences also became familiar with the J-E-L-L-O jingle.

Advertising agencies such as J. Walter Thompson seemed to be more inspired than deterred by the restrictions. They dreamed up schemes working sponsor names into talent acts. When vaudeville and radio merged, advertisers such as Eveready Battery, Gold Dust washing powder, and A&P groceries named performers after their brands. Each time the radio announcer thanked the Gold Dust Twins or the A&P Gypsies, he was plugging the product. More creative sidesteps were contrived. Jack Benny greeted his radio audience, "Jello again." It was just a matter of time before major ad agencies on Madison Avenue overcame broadcasting's barriers and created radio slogans with catchy jingles: "LSMFT—Lucky Strike Means Fine Tobacco," and "You'll wonder where the yellow went when you brush your teeth with Pepsodent."

When the Depression darkened America's mood, radio spots resonated with appeals to "fear, shame and blame [that] . . . conveyed a common message: 'If you don't buy this product, you will be sorry.'"[2] Commercial profits sank during the 1930s, and marquee ad agencies found it impossible to maintain their billings, the monetary measure of advertising sold. When sales dropped, salaries were cut, and personnel were laid off—these were dire times.

MA PERKINS. Radio programming and products merged during the 1930s, when Oxydol detergent sponsored the serial *Ma Perkins*. The genre "soap opera" took its name from this family melodrama, which featured actress Virginia Payne in more than 7,000 broadcasts.

Audience distrust of commercials grew in response to the malaise as well as the growing repository of dishonest ads. This dismal valley was charted by Ballyhoo, a 1930s magazine devoted to uncovering **puffery**, advertising claims that were just too wonderful to be true.[3] Yet, radio sales people persisted throughout this era in believing there was gold to be mined in broadcast time and they continued to prospect for it. At first, no one thought radio could sell commercials during daytime hours. That was until a Chicago ad agent came up with a plan to attract housewives in 1932. Frank Hummert drafted scripts for family melodramas while his assistant Anne Ashenhurst produced shows such as *Betty and Bob* and *Just Plain Bill*. It was Oxydol laundry detergent's sponsorship of *Ma Perkins* that gave the genre its name: soap opera.[4]

Saturday morning serials caught on with kids who tuned in for their favorite heroes. Tom Mix and the Ralston Straight Shooters kept cereal bowls full of Ralston's, while other youngsters gulped down Wheaties, the "Breakfast of Champions" during episodes of *Jack Armstrong, the All-American Boy*. For the girls, Ovaltine went well with *Little Orphan Annie*, and there were more comedies, musical variety programs, and quiz shows for the rest of the family—all with commercials. By 1938, radio had overtaken magazines' lead in total volume for advertising revenue. Clearly, the business of broadcasting was booming, while audiences became accustomed to a mix of programming and promotion.

RISE AND FALL OF SINGLE SPONSORSHIP

The magic that worked for sponsors in radio seemed to fare even better on television. The strategy of producers in the 1950s was to match product ads with customer tastes in programming. The humor of *The Ernie Kovacs Show*, for example, appealed to men who might want to light up a Dutch Masters cigar. American housewives would leave their kitchens to watch celebrities on *Betty Crocker's Star Matinee*. And *Texaco's Star Theater* was one of many shows to build a loyal following to the single-sponsor program. A singing quartet of Texaco attendants promised millions of Milton Berle fans each week that they could trust their car to "the man

Puffery Advertising exaggerations not intended to be taken seriously, but to make the messages more interesting and entertaining.

The business of commercial broadcasting is building audiences for exposure to advertising.

who wears the star." Soon, however, the harmony turned to discord when a certain type of TV program fell victim to corruption.

In 1958, news broke that *Dotto*, a CBS game show, was rigged. After the producers of other TV game shows, including *Twenty-One* and *The $64,000 Question* came before congressional hearings, the networks decided to remove most of their single-sponsored shows from the program schedule. Networks took over production, and began selling shows to multiple advertisers who participated in programs instead of sponsoring them outright. At the same time, the business of advertising was evolving in terms of strategy, organization, and resources.

Commercial Enterprise

It all begins with marketing, which is the business of bringing products to the attention of a desired audience. That cannot be accomplished, however, without getting to know the product first. Advertisers become personally acquainted with

Advertising begins with marketing—knowing the product, its benefits, and the target audience.

the product's strengths and weaknesses so they know how to persuade consumers to try it. The challenge becomes bringing product qualities out in advertising messages that will be remembered at the point of purchase.

The rules of the game for advertising on radio, television, and the Internet are familiar to the professionals who have made it their career, but mystifying to most everyone else. They range from how spots are packaged and sold to where they should be placed in the broadcast day. Advertising sales people perform a range of duties in a variety of offices—at a radio or TV station, an advertising agency, a network, or other firms that employ their own advertising staffs.

ADVERTISING FORCES

National businesses naturally rely on different media to get their point across to potential customers. Mammoth firms like IBM, General Motors, and Procter & Gamble often recruit agencies to draw up their battle plans in order to win their market share and keep on growing. They do this through a strategy called *media mix,* which identifies all of the channels used to attract new customers and build loyalty among old ones. Some firms employ in-house agencies to design commercials and campaigns. Regardless of where the advertising team is located—in-house or in Manhattan—winning players are well equipped to reach the right audience with the right message.

Advertising Agencies

Advertising agencies occupy the pinnacle of a much larger industry that criss-crosses into radio and TV production studios, web-design boutiques, and print shops. The full-service agency threads a number of duties: account management, creative tasks, media planning, and marketing research. These specialties report to a senior administrator, such as a vice president or president.

Account Management

Account sales personnel act as both consultants and facilitators. They learn fully the clients' promotional needs in order to intelligently advise them on the decisions to make with their ad dollars. They are paired with client accounts to build their business through effective marketing. They meet and consult with these people about their firm's goals and how advertising can achieve them. Once message objectives are drafted and target audiences are identified, an advertising plan can be roughed out for execution. The account executive then goes to the creative writers, artists, and media producers to vizualize the client's message in words and images, while consulting with media buyers about the proper mix for broadcast and print distribution.

Creative Ideas

The creative department is where imagination holds sway. It is there that ideas are born and sold. *Creative writers, art directors,* and *producers* are the artisans responsible for choosing words and pictures to create compelling images for the client. Radio and television *producers* are responsible for making the spots, which demands creative and administrative talents. Many details must be addressed in the production of any commercial—everything from selecting the talent and location to budgeting, scheduling, recording, editing, and eventually getting final approval of the finished spot.

Media Planners

Media planners and *buyers* are responsible for the planning and placement of advertising messages. They draw up and present the media mix as a strategy to fulfill the client's promotional goals. The agency's media buyer takes the finished campaign and distributes it in markets where it will do the most good, which often requires buying time with local, regional, and even national media.

Advertising Markets

The country is divided not just into cities but into markets that encompass more than one city. The term used to describe media locale is **designated market area (DMA)**. Nielsen uses the designation to describe an area of ratings competition between local television stations.

Marketing researchers inspect circulation figures and rating books to find the best stations and other media outlets for target audiences. Other forms of feedback, such as *focus groups,* may gauge a commercial's readiness for air before it's ever broadcast. Afterward, telephone surveys often check to see what type of impression the advertising spots have made. This research process is covered in greater detail in Chapter 9.

BROADCASTING FORCES

Inside the offices of local radio and TV stations, advertising and sales people work the telephones to make **cold calls**, where they introduce themselves to prospective advertisers or get in their cars to visit clients for the first time. The general sales manager leads the station's sales force, and is responsible for both local and national sales of commercial time. That individual acts as a team leader, motivating the sales staff of the station by drawing up a plan of advertising goals and the best ways to achieve them. Like a coach, the general sales manager tries to project a positive image for the station and build a winning team.

Local sales managers spearhead their forces in the city and surrounding vicinity. Their aim is to sell the station's commercial **inventory** to advertisers. They show AEs how to package and present the station's commercial time in order to maximize sales. Local sales managers also prepare presentations on audience ratings and research data on the competition, particularly cable and Internet advertising that has sliced away at the TV pie.

Designated market area (DMA) A. C. Nielsen's term for one group of communities where the audience receives the same radio and television signals.

Cold call Unannounced visit or phone call by an account executive to an advertising prospect.

Inventory Total budget of available ads for a radio, television station, or cable system.

Account Executives

The principal envoys and consultants in radio and TV stations, as in the advertising agency, are the **account executives (AEs)**. In local markets they begin their day interviewing potential clients. They make planned or cold calls, either seeking out prospects from a core list or finding new ones on their own. They discover who the advertisers are and, more important, what their needs are in terms of promoting new business. When successful, AEs bring back signed contracts and place spots in the inventory of their program log.

National Reps

National advertising sales usually are delegated to a firm outside the station. It simply is not practical for every radio and TV station to hire a sales person to represent the business in New York City or Los Angeles. That's the job of the national representatives who first made their appearance on the advertising scene more than 50 years ago. A revival of advertising swept the nation after World War II, and it gave birth to a new type of broadcasting sales person. Two radio salesmen, Ed Petry and John Blair, decided if they could represent several stations instead of just one, they could multiply their accounts and commissions. Blair and Petry found their inspiration was on target. It just made good business sense for commercial advertisers to turn the chore over to one firm instead of having to negotiate with hundreds of stations around the country. They were so successful, in fact, that the commercial networks took notice and opened their own national sales offices. The FCC, however, took a sideways glance at that move and, seeing a potential monopoly in the making, ordered networks to represent only their owned-and-operated stations, and leave the affiliates alone. About a dozen firms followed Blair and Petry into the

Account executives (AEs) Salespeople who serve as liaisons between agencies and clients, coordinating the specialists working on the account. Also, people in a broadcast or cable operation assigned to handle specific accounts and to act as marketing consultants.

sidebar
CABLE VERSUS BROADCASTING

No one likes to be told he or she is getting old—over the hill, no longer at the top of his or her game. Ask cable executives to describe the future of broadcast advertising, and you may find out why. In early spring each year, the cable and broadcasting executives make their respective pitches to the *upfront market* of advertisers in New York's Madison Square Garden. This is when the commercials are sold for the fall television season. In 2003, the seven broadcast networks (ABC, CBS, NBC, Fox, PAX, UPN, and WB) sold about $9.3 billion in commercial time, while all cable channels, including such familiar ones as MTV, Nickelodeon, and Comedy Central, took in about $5.3 billion. The following year, 2004, broadcast networks stayed about the same in the upfront market, but cable sales increased 13 percent from $5.3 billion to $6.1 billion.

Chalk up another round to cable. It has convinced advertisers that cable not only has the momentum, but that its audiences are more loyal and receptive to cable spots than broadcast audiences. In addition, cable viewers are growing faster in the key demo-

graphic of 18- to 34-year-olds. Broadcasters have not ignored these advertising challenges, but have fought back by emphasizing how individually no single cable program has the force and weight of viewership of, say, CBS's *60 Minutes* or NBC's *ER*.

How serious is this battle between cable and broadcast television for advertising dollars? Check out the websites of the Television Advertising Bureau and the Cablevision Advertising Bureau. You will read pointed headlines reporting that "Cable Penetration Hits Nine-year Low," and that the Nielsen ratings show 99 of the top 100 programs on broadcast networks while only 1 was on cable. Then go to the Cable Advertising Bureau web site and read how "Over the last 4 years, Broadcast's reach capabilities have greatly diminished while Cable's reach has grown exponentially to practically match or surpass that of Broadcast."

What makes all of this fighting seem strange is that consolidation in electronic media has placed many broadcast and cable channels under the same corporate umbrella. For example, broadcast networks CBS and UPN share the same board of directors with cable channels MTV and Nickelodeon, while NBC owns Bravo, USA, and Sci Fi channels. Sibling rivalries may be the fiercest of all.[5]

business of selling time for local stations to national advertisers—Katz and Telerep, among others.

National sales reps contract with local stations for two or more years, and ask for commissions in exchange for their national spot sales. Rep profits are calculated according to the charge per spot.[6] Two of America's largest advertisers, General Motors and Procter & Gamble, divide their multibillion-dollar budget for advertising several ways, and television gets about 40 percent of it. The remainder is invested in advertising on radio and in newspapers, as well as magazines and outdoor billboards. In terms of TV time, these huge advertisers spend about one-fourth of their money placing commercials in network shows, such as the evening news or situation comedies. They spend another 5 or 6 percent buying spots in syndicated vehicles—for instance, *Oprah* and *Wheel of Fortune*. Then they spend between 5 and 10 percent for cable channel placement. Even though a network television spot reaches about 200 affiliates, advertisers aim for audiences outside their reach of those network stations, and that's where national reps come in. They make the investments for spot buys—commercials purchased on a regional or national basis in markets around the country. Some have quipped that this is the "spray and pray" approach.

◤ Art of the Deal

When account executives seek to offer station time to a commercial client, they usually proceed in this fashion.[7] First, they secure the prospect's trust and respect by establishing an informal relationship. In the *approach,* AEs listen and learn of the client's business and marketing needs. Following that stage of the relationship comes the *discussion,* in which the account executive will determine if, practically speaking, the station, cable system, or website can effectively meet the prospective client's needs. If that conversation is productive, the AE and client will embark on the *negotiation* phase, in which advertising goals are discussed. If the client and sales person reach an agreement, there is what's known as a *closing,* where contracts are signed and commercials go into production.

NEGOTIATING THE SALE

In the negotiation phase, both AE and advertiser agree on the total *run* of commercials for the campaign. What that means is the number of spots to be aired over how long a period and in what dayparts (separate periods of the day). They look for time in the station's inventory of slots available for advertising. These **avails** are unsold commercial minutes where AEs can place ads for clients, including within programs and at hourly breaks. The AE then presents his or her plan in the form of a schedule or flight. A commercial *schedule* reflects dates and times when spots are scheduled for broadcast, whereas a **flight** suggests the entire period of time for a campaign. In some cases, AEs even have prospective commercials produced; these "specs" are samples designed to encourage potential advertisers to buy.

Advertising sales are made on the basis of time sold and audience reached—to put it in the terms of the trade, *frequency* and *reach*. **Reach** equals the *share* of the target audience that actually was exposed to an advertising message based on audience ratings, whereas **frequency** estimates the *number of times* the target audience was exposed to the commercial message.

The sales people and advertiser also decide if the spots are to run as adjacencies, which are commercials placed within or next to network programs. They also may choose to place spots in local programs at the **best time available (BTA).** If the spots are scattered throughout the day at the station's discretion, the advertiser

Avails Available time slots in a broadcast or cable schedule to be sold to advertisers.

Flight Commercial schedule from beginning to end when the advertiser places commercials with a station or cable system.

Reach Estimate of the percentage of the total audience exposed to a message, spot, or program at least once during its run.

Frequency Measure of advertising reinforcement—usually an estimate of how many times the audience has been exposed to a spot.

Best Time Available (BTA) Flight of commercials where the TV station or cable system promises to run the spots in the highest-rated time period available.

is getting the **run of schedule (ROS)** treatment. Most advertisers accept package plans offering discounts based on either the size of the package or the flexibility that is built into its schedule. One factor is *preemptibility.* Lower prices are charged for spots that can be moved with advance notice as opposed to those holding **fixed positions.** There are also different rates according to how much notice the station must give before preempting a spot—one day, one week, two weeks, or a moment's notice—they are all priced differently.

Naturally, rates vary by the time of day: The bigger the audience, the higher the cost for a daypart. The term *prime time,* for example, describes the evening daypart when more people are watching their television sets. If a sponsor wants to reach peak-audience numbers, the station charges its highest rate, often described as a *triple-A* (*AAA*) rate or by a similar term. On the other hand, if the advertiser is willing to accept a *fringe daypart,* when fewer audience members are tuned in, the cost for advertising will be reduced to a *B* or *C* rate.

Advertisers need to know of station discounts before buying a package of spots. Sales people benefit from extra commissions by selling longer schedules, and discounts are also given to long-term clients for special occasions such as holidays. Some stations use *rate cards* to show advertisers the base rate, which is often the highest price charged per spot, and lower figures with discounts included for special package plans. Some stations and networks do not release or even draft rate cards, but use a flexible scale for advertising negotiations. Another factor figuring in the cost is commercial length. Obviously, 30-second spots are more common and more expensive than 10- or 15-second commercials.

Stations also offer advertising deals with titles such as a *total audience plan (TAP),* which is a contract that spreads spots throughout the broadcast day, or an orbiting plan that runs commercials at slightly different times each day. Whatever the deal is called, the price of advertising for the client is gauged by the size and nature of the audience, the time of day they're reached, the number of exposures, and the flexibility of the schedule.

DATA FOR DOLLARS

The formula used for charging advertisers is calculated in several ways. The most common method is **cost per thousand (CPM)** viewers or listeners; the *M* is based on the Latin term *mille,* which is the Roman numeral for 1,000. The CPM, however, is not the only factor used to gauge cost. There is the **cost per point (CPP),** which is computed by dividing the total rating points by the cost of the spots. Then there is the **cume** (short for cumulative unduplicated audience), which is the estimate of how many people were exposed to the spot just once during a portion of its run (usually a week).

Other calculations weigh the size of the audience for the spot in terms of the impressions it has made. The **gross impressions (GIs)** indicate the total number of people reached by the entire run. The **gross rating points (GRPs)** or *grips* show the gross or total ratings points for the run of spots. A. C. Nielsen Company also counts the total households within a particular demographic group that the sponsor wants to reach. This formula is called **target rating points (TRPs).**

AUDIENCE PROFILING

Years ago, conventional wisdom held that advertising was about collecting as many eyes and ears as possible to bring to each spot. It didn't matter to whom the eyes and ears belonged—just so long as they were counted in the audience. That, of course, created "wasted circulation," advertising to people who were not interested. What businesses really wanted was a way to reach their target audience—people who would be more inclined to listen to the message in the first place, and do

---

Run of schedule (ROS) Preemptible package of spots that can be moved wherever a station or system desires and is sold at lower rates.

Fixed position Guarantee to televise a commercial at the precise time requested by the advertiser.

Cost per thousand (CPM) Advertising charge for making an impression on 1,000 members of the audience.

Cost per point (CPP) Advertising charge for reaching an audience equivalent to 1 percent of the population.

Cume Cumulative, unduplicated total number of people who pay attention to a radio or TV station for at least five minutes over a period of time, usually a week.

Gross impressions (GIs) Total number of people reached by all commercials over a specified period of time.

Gross rating points (GRPs) Total of all ratings points received for all spots over a specified period of time.

Target rating points (TRPs) Percentage of viewers who saw the message or program within the target audience.

protalk

Amy Menard
ACCOUNT EXECUTIVE

Advertising sales is a "people-person" job, says Amy Menard. While at the same TV station, Menard made the switch to advertising sales after working in the news and promotion departments of the ABC affiliate in Lafayette, Louisiana. Menard's typical day involves working with many people—business managers, retail merchants, and potential clients interested in generating new business through advertising. She begins at 8:00 A.M. with a sales meeting in the station, at which the general sales manager outlines goals for the day, the week, or the month, and even projects the quarter and year.

After the meeting is over, Menard begins making calls to advertising clients, including follow-up calls and cold calls for new prospects. Once she hits the streets, Menard has one idea in mind: "Sell, sell, sell!" At the end of the day, the station's general sales manager asks for her report. In that debriefing, she discusses the orders taken and the steps required toward filling those orders. That includes getting the commercials produced and giving instructions to the traffic department, which is the office responsible for scheduling programs and commercials.

Menard's primary job is working her account list, building relationships with her station's customers and clients. She says advertising sales people are really the eyes

> "There are plenty of businesses, who need the help of television advertising."

and ears of the station—the first to learn what people think of the programming and the talent. In return, she informs her clients of special events and programs on her TV station, anything that might be of interest to them for promoting their business. There is nothing like getting new business on the air, but there are times when the prospects seem lean and mean. She says, "Hearing NO never feels good. Let's just say you have to know how to move on!" Menard adds, "There are plenty of businesses [that] need the help of television advertising, [so] there is no point in wasting time with those who don't."

PRINCIPLE 3

The strategy of targeting steers commercials away from diverse audiences and toward homogeneous groups, creating niches.

something about it. Advertising deals worked out so that the client can pay according to the number of desired audience members who hear or see the spot. This strategy is sometimes called a *Volume Power Index (VPI)*.

In order to solve the puzzle of finding the right person to pair with the right ad, marketers use the tools of audience analysis and gauge demographics and psychographics. Demographics define audiences in fairly fixed categories, such as their age, gender, ethnicity, education, and income. Age, for example, plays the most significant role in advertising based on the conventional wisdom that 18- to 49-year-olds are the most profitable demographic. The reason is that such viewers and listeners are said to have more disposable income and yet are largely uncommitted in terms of their brand loyalties. As a result, advertisers generally pay more for commercial time that will reach young audience members in this group. That premise, however, has been challenged as some research shows older demographics are more desirable because of their more predictable spending habits.

Psychographics more precisely outline who the audience is in terms of their personalities, preferences, and lifestyles. Thus, the strategy of targeting steers commercials toward certain clusters of individuals who have much in common, but away from diverse audiences that tend to differ on product choices.

CLOSING THE DEAL

When it comes to closing the deal, there are a number of ways advertisers can pay for their spots. Cash plans, trade-out deals, and bartering are available, for example. There are also co-op deals that allow a manufacturer and its retailer to share the cost of commercials. There are even packages of spots based on viewer response called *per inquiry*. The emergence of cross-platform deals also has created a lot of interest. Major advertisers such as Kellogg's and Procter & Gamble have taken advantage of the growing concentration of media.

Cross-Platform Deals

Before Congress and the courts began to slow the trend toward media consolidation, *cross platform deals* were thought by some to be the wave of the future. When a CBS Viacom or an AOL signs a multimillion-dollar client capturing all of the advertising for cable, broadcast, and print media, it makes headlines. However, cross-platform buys that place advertisers in direct negotiation with corporations have not eliminated the so-called middle men—the national reps and media planners.

Fueling the drive toward cross-platform advertising are the parallel trends of consolidation and fragmentation—more channels owned by fewer companies reaching smaller audiences. Audience fragmentation is one of the defining realities of advertising in the twenty-first century, and so it seemed to be more efficient for a national advertiser to deal directly with one media conglomerate reaching multiple radio and television audiences. But the question remained: Was it the best advertising strategy? Cable systems and broadcast stations outside group-owned outlets may better fit the advertising campaign goals, and that's where ad agencies and media planners have a major role to play.

Bartering for Time

When airtime is exchanged for sponsors' products or services, the deal is called a **barter,** or, in the case of radio sales, a *trade-out deal.* Barter or trade-out deals work best when a broadcaster needs a sponsor's product, such as station vehicles or office furniture. The principal danger is that such deals can jeopardize the station's cash flow, undermining its payroll and schedule of debt payments.

Barter (or Trade-out) Deal for sponsors' products or services in exchange for advertising time.

protalk

Sean Walleck
KTXA SALES MANAGER

Broadcast advertising can be a roving career but also a rewarding one. Sean Walleck is the local sales manager for a television station in Dallas-Fort Worth. That's a position he has worked up to after stints at TV stations in Kansas City, Phoenix, and Chicago. Walleck, a philosophy graduate of Lake Forest College, did not use his diploma to get his foot in the door but rather his internship experience at a Chicago TV station. It was that experience that prompted Walleck to recommend a reality check of the industry for others seeking to follow in his footsteps.

Walleck's first word of advice for college graduates interested in advertising and sales: "Learn how to read a ratings book." Walleck is referring to those quarterly books from the A. C. Nielsen Company measuring a TV station's audience, which serves as the basis for selling commercials. The rows of numbers are extremely important, whether you're working in radio, television, a full-service advertising agency, or a time-buying firm. "It's about knowing how to read the Nielsens," he says, "and what each point means for both your sponsor and your station."

At the macrolevel, Walleck sees both advantages and disadvantages in the trend toward industry consolidation. Mergers have made sales more efficient—his station's

"Learn how to read a ratings book."

parent company is one example—but "people lose jobs" when those mergers occur. Economies of scale through consolidation make it easier to do business with a single media organization in a cross-platform buy (explained next). New forms of nontraditional revenues (NTRs) is another by-product of consolidations—especially as they apply to web promotions. Walleck's advertisers have the option to buy spots and border ads on his TV station's Internet site. "We sell spots and dots now," he says.

Walleck has a final piece of advice for aspiring professionals who want to keep their jobs in broadcast advertising after a merger: "Learn how to sell." In electronic media, that means learning how to become a careful consultant for your clients. This is a long-term relationship between the advertiser and the station, and no one wins by cutting corners just to make a fast buck.

Did you ever wonder where those spots for Smokey the Bear, McGruff the Crime Dog, or the Crash Test Dummies came from? Turn back to 1942, when the United States was at war on two fronts. One was in the battlefields against the Axis powers of Germany and Japan and the other front was here at home. At the White House's behest, the War Advertising Council was formed to fan patriotic flames through public service advertising.

Posters reminded citizens that "Loose Lips Sink Ships." When more help was needed in the factories, "Rosie the Riveter" recruited women to work at industrial jobs. In order to save the food supply for soldiers, public service advertising encouraged Americans to plant "victory gardens." Perhaps the War Ad Council's biggest victory came in raising $35 billion by promoting war bonds. After the war, President Harry S. Truman asked the council to continue its peacetime mission, and the National Advertising Council was formed.

In public service, the nation's top advertising agencies do the creative work *pro bono*—free of charge—while networks and local broadcasters join in by donating millions of dollars in free time. The National Association of Broadcasters is pleased to mention this fact whenever Congress takes up public-interest legislation threatening to impose new legal burdens on broadcasters.

The Ad Council receives hundreds of requests for campaigns each year but winnows out all of the ones that are political, religious, or commercial in nature. What's left are issues relevant to most Americans, from education to the environment, and the Ad Council creates about 40 campaigns each year to address those issues and concerns.

In television, nationally syndicated programs such as *Friends* or *Entertainment Tonight* may be bartered to local TV stations. **Barter syndication** means the broadcaster "pays" for the right to air the syndicated program by giving the show's owner the commercial time slots to sell to advertisers. *Barter-plus-cash* seals the deal by requiring an additional cash payment from the broadcaster for the license to air the show. The show's producers have already placed commercials in the program directed at a national audience, but there are empty slots available for the local stations to sell to their advertising clients as well.

Co-Op Deals

Cooperative advertising describes a method for paying for TV commercials where the local retailer and a national manufacturer share the expense. Co-op spots are more common to radio than to television. They usually feature a 50–50 split, dividing the cost between a national vendor, such as Coca-Cola or the Ford Motor Company, and a local retail outlet.

Per Inquiry

Sponsors buying **per inquiry (PI)** spots gauge the advertising rate by the number of calls received in response to the ads. Per inquiries invite people to order by phone any merchandise from compact discs to salad shooters. These spots usually call the audience to action with, "Call now! Operators are standing by," or words to that effect.

Make-Goods

Sometimes spots don't reach as many people as promised, and that calls for a new solution. When big stories break, such as the war in Iraq or a major hurricane,

Barter syndication Contract with local stations buying a syndicated program in exchange for an agreement to run spots it contains. Stations pay for the show in part or in whole by giving the syndicator an agreed-upon amount of commercial time to sell within the program.

Cooperative (co-op) advertising Agreement between local retailers and manufacturers to share advertising expenses.

Per inquiry (PI) Ads paid for by the number of units (products or services) sold by audience response to the ad. These spots use a phone number or post office box number for a direct response.

broadcasters may clear their commercial schedules for "wall-to-wall" news coverage. Those spots missing appointed times must be rescheduled as **make-goods.** The lost spots are broadcast at new times in agreement with the advertiser. These make-goods are called for when broadcasters schedule the commercial but something happens to keep it from running or a technical problem occurs. Make-goods are also slated to make up for lost rating points when shows fail to net as large an audience as the sponsor was promised.

Sustaining Time

If a program is *sustaining,* it means there are no commercials and the broadcaster is paying for the time. So, the broadcaster may choose to fill it with public service announcements (PSAs). The FCC encourages PSAs as a way of serving the public interest. These spots nurture causes such as the American Cancer Society and the United Way.

Advertising Strategies

The goal of advertising sales is a fairly simple one: Persuade the client to buy time and build up the company's image in order to generate more business. Some commercials are directed simply at enhancing the sponsor's image, which is *institutional advertising.* Other spots aim to increase customer use of a product or a service. There is an obvious reward for successful marketers, but to motivate the creative side of the industry, companies compete in contests each year and show off their best spots. Awards given by national advertising associations, such as the *Clios* or *Addies,* indicate creative excellence but do not always indicate how effective the ad may be in stimulating business.

BUILDING BRAND IMAGE

Whether selling a product or building a corporate image, **branding** is the name of the game—differentiating the advertiser and product from its rivals, giving it a personality and a name, and, most important, showing what is in it for the customers. Branding is mainly about images and perceptions. The reason Kellogg's Corn Flakes became such a popular item had as much to do with Tony the Tiger's claim that they were "Grrrrreat!" as their actual taste. Is there any real difference in the way Kellogg's Corn Flakes taste from other brands? Advertisers claim that what customers taste most is the brand's image, and they will buy products with what they perceive to be a superior image, so long as they can afford it.

PRINCIPLE 4

Consumers buy a product when they perceive its brand image to be superior to the image of its competitors.

UNIQUE SELLING PROPOSITIONS

Advertising success is tied to a second principle—**unique selling propositions (USPs).** This idea became the mantra of the 1960s after advertising executive Rosser Reeves convinced his colleagues that it was the best—if not the only—way to cut through ad **clutter,** the growing congestion of all media drawing on advertising for support. In his book, *Reality in Advertising,* Reeves argued, "The consumer tends to remember just one thing from an advertisement—one strong claim, or one strong concept."

Radio and TV producers interpreted Reeves to mean one strong benefit promoted in a catchy slogan or jingle.

Make-goods Spots added or rescheduled when a television station or network fails to air them properly or to deliver a CPM guaranteed to the sponsor.

Branding (or Brand imaging) Strategy to differentiate a product's name and logo from other competing brands in the market.

Unique selling proposition (USP) Facet of a product distinguishing it from rival brands and of benefit to its customers.

Clutter Excessive number of spots in a break, or general barrage of ads obscuring any single message.

"How do you spell relief? 'R-O-L-A-I-D-S.' "
"M&M's melt in your mouth, not in your hands!"
"Certs breath mints with a magic drop of retsyn."[8]

Branding and USPs ideally would place the product in a superior position to its competitors. Such theories are all well and good, but in order to put them into practice, advertisers need to know who the customer is, who the rivals are, and how to make the audience think the product is superior.

Advertising Convergence

The vision of someone avoiding TV commercials by either pushing the mute button on the remote or channel surfing during spot breaks has long been an industry concern. Now, the nemesis for advertising is the digital video recorder (DVR)—often known by the pioneer brand name TiVo. It allows viewers to skip past commercials altogether, and the industry predicts there will be a DVR in one of every five television households by 2008. About half of today's DVR owners jump past the spots when watching shows they've recorded, and as a result investments in TV commercials are subject to future cutbacks. Online advertising will reap some of the resulting benefits if that is the case, but network television is not giving up without a fight. TV's new weapon of choice is *product* or *brand integration.*

What is product integration all about? When Fox debuted its prime-time drama *24*, it ran no commercials in the program but did advertise spots at the beginning and end of the show. That was to keep from interrupting the fast-paced drama; however, audiences also may have noticed that the series star, Kiefer Sutherland, was driving a Ford SUV throughout the program. You also might not associate Paris Hilton and Nicole Richie with Burger King—that is, until you watched them in Fox's *The Simple Life 2*, where Burger King became part of the program. That's what the advertising deal specified, and Fox was happy to deliver. Anyone watching Fox's biggest hit, *American Idol*, will also notice the talent emerging from the Coca Cola "Red Room."

Whether it's called *product placement, contextual commercials,* or brand integration, the practice has generated a fair amount of discussion and criticism. Commercial Alert, an organization associated with Ralph Nader, has petitioned the Federal Communications Commission to probe the practice of product integration on the premise that it skirts the government rule requiring that advertisers reveal their identities over the airwaves. Commercial Alert calls this new form of advertising integration just one more example of how television has become one big infomercial.

INFOMERCIALS

Two forms of TV programming can fairly be called advertising: home-shopping networks and infomercials, which are formally titled **program-length commercials (PLCs).** Other names used to describe this format are *paid programming, long-form advertising,* and *documercials.* There are three ways to know you have stumbled on to an infomercial. First, it is trying to sell something by demonstration, display, testimonial, or a combination of three. Second, it is at least a half-hour long. And third, it invites a direct response of some monetary investment.

Infomercials once ran only in hard-to-sell time deep in the caverns of late-night TV, where insomniacs, the graveyard shift, and some bed-ridden patients resided. Now they occupy daytime hours on cable and most of the day on some LPTV

Program-length commercials (PLCs) Thirty- to sixty-minute programs, usually called *infomercials*, that promote a single sponsor or product.

stations. Not surprisingly, their growth parallels the proliferation of cable and satellite TV channels.

Infomercials Investigated

The FCC first turned its attention to infomercials in 1973 when it decided that the entire half-hour should be counted as advertising, and that the public interest was ill served by such a huge chunk of commercial time. The government reversed itself in 1984, finding nothing illegal about infomercials, suggesting that viewers need only be aware that host celebrities and "experts" endorsing products are paid by the sponsor to do so. After the FCC lifted its ban on infomercials, however, a wide assortment of advertisers from star-gazing astrologers to home appliance dealers began producing their PLCs.

HOME SHOPPING NETWORKS

Home shopping on television is a big business, and getting bigger all the time. The first major U.S. channel, HSN (Home Shopping Network), is also the most widely distributed global TV shopping network with outlets in Germany, Japan, China, Italy, and France. This multibillion-dollar enterprise also generates stable audience numbers despite what might be happening to the ratings of other network programming.

Shopping by television began nearly 30 years ago in Clearwater, Florida, as an improvised solution to an unpaid bill for a radio advertiser. A local AM station manager decided to auction off 112 electric can openers on the air to help the retailer pay his bill. The station sold every single can opener, and decided to start a regular call-in-and-buy show called *Suncoast Bargaineers*. A cable access channel picked up the idea in 1981, and before long it was a television hit. Barry Diller, former chairman of the Board of Paramount Pictures and Fox Inc., bought the Florida-based Home Shopping Network in 1993. Four years later, he combined e-commerce with television shopping through the web page HSN.com.[9]

CHANNEL SHOPPING. It's becoming a very big business—not just for the Home Shopping Network (HSN) of Tampa, Florida—but for a variety of Direct Retail Television (DRTV) clones. HSN is still the king though. During one day of the Christmas season, it racked up $30 million in sales.

HSN's nearest rival, QVC, is a relative newcomer. It began promoting Sears products in 1986 from studios in West Chester, Pennsylvania. Like HSN, it, too, integrated its TV and Internet commerce, and not surprisingly, others have joined the competition to merge shopping on television and online.

Advertising in the Internet Age

The Web began burgeoning into a billion-dollar enterprise in the 1990s, and produced a record $8 billion in e-commerce sales for the year before the dot-com implosion of 2001. By 2004, the Web had rebounded and was a stronger advertising medium than ever before. Forecasters predicted it would top its 2000 peak and generate more than $9 billion in on-line spending. It's become the central focus of both convergence and creativity. Websites have devised a new means of gauging the effectiveness of advertising based on the level of interactivity customers demonstrate while browsing on-line.

CREATING WEB ADVERTISING

The original format for web advertising was called **banners**—static boxes across the top of the page now considered somewhat passé. In the creative department, web designers recommended that banners be kept to seven words or less and feature small but colorful art. Jim Sterne, in *What Makes People Click*, says the boxes began as advertising billboards but were crowded out by banners everywhere and "traffic began going places where things were happening."[10] Next to come in vogue were *skyscraper* or *tower* ads down the sides of the web page or *wallpaper* in the background. Then came the pop-outs—ads springing out from the borders of the computer screen. After that, something called *interstitial* ads began fading in and out on web pages.

Debate has focused on the television spots that download while web users are reading a screen, and *pop-under exit panels* triggered by a user's attempt to leave the website or back out of it. These pop-ups have caused so much frustration

Banners Horizontal advertising blocks on the Web placed above content (like billboards).

even some advertisers want them scrapped. "When viewers make the decision to go off-line," says Mark Amado, an account director for a division of Initiative Media, "their mindset has already moved on to something else, and we should honor that decision by not hitting them with one more potentially out-of-context message."[11]

TRACKING WEB VIEWERS

Measuring web traffic has become a profitable industry for on-line audience analysis firms. Various software tracks every move a person makes on the Internet.[12] This technology consequently has raised privacy concerns among Internet observers.

Layers of Web Browsing

Web advertising fostered new economic models based on the layered nature of web browsing from one click to the next. There are four levels of audience interactivity measured on the Internet. Ad placement may be counted by the traditional CPM basis, or measured by the *cost per click (CPC)*, the *cost per lead (CPL)*, or the *cost per sale (CPS)*.

The first level, CPM, is based on impressions, and charges for website billboards on a cost-per-thousand viewer basis. More profitable are on-line ads that catch not only customers' eyes but their hands as well. Web sponsors pay more if viewers actually click on the link of the banner advertising to the website. This charge, the second level, is called the **click-through rate** or the cost per click.

The third level of advertising on the Internet is gauged by the number of times someone sees a banner or logo and clicks through to a link on the website. The person then decides to fill out a form. That is what is called a *lead*—no money has changed hands, no products have been purchased, but a potential customer has given his or her personal data on-line. Advertisers calculate the cost per lead. Many sites now require registration, and once viewers are interested enough to sign in on a website, the information they give creates a database of consumer targets for advertising.

The fourth level is measured by e-commerce sales. An example of interactive e-commerce would include the purchase of travel tickets by responding to a site knowing that it promoted the vacation resort and sold fares there as well. This is a measure determined as the cost per sale.[13] It is through these measures and others that Internet advertising has built its revenue based on how web users interact with the commercial message, and the degree to which they begin to negotiate on-line.

Web Entanglements

There are good reasons why the blue skies of web advertising clouded over in 2001. Hundreds of thousands of websites overwhelmed the average user who visited only a few sites each week. As users became more familiar with favorite web pages, they developed "banner blindness," an inattentive attitude toward the ads. And as more web users headed for popular sites, loading times increased and crashes occurred. Finally, web merchants noticed initial reluctance of users to give credit card numbers, for security reasons. E-commerce proponents argued that the likelihood of fraud is no greater on-line than off-line, and the reluctance began to diminish. Still, the reach of e-commerce is limited to those who have computers, have an Internet service provider (ISP), and are willing to trust an online merchant with their credit card.

PRINCIPLE 5

Internet advertising draws revenue based on the viewer's level of interactivity with the commercial message.

Click-through rate Frequency at which viewers move from the Web advertisement to the sponsor's site to place an order.

protalk
ZZ Mylar
INTERNET SALES COORDINATOR, MYSANANTONIO.COM

Every new advertising medium comes with a price tag, which raises the question of viability—Can it generate enough income to sustain its existence? ZZ Mylar has joined the growing ranks of advertising executives who have answered the question about the Internet's financial future. After working in sales and promotions for TV stations in Dallas, Houston, and San Antonio, Mylar made the transition to web advertising. Her website, MySanAntonio.com, merges media under the Belo corporate umbrella, and it brings an on-line presence to her former station, KENS-TV, by creating new promotions that combine the website with television programming. "It's an amazing product that has people very excited—very revolutionary," says Mylar.

Just as before, Mylar spends a good part of the day making sales calls and writing new contracts—that is, when she is not checking on current campaigns, making sure they have the impact intended. One thing she likes about on-line advertising sales is being on the cutting edge of innovation. "When an ad agency gives you the green light to contact their client directly, you know the information you have must be new."

The move from television to the Web has been an educational one for Mylar, and a lot of it is self-taught. She was able to break new ground in selling "News on Demand,"

> ## "It's an amazing product that has people very excited—very revolutionary, "

which is one way for MySanAntonio.com viewers to catch up with KENS-TV news while sitting at their computers. They click on only the stories they want to see.

There's a lot to like about on-line sales, says Mylar. "There is much more opportunity with the new inventory." Advertising selections can be made, different products can be sold, and presentations are no longer confined to a single set of spots available during the broadcast day. In other ways, however, it is quite similar to representing a TV station. She still relishes the satisfaction of making big sales and reaching new customers with a product or service that her clients have to offer.

Still, selling an innovative product has its risks. If a campaign fails, Mylar says, the client may jump to the conclusion that there is something wrong with the Web rather than considering how effective the creative approach was, or other possible factors. The bottom line for Mylar is that regardless of whether in television or on-line media, advertising consultants must be effective listeners. Their work is about discovering what a client's needs are and determining how to build that client's business. "It cannot be just about making the sale and getting a commission," she says; it has to be about giving honest advice. An attitude of service and the motivation of a self-starter who stays organized and up to date are prerequisites for a successful career in media advertising.

Future Growth

The plusses for web advertising are that it tends to reach 18- to 34-year-olds who are more affluent and well educated. It also reaches them at the point of purchase in the realm of e-commerce. In addition, web advertising allows sponsors to change their advertising easily. Some observers predict the Internet will become an even greater source of advertising once broadband technology converges with interactive radio and television. It already has made a difference in politics. In the 2004 election race, political spending leaned toward the Internet, where blogs and e-mails attracted a 1 percent increase in political advertising, whereas radio and outdoor billboards saw a decrease in ad spending.[14]

Promotion

It should come as no surprise that the merchants of media advertising spend part of their broadcast budgets promoting themselves. Radio, television, and web promotions sell a station or channel's personalities, programs, and brand image to the audience. The promotion can be as small as an icon on a web page or as large as a hot-air balloon. There are promotion activities designed for every advertiser and every audience. Although stations benefit from blowing their own horns, it also helps to employ other channels, and involve both *internal* and *external media*, particularly when inviting new viewers.

PURPOSES OF PROMOTION

Promotional spots can boost ratings if they are directed at the audience, or a sales promotion can encourage advertisers to buy more time. Promotional campaigns also unify sponsors and audiences through shared community projects or a particular theme. Themes, such as "We're the ones to watch" or "Appointment TV," reinforce both the channel's and the program's relevance to the viewer. Television news programs may use catch phrases, such as "Live, local, and late-breaking" or "On your side," to underscore their relationship to the viewer. Local stations also promote daily newscasts using avails to show clips of stories promised in the upcoming newscast. Series or situation comedies are highlighted with clips of an upcoming show's entertaining scenes.

CABLE, SATELLITE, AND INTERNET PROMOTIONS

Cable and satellite promos solicit subscribers for pay services by using a type of comparative advertising. DirecTV positioned itself as a better choice than cable by virtue of its crystal-clear digital channels in high definition. Cable systems fought back by promoting their bundled options of cable modems, telephone services, and video on demand. Those fights continue but on different terms. Other promotions spotlight sports and movie packages for viewers or feature discounts. It's not surprising to see endorsements, logos, and themes used for media promotions, especially when expensive accounts are involved.

It is difficult to overstate the promotional value of the Internet, as virtually all radio and TV stations have established some on-line presence accessible to web viewers. Websites also offer chances for promotional campaigns for certain causes. Sponsors generally like to be associated with doing good deeds in their community, and many stations take the opportunity to participate in campaigns such as "Partnership for a Drug Free America" or the "Susan G. Komen Race for the Cure" for breast cancer. They further advertise their support on station websites to promote both the station and the cause.

▀ Issues in Advertising

Critics of media advertising don't have to look far to find places where the loftier goals of informing the public and building business have somehow missed the mark. Debates rage over issues touching on the ethics and quality of contemporary advertising. Critics also are concerned with a range of advertising issues from subliminal ads to deception and fraud.

SUBLIMINAL ADVERTISING

Subliminal advertising, using barely perceptible words and images to move viewers, has been around since the 1960s. Yet, reliable evidence to confirm its effects is illusory. Jim Vicary brought it to the public's attention first in 1957 when he conducted an experiment by splicing the quick messages "Eat popcorn" and "Drink Coca-Cola" into a movie reel. He reported upsurges in concession sales, but later replications failed to verify his results. Vicary later admitted his initial study was a hoax. Nonetheless, the subliminal theory generated much interest and even one ardent proponent, Dr. Wilson Bryan Key.

In Key's first book on the subject, *Subliminal Seduction,* he observed sexual imagery rampant in subliminal advertising, noting that even Kent cigarettes were designed to appeal to women through its "strong masculine name, suggesting a solid distinguished WASP heritage." The fact of the matter was that the cigarettes were named after Herbert Kent, president of the company.

Key's analysis could be described as more artful than empirical, given the lack of confirming evidence for subliminal advertising. Nonetheless, the FCC imposed a

sidebar

POLITICAL ADVERTISING

Elections bring more than just voters to the polls. They also bring politicians to broadcast stations and networks with huge infusions of cash. Broadcasters are increasingly dependent on these "campaign buys" that make up for less reliable sources of revenue. In 2004, TV advertising for all political races was projected to bring stations $1.5 billion—an increase of more than 50 percent from the previous election year.

Political advertising is a two-edged source, however, due to its controversial nature. Debates rage over attack ads that smear a candidate's character, such as the spots by the so-called Swiftboat veterans attacking Sen. John Kerry's Vietnam record and award of medals.

In 2004, political heat was deflected from advertising by a film documentary, Michael Moore's *Fahrenheit 9/11,* which criticized the Bush administration. Members of the Republican-allied Citizens United, enraged by Moore's film, sought an opinion from the federal government that would remove its free-speech protection as a documentary and place restrictions on the commercials promoting it.

The Federal Election Commission's counsel supported this view and even drafted an advisory opinion indicating radio and television stations may be subject to restrictions on commercials promoting *Fahrenheit 9/11.* The government, however, has long granted broadcasters political exemptions for newscasts and documentaries, which includes its advertising and promotion. One thing became clear from this controversy: Any attempt to block ad revenue from reaching broadcasters during an election year would face a fight.

rule prohibiting stations from using the technique. Some broadcasters keep experimenting with the idea, nonetheless.

COMMERCIAL CLUTTER

How many TV spots are enough? For the first dozen years of commercial television, fewer than five minutes of ads aired per hour, and almost all spots were 60 seconds in length. Then, in the 1960s, networks began boosting their rates, and ad agencies answered the challenge by **piggybacking** spots, combining two product messages into one, 30 seconds in length. The half-minute spot became the standard unit of sale by the 1970s, but TV stations and networks sold them at one-minute rates—clients paid the same rate for the time, but only received half as much.

The American Association of Advertising Agencies (AAAA) found that for every 60 minutes of prime-time TV viewers watched in 2000, they spent almost 17 minutes viewing (or escaping) ad spots. The number of commercials has increased, but their length has diminished. Ten- and 15-second spots became commonplace in the 1990s. Time limits imposed on commercials have been matters of concern for professional associations. At one time, the National Association of Broadcasters asked radio stations to sell no more than 18 minutes of ads per hour, and recommended that television stations limit spots to 9½ minutes an hour or less during prime time and 16 minutes in other dayparts. In 1984, the Justice Department charged that the NAB's ad limits reduced competition, and in response, the NAB abandoned its restrictions on advertising time as well as its *office* for fielding complaints on that issue.

Now, network TV typically carries a 24-spot load per hour, and the commercial volume is even greater on cable, which often has 28 advertising messages or more per hour. A former television program director says the clutter "is killing the goose that laid the golden broadcast egg."[15] A leading trade publication predicted that if this is left unchanged, the television advertiser will eventually dominate the list of things most annoying to Americans (bested only by telemarketers).

Advertising and the Law

In terms of First Amendment freedoms, commercial speech gets only mid-level protection—more than obscene or indecent expression but less than political and private speech.[16] There are three principles of law that advertisers must follow: The content of ads must (1) be truthful and not misleading, (2) show evidence to back up its claims, and (3) be fair. Case law has given rise to several remedies when these principles are not met, including corrective advertising and counter-commercials. What this essentially means is that freedom of speech for advertisers is limited by their responsibility to substantiate commercial claims.

PRINCIPLE 6

Freedom of speech for advertisers is limited by their ability to substantiate the claims that their commercials make.

THE FTC

The Federal Trade Commission predates the Federal Communications Commission. Congress established the FTC in 1914 to ensure fairness in competition, and placed advertising under its purview in 1938 with the passage of the Wheeler-Lea Act, designed to eliminate "unfair and deceptive acts" of commerce.[17] The FTC's Bureau of Consumer Protection, as a result, is directly responsible for protecting citizens from false or misleading ads.

Piggybacking TV commercial practice in the 1960s, in which a single sponsor presented two products in a 60-second ad.

Deceptive Advertising

Broadcast "advertisers must have reasonable support for all express and implied objective claims." That is why TV spots no longer show Wonder Bread building bodies 12 ways, or making claims that Listerine not only makes your breath smell sweet but prevents colds and sore throats.[18] The laws against "misleading" commercials have been interpreted to mean both what is included in a message and what is left out, and in the bread and mouthwash cases, the claims were judged to be misleading. If the "consumer's decision to buy or use the product" is based on false information or reasonable inference; if an ad "contains a statement or omits information that is likely to mislead consumers," then the advertisement is against the law. It's that simple.

Comparative Advertising

For a time, TV networks rejected comparative commercials, but that practice was judged to be unconstitutional. Now, the FTC encourages honest comparative advertising with some limits. A competing brand cannot be called "brand X," for instance, and all claims of comparison must be documented. Whenever advertisers fail to meet their burden of proof, the government can apply a number of remedies, including cease-and-desist orders and corrective advertising.

Corrective Advertising

Historically, **corrective advertising,** used to redress false impressions, has cost millions of dollars. One remedy imposed by the FTC, for example, had Ocean Spray tell its customers that extra "food energy" promised in its cranberry juice was actually calories and not protein or vitamins. Profile diet bread had to correct the false impression that its calories per slice were fewer than the calories contained in regular slices of bread.

Safety

Do broadcasters have to make sure the spots their stations are running are safe? According to the FTC, if a spot "causes or is likely to cause substantial consumer injury then it may be judged unfair." If, for example, a food label failed to properly indicate its contents and caused a customer's allergic reaction, or if a child were harmed while playing with a toy he or she thought to be safe, then the advertising could be judged unfair.

The government's trade agency is not the only office that responds to complaints about advertising. The Federal Communications Commission has in the past reviewed allegations of unfairness and has even gone so far as banning some products from commercial broadcasting.

THE FCC

The Federal Communications Commission specifies a boundary between advertising and programming, and stations must identify the source of the advertising. This identification rule is rarely violated except during campaigns when political advocates could try to conceal their cause. Now, product integration raises new questions about sponsor identification, as noted earlier in this chapter. Advertisers that are abusive of product claims are likely to find themselves negatively listed with professional organizations, such as the advertising review boards of the Better Business Bureau.

For nearly four decades, the FCC tried to enforce fairness when controversial issues emerged as part of commercial advertising. The Fairness Doctrine, however, was abandoned in 1987 when it was found too difficult to enforce equitably. As a result, fairness is now principally a matter of professional ethics and responsibility.

Corrective advertising Correction of misleading spots by a new set of commercials that contain truthful information.

career focus

ADVERTISING AGENCY

Job opportunities in advertising are found in the offices of cable systems, radio and TV stations, as well as advertising agencies. Advertising agencies are just one part of a much larger industry composed of audio and video production studios, web design companies, and print shops, which act as suppliers to the industry.

The full-service ad agency typically defines its duties as servicing client accounts, creative tasks, media planning, and marketing research. These offices report to a senior administrator—a **vice president** or **president.** There also are **supervisors** or **directors of account management,** as well as **creative** and **media directors.**

Account managers monitor agency performance and handle long-term planning, personnel assignments, and revenue projections. They oversee account executives, who meet with agency clients in order to draw up advertising goals and strategies. **Account executives** are responsible for advising the agency on the client's product, the market, and relevant consumers. Their duties include client relations, project planning, and budgets.

Creative writers, art directors, and **producers** are responsible for composing the advertising message and choosing the right words and pictures to reach the target audience. **Radio and television producers** are responsible for recording and editing the spots, which demands creative and administrative skills. That includes everything from the selection of talent and location through budgeting, scheduling, recording, editing, and granting final approval of the commercial.

Media planners draw up and present the media mix, which is a plan proposing the specific channels to fulfill the client's promotional goals. They use ratings data and audience research to develop their strategy. They should have knowledge of the cost and effectiveness of print, electronic, and outdoor media available. Analytical skills based on reasoning from the ratings are necessary, since media planners must draft the media mix and defend it to both the client and the agency. Some agencies also employ **marketing researchers,** who analyze ratings data and conduct audience research.

After the client signs off on the media plan, **media buyers** make the purchases of the advertising space and time. For certain media, they use advertising rate cards. Newspapers, magazines, and some broadcasting outlets work this way. More often in broadcast advertising, particularly television, some back-and-forth bargaining over scheduling and pricing takes place.

Advertising or marketing researchers find, interpret, and evaluate the data about advertising and its audiences. These positions are in demand, drawing people from marketing research firms, government agencies, and media consultants.

Broadcast Sales Offices

General sales managers in radio and television motivate and lead the sales team, establish sales goals, and lead the station's effort toward increasing its share of the advertising revenue in the market. **Local sales managers** develop and direct the local advertising effort by working with account executives and keeping in touch with local clients and ad agencies.

Account executives serve as consultants for clients and sales people for the station. They make presentations and negotiate deals for advertising time. **Sales assistants** provide support from inside the station by taking orders for commercial time by phone either from account executives in the field or directly from clients over the phone.

COUNTER-COMMERCIALS

During the days of the Fairness Doctrine, broadcasters who held forth on controversial issues or aired "advertorials," pointed messages on those topics, were also required to vent opposing views. In one instance, smoking became the controversial issue, and the FCC deemed antismoking **public service announcements (PSAs)** an appropriate means for broadcasters to counter TV commercials for cigarettes.[19]

Public service announcement (PSA) Spot aired for free by a station or system promoting a nonprofit organization. PSAs are often used to fill unsold avails.

These counter-commercials to smoking were effective for a while—broadcasters and tobacco companies noticed a drop in sales by more than 2 percent—but they soon turned to a different solution. Rather than continue to sustain a loss in business from counter-ads, tobacco executives actually lobbied for the Federal Cigarette and Labeling Advertising Act banning broadcasts of cigarette ads on radio and television. The result was a shift in cigarette advertising from broadcasting to other media, including outdoor and print, especially magazines.

Double Billing

Co-op advertising has been subject to a particular fraudulent practice known as **double billing.** This usually occurs in radio advertising sales, when the station sales-person and a local retailer conspire to inflate the reported price for commercial time in order to bill the national manufacturer a higher price than what the spots actually cost. The conspirators then split the gains. This was once an offense to be investigated by the FCC, but during the deregulation days of the 1980s, it was removed from the agency's purview except as a possible issue for license renewal. That left it up to national vendors to seek redress through civil claims if a local station bilked them on the cost of spots.

Clipping Time

Some times, stations get into a jam by failing to run local spots at the right time, and may try to work their way out illegally—adding make-goods, where they do not belong. One form of chicanery, *clipping*, robs the network or syndicated programs of program time. This occurs when the local station decides to clip the closing credits of a network or syndicated program in order to put on a local spot. It also can involve replacing network or syndicated advertising with a local commercial run over it. This becomes fraud when the station signs a contract to air both network and syndicated programs—as well as the commercials at the time offered, but fails to do so.

Double-billing Illegal advertising practice in which a station fraudulently bills a retailer at twice (or more) the normal rate. The retailer then sends the bill back to a vendor who pays twice as much as his share to relieve the local retailer of the excessive cost of the commercials.

Now Legal

Summary

The commercial model of broadcast advertising that now pervades radio and television was held in check during the early days. Self-imposed limits on time and content were the rule of the day. Now, broadcast advertisers are surrounded by competition and are constantly seeking ways to survive. This competition begins with marketing—knowing the product, its benefits, and the target audience. The strategy of targeting started as a single-show concept with sponsors' names integrated into entertainment fare and evolved into participation divided among multiple advertisers per program.

The goal of commercial media is to create audiences who will meet the client's marketing needs. Fundamental to this aim is the relationship between account executives and advertising clients. In exchange for advertising dollars, clients expect to see a picture of their customers identified by demographic and psychographic data. The strategy of targeting steers commercials away from diverse groups and toward audiences who have at least one thing in common—an interest in the advertised product.

In their quest for advertising sales, broadcasters price their most valuable commodity—commercial time—by the number of spots sold, the time of day, the flexibility of schedule, and audience variables. Just as products are measured by their contents, advertising is calculated by several audience measures. There is the cost per thousand (CPM), the cost per ratings point (CPP), gross ratings points (GRPs), and the cumulative number of viewers (cume). Broadcast time is gauged and priced by these measures of the size of the audience.

Researchers understand how powerful the imagery of slogans and icons can be. Broadcast audiences respond to advertising (1) that demonstrates a unique personal benefit to them and (2) when they perceive a product's brand to be superior to that of its competitors. It is a specialized business at the national level, with major agencies and representative firms dividing up accounts with networks, cable systems, and broadcast stations. Advertising also is a team enterprise with different players in sales, marketing, creativity, and research. The Internet continues to attract attention and commercial revenue from firms that use a variety of formulas for charging clients for space on-line. Some on-line media gauge the cost by the viewer's response in terms of web interactivity. Commercial speech online or in any form of media is subject to legal and ethical scrutiny to test its truthfulness. Advertising that is fraudulent or deceptive is unethical and illegal.

Food for Thought

1. In the early days of radio, broadcast advertising met with resistance from those who thought it would lower the quality of the new medium. Would you defend or criticize commercial radio today based on the quality of its advertising? What examples would you give?

2. Commercials at the movies during the previews of coming attractions have become a new source of revenue. Do you find this type of advertising effective? Why or Why not?

3. Careers in advertising focus on sales, creative production, media buying, and research. If you were to pursue a career in advertising, what area would you choose, and why?

4. Some stations allocate time for public service advertising, while others run PSAs only as filler for unsold time. In your opinion, should broadcasters devote more time to public service spots to serve the community? Explain your answer.

5. For a short time, the FCC banned infomercials, suggesting that broadcasters would be spending too much time in marketing products, which would not be in the public interest. Do you think infomercials are in the public interest? Why or why not?

6. Advertising on-line has led to what some researchers have called "banner blindness" and commercial clutter on the screen. What type of Internet advertising do you consider to be most effective and why?

7. Do you agree with the remedy of corrective advertising prescribed by the government as a cure for deceptive or false advertising—or should it be just a matter of "buyer beware"? Explain.

8. Both alcohol and cigarette advertising are limited on television by either government or self-imposed restrictions. How do you feel about advertising tobacco and alcohol on radio and television? Are some forms acceptable, but not others? Support your answers.

Law

> *Restriction of free thought and free speech is the most dangerous of all subversions. It is the one un-American act that could most easily defeat us.*
>
> —U.S. SUPREME COURT JUSTICE WILLIAM O. DOUGLAS

Chapter 11

It was not your typical fall election in October 2003. Suffering under a huge amount of public debt, Californians went to the polls to recall their elected governor, and at the same time choose his successor from a field of more than 200 candidates. Given the Golden State's wealth of television and screen talent, it was not surprising that two of the gubernatorial hopefuls were actors: action film star Arnold Schwarzenegger and TV personality Gary Coleman of *Diff'rent Strokes.*

The National Association of Broadcasters quickly alerted its member stations that if they chose to televise either Coleman or Schwarzenegger's television shows and movies, the "equal-time" rule could kick in and cost them commercial revenue. If California broadcasters, for example, aired just one of Schwarzenegger's blockbuster films—*Terminator, Predator, Commando, Total Recall,* or *Kindergarten Cop*—then Schwarzenegger's rival candidates could demand they be given television time for free. As one observer put it, California television could become a non-stop political commercial. Schwarzenegger won the race without TV station managers losing their shirts.

Political advertising is one area of law where broadcasters face legal restrictions distinctively different from those imposed on other media, enforced by a federal agency that has been given the authority to uphold the public interest. The justification for this legal scrutiny is based on radio and television's use of the electromagnetic spectrum. Since the broadcast airwaves belong to the public and are in limited supply, the federal government decides who is to be licensed for their use, and how they should exercise this privilege.

Fewer media companies hold more and more of the broadcast licenses and that has raised concerns about the trend toward a monopoly. Citizen groups have asked the government to hold the rein on multi-billion-dollar media mergers. Special-interest groups also have asked broadcasters to curb excessive advertising and indecent content, while improving children's education with their programming.

Protections for the electronic media's rights are also founded in law. Although no permanent safeguard shields their intellectual property, they do have copyrights for temporary periods of time. Courts also uphold the rights of privacy, reputation, access, and confidentiality in media law. As in most legal disputes, judges and juries are asked to weigh the rights of opposing groups and their interests.

▐ Beginning with the Basics

The rule of law is a commonly heard phrase when an actor, celebrity, or politician finds his or her fate shifting in the scales of justice. Without this standard, the rights of the governed would be subordinate to those who govern them. **Jurisprudence** in U.S. electronic media puts this principle into practice through three branches of government: *executive, legislative,* and *judiciary.* The executive branch administers the laws that are drafted by the legislative and that are interpreted by the judiciary. What does it mean for electronic media? Because broadcast waves are a natural resource—one defined as scarce and limited—their use is guarded in laws established by Congress and enforced through an executive agency, the Federal Communications Commission.

Less room is available for broadcast channels than for people who wish to secure them, so the FCC faces a formidable task in assigning frequencies. The agency receives thousands of inquiries for broadcast licenses each year, and most of them must be turned away. In the past, broadcasters' resources and their promises to serve the community's interest helped the government decide between competing license applicants. Now, it's mainly a matter of money, although the government does have a system of credits for certain applicants. The Communications Act of 1934 stresses the national interest *"in communication by wire and radio so as to make available, so far as possible, to all the people of the United States a rapid efficient, Nation-wide, and world-wide wire and radio communication service with adequate facilities at reasonable charges."* So the ideas of public airwaves, spectrum scarcity, and local service are time honored, subject to the actions of Congress, the interpretation of the courts, and the decisions of the FCC. Broadcasting's legal history is less than a century old, but tracing its rules and procedures can be a tricky business.

PRINCIPLE 1

The government licenses broadcasters to serve the public interest.

WIRELESS RULES

Beginning the ledger of radio law is the Wireless Ship Act of 1910. The U.S. Navy requested a rule for the "apparatus and operators for radio communications on certain ocean steamers" carrying "50 or more persons."[1] The tragic oversight of this law was its failure to prescribe how many hours a day the wireless operators had to stay at their posts. On April 15, 1912, that omission came to light when, with no radioman on duty, a jagged iceberg sent the "unsinkable" *Titanic* into maritime history. Congress was roused to spell out a law requiring staffing of every ship's radio room both day and night.

Radio stations were silenced during World War I. Soldiers trained in wireless returned home and continued to tinker with homemade transmitters and crystal sets. Their huge clash of signals inevitably created interference because only one frequency was available. Early stations were assigned to broadcast on the radio dial at 360 meters, the length of the wave, but that channel grew so crowded that early pirates seized another channel that had been reserved for government weather and crop reports.

The radio spectrum was in such disarray that Herbert Hoover, U.S. Secretary of Commerce and Labor, summoned leaders to four national radio conferences to straighten out the mess. The 1922 session produced a new channel for broadcasting at 750 kHz. Still, two frequencies were not enough and some stations were forced to share their time on the air just to be heard.[2]

The spread of wireless sets inspired eager operators ready to broadcast across the national map. By 1923, the number of radio stations exceeded 560, and they had

Jurisprudence Theory or philosophy of law. Once referred to underlying principles or rationale for judicial actions.

THE *TITANIC*. The maiden voyage of the famous luxury liner in April 1912 had an impact on American radio law. If wireless operators had been required to stay on duty all night, more of the ship's 1,517 passengers might have been saved.

exhausted the government's file of three-letter combinations for call signs. Secretary Hoover needed new advice on how to nurture radio's continued growth. After three more meetings with hundreds of engineers and station owners from companies such as AT&T and Westinghouse, the government cleared more channels—from 550 to 1500 kHz. Telltale noise from radio interference had been alleviated, but enforcing the law became next to impossible.

Intercity and Zenith Cases

Secretary Hoover found his authority lacking when two court cases went against his department. The first ruling reversed an earlier decision to deny Intercity Radio Company of New York a license. Its broadcasts would infringe on government channels, so the secretary saw no option but to refuse the application. However, a federal court held that Hoover had no power to refuse a license.[3]

In the second case, a Chicago station trespassed on a wavelength reserved for Canadian radio. Zenith Radio's station, WJAZ, was denied its license, but the company president Eugene F. McDonald would not take no for an answer. He felt the government had slighted his company in favor of larger rivals, such as General Electric and Westinghouse. So he ignored the federal refusal and resumed programming at 910 kHz.[4] In both cases, *Hoover* v. *Intercity Radio* and *United States* v. *Zenith (WJAZ)*, judges reached fundamentally the same conclusion: The department's power to deny licenses was written nowhere in law. Congress had failed to empower Hoover to orchestrate radio frequencies, power, and hours of operation.[5]

Federal Radio Commission

After hearing from radio engineers and industry leaders in 1926, Secretary Hoover was ready to see something written in law that would give him authority over this chaos. The Radio Control Bill of 1927 granted his wish. Representative Wallace H. White (R-ME) and Senator Clarence C. Dill (D-MT) pushed through a bill legislating a new, *temporary* agency to remedy radio's problems. The Federal Radio Commission (FRC) would, "consistent with the public interest, convenience, and necessity, make reasonable regulations governing the interference potential of devices . . . emitting radio frequency energy."

Casualties in the Public Interest

There were new problems to be resolved beyond the noise of interference and unruly station owners. When word reached Washington of medical charlatans and evangelists hawking their wares by radio, the FRC stepped in to establish legal precedents. One huckster, John R. Brinkley, had signed on a remarkable station, KFKB (Kansas First Kansas Best) in 1923. The license for the station in Milford, Kansas, was granted to a hospital association. "Doc" Brinkley specialized in a cure for lost male virility—he surgically implanted goat glands in his male patients. Over his 5,000-watt station, Brinkley counseled midwesterners through the *Medical Question Box* show. He read their letters, diagnosed their symptoms, and prescribed cures—all sight unseen. Once the FRC got wind of his medical practices, it moved in with a second opinion—close down his station, KFKB.

Another medical marvel was broadcasting farther north, in Muscatine, Iowa. The voice of Norman "TNT" Baker rang forth from Know-The-Naked-Truth (KTNT) radio. "Doc" Baker sold his "penetrating oil" over the air for those suffering from appendicitis—whenever he was not shilling for an instrument with the "newest musical tone in 40 years," the *calliaphone*. His more orthodox colleagues took a public position against his radio carnival, incurring his wrath. After the American Medical Association (AMA) denounced TNT's sideshow, Baker charged that the AMA's members were the "Amateur Meatcutters of America," and said M.D. only meant "More Dough!"

The FRC, honoring free speech and its mandate not to censor broadcasting, was reluctant to turn off Baker's or Brinkley's radio transmitters. It did have the power to deny their petitions to renew their radio licenses, and did deny them in 1931. Undaunted, the mesmerizing doctors packed up their antennas and moved to warmer climes. Baker and Brinkley were heard broadcasting cures from Mexico, using the airwaves to reach desperate and naive patients.[6]

Perhaps more than medicine, religion persuaded radio listeners, but early evangelists found the government could draw the line on their sermons. The FRC took action in 1931 against Rev. Robert Shuler, a California evangelist famous for his attacks on the Roman Catholic Church. After hearing an appeal to recover his license for KGEF in Los Angeles, the Court of Appeals found "Bullet Bob" Shuler's incendiary broadcasts were not subject to First Amendment protection due to their defamatory nature. He lost the privilege of using the public airwaves.[7]

Another spiritual star of California radio was Aimee Semple McPherson. She preached from the Four Square Gospel church in Los Angeles, where her radio broadcasts wandered up and down the dial like a sojourner lost in the wilderness. She was warned in 1925 that KSFG had better find its rightful frequency and stop interfering with other stations; this only angered the charismatic minister. McPherson's telegram to Secretary Hoover stated: "Please order your minions of Satan to leave my station alone. You cannot expect the Almighty to abide by your wavelength nonsense."[8]

Emergence of the FCC

Congress never intended for the FRC to spring to full life for the future regulation of radio. There was just a one-year plan designed to allow the agency to "function only occasionally," then hand back its chores to the Commerce Department. That hand-back never took place, though. Instead, the FRC evolved to become the Federal Communications Commission. Congress extended the commission's reach over rule making in 1934 to encompass telephones, telegraphs, and all electronic media. The FRC was given a broad canvas, with the media landscape barely sketched.

Adversaries and Ascertainment

Because a democracy draws its breath from the consent of the governed, early attempts at dictating regulatory issues

PRINCIPLE 2

Public ownership of the spectrum requires government oversight over the use of its frequencies.

to broadcasters were ill-fated. One notable flop was an FCC policy statement attacking broadcasters for giving away too much air time to networks for their programs and advertising. The statement was titled *Public Service Responsibility of Broadcast Licensees,* but became known by its colorful nickname, the Blue Book, inspired by its binding and cover. Charles A. Siepmann, a British consultant to the FCC, composed the prescriptive commentary on U.S. radio, taking aim at the lack of public-affairs shows and "sustaining" programs free of commercial interruption. American broadcasters were livid. They found the tone and tenor of the Blue Book demeaning and its threat of government oversight intimidating. Leading the resistance was *Broadcasting Magazine,* which called Siepmann's work as "masterfully evasive as it is vicious."[9] The Blue Book was never fully enforced.

In 1960, the FCC took a new tack. It published a program policy statement endorsing programming it considered to be in the public interest. Fourteen favored items included community news of political affairs and programs on religion. There also were to be educational and children's shows, farm reports and business news, and editorials and minority access. This statement placed entertainment programming at the bottom.

This document also called for an *ascertainment* of community problems to be dealt with by each broadcaster through programming. By interviewing civic leaders and conducting surveys, broadcasters would discover issues of importance. Most important, this 1960 policy was enforced as part of broadcasters' renewal process for their licenses. That was true until the 1980s, when President Reagan's appointee to lead the FCC, Mark Fowler, called for an overhaul of such rules. In terms of the public's interest, he likened radio and television to a household appliance—"a toaster with pictures."

FCC SEAL. The Federal Communications Commission stands as the authority over licensing and regulation of electronic media in the United States. Although prohibited from censorship, the agency's authority does ensure that broadcast licensees serve the "public interest, convenience and necessity."

Deregulation

Under Fowler, the FCC found its ascertainment policy—and others like it—offensive to the First Amendment. He dropped the policy into the "circular file" of discarded rules, to be replaced by the marketplace's wisdom. First, commercial radio was deregulated in 1981, followed by commercial television in 1985. Congress joined in the retrenchment by reducing the size of the FCC from seven to five commissioners. Things changed during the 1990s, when deregulation was no longer supported as a one-size-fits-all solution for FCC problems. In 1996, Congress passed the Telecommunications Act, which lifted the limits on station ownership, but provided parental control of program content through the V-chip.

FCC STRUCTURE

The FCC is first responsible for administering and enforcing the laws passed on Capitol Hill. It does so by delegating its regulatory powers to its bureaus and offices. Each one functions under the agency's procedures for rule making and policy enforcement. The agency examines license applications and leads aspiring broadcasters through the maze of the federal bureaucracy. To understand this branch of administrative law requires knowing how it is organized.

The Commission

The FCC consists of five individuals appointed by the president and confirmed by the Senate. They exercise the will of Congress, which has given them quasi-judicial and legislative powers to draft and enforce rules governing electronic media. All commissioners are sworn to serve five-year terms, though few actually stay on that

When President George W. Bush first took office in 2000, he had to choose who would lead the FCC—a sitting commissioner, Michael K. Powell, or a friend from Texas. Bush chose the 38-year-old Powell to succeed William Kennard, making him the second African American to hold that position of leadership. As the son of General Colin Powell (later U.S. Secretary of State), President Clinton first summoned Michael Powell to the FCC as a Republican Commissioner in 1997. His experience and conservative philosophy encouraged President Bush to appoint him chair in 2001.

Powell's actions veered occasionally from traditional GOP policy. For example, he favored low-power FM radio—advocated by the more liberal commissioners—to give churches and schools a chance to broadcast. Yet, when asked by reporters about the so-called digital divide, the gap between the *haves* and *have-nots* in computer technology, he sounded a more cynical note: "I think there's a Mercedes divide. I'd like one, but I can't afford it."[10] Powell supported controversial changes in media ownership rules that permitted an even greater concentration of power among rich, media conglomerates.

Powell's distaste for dictating content to broadcasters simmered below the surface during an early indecency controversy—the first of many—during his tenure. The FCC issued a notice of apparent liability and forfeiture (in other words, a fine) against a Pueblo, Colorado, radio station for playing an edited version of one of Eminem's rap songs, *The Real Slim Shady.* That move echoed an earlier era when New York shock jock Howard Stern managed to tally up $1.7 million in fines for his brand of radio sleaze. Another FM station, KBOO in Portland, Oregon, was fined $7,000 for playing *Your Revolution* by rap artist Sarah Jones. In 2002, however, the FCC's Media Bureau reversed the indecency decision against KKMG-FM in Pueblo, Colorado.[11]

After half a million viewers took time to complain about Janet Jackson's breast-baring halftime performance at the 2004 Super Bowl, Powell joined the outcry of viewers—signing off a large fine against CBS stations that aired the incident. He urged broadcasters to revive a code of decency that would help them self-police the airwaves and provide more family-friendly viewing, and was compelled to respond to conservative pressures to crack down on indecency.

Just one day after President Bush's second inauguration in 2005, Powell announced his resignation as FCC chair "with a mixture of pride and regret," pointing with pride to a "bold and aggressive agenda" that focused on stimulating innovative technology in electronic media. He was replaced by Kevin J. Martin, a Republican from North Carolina trained at Harvard's School of Law.

long. The president appoints them, selecting three of them from his own party, and choosing one of the three to preside as FCC chair and to set the agenda.

Bureaus and Offices

The old slogan at the ballpark was "You can't tell the players without a program," and at the FCC you need an organizational chart to tell what's going on. Its bureaus and staffs are shuffled and reshuffled to match the priorities and policy views of each new chair. Chair Kevin J. Martin, for example, selected new staff members from the Wireline Competition Bureau. In 2001, Powell took audio and video services from the old Mass Media Bureau, and moved them in with Cable Services to form the Media Bureau. He created the Wireline Competition Bureau to replace the Common Carrier Bureau, as Figure 11.1 shows. Before him, Chair Reed Hundt saw the need for an international affairs bureau and another for wireless telecommunications.[12]

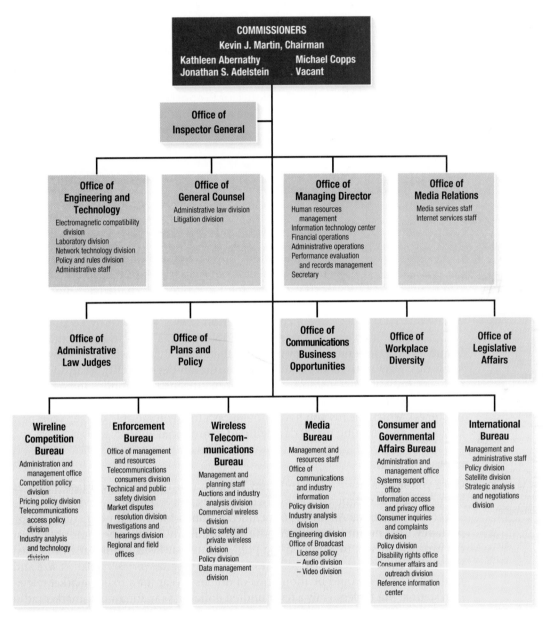

FIGURE 11.1 *Organization Chart, Federal Communications Commission*
As part of its reform plan to become more effective, efficient, and responsive, the Federal Communications Commission approved the reorganization of several of the agency's bureaus. In making these changes, the FCC was guided by the following principles: Develop a standardized organizational structure across the bureaus; reflect changes in regulation and workload; recognize that dynamic industry change will continue; and use the reorganization to improve the technical and economic analysis in decision making.
Source: FCC News, Federal Communications Commission.

The FCC's Watch

Congress delegates the powers of the Commission to ensure the technical quality of broadcast signals while preserving the U.S. stake in their use. To reach as many communities as possible, the FCC assigns call letters, monitors frequencies, inspects stations, and even sees that TV towers are properly lighted and painted.

Rule-Making Process

Electronic media change faster than new rules can be drafted or old ones revised, but the agency honors its procedures for changing rules or adding new ones. It begins by calling attention to a perceived problem. That issue is framed in a document called the **Notice of Inquiry (NOI).** The NOI alerts broadcasters by telling them that the FCC has deemed a question to be of special importance, and invites their comments on its answer. The next step signals that there's a good chance that a change of rules is on its way. The **Notice of Proposed Rule Making (NPRM)** will give broadcasters a second chance to speak their minds. When the agency arrives at the third stage, it publishes a **Report and Order (R&O),** announcing a decision. The R&O defines in detail how and why any new rule was adopted and whether amendments were tacked on, or may simply explain why the status quo was preferred. All R&Os are in the *Federal Register,* published by the U.S. Government Printing Office in Washington, DC.[13]

PRINCIPLE 3

The scarcity of spectrum space limits broadcast frequencies and station licenses.

Notice of Inquiry (NOI) Public statement by the FCC, designed to create awareness of a legal problem or issue and to invite informed comments on how it should be resolved. The Federal Register is responsible for publishing the NOI.

Notice of Proposed Rule Making (NPRM) The FCC takes this second step in the regulatory process to show it is planning to make or amend its rules and is inviting further comments.

Report and Order (R&O) FCC statement explaining new rules or changes in old ones. The R&O is published in the Federal Register.

Construction permit (CP) Authorization by the FCC to build a broadcast facility or to make substantial changes in an existing one.

Renewal Expectancy FCC's stated commitment to favor existing broadcast licenses over competitive applications at the time of renewal, unless given cause to do otherwise under the Telecommunications Act of 1996.

LICENSING PROCESS

The most basic power delegated to the FCC is the licensing of broadcast stations. If you sought to own a radio or TV station, the most rigorous challenge is finding an available frequency. To improve your chances, an engineer *and* a communication lawyer will help by keeping their eyes peeled for public notices when the next station or frequency is up for grabs. If they find one, the Media Bureau will have a few forms to fill out. New stations begin with a **construction permit (CP).** That document will confirm your citizenship (only American citizens may apply), the quality of your character (no convicted felons), and the depth of your pockets (sufficient resources), which means you have enough cash that you are able to keep your new station up and running for 90 days without commercial sponsors.

License Renewal

Once you have a license in hand, how long do you think that will last? Licenses are now held for eight-year terms before renewal is required.[14] Revisions to the Telecommunications Act in 1996 included a bonus for broadcasters: **renewal expectancy.** That meant licenses were safe and secure until someone could show that the broadcaster was treading on the public's interest. Even though the broadcast license is free, the government assesses an annual fee that varies by market size and medium. That fee would be thousands of dollars a year for a major market television station, but far less, only a few hundred dollars a year, for a small-market radio station.

Public Inspection File

Mounds of paperwork may have vanished thanks to deregulation, but one document folder still is required by the FCC's rulebooks: the *public inspection file.* These records are available for all to see, and include a station's license applications and reports to the agency, contour maps of the station's coverage area, letters and e-mails from the audience, and a list of local issues and of programs designed to address them. Failure to make the public inspection file available during business hours for anyone to thumb through—even rival broadcasters—can draw an FCC fine, or, as the agency calls it, a *forfeiture*. What's not included in the public inspection file are the station's financial documents, including profit and earnings statements, payroll records, tax forms, and other business records. The government respects the fact that broadcast stations compete as private enterprises.

Penalties

The agency may fine a station or revoke its license for putting *obscene or indecent language* on the air, *soliciting money under false pretenses, advertising illegal lotteries,* or other violations of the public trust. The FCC almost never revokes a station license; it renews licenses about 98 percent of the time. The few exceptions show the dark taint of misleading the government (lack of candor), often compounded by a poor record of public service. Beyond revocation or forfeitures, the agency employs other tools to uphold the law. License renewals can be okayed for the short term, until the station has found ways to solve its problems. Other FCC sanctions include letters of reprimand placed in the station's public file, and the less frequent cease-and-desist order used to halt a harmful station practice.

FEES, AUCTIONS, AND LOTTERIES

It sounds like a lot of money—*$280 million* a year—but to budget one agency overseeing all of U.S. telecommunications, it's not much. President Bush asked Congress to approve an FCC spending plan of *$280,798,000 in 2004*. An increase was needed to raise salaries and benefits and to pay for contract services, including engineering consultants. It is important to realize that the FCC rakes in more than its budget each year—a lot more. Some of it comes from a variety of fees. Satellite television, for example, contributes somewhere between $114,000 and $131,000 per satellite, cable systems pay about 66 cents to the government per subscriber for millions more, and radio and TV stations add hundreds of thousands in their annual fees.

The FCC has grown to become one of Uncle Sam's favorite nephews ever since *spectrum auctions* were established to raise money from a natural resource: the electromagnetic spectrum. Since a federal court ruled in 1993 (*Bechtel v. FCC*)[15] against the comparative-hearing process for broadcast licenses, billions of dollars have been raised for the U.S. Treasury by taking bids for licenses.

PRINCIPLE 4

Competitive bidding—not the promise of performance—determines who gains access to the telecommunications channels.

Comparative hearings had been held by an administrative law judge who chose between competing broadcasters by applying a public-service criteria. After the *Bechtel* case, the agency first moved on to lotteries for licensees before Congress gave the go-ahead to competitive bidding. Then, spectrum auctions threw open the door to virtually anyone who could submit an advance payment. In 1997, Congress expanded this money-making program by essentially doing away with the lottery as an alternative means.

■ Ownership: Concentration and Diversity

Multibillion-dollar enterprises owning dozens of radio and TV stations have been growing larger in size but fewer in number. When Ben Bagdikian wrote *The Media Monopoly* in 1983, half of the broadcasting, newspaper, magazine, music, film, video, and publishing groups were in the hands of 50 companies. In two decades, that number has shrunk to 5 media conglomerates: Time Warner, Disney, News Corporation, Viacom, and Bertelsmann. Looser controls on broadcast ownership took effect with the passage of the Telecommunications Act of 1996, which accelerated the trend toward media mergers. Consider this statistic: From March 1996 to March 2001, the number of U.S. radio stations grew by 7 percent while the number of owners fell 25 percent. In June 2003, the FCC moved to raise the limit on the number of television stations one company could own. At first, it proposed raising the maximum limit from 35 to 45 percent of the U.S. television households a group could reach, but a federal appeals court blocked this consolidation-friendly plan, and ordered the FCC to rewrite its ownership revisions. It settled on a compromise measure.

OWNERSHIP REGULATION

Republican leaders in the U.S. Senate in 2004 tried to clear the way for a vote on the FCC's new ownership cap, and offered a compromise figure of 39 percent. Pro-deregulation conservatives favored this move, which also had the support of the National Association of Broadcasters. However, a much larger number of special-interest groups representing such diverse viewpoints as the National Rifle Association and the American Civil Liberties Union stood in opposition to further consolidation, which this rule would allow, and so it failed at first to pass in the Senate. The Congress finally settled on the 39 percent limit in January 2004, following intense lobbying from the President and industry leaders.[16]

First Limits

The FCC placed its first cap on ownership of radio and TV stations in 1952 with the "rule of sevens." This simply meant that the maximum number of broadcast outlets for any one person or corporation to own was seven AM, seven FM, and seven TV stations. The rationale was that more broadcast owners meant more differences of opinion and more diversity in the marketplace of ideas, and that formula stood until 1984.

The nation's news media policy was founded on what the Supreme Court identified as an underlying assumption of the First Amendment: that "the widest possible dissemination of information from diverse and antagonistic sources is essential to the welfare of the public."[17] That thinking was challenged, however, during the deregulation era, when the government created a friendly environment for media mergers. The rule of sevens became the rule of eights, and then twelves, and then twenties, until it was expanded by Congress in 1996.

The Telecommunications Act of 1996 put in place an "all-you-can-own" rule for radio stations on a national basis, but it left in place some limits at the local level. There are rules governing the number of radio stations, TV stations, and newspapers one owner can control by applying a formula to take into account the varying sizes of media markets.

Duopolies, Triopolies, and Multiopolies

It used to be that one television group owner could not control 2 TV stations in one market; however, that rule against *duopolies* was relaxed in 1999 to accommodate more media mergers. Ownership of both radio and television stations in a single market is now limited by the total number of stations there. For example, a single owner can hold licenses to a *triopoly*, 3 TV stations—if the market has more than 18 stations and if only 1 of the 3 is rated among the top 4. *Duopolies*, ownership of 2 TV stations, are permitted in markets of between 5 and 18 stations, if only 1 of the 2 is rated among the top 4. For markets with 5 or fewer TV stations, the rule is only 1 TV station per owner.

Radio station rules are similar in that a sliding scale determines the numbers. The ownership limits hold that no more than 8 stations can be controlled in the largest markets by one owner (45 stations and up). The ownership limit is 7 stations in markets with 30 to 44 radio stations, 6 in smaller markets of between 15 and 29 stations, and 3 radio stations in markets with 14 or fewer stations.

CROSS-OWNERSHIP

The cross-ownership rule, forbidding partnerships between newspapers and broadcasters in the same city, was written into law in 1975. Over the years, new cable and satellite channels as well as websites were added. Network owners maintain that marketplace reality has done more to protect diversity and competition than outdated law ever did. Citizen groups agree that more channels are available, but contend they are often owned by the same huge corporations, and that the ones that are not so well supported discover it difficult to compete.

Newspaper companies such as Tribune, Belo, and Post-Newsweek owned both dailies and broadcast stations in the same cities before the 1975 cross-ownership rule was adopted. In most cases, the rule allowed them to hold on to those properties. Now, major media groups would like to buy new TV stations in other markets where they own newspapers or cable systems, and one court appears to be on their side. In *Fox Television Stations* v. *Federal Communications Commission* (No. 00–1222),[18] the District of Columbia Court of Appeals threw out the rule that barred one group from owning both cable systems and television stations in the same market.

As the rule stands now, there are no limits on cross-ownership in markets with more than nine television stations, but no single broadcaster can control an "inordinate share" of media in the community. In smaller markets, those with four to eight television stations, the FCC allows a single media owner to control one daily newspaper, a maximum number of radio stations, but no TV stations. A second cross-ownership possibility allows two TV stations, the maximum number of radio stations, but no newspaper. The third option allows one newspaper, one TV station, and *half* the maximum number of radio stations.

DIVERSITY IN OWNERSHIP

Before 1978, the government granted one-half of 1 percent of all broadcast licenses to minorities—only 40 stations out of more than 8,500. After 1978, the FCC deployed a tax certificate program designed to encourage the sale of cable systems and radio and television stations to minorities. The program did this by deferring or eliminating the capital gains tax that station sellers would have to pay on their profits from a sale. If you owned a broadcast or cable property and were willing to sell it to a female or minority, you would get a tax break. During the next 17 years, more than 350 broadcast and cable systems lined up to take advantage of that offer.

There was another push toward the diversity in ownership goal established during President Jimmy Carter's administration. The FCC encouraged distress sales of broadcast and cable properties to minority and women licensees by allowing the sellers to recover some of the market value of "intangible assets," a recovery of their investment not normally allowed under other circumstances.

In 1995, Congress and the courts did away with both programs. The result was a slip in the number of TV stations owned by minorities. U.S. Commerce Secretary Norman Mineta stated, "Clearly there is reason for concern."[19] A report prepared for the Office of General Counsel of the FCC called it a serious situation.[20]

MINORITY EMPLOYMENT

In the three decades the FCC's Equal Employment Opportunities (EEO) rules were on the books, minority employment in electronic media jumped from 9.1 to 20.2 percent. Women fared best, holding more than 40 percent of the available jobs, but only 15 percent of the women were serving as general managers.

The EEO rules encouraged broadcasters to consider minorities and women for job vacancies: "No person shall be discriminated against in employment by such stations because of race, color, religion, national origin, or sex." The rules did not set quotas or specify types of people to be hired. However, they did allow the FCC to pass judgment on hiring decisions after the fact.

A severe blow was dealt to the FCC's guidelines when the government lost two court cases. *Lutheran Church-Missouri Synod* v. *Federal Communications Commission*[21] involved two radio stations at Concordia Seminary in Clayton, Missouri. An appeals court overturned the ruling that their hiring of minorities failed to comply with EEO guidelines. A second case in 2001 held that a broadcast station's outreach program disadvantaged nonminorities.[22]

So, it was back to the drawing board for the FCC. A new plan was implemented in 2003. This policy no longer required hiring—or even reporting all employment records in terms of race and gender—but instead called for a "broad outreach," which the station could document in its files. The new policy asks for a wide dissemination of employment notices, including ones sent to local groups concerned about minority hiring practices. Broadcasters also are asked to participate in employment-related activities, such as job fairs and internship programs. Their annual reports for public inspection files and websites are required to document station policies on nondiscrimination in hiring. The difference is that no punishment is meted out for failure to hire a ratio of minorities and females.

Regulation by Delivery System

It's no surprise to TV station managers that few viewers receive their signal by rabbit ears or rooftop antennas anymore. Most U.S. households (68 percent) subscribe to cable and about 18 percent use satellite dishes. The broadband road to the digital future looks uncertain for over-the-air broadcasters. They will convert to digital frequencies and give up analog channels, while competing media and manufacturers join in the transition. Broadcasters say they want a level playing field, however, in order to face intense competition from satellite, cable, and the Internet.

DIGITAL TELEVISION (DTV)

A major magazine for broadcast journalists posed an intriguing question: "How and When Will HDTV [high-definition television] Affect Television News?"[23] The article was published 18 years ago, and so the answer would be "Not any time soon." Digital television (DTV) was finally approved for U.S. homes in 1997, but high-definition TV was left optional. Congress set a deadline at the end of 2006 for broadcast stations' conversion to DTV, but did not force their hand on high-definition conversion. The FCC assigned 6 megaHertz of extra spectrum for stations to establish new digital channels while continuing to broadcast in analog. As the 1997 law stated, TV stations would broadcast only in analog until 85 percent of the American viewers owned a DTV set.[24]

To speed the transition along, the government offered a new interpretation of the 85 percent rule. The FCC proposed to count all cable subscribers as digitally served—even if those viewers did not own a digital set or subscribe to digital cable. The problem with abandoning analog channels in order to move to digital frequencies was that it could leave many low-income viewers without television service—if it is done too soon.

The "Broadcast Flag"

Broadcasters and Hollywood were both concerned about the possibility of DTV piracy of high-definition movies and other digital broadcast content via the Internet. The FCC responded with a rule requiring TV set manufacturers to install the circuitry that would look for a signal in the video stream called a "broadcast flag." These DTV receivers would read the codes in the programs, and prevent swapping of DTV content over the Web. Opponents of the technology challenged the FCC rule in court, claiming the government lacked the authority to impose such a rule, especially one that arguably would affect consumers' home video recording rights. A federal appeals court was asked to decide if the flag would be legal before the rule takes effect in July 2005, and the comments from the bench indicated that the FCC might have gone too far.

CABLE REGULATIONS

Congress appeared to be whipsawed by rival groups when it drew up its cable regulation acts. First, there was the 1984 law the cable industry wanted to see enacted to keep franchising authorities from imposing outlandish service requests and levying exorbitant franchise fees. Then, Congress heard the protest of consumers in 1992, and served notice on the cable industry to begin offering rate relief. The Cable Television Consumer Protection and Competition Act sought to control the charges for both basic and expanded basic tiers of cable channels. Regulation then deregulation and back again have been part of cable's legacy in law for years.

Franchising the System

To get a better view of how cable systems differ from broadcasting, look at one basic difference: federal licensing does not apply to cable because it operates under agreements that bring it under local governing bodies. The franchise contract is adopted as an ordinance, a local law specifying length of the deal (10 to 15 years is standard); terms of service fees charged; number of channels offered; public, educational, government channels; and other technical issues.

The city or governing agency is entitled to ask for up to 5 percent of the cable system's gross revenues in fees, so there's a built-in incentive to select the operator most qualified to generate more revenue. In most instances, the process for selecting a cable system starts with a Request for Proposal (RFP). Then, hearings are held with representatives from competing companies before the local government reaches a franchise agreement with one of them. Some civic leaders forced rival cable systems into unreasonable terms for service contracts in exchange for awarding them franchises, which is why Washington stepped in.

1984 Cable Law

Congress passed the Cable Communication Policy Act in 1984 to set the ceiling for municipal franchise fees at 5 percent. The act also prevented cities from specifying just how much money a cable system could or could not charge its customers. It also added a new title to the Communications Act. Title VI coded into law most of the earlier FCC rules for cable, and gave the FCC power to issue guidelines for franchise agreements, including renewals and fees. It freed cable systems from aggressive franchisees hoping to increase their revenues through the franchise fees.

protalk
Richard E. Wiley
DEAN OF THE FCC

Commissioners of the FCC may come and go every five years or less, but one FCC chair has played a pivotal role in policy for more than three decades—that's three decades after the end of his tenure!

In an era of digital convergence, Richard E. Wiley led the nation's efforts to adopt a new digital television standard and make possible dramatically clear high-definition television. However he confesses that for a number of years, it was a bit frustrating. "At first, the transition was slow and uncertain. However, thanks to the recent leadership of the FCC (and, in particular, Chairman Michael Powell), and to the private sector's energetic response, things really have begun to accelerate."

The major development, as Wiley sees it, is the ever-increasing HDTV programs being provided today by broadcast, cable, and satellite networks. Such programming "gives people the incentive to purchase new digital television receivers, especially with prices falling over 75% since they were first introduced in 1998."

Wiley also cited with approval recent FCC decisions to require digital tuners in larger DTV sets, to make possible integrated "plug-and-play" receivers (not requiring set-top boxes), and to prevent widespread distribution via the Internet of copyrighted digital programs. "All of these actions," he says, "will give further momentum to the

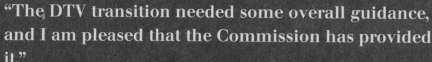

"The DTV transition needed some overall guidance, and I am pleased that the Commission has provided it."

digital transition. What's left now is for the Commission and Congress to determine when the transition should end and when conventional television, as we have known it for over 50 years, is phased out."

Wiley also has been involved in the FCC's effort to develop new broadcast ownership rules, representing various newspapers and broadcast clients before the Commission and the courts. Some of the agency's decisions have placed broadcast networks at odds with local stations. The former FCC Chair, and now head of Washington, DC, law firm of Wiley, Rein & Fielding, mused, "I have clients on both sides of that issue, so let me just say that I am with my clients!"[25]

On the other hand, some cable operators saw the 1984 act as a green light to begin raising their rates for basic and premium tiers. The General Accounting Office (GAO) measured a 43 percent hike in the average cable bill from 1986 to 1989. So Congress stepped in again, this time drafting a plan with the consumer in mind.

1992 Cable Law

The 1992 Cable Act put a cap on local cable rates, and ordered every system with 36 channels or more to set aside at least 3 channels for lease to outside parties. This **leased access channel** provision gave power to the franchising authority to set maximum rates for special lease customers. The Cable Act (1992) allowed competition among cable systems, and even allowed "overbuilds," in which two cable lines pass the same neighborhood, which is still a relatively rare occurrence.

Congress asked the FCC to cap the number of cable systems one owner could control. The FCC adopted a rule barring any company from acquiring more than 30 percent of all cable subscribers in the United States. A federal appeals court rejected that action, holding that the 30 percent limit was at cross-purposes with the First Amendment.

Must-Carry Rules

The 1992 Cable Act also took a new turn with regard to the "must-carry" provisions for local broadcast channels. You may recall from Chapter 3 that the **must-carry** rule requires cable operators to carry every TV station within a certain radius of their systems. The act had barely become law when Turner Broadcasting, joined by several cable groups, appeared in court to challenge the law's provisions. In two rulings, dubbed as *Turner I* and *Turner II*, the U.S. Supreme Court held the must-carry rule was acceptable because it protected noncable homes from losing broadcast TV stations that might be forced out of business if taken off cable's basic tier of channels. It also protected fair competition and the diversity of program sources.[26]

A broadcaster could choose to negotiate a "retransmission consent agreement," which charged the cable operator a fee for carriage. Except for a few major cities, most cable systems were unmoved by this option. Television stations needed the exposure on cable anyway and, in most cases, accepted must-carry without imposing fees. This meant the stations' signal would be carried for free, as usual.

In 1998, the Federal Communications Commission began the process of rewriting its rules to afford television stations a second channel on local cable systems as part of the transition plan to DTV. Cable operators protested because they claimed that digital must-carry rules would force them to eliminate other cable channels (C-SPAN had been mentioned) to make room for extra broadcast channels. Broadcasters countered that would not be necessary because DTV channels were compressed, but early in 2005 the FCC sided with cable and voted not to impose any must-carry obligation on cable for broadcasters' extra DTV channels.

Plug-and-Play Sets

The government moved ahead with plans to allow cable customers to plug a cable directly into digital TV sets without requiring the use of a set-top box. The FCC's rules regarding plug-and-play, cable-ready digital-TV sets raised a question of protection for film distributors and television content producers, who were concerned about digital piracy of movies and other content. The industry asked for a technical solution called "selectable output control," which would have permitted the studios to turn off certain outputs deemed insecure. The FCC generally rejected that proposal, but did adopt "encoding rules" that set a ceiling for cable or satellite copying: Broadcast programming is freely "copyable," cable programming is limited to copy-once, and video-on-demand and pay-per-view content has all copying restricted.

Leased access channels Legal requirement that cable systems reserve some channels for leased commercial use for parties not affiliated with the TV system.

Must-carry Rule requiring that broadcast stations be carried on cable, satellite, or other subscription video service.

SATELLITE REGULATIONS

If radio and TV stations are licensed and cable systems are franchised, then does satellite television operate without controls? Satellite systems are licensed, and the Cable Television Consumer Protection and Competition Act of 1992 granted satellite television equitable access to all cable channels available. Congress acted to advance satellite television in 1999 with the Satellite Home Viewers Improvement Act, a law that gave DISH and DirecTV the right to carry local channels.

TELCO LAWS

After Bell Telephone was broken into seven regional Bell operating companies (RBOCs), or "baby Bells," in 1982, a new playing field for telecommunications emerged. First, the FCC's video dial tone (VDT) decision of 1992 opened the door for phone companies to common-carrier video service through their wires. Then, the 1996 Telecommunications Act lifted the cross-ownership ban on telephone and cable services. That act allowed the baby Bells to begin offering video programs, while the cable company began selling telephone service to its subscribers, and began offering Voice-over Internet Protocol (VoIP).

The FCC now oversees what are called open video systems (OVS) for phone companies. These systems can become a common carrier for others to use or a multichannel television service of its own—two means by which telephone firms may compete in the video marketplace. Where fiber-optic lines have been installed, telephone companies deliver data and video through broadband networks. These networks require certification by the FCC, however, and may be subject to a fee charged by local governing bodies. In fact, a federal appeals court held that the city of Dallas, Texas, could subject a phone company, Southwestern Bell Video Services, to the same franchise fees and rules it had imposed on cable television.[27]

SMATV AND WIRELESS SYSTEMS

Private apartment complexes and some residential communities draw on satellite master antenna television (SMATV), sometimes known as *private cable*. Because SMATV systems usually do not cross public right-of-ways or use broadcast spectrum, they are subject neither to licensing nor to franchise agreements that cable companies are required to sign. The FCC has held these private cable systems are also free of state and local governance.

Microwave Television

Microwave television channels, broadcast by multipoint microwave distribution systems (MMDS), use the public airwaves and require a license from the FCC. Dubbed "wireless cable," MMDS uses microwave frequencies between 5 and 10 GHz, and serves about 200,000 customers in the United States. First sold as an alternative to cable programming, MMDS began to promote its broadband Internet services in the late 1990s. In 1998, the FCC sought to encourage growth in the wireless cable industry by allowing MMDS to compete with Internet access and videoconferencing for its customers.

 # Content Standards

As practiced in the United States, media law may be divided into two areas: structural and content regulation. Discussion of mergers and acquisitions reflects a view of the industry's structure, in terms of legal administration and media ownership.

We take a look now at matters of content, beginning with two rules sometimes erroneously tied to the term *equal time.* Both rules were meant to ensure balance and fairness in terms of the broadcaster's obligations to citizens, politicians, and issues of public importance. Neither rule, however, guaranteed *equal time.* The third area of content law covers children's programming.

FAIRNESS DOCTRINE

The **Fairness Doctrine** grew out of the FCC's longstanding commitment to "the free and fair competition of opposing views."[28] The struggle to define *fairness* goes back to a report issued by the agency's predecessor, the Federal Radio Commission (FRC). In the *Great Lakes* case of 1929, the FRC emphasized the fairness principle, applying it to addresses by political candidates and to all issues of importance to the public.[29] Twenty years later, the agency gave seed to the Fairness Doctrine in a special policy statement titled *In the Matter of Editorializing by Broadcast Licensees.*[30] It repealed the FCC's pre–World War II ban on editorials, replacing it with an affirmative obligation to cover "public issues in the community." As the doctrine evolved, it spoke forcefully to debates about the dangers of nuclear energy, pollution of the environment, and even cigarette smoking. Yet, after decades of contentious hearings, contradictory decisions, and uneven enforcement, the FCC abandoned it. The commission's Fairness Report in 1985 deemed the Fairness Doctrine a failure. Congress disagreed and tried to codify the doctrine into law. President Reagan's likely veto and radio talkshow host Rush Limbaugh were all that stood in the way—but they were enough. President Clinton invited a new Fairness Doctrine bill, but Congress lost both the initiative and the votes needed to enact one. Finally, with the blessing of two lower court rulings—one involving a nuclear plant in Syracuse, New York, and the other a labor dispute in Arkansas—the FCC stopped enforcing "fairness."[31]

Some loose ends of the doctrine were later addressed in court. A federal panel ordered the FCC to rescind the doctrine's personal attack and editorializing rules because they "entangle the government in day-to-day operations of the media." The personal attack rules put broadcasters in the position of having to notify any individual or group who had been personally maligned over their station's airwaves, and offer them free time to reply to the attack. Similarly, the political editorial rule stated that a broadcaster who endorses a candidate for political office would have to surrender free air time to the candidate's opponent in the race. In 2000, the Radio-Television News Directors Association succeeded in getting both rules overturned.[32]

SECTION 315—EQUAL OPPORTUNITIES

The second rule, **equal opportunities** affects the election campaigns of members in Congress. It holds if a station permits any person "who is a legally qualified candidate for any public office to use a broadcasting station, he shall afford equal opportunities to all other such candidates.[33] There is also a candidate access rule, Sec. 312(a), that requires broadcasters to sell advertising time to candidates running for federal office.

At the dawn of radio regulation, the bill that enacted the Federal Radio Commission advised broadcasters they must afford equal opportunities to all candidates *for campaign advertising.* The law added that stations "shall have no power of censorship over the material broadcast." Congress moved the rule to Section 315 and the 1934 law put some teeth in its enforcement. Radio licenses could be revoked if stations were caught willfully and repeatedly failing "to allow reasonable access to . . . a legally qualified candidate for Federal elective office."[34]

Section 315 prompts a number of "frequently asked questions" (FAQs). Broadcasters, for example, ask how much to charge politicians for their campaign spots.

Fairness Doctrine FCC policy that evolved as a rule requiring broadcasters to report and discuss controversial issues of public importance and to present opposing viewpoints on those issues. The FCC rescinded parts of it in 1987 and 2000.

Equal opportunities Federal requirement that candidates have the same degree of access to local radio and television stations, including advertising rates and time available prior to an election.

That answer is found in the "lowest unit charge" rule, which supposedly gave candidates the bargain-basement rates offered to a station's most favored advertisers. Stations routinely skirt that rule, however. If a candidate wants a particular time slot with a substantial audience—say, adjacent to the local newscast—the station can and often does raise the rates so that the spot won't be preempted. One study of 17 media markets found the price per political spot tripling in the three months prior to the election—the closer to the ballot box, the costlier the airtime."[35]

The next FAQ: Can a campaign spot for a candidate be edited? Section 315's no-censorship clause has created a few problems for broadcasters. In 1972, for example, the gubernatorial race in Georgia featured a white supremacist candidate who spewed hatred against blacks in his radio spots. Broadcasters were repulsed by J. B. Stoner's campaign ads and challenged the rule in court, but to no avail. The political spots posed no "clear and present danger of imminent violence," the court held, and had to be broadcast without censorship.[36]

Hustler magazine publisher Larry Flynt threatened to push the limits of the no-censorship clause even further by promoting pornography in his campaign for U.S. president in 1984. The FCC answered that challenge by advising stations that its rules against obscenity and indecency should be honored even if it meant censoring Flynt's ads. The legal definitions of those terms are discussed later in this chapter.

When antiabortion activists asked to show dead fetuses as part of their campaigns, television broadcasters faced another dilemma. Should the gruesome pictures be censored or broadcast without edits? WAGA-TV in Atlanta elected to channel the graphic spots to "safe-harbor" time slots after midnight, and a U.S. District Court backed up that decision.[37]

Then came the question of newscast coverage: Was equal time needed for every minute radio and TV reporters covered political incumbents? That certainly would have prompted some news directors to ban most political reporting during the campaign season. So the law was amended in 1959 to exempt *bona fide* newscasts, news interviews, news documentaries, and spot news coverage.[38] As a result, political candidates not only appear on the morning news shows but they are also seen on talk shows, which some critics argue stretches the definition of a bona fide news interview. For example, the FCC gave a news interview exemption to shock jock Howard Stern so that he could visit on air with Arnold Schwarzenegger during his

sidebar

CAMPAIGN REFORM

Here's a riddle for you: How does a federal candidate get elected without broadcast advertising? Answer: Only by running unopposed. From 1980 to 2000, the amount of money invested in political advertising in the United States quadrupled from $200 billion to $800 billion.

In 2004, Madison Avenue predicted more than $1.5 billion would be spent on campaigns to win the White House and seats in Congress. Some of the advertising proved controversial, given the attacks were waged early in the presidential campaign by so-called 527 groups, a title taken from Section 527 of the Internal Revenue code. Political activists, such as the Swift Boat Veterans for Truth and Moveon.org, raised the temperature of the heated race with allega-

tions of moral misconduct hurled at both President George W. Bush and Senator John Kerry. Suits were filed against the Federal Election Commission for not disciplining the 527 groups who were spending money on attack ads in key states where voters were undecided. Broadcasters were not complaining, however, and neither were cable systems and Internet organizations, which rely on this uptick in commercial spending.

Even though "soft money" (cash contributions falling outside federal election laws) may be one day eliminated from the campaign process through congressional reform, the need for money to buy commercial time is a hardy perennial. Some reformers argue that that won't change until the United States joins the rest of the world's democracies in offering *free* television time for candidates to discuss political issues.

2003 campaign to become California governor without inviting his opponents to share equal time.

As for political debates and press conferences, a special exemption was made for the famous Kennedy and Nixon debates in 1960. Since that time, however, political debates have been considered fair game for equal-opportunity requests. Congress eventually understood that if stations were supposed to cover political news to the "fullest degree," debates and press conferences would have to be added to the exemptions, which they were in 1975.[39]

One final question: When candidates log an appearance in a situation comedy or a televised motion picture, does it trigger equal opportunities for their political opponents? Ronald Reagan's movies were routinely yanked from TV program schedules while he campaigned for the White House in the 1980s, and, as noted, California's TV stations could not telecast movies such as *Predator*, starring Arnold Schwarzenegger during his campaign. The FCC usually interprets "use" by a candidate to mean the campaign's materials or the showing of earlier films and TV appearances.

CHILDREN'S TELEVISION

Rules for children's television have long been of interest to parents and politicians. Action for Children's Television (ACT) was formed in the late 1960s in response to cartoons with host characters gulping down brand-name cereals, and with action heroes teaching youngsters violent solutions to dramatic dilemmas that could influence a child's social interactions.

Action for Children's Television had entered the arena after it saw that television had veered from the vision articulated by FCC policy statements. In 1974, the FCC urged broadcasters to provide airtime for youth-oriented educational shows and to be wary of how advertisers approached young viewers. Five years later, the commission checked in to see how its *Children's Television Report and Policy Statement* had fared with broadcasters.

The verdict was mixed. Broadcasters were curbing commercial time, but the educational content of children's programming left something to be desired. TV station managers described *Leave It to Beaver* as a learning experience. At that time,

Few people are confident enough to believe they can make a difference in the code books of federal law. Those willing to work at it often find that they can. A group of Boston area mothers formed Action for Children's Television (ACT) in 1968. The women were inspired to close ranks in support of *Cap'n Kangaroo* after a local TV station announced it would be cutting his morning show in half. The ACT group initiated

a letter-writing campaign. Some 2,500 letters later, Cap'n Kangaroo had his full hour back in Boston.

Action for Children's Television then trained its sights on the nation's capital. The FCC had no detailed policy for children's programming, so ACT leader Peggy Charren and her friends in Newton, Massachusetts, decided to help the agency complete its homework. For more than a year, they prepared studies and talking points on what could be done to improve children's television. Charren led her group to Washington, DC, to meet with the FCC commissioners. They were impressed with ACT's depth and understanding of the broadcast issues.

At first, ACT attacked excessive violence in children's programs but soon realized that raising the flag of censorship might jeopardize their cause. They chose instead to focus on TV advertising, arguing that children should not be "dismissed by the medium simply as a market—a group of naive little consumers."[40] The membership of ACT grew slowly and steadily while its leadership developed a sense of what would and would not work in dealing with lawyers, network executives, and Congress. The group secured backing from the Ford and Carnegie foundations, allowing it to get better foothold in Washington.

The pinnacle of ACT's record of accomplishments came with the passage of the Children's Television Act of 1990. Soon after that event, ACT closed its offices, but Charren remains outspoken on children's issues. Her efforts include a campaign to direct indecent content to "safe harbors" (time periods when children are not expected to be watching television).

---

deregulation of broadcasting was reaching a zenith in Washington, so the FCC looked the other way. Alternative media, cable, and videocassettes would help mold young people's minds, reasoned broadcasters.[41] ACT refused to accept this rationalization and went to court to make its case. The mothers' group lost its first battle over children's programming in 1983, but mounted additional charges that met with success.[42] The FCC finally reversed its position by drafting new controls for children's television. Meanwhile, Congress drew up the Children's Television Act and passed it in 1990.[43] Six years later, the FCC amended its rules to strengthen its enforcement of the act.[44]

The new children's TV rules limit advertising time to 12 minutes per hour during the week, and 10½ minutes per hour on weekends. Commercials must be separated from program content, and advertising must not appear within the context of the children's programming. Television stations also must locate three hours in their weekly lineup for educational content that serves the intellectual, emotional, and social development of children, defined as 16 years old and younger. Licensees failing to comply

PRINCIPLE 6

Advocacy groups influence policy and law for electronic media, including children's television.

are subject to explain why, improve their performance, or pay thousands of dollars in forfeitures.

Copyright and Promotions

A media lawyer recalls that when he began his career, it took about 15 minutes a week to deal with *copyright* litigation. Now, this type of litigation fills most of his working days. In the digital age, no avenue of law has seen as much new traffic as intellectual property rights. In light of the ease with which any computer can transmit words, pictures, and sounds across global borders, it's easy to understand why. Digital convergence of media has made it easy to copy and distribute material, and the owners of creative property have gone to court to protect their rights. This episode culminated in a period of legal struggles undertaken to keep creative content in the care of its authors or owners.

COPYRIGHT ORIGINS

The eighth section of the first article of the U.S. Constitution delegates to Congress the power to "promote the progress of science and the useful arts." Congress has enacted laws to secure for authors and inventors exclusive rights (for limited periods) to their own writings and discoveries. The U.S. government recognizes three areas of ownership—slogans, logos, and brand names—as suitable for trademark protection. Inventions are eligible for *patents,* and copyrights are reserved for a "fixed tangible medium of expression." This could mean a play, news story, record, or script, all of which can be copied in some fashion. The fixed expression must be both original and creative; although it may be considered intellectual property, the ideas themselves cannot be owned.

In 1976, a patchwork of cases and acts converged in a single law to protect ownership rights for creative artists in seven areas: (1) literature; (2) musical compositions; (3) drama; (4) choreography and pantomime; (5) pictures, graphics, and sculpture; (6) films and audiovisual works; and (7) sound recordings.

In order to ensure that a work is protected by copyright, the creator should follow a few rules: Label the work with the author's name, provide the date of publication, and use the symbol for copyright or just add the word *copyright* to the work. Original works need not be registered to receive legal protection. However, registering the work secures an advantage in court when an artist seeks to recover costs or **damages** for infringement.

The law defines one pertinent exception to protection: A "work made for hire" falls outside the bounds of copyright ownership. "For hire" describes any creative work within the scope of an artist's paid employment. Ownership then belongs to the employer.

> **PRINCIPLE 7**
>
> There is no absolute protection for one's ownership of original and creative content.

FAIR USE

Sometimes it's legal to use other people's "fixed expressions." **Fair use** is the term for a privilege to use copyrighted material in a reasonable manner without the owner's consent. Before the copyright expires and the creative expression enters the public domain, scholars, journalists, and other artists can make *limited use* of the material as long as doing so has some *useful literary purpose.*

The Copyright Act of 1976 gave the courts four criteria by which to tell the difference between fair use and infringement: (1) Did the original work fall under copyright protection? (2) How much of the substantive content was exploited? (3) Was that secondary use intended to make a profit? (4) Will that use affect future profits

Damages Monetary compensation a person receives for injury to his or her property or rights because of the willful act or negligence of another.

Fair use Limited legal permission to use copyrighted materials without the owner's consent, particularly when news, criticism, or scholarship is involved.

BETTE MIDLER. Celebrities have a right of protection to their personal talents and images. When Ford Motor Company used a sound-alike of Bette Midler in a TV commercial, the court ruled in her favor in order to protect her copyright ownership.

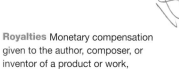

Royalties Monetary compensation given to the author, composer, or inventor of a product or work, calculated according to a percentage of its sales receipts.

for the copyright holder(s)? The courts have been reluctant to permit fair use for certain types of commercial spin-offs, such as when a singer imitates Bette Midler on behalf of Ford Motors or when Samsung Electronics uses a Vanna White robot to sell its products.[45]

ROYALTIES AND MUSIC

Radio and TV stations don't play their music for free. They negotiate a license fee with the copyrights-owners, based on the station's audience size, how many times it airs a song, and the music selected. The license requires that **royalties** be sent to either the American Society of Composers, Authors, and Publishers (ASCAP); Broadcast Music Incorporated (BMI); or the Society of European Stage Actors and Composers (SESAC). Not everyone is willing to pay ASCAP, BMI, and SESAC royalty fees, or even pay for the music they listen to for personal enjoyment, which has cost the recording industry billions of dollars in royalties and profits.

Digital Pirates

Copyright pirates have been busy swapping digitally transmitted music over the Internet. The recording industry determined to capture the pirates and bring them to justice. Congress responded in the 1990s by passing two major laws. The Digital Performance Right in Sound Recordings Act (DPRA) was adopted in 1995, giving copyright holders royalties from the digital performance of their records over satellite and pay cable services.

Three years later, the Digital Millennium Copyright Act (DMCA) extended music ownership protection to websites. The DMCA also spread the ban on illegal distribution to motion pictures, software, and magazine photos.

The law also lengthened copyright periods through the Sonny Bono Copyright Term Extension Act—named after former California congressman and musician Sonny Bono. The Bono provision extended property ownership from the author's life plus 50 years to life plus 70 years. The Bono Act was advocated by the Disney Corporation, which was seeking to maintain control of Mickey Mouse. The American Library Association challenged it on the basis that it prevented "an extraordinary range of creative invention" from entering the public domain just when the Internet was enabling people to draw on such creative works. Critics further objected to the Bono Act's retroactive application that did little to stimulate new creativity while doing, much to protect Disney's financial future. In 2003, the U.S. Supreme Court ruled in the law's favor holding that while the extension might be viewed as poor public policy it still was within the power of Congress to enact such a measure.[46]

The DMCA also covered webcasting music on-line, which was badly in need of a royalty system to compensate copyright holders. The government designated the Recording Industry Association of America, as the group responsible for collecting and paying fees for webcasting. In addition, the DMCA freed Internet service providers (ISPs) from corporate liability stemming from copyright abuses by subscribers illegally downloading and sending music over the Web. The DMCA, however, required ISPs to take action once they became aware of this type of music piracy.

Napster and Friends

Napster, a website created and incorporated in Boston by a former Northeastern University student, Shawn Fanning, attracted multiple legal claims of copyright in-

Fair use has been taken to mean that a parody of original content does not infringe copyright if permission is obtained in advance. The musical comedian Weird Al Yankovic parodies pop artists, but not before getting their permission to do so. Appropriating an artist's music and using it for profit without permission can lead to disaster in copyright court—but doesn't always.

When members of the rap group 2 Live Crew wanted to record a parody of Roy Orbison's rock classic, *Oh, Pretty Woman,* they tried but failed to get permission from Acuff-Rose Music, Inc. So the rappers decided to record their parody without permission and put it on the album *As Clean as They Wanna Be.* The piece featured lyrics such as "Baldheaded woman you got a teeny weenie Afro." Acuff-Rose

OH, PRETTY WOMAN CASE. 2 Live Crew parodied Roy Orbison's *Oh, Pretty Woman* without permission, and was sued for copyright infringement. The rap group prevailed in court because of the satirical nature of its song.

ACHIEVING MUSICAL PARODY

Music sued for infringement. The U.S. Court of Appeals ruled in 2 Live Crew's favor, concluding that its musical parody was "like less ostensibly humorous forms of criticism, is [of] social benefit, by shedding light on an earlier work, and in the process, creating a new one.[47]

A different outcome developed when comedian Joe Piscopo performed a parody of a rap group known as "The Fat Boys" for a beer commercial. The district court held that Piscopo's act did not qualify for fair-use protection, since it satirized no original work and violated the Fat Boys' right of publicity.[48]

fringement. Fanning had composed an index of titles allowing web users to browse and then download records from other members' computers. New artists appreciated this chance to distribute their songs without record companies intervening in the process. Established musicians and record companies, however, considered the downloads to be illegally depriving them of their royalties.

In a San Francisco courtroom, Chief District Judge Marilyn Patel dealt with the Napster litigants. Napster had hoped the Audio Home Recording Act (AHRA) of 1992 would allow its web-swapping service to do business. They pointed to the U.S. Supreme Court's ruling in the 1984 *Betamax* case that held VCRs were a technology capable of noninfringing uses.[49] However, the AHRA prohibited serial copying—making multiple duplicates—but did not cover computers at all. When a three judge panel from the Ninth Circuit Court of Appeals ruled early in 2001 against Napster, the writing was clearly on the wall. Napster effectively shut down its service that summer.

In June 2002, Napster filed for bankruptcy protection, listing nearly 8 million in assets but more than $100 million in debts. This allowed one of its largest creditors, Bertelsmann Music Group, to take over Napster's operations.

Before Napster actually collapsed, several Napster-styled companies developed new systems for swapping MP3 music files. KaZaA, Grokster, and Morpheus, among others, used file-sharing software that enabled their customers simply to access each other's computers rather than routing to a central server where songs were indexed as Napster had done. A federal judge in California ruled that this effectively shifted the infringing activities from the software manufacturer to the users themselves.[50] The only recourse left for the RIAA then was to go after the MP3 file swappers on an individual basis, which it did. News reports indicated college students were paying between $12,000 and $17,500 to settle outside of court with the music industry.

PAYOLA OR PROMOTIONS?

During the "Happy Days" era of the 1950s, when rock 'n' roll radio was booming, a sleazy business in darkened control rooms came to light. News broke of record labels trying to manufacture hits for their artists by slipping cash under the turntable to disc jockeys. The deal was that a radio station would play an artist's record so listeners would want to go out and buy it. After this practice, known as *payola*, was revealed in 1959, Congress moved to amend the Communications Act with a section (#508) prohibiting it. The story does not end there, though.

In 2002, the recording industry called on the U.S. government to crack down on a new method for pushing records onto the airwaves. An independent music promoter, or "indie," pays a radio station for the right to represent it exclusively. The promoter's fee paid to the broadcaster can range between $100,000 and $400,000. That would be a big sum for one person to pay, but the indie easily recovers that amount by billing record companies for each song added to the station's playlist. The recording industry estimates that major record companies pay promoters millions of dollars each year in such "promotion fees." Lawmakers in Congress drafted legislation to end the practice, but a partisan split stalled the issue.

When the largest radio group in the United States, Clear Channel Communications, announced that is was going to deal exclusively with "indies" and not others, things began to happen. The Recording Industry Association of America (RIAA) and nine other artist groups and unions called on the FCC to investigate the practice surrounding indie promotions, record companies, and radio stations.

▋ Free Speech and Free Press

Tort A wrong that breaks the implied social contract of behavior toward others. Torts include libel, slander, and invasion of privacy. Relief is usually sought in the form of monetary damages.

Civil suit Action involved in petition for damages as the result of a personal wrong or injury. It differs from criminal prosecution that seeks justice for crimes against society.

Slander False expression in speech, gestures, or signs that harms the reputation of an individual. May be applied to broadcasting in some states.

Libel Published information that is false and injures a person's reputation. If it is an obvious libel, it is called *libel per se*, but if it injures only by implication, it is *libel per quod*.

The Bill of Rights, the first 10 amendments to the U.S. Constitution, was adopted four years after the document was drafted. The First Amendment set forth the legal principle for communications law by stating in part that "Congress shall make no law . . . abridging the freedom of speech or of the press." The First Amendment protects free speech, but lawmakers in Congress have wrestled with where to draw the lines on indecency, defamation, and privacy. Meanwhile, electronic journalists seek to gain greater access to information, including courtrooms to report on trials using the tools of their trade—cameras and microphones.

DEFAMATION

Not all free speech deserves protection, since it must be balanced with other rights, including the right to personal reputation. *Defamation* in modern times branches out to form two types of **torts (civil suits). Slander** refers to utterances, and in some states, defaming a person over the air by radio or television is treated as slander. **Libel** refers to printed publications, and suggests that something written is responsible for the injury to reputation. Whether a statement amounts to a "publication" depends principally on the message's permanence and whether it may be passed on to others. Most states have adopted the judicial definition of *libel* for broadcasting.

protalk
Heidi Constantine
MEDIA LAWYER

Heidi Constantine enjoys being on the cutting edge, and in her law office in Manhattan that's where she is, working cases in new media, intellectual property, and "Ibusiness." Her communication studies at the University of Louisiana at Lafayette sparked a special interest in media law, but it was while interning with a Hollywood film and television studio that she found her niche in intellectual property rights. "It was then I realized that this was an area that would never get boring for me due to the constant evolution within the field," says Constantine.

Professionals in electronic media work at the intersection of technology and creativity, and can easily lose the fruits of their labor due to ignorance of property law. "Knowing how to protect your ideas is key to success in today's economy," Constantine says, and that's why she works to keep her clients from losing the benefits of their ideas.

Constantine predicts that the law will continue to evolve toward favoring "big business" through the protection of copyrights, trademarks, and patents. She sees the Sonny Bono provision as a good example. Its extensions of copyright time periods will keep U.S. copyright laws on par with those of the European Union. Constantine

> "I don't think the First Amendment gives us the right to exploit the ideas of others without compensating them."

supports this, seeing no reason why creative people (or their heirs) should not benefit from their work. "I don't think the First Amendment gives us the right to exploit the ideas of others without compensating them," she says.

Intellectual property law will continue to progress in terms of use of the Internet as test cases unfold. "These cases are important because they will set our boundaries for use of the Internet," Constantine says.

Media law provides graduates with the opportunity to combine their interest in communication with a study of our legal system and perhaps effect change in society. Constantine enjoys the perks of working in a fairly stable profession in which she can learn from bright people. That makes the long days filled with writing briefs, letters, and settlement agreements all worthwhile.

This view holds that speech modulated by electronic media is a type of publication, and consequently warrants heavier penalties than slander does.[51] Either category of offense falls mainly under civil law. Criminal libel statutes remain on the books in a few states, but rarely are enforced.

A plaintiff seeking to restore his or her good name through a libel action must show the court that the publication did falsely malign the plaintiff through ridicule, contempt, or scorn in a way in which his or her identity was discernible. Even if a name is not mentioned, if the subject's identity can be detected, then the statement—if both false and defamatory—is libelous.

Libel Defenses

When media lawyers are asked to defend a broadcast journalist against a claim for defamation seeking to recover large sums of money in damages, they begin by asking the **defendant** several key questions. First, they need to determine whether the offending statements that were broadcast could be considered true. Truth is not an absolute defense, and it is often difficult to prove in a court of law. The second question, then, concerns the defendant's source(s) for the information. Did it originate from reliable sources, or was it just secondhand gossip and rumor? Third, lawyers need to discover if the broadcast information was *privileged,* which means the words were first spoken or written in a public forum where reporting such material is permissible. Statements such as these give the reporter a *qualified privileged* to quote, particularly if they are part of a public record, such as debates in the halls of Congress or on the floor of a courtroom. Finally, did the broadcaster simply state his or her opinion or was it an assertion of fact? If it is a statement of opinion, the offending words may be protected as **fair comment and criticism** in the case of opinion. Fair comment permits some types of satire or humorous exaggeration if the audience would not reasonably infer the comments to be a fact.

One example of humorous exaggeration is the Carl Sagan case. The famous television astronomer learned that Apple's engineers had code-named one of their new personal computers after him, called the Sagan. He did not want that honor and fought it in court. So, as a joke, Apple engineers dubbed its new model the BHA for "butt-head astronomer." Sagan sued but the ruling went against him since the slur did not imply an objective fact.[52] Therein lies the key to whether humor crosses the line between fair comment and defamation. If it does imply a fact, it crosses the line and can be actionable. In the famous *Hustler Magazine* v. *Falwell* case,[53] an advertisement parody indicated the Virginia evangelist had a drunken rendezvous with his mother in an outhouse. The court held the advertisement could not be taken seriously. Courts have also allowed room for rhetorical hyperbole and refused to grant damages when the speaker was clearly exaggerating.

After the defense has addressed the key questions, it is time for the plaintiff to state his or her case. The plaintiff's lawyers must first ask if their client is a public figure or a private citizen. If the plaintiff is a private citizen, the bar of evidence is lower for the attorneys seeking damages. All they have to show is the media defendant was negligent in their investigation of the alleged facts. On the other hand, if the plaintiff is found to be a public figure, the burden of proof is higher. The plaintiff's attorney must show **actual malice** was involved in order to secure an award of damages. *Actual malice* is not defined as hostility or a vengeful attitude. It describes circumstances where the journalist or announcer knew beforehand that the defaming information was false, or at least showed a reckless disregard in determining its truth.

The landmark case in this area, *New York Times* v. *Sullivan* (1964), involved a political advertisement. It appeared in the *Times* on behalf of Rev. Martin Luther King Jr.'s legal defense fund.[54] The U.S. Supreme Court agreed that Police Commissioner L. B. Sullivan (though not identified by name) did represent the Montgomery police. The ad falsely accused him, in that official capacity, of trying to

Defendant Person who is the object of either a criminal or civil action. The defendant must answer the charges in criminal court or the plaintiff's complaint in civil court.

Fair comment and criticism A traditional defense in libel or slander suits involving opinions, dating back to the common law.

Actual malice Defamatory statements made with knowledge of their falseness or with reckless disregard for their truth or falsity.

"starve students into submission" at Alabama State College. In his opinion, Justice William Brennan declared that a public official should not recover damages for a libel directed at his official conduct unless the statement was made with actual malice—the ruling thus placed libel against public officials on a different level from libel against ordinary citizens.

Public Figures

Some people may lack formal titles or standing in public office, but they are viewed as *public figures* in the eyes of the law. Las Vegas entertainer Wayne Newton, for example, brought suit against the NBC television network for a report implying that his purchase of the Aladdin Casino involved Mafia money. At first, a Nevada court found that Newton, a public figure, was owed $5.2 million in damages because the network television story had clearly harmed his persona. NBC appealed the verdict to a higher court, which reversed the decision. The Nevada jury's verdict in favor of its "hometown hero" was overturned because actual malice had not been shown.[55]

PRINCIPLE 8

Public officials and figures receive less protection for their reputations than private citizens do.

PRIVACY

When celebrity gossip passes for news, journalists risk intruding on personal privacy. Two Boston attorneys writing in a law journal expressed that sentiment more than a century ago. Samuel D. Warren and Louis Brandeis claimed that stories of personal gossip had crowded out news items that mattered, and consequently were destroying readers' "robustness of thought and delicacy of feeling."[56] Warren's wealthy family, in particular, had been subjected to some unwanted coverage in Boston's dailies. In an 1890 edition of the *Harvard Law Review*,[57] the two men proposed a law to protect citizens from embarrassing intrusions into their private lives.

It is interesting to note that their concerns with privacy was provoked in part by new technology—that is, by 1890 standards: "Instantaneous photographs and newspaper enterprise have invaded the sacred precincts of private and domestic life; and numerous mechanical devices threaten to make good the prediction that 'what is whispered in the closet shall be proclaimed from the housetops.'"[58] In 1916, President Woodrow Wilson appointed Brandeis to the U.S. Supreme Court, where he sought to lay the legal foundation for privacy protection. He ultimately succeeded in 1939 through the Restatement of Torts, where the harm of privacy invasion was formally outlined.

Offenses and Defenses

Courts consider four circumstances to be legitimate invasions of privacy. First, celebrities who find their personal images or likenesses used to sell a product without permission may sue for their right of publicity, or as it is sometimes called, *appropriation*. Second, paparazzi who trigger a barrage of camera-shutter clicks toward public figures on their private property are *trespassing on personal solitude*. Third, disclosure of *embarrassing private facts* is an unjust invasion of privacy. Finally, if the media place someone in a *false light*, a jury can award damages based on privacy invasion.

The two defenses against claims of privacy invasion are proof of newsworthiness and/or consent. Celebrities such as Clint Eastwood, Christie Brinkley, and Cher have won suits when invasive media failed to prove either newsworthiness or consent.

JUSTICE LOUIS D. BRANDEIS. This Supreme Court Justice, considered one of the architects of American privacy law, coauthored a journal article describing the necessity for personal protection from intrusion by the news media.

Cyberprivacy

On the Internet, privacy has moved to center stage in cases involving e-mail searches without warrants and the use of pseudonyms, among other areas. When the Georgia legislature in 1997 tried to prevent people from using nicknames in their e-mail addresses, the American Civil Liberties Union moved to protect such anonymous communication. A federal district court overturned the Georgia statute for being vague and overly broad.[59]

Congress stepped in to stop unreasonable searches and seizures of e-mail messages found on a personal computer's hard drive. The Electronic Communication Privacy Act (ECPA, 1986) held that such computer evidence requires either a search warrant or a court order before police may seize it. The ECPA also keeps law-enforcement agents from intercepting e-mails without a magistrate's signature. A petty officer in the United States Navy, Timothy McVeigh (not the Oklahoma City bomber), sued and won under the ECPA when the government, without a magistrate's approval, investigated his sex life through an America Online e-mail account.[60]

CAMERAS IN THE COURTROOM

The question of television coverage in courtroom trials invariably summons to the fore two fundamental rights: one exercised by the free press to cover public trials, and the other owed to the defendant for a fair hearing in court. At least four times, the U.S. Supreme Court has taken up the issue of television publicity and trials with one common conclusion: The trial judge must protect the courtroom and the defendant's rights from the excesses of TV publicity.

Trials Under the Big Top

Some judges are happy to grant access to photographers and broadcast journalists, whereas others are not so inclined. Those opposed cite trials in which judicial decorum was destroyed by media frenzy. Throughout the twentieth century, the Sixth Amendment's guarantee to a fair trial came into conflict with freedom of the press, and overturned convictions were sometimes the results.

HARRISON FORD IN THE MOVIE VERSION OF *THE FUGITIVE*. The conviction of a Cleveland physician for the murder of his wife inspired both television and motion-picture portrayals of the "Fugitive." To journalists, however, the 1954 trial of Dr. Sam Sheppard was notorious for the circus atmosphere created by reporters covering it.

Dr. Sam Sheppard's famous murder trial in Ohio inspired the dramatic TV serial and motion picture, *The Fugitive*. The case now stands as an argument against pre-trial publicity and cameras in the courtroom. Dr. Sheppard was convicted in 1954 of murdering his wife. The U.S. Supreme Court faulted the trial judge for failing to maintain dignity and order, and overturned Dr. Sheppard's conviction. He was later retried and convicted for the crime. The judiciary has viewed the news media as a threat to a defendant's right to a fair trial, not only during the prosecution of a case, but beforehand as well.

Cameras Banned

A family tragedy involving the great aviator Charles Lindbergh had a direct bearing on the broadcast coverage of trials. In 1937, the American Bar Association, reacting to the sensational trial of Bruno Hauptmann for the kidnapping and murder of the Lindbergh baby, moved to prevent courtroom access to photographers and broadcasters. That ban on access was drafted as Canon 35 of the ABA Code of Conduct. The canon came under scrutiny after *Chandler* v. *Florida* was appealed in 1981.[61]

In that case, two Miami police officers were charged with burglarizing a restaurant. The city's news media followed the case from arrest to prosecution. TV cameras covered the trial, and although only a few minutes actually appeared on the local newscasts, the defense claimed the coverage had interfered with justice. The U.S. Supreme Court reached a different conclusion and ruled that the presence of cameras was not enough to deny the officers their right to a fair trial. Afterward, the ABA dropped its ban on cameras from its code. Today, 38 states allow televised trials and 48 states permit judges to decide when access by electronic and photographic media is warranted.

PRINCIPLE 9

In granting electronic media access to the courts, jurists balance the right to a fair trail against the freedom of the press.

OBSCENITY AND INDECENCY

This is a difficult area of law. Should words and pictures that incite lust ("prurient interest"), offend the senses ("patently offensive"), and show little in the way of substance ("lacking in serious literary, artistic, political, or scientific value") be prohibited? This criteria for obscenity emerged in the **precedent**-setting case of *Miller* v. *California*.[62] There are three sections of federal law where electronic media may be found in violation for obscene or indecent content.[63] The difference between obscenity and indecency is the latter may have some redeeming values, but according to the law, obscenity has none.[64]

The Carlin Case

On an October afternoon in 1973, a father was driving with his young son around New York City. He had tuned the car radio to Pacifica Foundation's WBAI-FM. A routine by comedian George Carlin featured a litany of profane and indecent terms that supposedly had been banned by the FCC. He presented them in a type of comic free verse titled "Seven Dirty Words."

The father was a member of Morality in Media (MIM), and he wrote a letter complaining of that experience to the FCC. The agency agreed that Carlin's monologue, broadcast at that time of day, was a faux pas of legal magnitude. Rather than slap a fine on WBAI-FM, however, the agency issued what amounted to a warning—a "declaratory order." Pacifica Foundation appealed that sanction, and the FCC welcomed this challenge in hopes of gaining some judicial guidance.

What came down from the U.S. Supreme Court affirmed the FCC's opinion that radio references to "excretory or sexual activities or organs" were patently offensive during daytime hours. The court ruled that such expressions were a nuisance rather than a high crime, and declared them best channeled to a "safe harbor" in the

Precedent Decision that serves as the authority for subsequent cases in an area of law.

GEORGE CARLIN. The king of iconoclastic comedy has come a long way since 1973 when the case arising from his "Seven Dirty Words" monologue established indecency standards for broadcasters. The former disc jockey was cited for lifetime achievement at the American Comedy Awards in 2001.

PRINCIPLE 10

Obscenity always is illegal, but "indecent" communication is permitted when children are not in the audience.

broadcast schedule—a period when children normally would not be found in the audience.[65]

FCC Crackdown

In 2004, a firestorm erupted over Janet Jackson's wardrobe malfunction at the hand of Justin Timberlake during the Super Bowl half-time show. More than 531,000 viewers complained about the sexually laced performance, and as a result, the FCC proposed a record indecency fine against stations affiliated with CBS-Viacom, which orchestrated the event through its MTV producers.

Earlier in the year, the agency repeatedly sanctioned the nation's largest radio chain, Clear Channel Communications. *Elliot in the Morning*, a Washington, DC, radio show, was cited a $247,500 fine for nine graphic and explicit sexual references, and the host of the *Bubba the Love Sponge* show in Florida was fired by Clear Channel, after a record $755,000 fine for his show.

To understand the differences between *indecency* and *obscenity*, we need to recognize the standards distinguishing the two offenses: *Obscenity* is defined by a three-part test that asks if the questionable material appeals to the "prurient interest," if it depicts sexual conduct in a "patently offensive" way, and if it lacks "literacy, artistic, political, or scientific value." *Indecency* is defined somewhat differently by the FCC as "language or material that, in context, depicts or describes, in terms patently offensive as measured by contemporary community standards for the broadcast medium, sexual or excretory activities, or organs." The key difference is that obscenity is always illegal, but indecent material may be broadcast during the "safe harbor" of hours between 10:00 P.M. and 6:00 A.M.[66]

Indecency and Cable

The Cable Act of 1984 gave franchising cities the power to ban shows criticized as either obscene or indecent. A cable subscriber in Miami, however, successfully had that law overturned. His complaint was based on the fact that cable subscribers are inviting particular channels into their homes, and therefore should be entitled to watch what they want.[67]

Similarly, the courts nullified two of three provisions in the Cable Television Consumer Protection and Competition Act of 1992 on First Amendment grounds. The Supreme Court prohibited cable operators from either censoring public access channels or segregating indecent programming to certain channels and blocking access to them. The high court, however, did allow cable systems to prohibit indecent programs reserved for lease to third parties. (Part of that 1992 cable act required system operators to lease their channels to other uses, such as home security services, computer data exchanges, or other businesses not to be viewed by cable subscribers.) The law with respect to the Internet and indecency is a different story.

Internet Indecency

When Congress passed the Telecommunications Act of 1996, the Communication Decency Act (CDA) was attached to keep any form of interactive computer or cable

COMMUNICATIONS CONTRACT

The transforming influence of telecommunications has created a high demand for attorneys specializing in information and communications law. Career opportunities are expanding in the Federal Communications Commission, the Federal Trade Commission, and a host of other offices in the federal government. New career paths also are available in state agencies involving utilities that deal with telecommunications, as well as in firms specializing in information and communications law.

Law schools intent on meeting the needs of the information economy are adding study "concentrations" in communication, entertainment, media, and information law. Indiana University, for example, offers a degree in information and communications law.

Fortunately, people who pursue law as a profession are generally interested in communication. Many obtain undergraduate degrees in that field before going to law school. They thus have taken courses relevant to a legal career, including public speaking, media law, debate and rhetoric, persuasion, and ethics. Later, as lawyers, they must extend and refresh their education, partly through continuing legal education (CLE) courses. Because the laws regarding computers, digital networks, and all electronic media are dynamic, continuing education is particularly important in media law.

The law school graduate specializing in some aspect of communications law can find work in major law firms, the U.S. Patent and Trademark Office, the FTC, the FCC, or other communication-related state and federal agencies. Legal jobs are identified by a variety of titles, including **attorney in communications and media law,** which could focus on public utilities, telecommunications, and administrative law. Another title in this field is **telecommunications associate,** which also would cover regulatory communications and special knowledge of the technology and technicalities of the industry. Careers in law and communication also include **legal researcher, paralegal, legal secretary, legal reporter,** and **legal educator.**

services from transmitting "obscene, lewd, lascivious, filthy, or indecent" material. Even before the digital record had been made of President Clinton's computerized pen signing the measure, the American Civil Liberties Union (ACLU)—joined by a host of publishers, citizen groups, and librarians—challenged the CDA's constitutionality.

The Court of Appeals in *ACLU* v. *Janet Reno* (then the U.S. attorney general) ruled that the CDA was a content-based restriction on speech and this was impermissible. The court said that whether "indecent" or "patently offensive," speech was entitled to constitutional protection. Justice John Paul Stevens delivered the majority opinion in a 7–2 ruling that held that the CDA placed "an unacceptably heavy burden on protected speech." The ruling also said the act was not worthy of rewriting to save an otherwise "patently invalid unconstitutional provision."[68]

Congress responded by passing another bill in 1998 aimed at the Internet, this time designed to protect children. The Child Online Protection Act (COPA) banned content harmful to minors based on the obscenity definition rendered in *Miller* v. *California.* However, COPA fell to the same fate as the CDA after the ACLU and digital activist groups filed suits against it. The Third Circuit Court of Appeals held that COPA's reliance on community standards in cyberspace was impractical, and pinned its ruling to the belief that computer technology blocking access to minors would be a better solution than government intervention into free speech.[69] A third attempt by Congress to control Internet pornography, the Children's Internet Protection Act of 2001, was affirmed by the Supreme Court. This law was aimed at public libraries, and simply ordered them to install anti-porn filters on their computers or risk the loss of federal funds. U.S. taxpayers spend about $200 million a year funding access to computers and the Web in public libraries, and a majority of the justices agreed that filters were the best means to keep sexually oriented materials from reaching adolescents there.

Summary

Electronic media law is founded in two contrasting traditions: one encompasses the freedoms protected by the First Amendment and the other is an obligation to use the airwaves in the public's interest, convenience or necessity. In the 75 years since this broadcasting standard was incorporated into law, radio and television stations have seen an influx of competitors from cable, satellite, and Internet companies. This new playing field has shifted the debate from discussing how the public interest standard is to be enforced to how the law can successfully adapt to the broadband future of delivery systems that merge both technology and corporate headquarters. Broadcasters are hoping to see rules that will allow them to compete with their rivals who are less constrained in the regulatory realm.

The debate over ownership restrictions has so far tested the idea that cross-ownership and duopoly rules will still afford a diversity of voices, or are no longer necessary in the age of digital convergence. Rules regarding political communication, indecency, and even children's television show how important it is for broadcasters to recognize their special obligations based on public trust. As long as the number of radio and television station applicants exceed the number of channels available, and free radio and television broadcasting prevail, there will be public interest in the performance of licensees. The FCC has replaced its rationale for making those decisions from government evaluation to renewal expectancy, while maintaining the key principles undergirding the 1996 Telecommunications Act.

In the digital future, new legal dilemmas will pose new challenges, particularly in the areas of intellectual property rights. Copyright protection demands solutions that will keep digital pirates from making money off the creative works of others. New technology will call for new laws to protect the old principles of freedom and democracy.

Food for Thought

1. The FCC has moved toward selling spectrum in auctions rather than giving it away. Do you think that this system unfairly advantages the rich and powerful in electronic media? Why or why not?
2. The FCC has proposed new guidelines to increase diversity in employment by having stations keep records and reporting the ratio of minorities and women to Congress. Do you think that it will be effective in increasing equal employment opportunities? Why or why not?
3. The FCC's crackdown on televised indecency has been applauded by some and attacked by others. How would you solve the problem in terms of new rules and policies?
4. Do you think that cable companies should be required to pay broadcasters for carrying their channels to cable subscribers? Why or why not?
5. Do you think the V-chip has been effective in curbing children's viewing of violent or sexually oriented programming? What alternative solutions would you propose?
6. Napster's music swapping service is now a matter of history, but would you be willing to subscribe to a pay-service for computer-downloads of music? Support your position.
7. If you owned a record company, would you allow new artists to post their music for free on the Internet? Why or why not?
8. Do you think cameras in the courtroom hamper a defendant's ability to get a fair trial? How would you balance the competing interests in this matter?

Professional Ethics

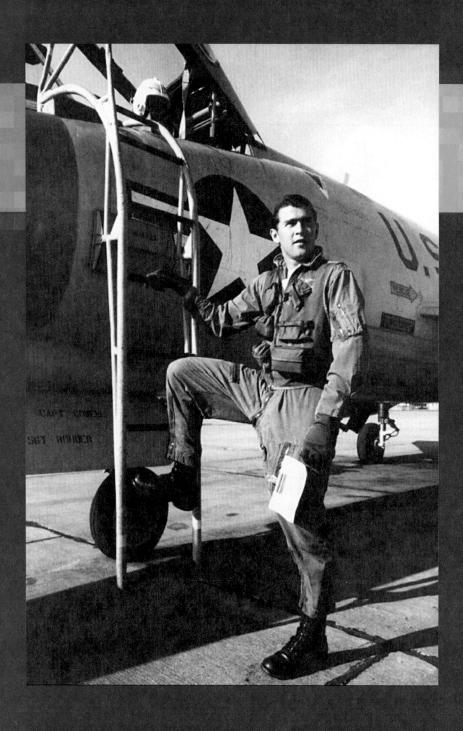

Chapter 12

For several weeks during the 2004 presidential campaign the nation was transfixed by an old story—one that had been around for years—but this time it was different: A major news organization had added its weight and credibility to one side of the controversy. The story was about whether the President of the United States had during the Vietnam era failed to fulfill his duties in the Texas Air National Guard. The disputed claim was that George W. Bush had not only failed to show up for a physical examination in 1972 but he also missed training activities in Alabama for several months. The White House acknowledged that the president failed to appear for his medical checkup, but denied any further lapses in his military record, adding that he was proud of his service. If the future president had *not* appeared for several months of duty, then why was he not reprimanded and how could he receive an honorable discharge instead?

CBS's *60 minutes* anchored on Wednesdays by Dan Rather, came forward to address those very questions. It had discovered documents—memoranda—attributed to Bush's commanding officer, the late Lt. Col. Jerry Killian. The memos' author complains of being pressured to "sugar coat" Bush's performance evaluations, and added that the future president had been "talking to someone upstairs" in order to get favorable treatment. Within minutes after CBS News aired the story, the memos were challenged on the Internet as forgeries based on their content and style. One blogger named "Buckhead," identified as a Republican lawyer in Atlanta, noticed the memos used a type of superscript and proportional spacing that appeared to be computer generated. Another blogger saw that the memos' wording relied on Army and not Air Force jargon.

What made matters worse for CBS News was that its efforts to demonstrate journalistic integrity after the broadcast seemed to backfire. The network's sources who were needed to verify the memos equivocated, or claimed that they had expressed personal doubts about their authenticity. Lt. Col. Killian's secretary, Marian Carr Knox, knew that she had *not* typed the specific memos that appeared on the *60 Minutes* segment, although she did recall her boss making similar statements about Bush's record.

It was not the memos' content but their authenticity that became the pivotal issue. The question was no longer whether the president fulfilled his duties to the National Guard, but how could CBS News commit such a journalistic blunder—one for which the network had to apologize—costing four employees their jobs and resulting in a lengthy 228-page investigation conducted by a former attorney general under President George H. W. Bush, and a former Associated Press executive.

Other national news organizations had reported similar allegations pertaining to Bush's military record in Alabama. Yet, CBS News claimed it had the evidence, and was caught in the eye of the storm with a story that appeared to be flawed. Within two months of the apology, Dan Rather announced he was leaving the anchor desk, and would serve only as a correspondent for *60 Minutes* airing on Wednesdays.

The job of the journalist is to seek truth, to verify sources, and above all, get the story right. When in error, the only solution is to correct it, which is what CBS News tried to do by appointing an independent commission to investigate its reporting process.

Regardless of the question involved, there can be no avoiding tough calls on deadline, even controversial stories that may influence the outcome of a national election. Each tough decision is made in full view of the audience, which produces reactions over the Internet and radio talk shows. Those voices may resound louder than the original message. As a result, broadcast journalists must know how to handle ethical questions that reflect their organization's standards and practices. Studies of professionalism draw on principles that point toward ideals such as fairness and truth considered by ethicists to be moral imperatives. This chapter will reflect on such ethical prescriptions for professional conduct.

Why Study Ethics?

There is a school of thought that believes principles and ethics are an indelible combination of childhood rearing, social influence, and cultural mores—so why bother teaching them if they are instilled there in the first place? Another school holds that principles and ethics are irrelevant to professional conduct—easily forgotten because everyone does what they must do to survive. If that means sacrificing the moral high ground—then so be it.

Both viewpoints discount the part that ethical reasoning and professional standards play in measuring conduct. We are committed to the idea that those who achieve a pinnacle in their profession by unscrupulous means are exceptions to prove the role. True success is not contradictory to character and conscience. This chapter will examine the evidence and explain our ethical heritage. After all, what contemporary professionals have to say about tough calls and the standards they rely on indicates what is useful to know to build a future career.

ETHICS DEFINED

Conventional wisdom differentiates the study of ethics from the study of law by noting how the law must enforce what a person *cannot* do but a person's ethics guides what he or she *must* do. Look up the word *ethics* and you will find that it goes deeper than that—the Greek word *ethos* denotes custom or character. Ethical choices become customary by what one practices every day. As the ancient philosopher Aristotle put it, "We are what we repeatedly do. Excellence then is not an art, but a habit." *Ethics* is actually what happens in one's thoughts, whereas *morality* is what others see on display. One simple question sums it all up: What is the best way to think and behave in a particular situation?

Generally, ethics fall below the level of law, although some ethical choices do prompt litigation, particularly when personal wrongs are alleged in civil court. Both areas—ethics and law—deal with questions of right and wrong. The origin of humanity's legal and ethical heritage springs from the same sources: ancient and medieval scholars, science and the scriptures, even the cultural and moral trends of the day.

The journey usually begins in Athens, where Greek philosophers tried to find the proper balance of life in terms of aesthetics, epistemology, and ethics. Western philosophers look to the writings of Aristotle (384–322 B.C.), who assembled his thoughts on virtue for his son, Nichomachus, and whose writings became known as Nichomachean ethics. Aristotle believed that the course to human happiness emerged through the realization of personal good. When that virtue is realized, then the person can begin to flourish. *Flourishing* requires determining how best to achieve a higher standard of ethical living. The way to accomplish this goal could lead one to choose role models, or in Greek terms, heroes. Patterson and Wilkins observe, "An Aristotelian might well consult this hero as an expert when making an ethical choice."[1]

ARISTOTLE, GREEK PHILOSOPHER AND ETHICIST.

TELEOLOGY

Many theories shape what a person knows and thinks about professional ethics, but we will focus on two prominent ones. The first one is called *telos* from the root word of **teleology**, and it suggests an *end* or *purpose.* In other words, what are we trying to accomplish here? Aristotle was a teleologist because he taught that people must initially discover their purpose before they can make wise decisions. In order to find your *telos*, you must make choices by reasoning from principles. One principle helps to narrow the options. Aristotle called it the *golden mean*—the middle point between extremes. The goal of moderation avoids extremism, advocating self-discipline over self-denial. Aristotle's lessons were instructions for virtuous living, which is why his lessons are called *virtue ethics.* Temptations to sloth, fraud, theft, or violence would never enter the discussion since, by definition, they are *vices.*

PRINCIPLE	1

The golden mean recommends a middle ground between extremes.

Even more central to Aristotle's philosophy is the work of ethical analysis. People reason from principles in order to make good decisions. Rival Greek philosophers, the "Sophists," promoted more attractive solutions. Their philosophy, known as *sophistry,* describes attractive but *not* well-reasoned choices.

DEONTOLOGY

A second foundation theory in ethics is rooted in the Greek word *deon,* meaning duty, which is one of the essential moral values ensuring happiness. **Deontology** elevates *duty* over *purpose,* and originates in the scriptures and commandments that recommend service to a higher authority. Deontology is how a person often defines his or her heroes. When New York City police and firefighters rushed into the World Trade Center on September 11, 2001, to save lives at the cost of their own, it was a sense of duty summoning them onward.

Two millennia after the Greeks, German idealists put a new face on deontology. Their quest was for philosophical ideals that would show how each person could embrace the good life by acting on principle, which is why they were called German Idealists. One of their leaders, Immanuel Kant (1724–1804), searched for universal maxims known as **categorical imperatives.** Unlike Aristotle, who focused on the individual as an ethical actor, Kant responded to the nature of one's moral *actions* that bode either good or ill. These would include ethical obligations to be fair and truthful, the duties of merit. He also noted the *strict duties* require people to avoid moral wrongs: lying, stealing, cheating, and doing harm, for example. He believed these duties transcend an individual's circumstances. In Kant's view, correct motives

Teleology Ethical theory based on *telos* for end or purpose. Assumes the purpose of our lives is discernible and is ordained either by divine powers or by our own will.

Deontology Ethical theory based on *deon* for duty, which is the essential moral value (rather than happiness or virtue).

Categorical imperatives Ethical theory based on discoverable principles or laws that can guide personal behavior.

PRINCIPLE 2

Categorical imperatives such as fairness and truth guide ethical decisions.

will ultimately produce happiness, regardless of one's success—like the old saw about how winning or losing is not as important as how you play the game.

In terms of ethical theory, the *deontological* approach actually forms no basis for resolving conflicts between *loyalties,* the "willing and practical and thoroughgoing devotion of a person to a cause." Josiah Royce wrote in 1908, "The whole moral law is implicitly bound up in one precept: Be loyal."[2] Journalists, for example, may find their professional loyalty at odds with their general loyalty as citizens. What happens then? For example, a reporter was writing about a struggling immigrant in the United States saving up his earnings to send back to his family in Mexico. The reporter's story tipped off authorities that this individual was an undocumented alien, and brought about his deportation. Should the journalist have concealed the man's illegal status and ignored federal law? If your answer is, "Do your duty," the next question becomes, "Which one?" Ethicists suggest that a reasoned *examination of loyalties* based on principles and consequences will reveal the correct *decision.*

CONSEQUENTIALISM

An offshoot of teleology, which focuses on purpose-shifted attention to the impact of ethical choices on the end result, is called **consequentialism.** The writings of Jeremy Bentham and John Stuart Mill, nineteenth-century British philosophers, are cited as sources of this perspective. What Mill and Bentham suggest is that decisions must be made for the greater good. In other words, ethical choices are the ones that hold the greatest promise for helping the most people. and therefore have the most *utility.*

Utilitarianism focused on the consequences of action in order to determine whether the actions are ethical or not. If ethical actions produce the greatest good, unethical decisions are the ones that benefit the fewest at the expense of the many.

Consequentialism Ethical theory that holds that the correct moral choice is the one that produces the best results.

GERMAN IDEALIST IMMANUEL KANT.

UTILITARIAN JOHN MILL.

Patterson and Wilkins argue that utilitarianism "can lead to ethical gridlock, with each group of stakeholders having seemingly equally strong claims with little way to choose among them."[3]

Mill's classic treatise, *On Liberty*, spelled out another ethical notion—the greater the freedom, the greater the happiness. However, he argued that because personal liberty expands people's choices, it gives them more opportunity to act for the greater good. In Mill's **utilitarianism**, there is a strong sense of moral duty. Twentieth-century pragmatists Bertrand Russell and John Dewey agreed with Mill that considering outcomes is essential to making moral choices. However, they disagreed about any search for a single truth. Truth is a matter of who is asking the questions, so to them, truth was a relative value. This relativism also relates to two other ethical theories: egoism and situationism.

Egoism asks first what might be in your best interest, and beckons self-interest instead of sacrificing for some greater good. **Relativism** and **situationism** place emphasis on personal goals and circumstances. Such perspectives minimize the role of duty, and "can lead to moral anarchy in which individuals lay claim to no ethical standards at all."[4] If there is no objective reality, then people fall into the mental quicksand of what is called *solipsism*—a type of thinking that places individual thoughts, feelings, and perceptions above all else.

> **PRINCIPLE 3**
>
> Ethical decisions are based on an assessment of duty, purpose, and consequence.

Civilized society cannot function under such tenets. People must ask themselves, is it necessary to choose between ethical theories—in order to select one perspective based on a single set of duties or purposes above the rest? Ethical dilemmas call for such an assessment, as well as moral responsibilities and consequences. If one's reasoning is based on sound principles, then finding a solution is within one's reach.

▌Professionalism

When it comes to choosing a career in the media, the question becomes whether one is entering a profession, craft, or trade—and it is far from settled. The answer usually depends on whom you ask, the career chosen, and how *professional* is defined. If one describes a *professional* as someone holding formal degrees with some system of certification, as in medicine or law, then a career in electronic media probably would not qualify. However, if a *profession* was defined by a set of standards or ethics, then many careers in electronic media would fit that description.

Generally, volunteer members of professional associations draft or revise codes of ethics in committees, which hold little or no power to enforce them. The codes are general statements of principle designed to encourage *professional* conduct. Lawyers may discourage radio and television stations from defining their ethics for fear of having to defend them in a court of law. However, dedicated professionals do not shy away from stating their ethical beliefs simply because lawyers warn them it could be used against them in a lawsuit. Ethical codes are one element of professionalism, and education is another.

HISTORICAL AND CONTEMPORARY PERSPECTIVES

When famed newspaper publisher Joseph Pulitzer exchanged letters in 1904 with Harvard University's president on the subject of teaching ethics as a prerequisite for a professional career in journalism, the two men did not quite see eye to eye. President Charles Elliot was enthusiastic about teaching students the journalist's duty to the public. He proposed classes at Harvard that would be designed to show how editors and publishers influence public policy. Pulitzer preferred another approach.

38

Utilitarianism Ethical theory that determines the correct moral choice based on decisions that produce the greatest good for the most people.

Egoism Ethical theory holding that the primary beneficiary of an action should be the one who takes the action, and that sacrifice is not necessary.

Relativism Ethical theory based on principles accepted according to circumstance rather than on universally applicable principles.

Situationism Ethical theory that approaches each situation as unique and holds that absolutes are too inflexible.

"Ideals, character, professional standards . . . a sense of honor should be the motif of the whole institution," he replied.[5] Pulitzer shifted the emphasis from the newspaper office (in Elliott's proposal) to the world at large.

ETHICAL APPLICATIONS

Regardless of the source—the Bible, the Quran, family instruction, or cultural mores—**normative ethics** are the general principles governing *decisions*. They are based in the philosophies and translate virtues such as *charity, fidelity,* and *truth* into everyday choices. One such moral principle is **egalitarianism**—the idea that people are entitled to equal rights and privileges in a just society.

Political theorist John Rawls was interested in achieving *distributive justice* for all *by applying* what he called "the veil of ignorance."[6] This intellectual approach suggests that questions of justice could be resolved if all decision makers would become blind to individual interests, and act as any one of the stakeholders in a conflict. For the media, this *veil* might be viewed as something like the visage of the goddess, Justice, who weighs only the issues relevant to a dispute so that all people are treated as equals.

Normative or **prescriptive ethics** have been translated into codes developed by associations and professional groups. The National Association of Broadcasters (NAB), the Society of Professional Journalists (SPJ), and the American Advertising Federation (AAF), among others, present their members with these guides of professional behavior. They contain duties that are both prohibitive and affirmative in nature, and point members toward professional action.

PRINCIPLE 4

Prescriptive ethics draw on higher principles to point us toward actions.

CODES OF ETHICS

The first National Association of Broadcasters Code was adopted in 1929 and consisted of just eight rules, half of which were designed to alleviate concerns about the commercialization of radio. Broadcasters, for example, who subscribed to the code promised to air no promotion that was "fraudulent, deceptive or obscene." Over the years, the NAB Code was expanded to address a variety of issues in news, politics, religion, and children's programming. Not all broadcasters subscribed to it, but for those who did, NAB staff members were assigned to investigate complaints made against them or their station.

In terms of advertising, the early code banned radio commercials between 7:00 P.M. and 11 P.M. to make room for "relaxing nighttime listening, often as a family." That ethic only lasted until 1937, when nine minutes of commercial time were allowed per daytime hour, and only six minutes at night.

Violations of the NAB Code carried only one penalty: removal of the NAB seal from the wall of the broadcast station office. The Code was revised in 1952 to include television, but 30 years later was eliminated altogether. The scene of its demise was a court battle over the *piggybacking* ban, preventing a sponsor from squeezing two product messages into a one-minute spot.

Alberto-Culver, a cosmetic firm, complained that this restrictive rule kept it from getting the commercial schedule it needed for its advertising. The NAB asked the Department of Justice if the code violated antitrust laws, since the ban on piggyback spots was in restraint of trade. The government affirmed that position, and a federal court judge agreed. In November 1982, the 53-year-old NAB Code was erased.

For eight years, there was nothing on the books resembling a statement of good behavior at the NAB. In 1990, the NAB's executive committee, working in concert with its attorneys, came up with a voluntary statement of principles covering just four areas: *children's television, indecency and obscenity, violence,* and *drug abuse.* Gone

Normative ethics Branch of ethical theory that determines which human actions are based on moral values, principles, and conduct.

Egalitarianism Ethical theory based on the notion that people are entitled to the same rights and privileges without respect to extraneous factors such as religion, gender, or race.

Prescriptive ethics is a normative approach and involves taking a stand about what standards and principles ought to govern behavior.

were all of the rules and prescriptions added to the old NAB Code replaced by a statement of the broadcasters' rights under the First Amendment, and "the desires and expectation of its audiences and the public interest."[7] This "general and advisory" statement asked broadcasters to "exercise responsible and careful judgment" when considering programming dealing with those issues. Washington lawmakers and one member of the FCC, Michael Copps, asked for a new NAB Code in 2002, but the response from the broadcast community was decidedly negative.

BROADCAST JOURNALISM ETHICS

Among careers in electronic media, it is broadcast journalists who most often confront ethical issues and dilemmas when making decisions. They have relied on their own codes of ethics for over half a century. In 2000, the Radio-Television News Directors Association (RTNDA) revised the set of normative values for its code that included "public trust, truth, fairness, integrity, independence, and accountability."

When first meeting as the National Association of Radio News Directors (NARND) in 1946, broadcast journalists approved the guideline to "accurately and without bias . . . within the bounds of good taste" report the news. They further endorsed a reporter's duty to "journalistic principles and ideals," and duty as well to the general manager of the station.[8] In its postwar resolve, the NARND vowed to move radio reporting beyond its "rip 'n read" phase, petitioning station managers to hire independent journalists to cover local news.

Sensationalism

Just two years after that first code of ethics was drafted, broadcast journalists became concerned with what might be called *sensationalism*. Excessive play of the terms *bulletin* and *flash* on radio was the problem then. News directors called on their colleagues to avoid this practice in reporting, writing, and announcing the news. Today, the association's professional standards reject all reporting "that fails to significantly advance a story, place the event in context, or add to the public knowledge."

Sensationalism, however, is defined in different ways. The RTNDA advises anchors and reporters to avoid "techniques that skew facts, distort reality, or sensationalize events." The code of ethics for the Society of Professional Journalists (SPJ) asks reporters to "show good taste. Avoid pandering to lurid curiosity." The National Association of Broadcasters advises programmers "to avoid presentations purely for the purpose of sensationalism or to appeal to prurient interest or morbid curiosity." The Advertising Principles of American Business also calls for ads "free of statements, illustrations or implications, which are offensive to good taste or public decency." The term *sensationalism* also raises concerns about exaggeration and distortions as well as decency and taste.

Stereotyping

Ethical codes have spoken to the rights of minorities by reminding journalists to gather a diversity of opinions from informed members of the community and, in the name of fairness, to avoid stereotyping minorities. In 1950, broadcast journalists codified their concern with stereotyping. It became the eighth standard in the revised code, that "the race, creed, color or previous status of an individual in the news should not be mentioned unless it is necessary to the understanding of the story."

Today, the SPJ Code is even broader and more defined in attacking prejudice in reporting. It asks the press to "avoid stereotyping by race, gender, age, religion, ethnicity, geography, sexual orientation, disability, physical appearance or social status." The SPJ ethic further advises its journalist members to "be a voice for the voiceless."[9]

Judicial Decorum

When the RTNDA revised its code in 1966, broadcast journalists were asked to "conduct themselves with dignity" in court and to "keep broadcast equipment as unobtrusive and silent as possible." That action followed two sensational court trials—one of a doctor accused of murdering his wife in Ohio, and another involving the fraudulent sale of fertilizer in Texas. Reflecting on the carnival atmosphere created by such trials, RTNDA members vowed to place greater emphasis on each citizen's right to a fair trial.

Privacy

The RTNDA's rewrite of its code in 1966 also marked the beginning of its professional concern with the privacy of individuals in the news. Broadcast reporters were pledged to "display humane respect for the dignity, privacy and well-being of persons" with whom the news dealt. In addition, the amended code called for a special sensitivity to the privacy of children.

The SPJ Code calls for "compassion to those who may be affected adversely by news coverage" and the recognition that gathering and reporting the news "is not a license for arrogance." Only an overriding public need "can justify intrusion into anyone's privacy."

Dividing News and Advertising

The first code of standards for radio news recommended a wall of separation within stations, between its newsroom and the sales office. The code said commercials must be kept separate from news content and required someone other than the newscaster to deliver them. This separation was endorsed by the Society of Professional Journalists, which saw hybrids of advertising and news as violations of the *truth* principle and thus to be avoided. The SPJ ethic urges journalists to "distinguish news from advertising and shun hybrids that blur the line between the two." The Advertising Principles of American Business includes eight ethics, seven of which deal with truth and verification in commercials.[10]

▛ Prescription Truth

CBS Newsman Edward R. Murrow put it this way: "To be persuasive, we must be believable. To be believable, we must be credible. To be credible, we must be truthful."[11] University of Illinois ethicist Clifford Christians says that all principles of ethics begin with a promise. In electronic media the most fundamental promise is to tell the truth to the audience. *Truth* naturally summons to mind the need for accuracy and diligence in reporting, but it also suggests the danger of distortions, plagiarism, and false or misleading information. Over the years, hasty mistakes have cost more than one professional member of the electronic media a loss in credibility.

OUTFOXED?

Fox News's Geraldo Rivera's swashbuckling style of journalism has proven to be controversial. When Rivera appeared on Fox News late in 2001, claiming to be near Kanndahar in Afghanistan at the scene of a shooting where he said American soldiers were killed, he confessed to becoming choked up after saying the Lord's Prayer over the "hallowed ground" where "friendly fire took so many of our men and the mujahedeen yesterday." A reporter for the *Baltimore Sun*, however, observed that Rivera's location was actually hundreds of miles away from the site of the actual shooting, and

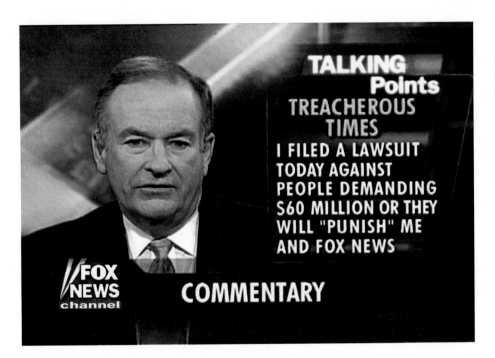

TALKING Points
TREACHEROUS TIMES
I FILED A LAWSUIT TODAY AGAINST PEOPLE DEMANDING $60 MILLION OR THEY WILL "PUNISH" ME AND FOX NEWS
FOX NEWS channel   COMMENTARY

FOX NEWS PHENOM. Fox News Channel's popular talk-show host Bill O'Reilly rode a smackdown style of interviewing and commentary to success in the ratings. Despite a sexual harassment suit from one producer, O'Reilly's audience remained loyal to an edgy brand of news and opinion.

the war correspondent explained himself by pleading confusion. Rivera's employer, Fox News, called it an "honest mistake," and expressed full confidence in his "journalistic integrity." The director of the Poynter Institute's ethics program, Bob Steele, was not so sure that Rivera's work at Fox News should be classified as journalism: "Is Geraldo Rivera a journalist or an entertainer? Is he a correspondent or a bin Laden bounty hunter? Is he a news reporter or a fiction writer?"[12]

Bias is another charge media professionals may encounter from all sides of the political spectrum, which underscores the importance of impartial reporting. That journalism practice has been seriously challenged by a media world awash with political pundits, raging bloggers, and spinning talk-show hosts. One national survey showed that more than one-fifth of U.S. people were turning to talk shows and other sources of mayhem and mirth for their daily diet of news.

Outfoxed: Rupert Murdoch's War on Journalism, a documentary by Robert Greenwald, charged the cable network, Fox News, with failing to honor its journalism ethic and trademark motto to be "fair and balanced." In the documentary, Fox News reporters were warned not to "fall into the easy trap of mourning the loss of US lives" in Iraq, and asked to avoid turning the 9/11 Commission's investigation into another Watergate. One former Fox reporter said any off-hand comments that made Democrats look stupid and Republicans look smart were especially appreciated. Greenwald's film made its case by using the words of former Fox News employees, executive news memos, and videotaped segments of Fox News programs, which were recorded without permission, according to Fox News executives.[13]

WAR PROPAGANDA

President Bush's administration announced in 2002 plans for two offices that would be dedicated to propaganda. The Office of Strategic Influence would operate in the Pentagon, and, reported *The New York Times*, it would rely on both information and *disinformation* for its propaganda campaign.[14] A second office of global diplomacy was to be dedicated to "intense shaping of information and coordination of messages." The reaction by the press to this announced betrayal of the truth principle was decidedly unfavorable. Texas columnist Molly Ivins wrote that you do not have to go "very far out on a limb to predict this will be a disaster. It will

What happens when the government pays commentators to advance its causes—are any ethical lines trespassed? In 2005, conservative news commentator Armstrong Williams admitted he had been quietly accepting federal money to promote President Bush's education program. Williams received $240,000 from the Department of Education in exchange for promoting the No Child Left Behind program, the center-piece of Bush's education agenda. The president said he was unaware of the payments and the Department of Education defended its deal. It claimed its contract with Williams was a permissible use of taxpayer funds; others charged it was against the law.

Tribune Media Services announced that it considered Williams's actions to be ethically out of bounds, and stopped distributing his columns, while the nation's largest African American journalists' organization asked other media outlets to resist this type of paid propaganda.

wreck our credibility in no time. . . . As any journalist can tell you, when you put out misinformation, all it does is poison the well of public debate."[15]

Shortly after the news broke of the military's plans to plant fake stories overseas, the Hollywood press reported Washington would be teaming up with one of its top action movie producers to produce a 13-episode "reality" series on ABC to profile its troops abroad. An ABC Entertainment executive claimed the Pentagon was eager to "produce what Americans want to see" because they regard it as an Army recruiting film.

PRINCIPLE 5

Without truth there is no trust, and without trust, meaningful exchanges of information are lost.

UNREAL WORLDS. Like most reality shows, CBS's *Big Brother* has a website with juicy gossip to satisfy viewers' thirst for news about reality-show actors.

INDECENT CONTENT

Lurid, indecent, and sensational content has provoked lawmakers and watchdog groups to recommend changes in television commercials and programming. FCC Commissioner Copps and Senator Robert Byrd (D-WV) criticized broadcasters for violating standards of good taste by airing objectionable programming. Broadcasters countered by claiming the commissioner was trying to bully them, and announced their hopes that Senator Byrd's complaints would fall on deaf ears at the FCC.

Why all the fuss over television decency? The Parents Television Council (PTC) conducted a content analysis indicating sexual activity was more prevalent and raunchier than ever before on network television. Television's level of crudity increased in 2003 with about 14.5 examples of bad language or sexual content per hour. The conservative media watchdog group said its vulgar meter for reality television surpassed scripted shows by an average of 4 incidents of foul content per hour and was rising fast. The PTC study found that the worst offenders were relationship-based reality shows, such as *The Bachelor* and *Average Joe.*

Given that parents are asked to monitor children's viewing habits, animation shows appear at first glance to be a safe bet. Unless it is what NBC billed as an "adult cartoon." called *Father of the Pride,* which the PTC noted "looks like a Shrek but sounds as sad and horny as last year's failed Whoopi Goldberg sitcom."[16] A RAND Corporation study found that all of this focus on sex was having an impact on children. Young people surveyed were more likely to become sexually active based on their television diet. If they were among the top 10 percent in viewing sexually related scenes, they were twice as likely to engage in sexual intercourse.[17]

Are broadcasters just giving the audience what they want? "Look, we're not saying we love the fact that kids are exposed to a lot more adult concepts and language a lot earlier than they used to be," editorialized *Broadcasting & Cable.* "But whether it is the inevitable and healthy maturing of the culture or an unseemly coarsening of the fabric of society, it is a fact of life." Senator Joseph Lieberman's (D-CT) advice to parents: "turn off the television if you find [shows] inappropriate for your children or offensive to their values."[18]

■ Fairness for All

The study of professional standards is important because in a free country, self-governance determines action more than the law does. Radio and television journalists are called on to protect people who are subject to the ravages of prejudice and stereotypes by fostering an atmosphere of tolerance and respect.

DIVERSITY DEFERRED

There is the expectation that the principles of equality and justice will prevail in the United States, but, by nature, people tend to think first in terms of their own group, gender, and ethnic origins. So, diversity has become a greater challenge for media writers and producers. A 2004 study of television shows found that U.S. viewers would be twice as likely to meet a Latino in real life, as they would see a Latino on television. According to *Fall Colors: Prime Time Diversity Report–2003*, released by the group called "Children Now," that is an improvement. Hispanic Americans represent 12 percent of the U.S. population, but appear in less than 6 percent of the shows on television. The study also found that even though there were more Latino characters than in previous TV seasons, the actors were often viewed in low-paying jobs and rarely viewed in a flattering light. Unfortunately, the same study found that almost half of all Middle Eastern characters (46 percent) on television were portrayed as criminals, about 15 percent of Asian/Pacific Islanders and Latinos played the bad guys, 10 percent of the villains were African Americans, and only 5 percent of the bad guys were whites.[19]

ABC News correspondent Elizabeth Vargas believes it is especially important to counter negative stereotypes where subjects of color are depicted as criminal suspects hiding their faces. Vargas said one ethical remedy would be to interview role models from the minority community—for example, to seek out physicians of color for health news "to show that, indeed, blacks and Hispanics are well represented in the successful, affluent, educated part of our society."[20]

It would be reassuring to believe major strides toward perfecting a diverse picture of the United States have been made. Some 55 years ago, Robert L. Hutchins of the University of Chicago convened a commission of 13 scholars and professionals to investigate the ethical performance of media. The commission felt that journalists needed to

SALSA SHAKE. Latin singer/dancer/star Jennifer Lopez reinvented herself as Jennifer Anthony after her marriage to celebrity Marc Anthony. Her cross-cultural appeal produces successful records, concerts, and movies.

PRINCIPLE 6

Being fair means being blind to individual differences that are irrelevant to the issue.

draw a "representative picture of the constituent groups of society." Other media obligations, the commission said, included giving truthful and fair accounts of the day's events, providing a forum for comment and criticism, reflecting the goals and values of society, and offering full access to the day's intelligence.

Independence and Self-Interest

Journalists are often subject to pressures from both news sources and advertisers. Professional ethics require them to report stories in an impartial and disinterested manner. Notice the term is not *uninterested* but *disinterested*. A vexing challenge arises when journalists are pressured to hold back on a story, either by advertisers or by their own bosses.

ADVERTISER PRESSURES

Television members of the Investigative Reporters and Editors (IRE) organization have found that advertisers carry a big stick but do not always speak softly. One survey asked those journalists if TV sponsors had ever "tried to influence the content of news on local commercial stations?" Almost 75 percent of the sample said "yes," and more than half had seen sponsors try to kill a story. The IRE survey revealed that 54 percent of the respondents said they knew of businesses that had withdrawn their advertising because of the content of news reports.[21] Dollars are potent weapons that advertisers may deploy against journalism.

Job Security?

Television reporters who tackle tough consumer topics—especially issues directed at automobile dealers—may even risk their job security. Consumer reporter David Horowitz was released from KCBS-TV in Los Angeles after raising the question of automobile safety one too many times.[22]

A husband-and-wife investigative team working in Florida found their coverage of Monsanto's synthetic growth hormone for cows upsetting to lawyers for the station's parent company. Their reporting for WTVT-TV (Fox) in Tampa Bay raised the possibility of carcinogens in the milk. The problem was the report never aired, and the couple lost their jobs. Steve Wilson and Jane Akre alleged in their suit that management had intervened on their news judgment.[23]

Covering the Mouse

The question of corporate influence over news policy came into play when ABC News stopped the broadcast of a *20/20* segment in 1998. In the story, investigative reporter Brian Ross described Disneyland's hiring practices, raising the question of whether pedophiles were effectively screened out by the Disney Company, which owns

CORPORATE ETHICS. As a member of the Disney family, ABC News has had to make difficult decisions on what stories to cover about Disneyland and other corporate properties.

protalk
Ernest Sotomayor
LONG ISLAND EDITOR, NEWSDAY.COM

Professional reporters and editors know how important the ethics of truth and accuracy are to getting their story right, but what concerns the Long Island editor of Newsday.com is that journalists may not be getting it right for everyone in their audience. "We talk a lot about ethics, and we tell students you have to be fair and objective and honest in your reporting—you don't plagiarize," noted Ernest Soto-mayor. "What we don't teach very much—and we don't have editors instilling in their managers—is that we have to have people who are going to ask questions that affect all communities."

Ernest Sotomayor's journey to the diverse world of New York journalism began far away from the boroughs. He took his first reporting job at the *El Paso Herald-Post* in 1976, several years after joining the business in an unconventional manner: "My mother made me get into the business," he says jokingly. She advised him to take a journalism class in the seventh grade because his cousin Frank had landed a job at the *Los Angeles Times.* "If it is good enough for Frank, then it is good enough for you," she told Ernest.

After working in El Paso and at the *Dallas Times Herald,* Sotomayor took a position with *New York Newsday,* where he became Brooklyn/Queens Editor, and

> Media must measure their quality in terms of how well they reflect people of color, who will one day represent the majority of Americans.

later he was named deputy business editor for technology at *Newsday.* He transitioned to on-line journalism in 2001 as Long Island Editor for Newsday.com, the newspaper's Web version.

Sotomayor knows how important it is to network with fellow professionals in order to reach the goals of diversity. His leadership in the National Association of Hispanic Journalists (NAHJ), and now as president of UNITY, a national coalition dedicated to journalists of color, placed him on a mission similar to that of higher education. It requires applying UNITY's resources to attract the best experts from print, broadcast, and Internet media, where journalism is practiced for both mainstream and communities of color.

Judging by the current state of affairs in the nation's newsrooms, Sotomayor knows his mission is not yet accomplished. "What happens in too many newsrooms is that what you have is this homogeneous group of people, and you don't have any diversity of thought," he said. Even though some media companies have made efforts to recruit from communities of color, they often fail to realize the job does not end there. "That is just the beginning point. Even within those groups you ought to have diversity, and they ought to understand that diversity just doesn't mean race; it means having diversity in the sources they are using in collecting information, and reporting stories that are broad based in their viewpoints," said Sotomayor. Demographics are changing; it is no longer smart for newspapers, broadcasters, and the Web to gauge their success purely in terms of circulation or professional awards. Media must measure their quality in terms of how well they reflect people of color, who will one day represent the majority of Americans.

ABC. Network news managers defended their decision to kill the story, denying that corporate pressure was involved. ABC News had covered negative stories before about its parent company. Eileen Murphy, ABC News spokeswoman, related that reporting on one's corporate bosses is never an easy task. "Whatever you come up with [regarding the parent company], positive or negative, will seem suspect."[24]

Another one of the Disney empire's properties, Miramax Films, had agreed to distribute Michael Moore's politically charged documentary *Fahrenheit 911* that assails the Bush administration for its War on Terrorism. Disney executives apparently decided that the company had more to lose than to gain by releasing the inflammatory film and backed away from it, thus avoiding the controversial questions it raised.

HYBRID CONTENT AND COMMERCIALS

Advertisers also try to blend in their products with news coverage, leaving journalists to decide the ethics of covering stories saturated in commercial content. Tough decisions over what to do about video news releases (VNRs), product placement, and the use of digital technology to rent part of the screen for marketing purposes are not uncommon. There is also the temptation that celebrity journalists face to become a promoter for commercial advertisers, although most reputable news organizations have rules prohibiting it.

Video News Releases

Early in 2004, television news directors around the country received a package in the mail: a videotape news story extolling President Bush's Medicare prescription drug bill. The video news release used actors in the role of journalists to report on the benefits of the White House plan. Was this the right thing to do, or did it cross the fine line between promotion and propaganda? Should taxpayers be supporting one party's efforts to influence the electorate by way of VNRs? Both President Bush's and President Clinton's administrations used VNRs to sell their agenda to the public by way of the nation's local television newscasts. The VNR is also a familiar practice to politicians with an eye on reelection, and for advertisers interested selling their products over television. A typical VNR release runs about a minute and a half in length and appears to be a typical reporter's story. Video news releases are produced for broadcast on local TV stations during their evening newscasts.

These informational pieces usually come prepackaged as reports on health tips, consumer affairs, and even political events. Fortune 500 companies sell products with them, political candidates use them, and they continue to grow in popularity as a public relations tool. If done well, these news/promotion pieces are hard to distinguish from a regular TV story—that is, unless the station chooses to identify the source.

So, where is the problem? Video news releases are neither created by journalists nor intended to cover the news from an independent and impartial viewpoint. If TV stations choose to use video provided by outsiders, such as public relations or advertising firms, the truth standard requires that the stations disclose that source. Some TV stations simply shovel VNRs to the audience without checking either their facts or their sources. In so doing, they treat their responsibility to the public's trust lightly and risk damaging news credibility.

"Newzmercials"

Another type of broadcast production that exudes a commercial scent is often mistaken for journalism. A station's employees produce these "newzmercials," which seem designed to please clients or sponsors, with little or no news value. University of Miami Professor Sam Roberts noticed one example of this disturbing trend while viewing a story on wine tasting. The expert interviewed praised the quality and price of a particular wine. The station cut to a commercial, and lo and behold, it was for that very wine sold by a local wine merchant. "It was a blatant infomercial disguised as a news story," says Roberts.

In New York City, a reporter produced a humorous piece about a portable chair to use at outdoor events such as golf tournaments. It seemed a harmless feature story until the anchors tagged it by reporting a toll-free number that viewers could call to order the chair. "Again, a shameless infomercial disguised as a news story in the middle of a newscast," observed Roberts.[25]

Commercial Ties

Fox News host Bill O'Reilly plays a dual role as a talk-show host on both radio and television, but he is a commercial announcer on just one medium—radio. O'Reilly recorded a commercial for a mattress company that aired on his radio talk show, but when attorneys for Fox News began to hear his voice extolling the virtues of the mattress on other radio shows, they wanted it stopped. Fox's lawyers sent a letter to Westwood One to keep the radio distributor from using O'Reilly's voice on other radio shows. Fox News had no problem with him reading commercials on his own show, but FNC drew the line there, apparently reasoning that O'Reilly's advertising mattresses on other radio shows would leave listeners with the impression that he is more of a pitch man than a news man.

Digitizing the Screen

Baseball fans tuning in to watch their favorite teams have noticed how the view from behind the pitcher's mound keeps changing. Networks have digitally overlaid part of the screens with advertising and promotions for its lineup of new programming. The commercial invasion does not stop there, though.

After its makeover in 2001, CNN's *Headline News* began broadcasting corporate logos on the lower half of the screen during business, sports, and weather updates. The President of Sales at CNN, Larry Goodman, said the combination of the screen logo and spot sales gave *Headline News* a competitive edge that it needed. He dismissed the idea dividing the screen between news and advertising interferes with journalistic integrity. "Viewers understand that television is ad supported," he says. "They are solely sponsors and don't have anything to do with editorial content."[26]

Embedded Commercials

Advertisers pay huge sums to have their products displayed in the actual story line and scenes of entertainment programs. Product placement has become but one means for circumventing digital video recorders like the ones made by TiVo that allow viewers to zip past conventional ads. Network advertisers call it *product integration* and have started placing commercial items on the set or even in the plot of TV programs. *The Christian Science Monitor* reported how a certain brand of doughnuts had found a home on the set of *Sex and the City*, and *American Idol* judges were sipping cups of branded soda. Product integration also made its way into such popular soap operas as ABC's *All My Children*, where actress Susan Lucci's cosmetics were featured on the series. Digital technology even allows products to be placed in programs after their first run when they've made the jump to syndication.[27] Consumer advocates argue that a line should be drawn between advertising and content.

Web-Weaving Ads

The programming language of the Web enables particular words to be underlined or highlighted so that viewers who click on them can be seamlessly transferred to another website. Such is the beauty of hypertext markup language (HTML), and advertisers have shown their appreciation by investing in such links. What happens

protalk
Bob Steele
POYNTER INSTITUTE

When journalists look for answers to on-the-job dilemmas, they often come to the Poynter Institute for Media Studies, where they can get additional training from professionals such as Bob Steele. Steele became interested in how good ethical decisions were made as a reporter, producer, and news director for television stations in Maine and Iowa. He chose ethical decisions as the topic for his Ph.D. dissertation, and examined both the principles of ethics and the process news people follow in making their decisions. Steele takes the position that ethical decision making is a learned craft, much like writing, reporting, editing, and other media skills.

Electronic media play a special role in society, and obviously professionals in the field have duties reflecting that mission. The freedoms enjoyed under the First Amendment, Steele says, carry "both significant rights and great responsibility."

Digital convergence in media has increased both flow and speed of information across multiple platforms. Steele says that makes it even more important for journalists to distinguish their work ethics. "It calls for quality craftsmanship and strong values," he says.

"There are lots of ethical land mines in the digitally convergent environment," says Steele. Some of these have to do with technology that allows on-line media to link to

"There are lots of ethical land mines in the digitally convergent environment."

sources of questionable credibility. Other concerns have to do with the commercial side of the business—drawing a clear distinction between advertising and editorial content.

Some would argue that professional ethics are too costly in a competitive world where speed and impact are at a premium and quality control is not. Steele would disagree. He does not regard "cost" as part of taking the high road. "I believe it's our duty to honor important principles that guide our work as journalists in service to citizens and community," he says. The "reward" for high ethical behavior may be intangible, measured mainly in the pride of rendering excellent service to others. "Granted, this journalism can also be financially lucrative to those who run media organizations," Steele says, "but that shouldn't be our driving force."

when on-line news sources decide to start linking words in their stories to advertising messages? Forbes.com became the first on-line news site to do it. Such advertising links could possibly alienate readers, or worse influence journalists, who might place advertising interests above news judgment.[28]

Advertisers also have tried *bridge pages* in order to link viewers from a radio or TV station's website to the advertiser's homepage. How do the links work? Suppose a TV news viewer enjoys a regular travel feature but becomes curious about the added information promised on the website. When the viewer goes on-line to get the information, he or she sees a travel agency's bridge page, a promotional advertisement that appears before the link to the news material.

POLITICAL ADS ON-LINE

On-line news sites attracted by large cash sums to be made in political advertising began serving up campaign ads in unprecedented ways in 2004 and creating tough ethical situations. How should a news site run political ads? Should it pair them on the same page as political news (a practice that alarms some professionals)? "If you were doing a story about the financial plight of a major airline . . . you probably wouldn't allow a big ad for that airline to run on the same page," said Merrill Brown, a former MSNBC.com editor-in-chief.[29] The will of advertisers may conflict with what a website would prefer to do unless an ethical policy has been established. *USA Today*'s website, for example, refuses to run political ads on its politics or elections pages, and other news sites refuse to place political ads except on non-news pages, although the practices widely vary from news site to news site.

CONFLICTS OF INTEREST

Broadcasters are asked to govern their lives in such a way that charges of conflict of interest, real or apparent, cannot be justly made. Turner Broadcasting System hands its employees a manual of ethics aimed at preventing them from publicly promoting, advertising, or endorsing "any product, service or organization without the prior written consent of an executive vice president, or the president, of CNN."[30]

Political Activism

Local TV stations uphold similar policies, sometimes to the surprise of their own personnel. Kelly Harvey was the weekend anchor at WTKR-TV in Norfolk, Virginia, when she found out about her station's conflict-of-interest policy. She arrived at work one day to discover she had been relieved of her anchor duties, placed on two-week suspension, and reassigned to the early-morning shift that began at 4:00 A.M.

The reason? She donated $1,000 of her own money to Democratic Senator Chuck Robb's reelection campaign. WTKR-TV's policy held its employees accountable for any "conflict of interest, or appearance of a conflict of interest that may cause any public question about the journalistic integrity" of the employee or the station. The news director said ignorance of the rule was no excuse, since this policy of professional conduct was fairly consistent among television stations.[31]

PRINCIPLE 7

A journalist who shows partisanship compromises his or her own credibility.

Models of Moral Reasoning

Professional ethicists hand out lists of questions to ask before making tough calls. The Poynter Institute, for example, which serves as a school of higher learning for

FIGURE 12.1 *Potter's Box*

journalists, lists 11 questions for photojournalists to answer. They include "What are my ethical concerns?" and "What is my journalistic purpose?" These questions can be broken down into basic areas dealing with purpose, ethics, and outcome. We offer here a general model for handling tough calls, but begin with a traditional model of moral reasoning called *Potter's Box*.

POTTER'S BOX

Ralph M. Potter was a doctoral student at Harvard who based his dissertation on a step-by-step procedure for making moral decisions. He began by first defining the situation, assessing the particular *values* and *principles* involved, then reaching a *decision* based on an assessment of *loyalties* (see Figure 12.1). Potter's Box has been used to resolve ethical dilemmas, although it fails to deal directly with issues of consequence. The emphasis is on loyalties rather than weighing the potential outcome of alternatives (as a teleologist might propose).

FIGURE 12.2 *Davie-Upshaw PEACE Model*

PEACE MODEL

What happens when journalists fail to heed their professional duty? Ethical accountability is usually a personal matter. Rather than censure violators, media organizations simply encourage observance of the professional codes. As one RTNDA leader, Jeff Marks, put it, "There are, indeed, many cases in which we would like to reprimand our colleagues publicly. However, in practice, a voluntary membership association cannot act as a prosecutor of its members."[32] We propose an original approach aimed directly at media decisions that embrace the ethical perspectives we have discussed. It forms the acronym PEACE, and it is based on questions of *Purpose, Ethics, Alternatives, Consequences,* and *Execution* (see Figure 12.2).

Success Principles

Professionalism is more than just making the right choices, although that is certainly a part of the equation. It is also about developing habits of success in one's chosen field. We consider now the work ethics that leaders in the media professions claim to be important. Clichéd as it might sound, competition is often called "cut-throat" by media veterans in describing their profession. A former president of CBS News, Van Gordon Sauter, characterized network news rivalries as "trench warfare, only without the mustard gas." Leaders in the media say their success is owed to knowing how to win by capitalizing on their talents, opportunities, passions, and principles of conduct. What it takes to stay on top of the game is also ethics.

SEIZING THE MOMENT

It almost became a slogan for the 1980s after the release of a popular movie about a New England prep school, The *Dead Poets Society.* In the film, Robin Williams plays a poetry teacher who wants his students to realize the importance of seizing the day and has them learn the Latin phrase, *carpe diem.* This same lesson served well the leaders of many industries, particularly the electronic media.

protalk
Deborah Potter
NEWSLAB

Deborah Potter believes the practice of ethical journalism is not an oxymoron. "It's essential, in fact, for journalists to ensure the credibility of their work, without which journalism itself is pointless," she says.

As a journalist and educator, Potter spent 16 years as a network correspondent for CBS and CNN before taking charge of the NewsLab, a nonprofit organization in Washington, DC, which was where she began helping TV journalists learn to tell difficult stories and to make tough ethical calls. In her role at NewsLab, she coordinates training programs, seminars, and scholarship support and research in areas of concern to electronic news professionals.

"Ethical journalism means excellent journalism," Potter declares. "It's difficult for one to exist without the other, because no matter how extraordinary a story may be, unethical behavior by journalists in developing that story will call into question the end result." That's why she finds the truth principle to be basically nonnegotiable. "Journalists who lie or dissemble to gather information may get what they're after, but at what cost to their credibility?" Readers and viewers may legitimately wonder: If a journalist would lie to a source, why wouldn't he or she lie to the audience?

"Ethical journalism means excellent journalism."

So Potter teaches journalists how to navigate the ethical minefields that await them in their careers. She helps them develop and follow their own guidelines for making sound ethical decisions, especially on deadline. "Guidelines work better than rulebooks," she believes, "because they allow journalists to consider individual circumstances" and to discuss the problem among themselves rather than blindly following any rules.

Some professionals think they can trust their "gut" to know the right thing to do in all circumstances, but Potter doesn't trust that type of decision making. "It is better for journalists to reason their way through an ethical dilemma," she says, "and be prepared to justify their actions, not just to themselves and their bosses, but also to the public."

CNN'S JUDY WOODRUFF.

CNN anchor Judy Woodruff was majoring in political science at Duke University when she asked the news director of the ABC affiliate in Atlanta for a job interview. That was in 1968, and after a brief visit about her interest in political science and mass communication, she was ready to leave the news director's office. Woodruff had moved to the door when the news director remarked, "Well, I think we can work something out. Besides, how could I turn down somebody with legs like yours?" She just gulped and said, "Thank you. I'll look for your letter in the mail," and left. Woodruff was stunned.

She accepted the job as a secretary, and for more than a year did everything from making coffee to cleaning film. This was not her dream. "I was chomping at the bit to get some reporting experience," she said, but the only on-air job available was a part-time slot doing Sunday-night weather. She had decided to pass on that opportunity until the news director pulled her aside and said, "If you're serious, you better get some experience. This is the way the real world works." Woodruff took the job.[37]

Reality-show producer Mark Burnett, who changed the television landscape through such program hits as *Survivor* and *The Apprentice,* says his basic philosophy is to just jump in. "You can't figure it out all in advance. The best you can hope for is that you'll have five out of 10 things figured out and you jump in, even if you can't swim, and you know you'll sort the rest as you go. And you fake it till you make it!"

Burnett also warns against overanalyzing situations to the point of paralysis. He credits his success to a personal willingness to accept defeat in the pursuit of a larger goal. "Anybody who totally expects to succeed on every attempt and needs to succeed is crazy," the Australian-born producer said. The scariest day of his life was when CBS gave him a thumbs-up sign for his first *Survivor* series, but he knew that he would eventually figure it out along the way. Once shooting got under way, and Burnett realized his vision would succeed, he learned another principle from the show's first winner, Richard Hatch. Hatch made it clear to Burnett how important it was to form alliances by telling the truth.[38]

TEAMWORK

This maxim may sound overly familiar, but hidden behind its well-worn exterior is an essential truth: Teamwork either makes an organization a success or creates a failure in its absence. ABC News Anchor Ted Koppel offers an anecdote about how teamwork contributed to the success of *Nightline.* At some point, a planned story for one evening's program showed little promise of development, and everyone seemed to know it. So, all members of the *Nightline* staff were summoned together to see what could be done about it. They each took a piece of paper and jotted down a new idea for the evening's topic.

Each staff person deposited his or her slip in a hat and waited for the winning idea to be announced by the executive producer, Rick Kaplan. The winner was a suggestion to cover a play, accused of racism, that was based on Mark Twain's novel, *Huckleberry Finn.* The winning idea came from the show's researcher, who was excited to hear her idea chosen, though her mood changed somewhat when she realized she was the one expected to follow through and help produce the show.[39]

NEVER STOP LEARNING

Some creative talents in television take the do-it-yourself approach by reading everything they can about their profession. Brannon Braga knew what she wanted to become at the age of 10, and so she decided to do something about it. She had no idea how to become a "Hollywood professional," but she devoured every book she could find on the subject and began writing screenplays. She went to film school in California, and, most important, kept practicing her craft by working on scripts every day.

After being turned down for a scriptwriting internship with Paramount Pictures, she applied again the following year and found that her persistence paid off. Like so many others, she converted her internship into a successful career, moving from writer to story editor to executive producer for *Star Trek Voyager.*[40]

Seizing a pivotal role in the planning of one of America's most popular news magazines requires an attitude of learning. Brad Bessey became coordinating producer for *Entertainment Tonight* after learning from his mentors. "Everyone you meet along the road can and will be a mentor to you," says Bessey. He singles out *ET*'s executive producer Linda Bell Blue as a "mentor, friend, and inspiration."[41]

CBS correspondent Susan Spencer found useful learning experiences both in college and on the job. She graduated with a double major in German literature and television/radio production at Michigan State University before tackling a master's degree in journalism at Columbia University. With two degrees from respected institutions in her field, it would seem that Spencer could write her own ticket, but that was not the case. She went to work first at a film production house in Louisville, Kentucky, and wrote advertising copy. Then a local public television station hired her to work on a public-affairs program. Spencer shot film, wrote and voiced copy, and covered hard news and features. It was enough to give her the training she needed, and she became co-anchor at WCCO-TV in Minnesota. Five years later she was hired as a correspondent at CBS News.[42]

What does it take for a woman to achieve the pinnacle of leadership in network television? Ask Jane Cahill Pfeiffer how her education evolved. She became president at NBC by learning all she could from each move she made in college and in her career. Cahill Pfeiffer credits excellent professors and her desire to participate every where she could on campus. "In the theater department, I worked diligently—directing, writing, and even acting." Upon graduation, she sought a life of religious devotion, but after a year in a convent, she abandoned her ambition to become a nun. Instead, she moved up the corporate ladder to hold directorships for oil, paper, insurance, and retail companies. After turning down an offer from President Jimmy Carter to join his cabinet, Cahill Pfeiffer was invited to join NBC by President Fred Silverman in 1977.

There, she made important strides in NBC's news coverage, improving the quality of the programming and making the management system more efficient. Cahill Pfieffer advanced to network chair at a time when women were still not accepted in the head office. She gives advice to others entering the field, saying, "Recognize you'll always be learning, because technology is driving so much change."[43]

PASSION WINS

In the world of advertising, Ted Bell is no lightweight. As former vice chairman and creative director of Young & Rubicam worldwide, one of the largest ad agencies in the world, he developed the heart to win in a fiercely competitive business. He advises other newcomers, "Be the most passionate person in the room. Not the smartest, not the cleverest, but the most passionate. Care more than anybody and you'll be the one that wins."[44]

Those who know Bell say his passion translates into a love for his work. He has loads of self-confidence, and his enthusiastic vision for projects seems to infect his colleagues. Bell is quick to explain that you can't be stingy with your time or your concern for others and hope to make it. His colleagues felt Bell's desire to work with them and they naturally responded with positive support and enthusiasm. Creativity and talent help, of course, but passion is the key. Small wonder that Bell

For the American-led coalition's war on terrorism, the news was particularly grim in the spring of 2004. During a two-week period, the world's news media were compelled to handle two sets of brutal images from Iraq: a series of photos showing naked Iraqi inmates suffering abuse at the hands of U.S. soldiers, and the beheading of an American hostage, Nicholas Berg, appeared on the Internet. Both sets of pictures reflected the brutal realities of this conflict, but also raised ethical questions: How should the images be conveyed to the audience, given the role of the U.S. forces in Iraq and the terrorists' objectives of striking fear in the minds of viewers?

Showing Pfc. Lynndie England smiling and pointing to naked prisoners with a leash in hand would indicate U.S. soldiers were inflicting sexual humiliation on the people they were liberating. The terrorists, on the other hand, sought to send out the images of their atrocities to as wide an audience as possible. To examine these two scenarios from an ethical perspective, the steps of the PEACE model should be used to guide the reasoning.

Purposes: Here is where we determine what the media are trying to accomplish in their coverage, and what the news actors may be hoping to achieve through their participation. It is also important not to overlook the audience's interests in seeking out the news. The media could hardly ignore the images once they were widely disseminated, and the audience would expect news coverage. Stories of prisoner abuse in Iraq had been reported, but attracted little notice until the Abu Ghraib pictures broke on CBS News's *60 Minutes.*

The terrorists wanted to propagate their grisly beheading of American businessman Nicholas Berg because for "young Muslim men, the image could be a recruitment poster designed to inspire excitement. To the terrorists' other target audience—Americans, Europeans, Asians or anybody else who may dare to stand against them—the image is a nightmare scenario designed to incite fear," said Tom Maurstad of *The Dallas Morning News.*[33] The audience, however, was growing weary of violent images. The Pew Internet and American Life Project discovered that between 40 and 49 percent of the public disapproved of such images being made available.[34]

Ethics: In each step the media should analyze the moral principles involved in both cases. What professional guidelines, station, or network policies can be consulted? For journalists, the terrorists' images of murder on the Web and in video represented a case in which key principles were in conflict. Both the Society of Professional Journalists and the Radio-Television News Directors Association advocate diligently and thoroughly investigating all sides of a controversy but warn against exploitation. The National Press Photographers Association's Code of Ethics states, "Treat all subjects with respect and dignity. Give special consideration to vulnerable subjects and compassion to victims of crime or tragedy. Intrude on private moments of grief only when the public has an overriding and justifiable need to see."[35] Here, professional duties must be weighed against concerns for the individuals involved. What harms will result? Discussion among members of the media organization is a must here.

Alternatives: In both examples, the choices involved determine how much to show the audience over how long a period of time and how prominently to display the images. The media, for example, had to choose how much of Nicholas Berg's fate to show—if any at all—and how to present the images from Abu Ghraib prison. The pictures tell the story with an impact that goes far beyond what words can describe. Images tend to create an emotional link with the audience as they look for people to identify with in the pictures. Berg's beheading would be shocking and yet strike a sympathetic chord for Americans; however, views of U.S. soldiers abusing Iraqis would prompt an entirely different response.

In this scenario, the questions involved how much of the images should be either blurred or abridged. *NBC Nightly News* and ABC's *World News Tonight* decided to stop the tape just as the killer drew his knife; however, CBS News decided to venture further into the violence of the act. It showed the killer grabbing Berg by his hair and slamming him to the ground before putting the knife to his neck. CBS News reasoned that it was important for the viewers to see as much as they can because the extra violence made it clear what evil the terrorists were willing to perpetrate.

Consequences: The fourth step is necessary to determine both long- and short-range outcomes. It calls for a reasoned prediction of the impact each

CHAMBER OF HORRORS. Abu Ghraib's prisoner abuse scandal inflamed the public and moved a military investigation from brief mention to lead-story coverage after sensational photographs were circulated. Lynndie England's poses with Iraqi inmates became iconic reminders of the evils of war.

possible choice might have on all parties involved. These effects would include the actors of the event, the media participants, and the audience. In considering the prison photographs, the question becomes, What parts of the images should be blurred in order to avoid offense? In the case of Berg's murder, Do the media fulfill terrorists' goals by showing events leading up to the beheading? Newspapers and networks tread carefully, sorting out how much information that Americans should have about U.S. soldiers in times of war, as well as their enemies.

At what point does the audience become either callous or sickened by the repeated showings of such imagery? Different treatments will have different impacts, and each choice holds the potential for either harm or good as far as the viewer is concerned. The Poynter Institute's ethicists advise that once a story's bell has been rung, it cannot be "unrung," but the decision becomes how long to keep ringing that bell.[36]

Execution: Finally, in this step a decision must be made and action taken but not until a rationale is fully developed and explained. This last step serves to recheck all the previous decisions in terms of their appropriateness and their presentation. Here, gatekeepers must consult with each other and test the reasons given. New dilemmas require new decisions in the media each day, and the only way to resolve them is for professionals to keep talking among themselves and with their audience. Most important, these questions must be asked repeatedly: What is our purpose here? What ethics (harm or help) can we support in making our decision? What are the other available alternatives? and What will be the impact of our eventual decision?

admires another passionate executive—Sumner Redstone—who rose to the top of one of the largest media conglomerates in the world.

In his seventies, Redstone stands among the wealthiest and commanding figures in U.S. media. He engineered Viacom's takeover of Paramount, Blockbuster, and CBS. He controls interests in publishing giant, Simon & Schuster, and in New York's professional hockey and basketball franchises. A self-made man? Perhaps, but Redstone is careful to credit his education first. Trained at Harvard and the famous Boston Latin School, he is one of the finer intellects in global media's corporate offices. One other facet of his career that stands out in an uncommon way: His determination brought him through a harrowing brush with death.

In 1979, Redstone was awakened by a fire in a Boston hotel and had to crawl out on a ledge to escape. He survived the midnight blaze only by clenching a windowsill with one hand while flames scorched the other hand and arm. After being lifted down by firefighters, he spent 60 hours on the surgery table to repair half of his body using skin grafts from the other half.[45] One might expect such a fearsome encounter could change a person's perspective on life. "Nonsense," says Redstone. The same man who Boston firemen lifted off that hotel ledge continues to lead his enterprise each day as the head of CBS/Viacom. "I have always been driven. I have a passion to win, and the will to win is the will to survive."

Summary

The professionals who lead electronic media vary somewhat in their standards and practices but their personal ethics are subject to the same principles of truth, fairness, and responsibility that guide other careers. Your personal blueprint as a professional will mark your success in the end. Textbooks and teachers, family and friends, corporate manuals and common sense—all will help you draft that blueprint. After that, the nature of your choice will direct the future and define your character. This chapter has showed how important it is to consider your duty to principle, your sense of purpose, and your reasoning about consequences in those defining moments.

Success is one of those funny terms that almost defy definition. Some professionals define it by status, personal wealth, or respect from their peers. However, success is more fundamentally *what you think of yourself at the end of the day*. Media success may be defined in corporate terms by ratings and dollars, but after the accountants have gone home and the Nielsen numbers tallied, there is another inventory to take. This personal accounting evaluates whether or not you've done the right thing that day. Such moral "gut-checks" are really what success is all about.

Food for Thought

1. If you were to identify two students cheating in class, and were considering whether to tell the professor, what principle would you follow:

   Duty. Tell the teacher to preserve the honor and integrity of the test.
   Purpose. Tell the teacher to avoid other students' getting an unfair grade.
   Consequences. Don't tell, to avoid appearing as a snitch and keep favor with other students.

2. Someone argues that codes of professional ethics are unnecessary because the only ones that matter are personal ethics anyway. What is your opinion?

3. Do you think government is ever justified in distributing dishonest propaganda, and if so, under what circumstances?

4. Do you think that television does a poor, adequate, good, or excellent job in portraying minority members of U.S. society? Please explain the reason for your answer.

5. Some citizens groups advocate boycotts against commercial sponsors of programs that violate standards of taste and decency. Would you ever participate in such a protest, and under what circumstances?

6. If you were a broadcast journalist and put under pressure by an advertiser to change or kill a story, how would you reason your response to the problem, using the PEACE model?

7. Discuss a time when you saw an opportunity to achieve a goal and seized the opportunity to your advantage, or perhaps had such a chance but failed to recognize it.

Theory *and* Research

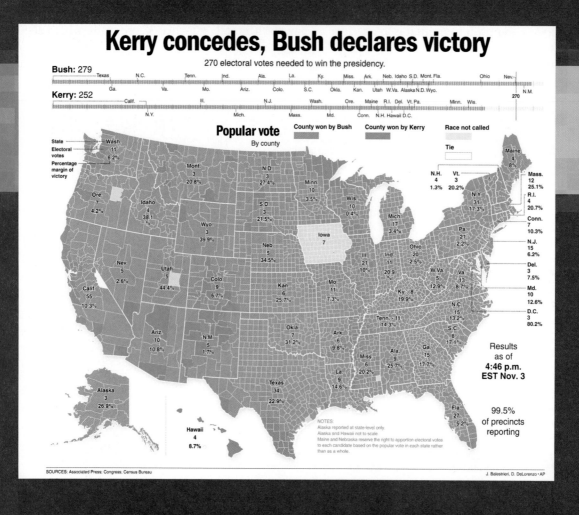

Kerry concedes, Bush declares victory

270 electoral votes needed to win the presidency.

Bush: 279

Texas   N.C.   Tenn.   Ind.   Ala.   La.   Ky.   Miss.   Ark.   Neb. Idaho S.D. Mont. Fla.   Ohio   Nev.

Ga.   Va.   Mo.   Ariz.   Colo.   S.C.   Okla.   Utah W.Va. Alaska N.D. Wyo.   270   N.M.

Kerry: 252

Calif.   Ill.   N.J.   Wash.   Ore.   Maine R.I. Del. Vt. Pa.   Minn.   Wis.

N.Y.   Mich.   Mass.   Md.   Conn.   N.H. Hawaii D.C.

Popular vote By county   County won by Bush   County won by Kerry   Race not called

Tie

State ······· Wash. 11 6.2%
Electoral votes
Percentage margin of victory

Mont. 3 20.8%

N.D. 3 27.4%

Minn. 10 3.5%

Ore. 4 4.2%

Idaho 4 38.1

Wis. 10 0.4%

Maine 4 8%

N.H. 4 1.3%

Vt. 3 20.2%

N.Y 31 17.3%

Mass. 12 25.1%

R.I. 4 20.7%

S.D. 3 21.5%

Wyo. 3 39.9%

Mich. 17 3.4%

Pa. 21 2.2%

Conn. 7 10.3%

Nev. 5 2.6%

Utah 5 44.4%

Neb. 5 34.5%

Iowa 7

Ill. 21 10%

Ind. 11 20.9%

Ohio 20 2.5%

W.Va. 5 12.9%

Va. 13 8.7%

N.J. 15 6.2%

Del. 3 7.5%

Calif. 55 10.3%

Colo. 9 6.7%

Kan. 6 25.7%

Mo. 11 7.3%

Ky. 8 19.9%

Md. 10 12.6%

D.C. 3 80.2%

Ariz. 10 10.8%

N.M. 5 1.7%

Okla. 7 31.2%

Ark. 6 9.8%

Tenn. 11 14.3%

N.C. 15 13.2%

S.C. 8 17.1%

Alaska 3 26.9%

Miss 6 20.2%

Ala. 9 25.7%

Ga. 15 17.7%

La. 9 14.6%

Texas 34 22.9%

Hawaii 4 8.7%

Fla. 27 5.2%

Results as of **4:46 p.m. EST Nov. 3**

99.5% of precincts reporting

NOTES:
Alaska reported at state-level only.
Alaska and Hawaii not to scale.
Maine and Nebraska reserve the right to apportion electoral votes to each candidate based on the popular vote in each state rather than as a whole.

SOURCES: Associated Press; Congress; Census Bureau

J. Balestrieri, D. DeLorenzo • AP

Chapter 13

"One citizen—one vote." It's the democratic ideal; however, a Florida bumper sticker quipped that it "may *not* apply in certain states."

Political humor aside, the fact that voters regularly make decisions about which candidates to elect is no small matter to the electronic media. In 2004, more money than ever before was invested in selling presidential candidates to the electorate—an estimated $1.6 billion—most of it spent on television advertising. Why? Television is more effective in reaching *undecided* voters, who are less engaged politically and less likely to turn to other sources for news of that nature. The citizens living in areas of the map not colored Republican red or Democratic blue are the desired viewers in what pollsters have labeled "battleground states." That's where the big money is spent during campaign seasons.

Yet, not everyone is clear on how pollsters go about drawing these partisan boundaries, or even what they report after taking their surveys. The 9/11 Commission, for example, published this blooper in 2004:

As best we can determine, neither in 2000 nor in the first eight months of 2001 did any polling organization in the United States think the subject of terrorism sufficiently on the minds of the public to warrant asking a question about it in a major national survey. (p. 341)

This congressional body overlooked important surveys taken in the spring and summer of 2001 by leading polling organizations and by members of the American Association of Public Opinion Research, including the Pew Research Center, NBC/*Wall Street Journal,* CBS/*The New York Times,* and the Gallup Organization.

In May 2001, for example, the Pew Research Center and the Princeton Survey Research Associates found that 64 percent of the American people viewed international terrorism as a major threat and 27 percent felt it was a minor one.[1] That same month, Fox News/Opinion Dynamics found that only 33 percent of Americans would be willing to give up personal freedoms in order to reduce the threat of terrorism, whereas 40 percent said they would not.

Other surveys prior to September 11, 2001, pursued the subject of terrorism in the wake of U.S. embassy bombings in Africa and of the *U.S.S. Cole* in Yemen. Although members of Congress may have missed these surveys, the electronic media pay more attention now to survey data than ever before in order to build their businesses, win advertising dollars, and direct the growth of their station, network, or television system.

This chapter approaches electronic media from a researcher's perspective in order to understand the impact radio, television, and the Internet have on people and society. Communication models take into account media sources and respective audiences and help us answer questions about the media's credibility, its impact, and its influence. Media researchers evaluate people's motives for television viewing, radio listening, and web browsing. And they have taken note of how all these electronic media shape the public's agenda as well as the public's social and political perspectives.

■ Purposes of Research

Very little can be taken for granted in the age of digital convergence, but one thing is certain: Major decisions are not made—whether about merging corporations or making new laws—without surveying the data. Media problems are solved not by guesswork and hunches, but by the best predictions available. Before choosing whether to make major investments in broadcast or broadband media, executives pore over spreadsheets to see *what the data show*. It's about reducing uncertainty and making informed and rational decisions. So important is this business of research—collecting data, analyzing them, and interpreting them—often the best jobs in radio and television go to researchers.

The systematic structure of mass communication research grew out of academic laboratories as well as media firms, where scholars and consultants apply the scientific method to develop theories and principles. **Research** is about answering questions and solving problems. It also involves testing theories about the way things work and predicting likely outcomes. So, it's all just based on common sense—correct? Not exactly. Albert Einstein observed, "Conclusions obtained by purely rational processes are, so far as reality is concerned, entirely empty."[2] Building a good theory is about discerning what happens *beyond* your perceptions.

So, how can you tell whether a theory is good or bad? First, you must ask yourself several questions: Does it build on what is already known to be true? Can it be proved false? Is it simple and practical enough to be tested? What researchers try to avoid is testing a theory that *reinvents the wheel* and fails to push back the frontiers of new knowledge.

A previous chapter on ratings offered a brief overview of applied audience research. This chapter focuses on academic studies and gives some attention to what is often called *administrative research* in which private consultants strive to yield commercial profits for the companies they serve. Scholars test theories that yield more in understanding than in revenue. In so doing, they reveal media processes and **effects.**

What are the influences of *propaganda,* and how does this process of persuasion work? What are *gatekeepers* thinking when they allow certain messages into the channels of radio and television while holding others back? How does media violence affect children's minds and shape their worlds? Do television shows function to gratify the audience, or do they create instead an unhealthy dependence on them? These are just a few of the fields of inquiry that scholars have tried to resolve in theory and research, and in the digital age new questions arise and beg for answers.

COMMUNICATION MODEL

It was not social scientists, but an engineer and mathematician who gave the early study of **communication** its first structural model. Claude Shannon at Bell Labs and Warren Weaver at the Sloan-Kettering Institute for Cancer Research collabo-

Research Process of systematically observing and collecting evidence in order to answer specific questions in a scientific manner through controlled and repeated measures.

Effects Approach to mass communication theory and research that measures the impact media have on members of the audience, either collectively or individually.

Communication Gathering, ordering, encoding, and transferring information from one source to a receiver.

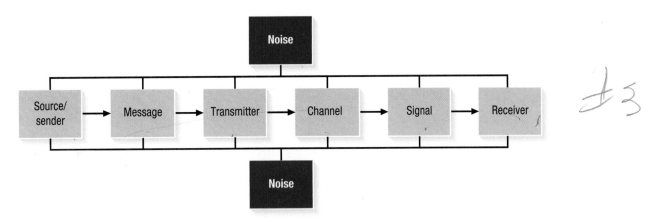

FIGURE 13.1 *Communication Model*
Shannon and Weaver's model of communication is the basis for a wealth of research. It was the first attempt to schematically illustrate the communication process and has been revised and expanded over the years.

rated on technical questions concerning the communication phenomenon. Could there be a logical way to trace its distinct parts and evaluate its processes? Shannon and Weaver drew up a mathematical model resembling an engineer's schematics, which became "widely accepted as one of the main seeds out of which Communication studies has grown."[3]

It was proven relevant in the digital era of convergence. Shannon and Weaver's model has addressed technical questions concerning how many bits per second are needed to transmit audio and video messages to computer screens. The influential model begins with an information source selecting a message, encoding it, and then sending it through a **channel** to be received as a signal and decoded back into a message.

The model (Figure 13.1) looks simple enough, but *noise* enters the process and threatens the message. Noise is basically anything that interferes with the **source/ sender**, the *receiver*, or interrupts at points in between. Noise may be generated by people, machines, and sometimes both.

Later researchers felt the model was missing one element in the communication process: *feedback*. A scientist from Massachusetts Institute of Technology, Norbert Wiener, added the feedback loop while developing weapons systems for the U.S. Defense Department. He saw it as necessary to either correct or confirm that the sender's signal was properly received. So how is feedback different from a new cycle of communication? It is focused on the initial message; only when a new subject is introduced does a new cycle begin. Notice that this model does not attempt to define *meaning*. Meaning has been of interest to theorists who have investigated how people initiate and respond to certain ideas and discard others. This research has been especially important during times of war, when messages are used to motivate people to fight for a cause, or perhaps even to surrender.

PROPAGANDA STUDIES

When U.S. military planes prepared for deployment over Iraq in 2003, their payload drop included more than explosives and missiles. Stacks of leaflets were placed on board the planes to warn the Iraqi people—especially soldiers still loyal to the regime of Saddam Hussein:

PRINCIPLE 1

The structural process of communication must be understood before the meaning of content is considered.

Channel Means used to carry a message from a source to a receiver, including electromagnetic waves, light, sound, or other media.

Source/Sender Message originator who encodes information and prepares it for transmission.

Iraqi Commanders and Soldiers: Show that you will not resist Coalition Force . . . Make Clear Your Intentions . . . Comply with our instructions and you will not be destroyed.

Another flyer read:

Coalition air power enforces the No-Fly Zones to protect the Iraqi people. Threatening these Coalition aircraft has a consequence. The attacks may destroy you or any location of Coalition's choosing. Will it be you or your brother? You decide.[4]

Pictures of soldiers dispensing coloring books and crayons to Iraqi children popped up on the Central Command's website for all coalition forces or Iraqi citizens, but no mention was made of the "collateral damage" of noncombatants killed by U.S. missiles, or other horrors of war.

It was not the first time mass distribution of propaganda was used as a means of persuasion. The early impetus for investigating communication activities arose from the national interest during a war 90 years earlier when Americans needed to know how to win both on the battlefield and in the hearts and minds of people. A student at the University of Chicago in the 1920s became curious about how propaganda had been used to help win World War I. Harold Lasswell studied the leaflets dropped from planes flying behind German lines to see how they encouraged enemy soldiers to surrender.

Lasswell considered this form of persuasion to be effective, and later scholars attributed to him the **"magic-bullet"** or "hypodermic-needle" theory. This idea suggests that a stimulus message will be followed by a direct response. Just as a bullet pierces barriers or a needle injects something under the skin—a message is the active ingredient persuading an audience to accept its "truth." The trouble was that it failed to account for the fact most people simply do not accept messages under their skin without challenging them first.

So, did Lasswell truly believe in the bullet theory for which he was credited? Communication scholar Wilbur Schramm doubted whether Lasswell or any other professor ever put much stock in it: "I have never known a serious scholar to endorse or make research use of the so-called Bullet Theory of communication effects." It may have been part of the popular wisdom of the era, but Schramm added dryly, "If any scholar did make serious use of this concept [of information], it signaled the childish side of the act of passage."[5]

Lasswell's propaganda work was published as a book in 1927,[6] drawing details from interviews with European leaders in World War I and using a form of content analysis. In it, the researcher defined four goals for propaganda: galvanizing hatred against the enemy, strengthening ties to allies, recruiting the cooperation of neutral parties, and demoralizing the enemy. He did believe **propaganda** was effective for the "management of collective attitudes by the manipulation of significant symbols." Lasswell viewed it, though, in neutral terms—simply as a strategy to shape people's personal convictions through mass media and socially derived messages. Not everyone agreed with his take on the matter.

NAZI PROPAGANDA. Wartime propaganda has produced landmark studies in media research. Communication scholars and social psychologists have used experiments and content analyses to measure the effects of propaganda on the audience.

Magic bullet (Hypodermic needle) Metaphors to describe how persuasive messages will strike target audiences with such dramatic impact to penetrate conflicting attitudes like a bullet or hypodermic needle.

Propaganda Spreading viewpoints or doctrines by government authority through the use of media.

Public relations pioneer Edward L. Bernays referred to *propaganda* as a means for manipulating and maintaining control of the masses. After the Nazis repeatedly broadcast the "Big Lie" (that the Jews were to blame for Germany losing World War I, and that pure Germans were a master race) through the late 1930s and early 1940s, propaganda took on even darker connotations. One book reviewer, Foster R. Dulles, labeled Lasswell's book a "Machiavellian textbook which should promptly be destroyed!"[7] Regardless of how this book was perceived, propaganda and its wartime deployment inspired much early study in mass communication.

Lasswell posed another key question for mass communication theorists to pursue. It contained five key elements: "*Who* says *what* to *whom* through what *channel* with what *effects*?"[8]

PERSUASION STUDIES

Just as Lasswell opened the door to the study of World War I propaganda through his use of content analysis, Carl Hovland, a Yale social psychologist, charted the impact of World War II propaganda through his experiments. The emphasis shifted from social processes of propaganda to the psychological processes of persuasion. Hovland and his fellow psychologists manipulated messages to see how they might reinforce or change an *attitude, belief,* or *behavior* that had affective (emotional) and cognitive (thought) components. He believed they could be altered through learning.

Selective Barriers

Persuasion theorists have explored the nature of *cognition*, which is how personal knowledge can influence electronic media effects. They have tried to determine how the behaviors leading to knowledge—**selective exposure/attention/perception/retention**—influence the impact of messages. *Selective exposure* is the tendency for people to expose themselves to television, radio and Internet content in agreement with their personal beliefs and to avoid disagreeable messages that cause dissonance. *Selective attention/perception* suggests an attraction to news and talk shows that reinforce strong convictions and, conversely, rejects contrary programming. *Selective retention* refers to how the memory functions in this process of interacting with electronic media—calling to mind supporting ideas from audio and video experiences.

Scholars have organized these mental defenses against electronic media into four barricades. Selective exposure serves as the outer barrier. Working inward, attention is next, then perception, and finally selective retention is the most intimate, personal barrier. They all work to maintain cognitive consistency by filtering out radio, television, and other media that create what social scientists have called **cognitive dissonance.** That's when people try to maintain their personal balance in thinking by avoiding or reordering media content that contradicts their beliefs or behaviors.

Why We Fight Research

The U.S. War Department, needing advice about how best to train soldiers for the mental rigors of combat, hired Carl Hovland as its chief psychologist and director of experimental studies in 1942. Hovland studied the influence of Army training films on U.S. soldiers who were preparing to go to war. He experimented with the *Why We Fight* series created by Hollywood director Frank Capra. The Army wanted to know if these films increased soldiers' awareness of events leading up to the war. How long would this learning last and would the films motivate soldiers to fight?

Hovland discovered that the 50-minute documentaries did heighten soldiers' awareness of events leading up to the war, but the films failed to motivate men to fight or even hate the enemy. Mere exposure to a documentary on the evils of Nazi Germany produced only the first step in a persuasion process. Starting at awareness, the soldiers took a cognitive journey that ended with a shift in attitude and behavior. This became known as the *hierarchy of effects.* Because persuasion is a process that occurs in phases, each step must be completed separately and successfully before the next one can occur. In the initial phase of knowledge acquisition, a viewer's *attention, comprehension, acceptance* and *retention* must be obtained, each in its turn. To complete his theory of persuasion, the quiet scholar from Yale added other key components, including the *sender's characteristics* (credibility), the *qualities of the message* (persuasive appeals), and the receiver's *cognition.*

Selective exposure/attention/perception/retention Tendency to organize and interpret messages to establish a coherent perspective consistent with preexisting beliefs and values.

Cognitive dissonance Research theory predicting human reaction to information that is contrary to felt attitudes and convictions.

CREDIBILITY RESEARCH

PRINCIPLE 2

The basis of credibility is formed by audience perceptions of expertise, honesty, and sincerity.

The ancient Greek rhetorician Aristotle spoke of **credibility** as the speaker's *ethos*. He noted that rhetorical appeals to emotion (*pathos*) and to reason (*logos*) contributed to a speaker's success. Audiences believe credible speakers, but tend to discount those who lack knowledge, goodwill, or status. Such judgments are based on perceptions of the source, which, to be credible, must be viewed as believable, knowledgeable, and sincere.

Early research in persuasion theory revealed a **sleeper effect,** in which arguments actually produce a stronger impact after the source is forgotten and credibility is no longer relevant. The audience begins to trust the information a few weeks later—after forgetting where it came from. Pitches for commercial products benefit from this phenomenon.

Message Appeals

Hovland experimented with other persuasive appeals, such as varying levels of fear. His aim was to find out which ones would be the most effective—*low, moderate* or *high* doses. He conducted experiments yielding surprising results. He measured human response to dental hygiene messages, and discovered *high*-fear appeals were actually less effective than *minimal fear* messages. Subjects told to brush daily after seeing graphic examples of badly rotted teeth were less likely to comply than subjects urged to brush after viewing x-rays and drawings of cavities.[9]

Later research in this area by Hovland and his associates favored a *moderate* rather than *minimal* approach to fear appeals. Subsequent studies formed the basis of what is called *protection motivation theory*. It suggests audience attitudes shift in response to the magnitude of the threat, its likelihood of occurring, and the appropriateness of the prescribed defense. Politicians have used this research to mount media campaigns, and health professionals have applied it to public warnings concerning disease outbreaks.

Recall Tests

Researchers believe that memory holds the key to attitude change and have a variety of methods to test memory's impact on media consumption. Surveys have identified which types of radio and television messages are easier to recall. **Recall tests** ask audience members which commercial appeals come to mind. **Random samples** are drawn to give all potential respondents an equal chance of selection.

Archibald M. Crossley and his Cooperative Analysis of Broadcasting (CAB) conducted the first radio surveys. As noted in the chapter on ratings, Crossley's interviewers typically asked a family whether its radio was tuned to a *sponsored* show on NBC or CBS. That was because the Association of National Advertisers was paying for the research. The advertisers needed information to figure where to buy commercial time. Consequently, the questions veered away from noncommercial (sustaining) programs.

TWO-STEP FLOW

One of Paul Lazarsfeld's contributions to media research grew out of a series of voter studies in the 1940s. He wanted to find out how much impact news coverage of presidential campaigns had in making up the minds of voters. Were radio and newspapers as influential as people seemed to think?

A panel study of the 1940 presidential race in Erie County, Ohio, between Franklin D. Roosevelt and Wendell L. Willkie provided the data for Lazarsfeld's conclusions in *The People's Choice*. The book showed how powerful people were in filtering the media's impact on voting decisions. Lazarsfeld and his team noticed a tendency

---

Credibility Measure of trust an audience invests in a source, based on perceptions of character quality, including honesty, expertise, and sincerity.

Sleeper effect Delayed impact persuasive messages have on attitude formation, based on time needed for arguments to be "awakened" in the subconscious after the source is forgotten.

Recall tests Means of gathering data by asking members to remember information disseminated by a media source, usually within a brief period afterward.

Random sample Group of people or items selected for study in which every member of this portion of the larger group has an equal and independent chance of selection.

for some citizens to base their ballot choices not on what they learned directly from radio and newspapers but on secondhand interpretations by friends and family. The Lazarsfeld team called this a **two-step flow** from the media to "opinion leaders" and then to primary social groups.

As with any discovery challenging the existing order, this theory provoked a fair amount of criticism. Critics charged that Lazarsfeld failed to properly measure the direct flow of news and that he overemphasized the role of so-called opinion leaders in the process. Lazarsfeld bolstered his argument for a *limited-effects* perspective in mass communication research based on data collected in Decatur, Illinois, during the 1948 race between Harry S Truman and Thomas E. Dewey. *Voting: A Study of Opinion Formation in a Presidential Campaign* also concluded that opinion leaders were interpreting news for primary social groups—friends and family.

PRINCIPLE 3

The two-step flow theory says that opinion leaders interpret news for others, thereby limiting the media's direct impact.

Critical Studies

What do you know about people who differ from you—and what do you merely believe? Early critical theorists shaped contemporary studies about the prevalence of racial and gender stereotyping, and about the western media's influence over societies and cultures. Critical perspectives have led the way in European scholarship of media and have grown more popular in recent times.

FRANKFURT SCHOOL

Lazarsfeld's faith in **empiricism**—gathering data from people in order to test theories and predictions—stood in stark contrast to the critical ideas that his European colleagues preferred. Marxist scholars Max Horkheimer, Theodor Adorno, and Herbert Marcuse were trained in philosophy and political economy in Austria and Germany. They formed the Institute for Social Research at the University of Frankfurt. Like Lazarsfeld, their religious and intellectual heritages forced them into exile when the Third Reich began to tighten its grip. They fled to New York City but continued to publish articles that challenged capitalist-based systems.

Lazarsfeld tried to persuade one of his comrades, Theodor Adorno, to join him in conducting statistics-based research, but failed. Instead, the Marxist scholar went to work on a popular-music project at CBS, and announced to anyone who cared to listen that he detested "jazz" and believed that reporting radio listeners' likes and dislikes was basically a waste of time. American audiences should be taught to appreciate serious music, Adorno felt; treating art as a commodity was just absurd.

Critical scholars from the Frankfurt tradition saw empirical research as a servant to the status quo. They challenged the empiricists' lack of concern over who owns the media, the nature of their goals, which ideas they promote and which ones they suppress—all evidence of submission to the capitalistic elite. What difference does it make what radio programs people choose to listen to if there is no choice about who produces them? Lazarsfeld responded to the attacks written by Horkheimer in *Traditions of Critical Theory* with his own case in favor of "administrative communication research." It basically supported the idea that audience opinions do matter. He believed both traditions, administrative and critical studies, should be nurtured in U.S. colleges and universities.

Two-step flow Model of opinion flow devised by Lazarsfeld, based on presidential election studies of 1940 and 1948 finding that media information flows from opinion leaders to individuals.

Empiricism Foundation of science based on seventeenth-century philosophy that our knowledge of the world could be accurately measured by our sensory experience.

JAG. Some American television programs are more successful overseas than others. *JAG*, for example, found a niche with international audiences who responded favorably to the military setting and its courtroom drama.

CULTURAL IMPERIALISM

The globalization of electronic media has fostered a critical perspective that tends to divide the world into exporting and importing countries. **Cultural imperialism** suggests that western values have dominated the world's cultures through exported radio and TV programs. Underdeveloped countries have trouble competing with the sophistication of western media, so the issue boils down to which country has the newest technology and the highest-quality production. This argument is countered, however, by evidence that even third-world countries prefer programs produced by their citizens in native languages, although the western model of television dominates—producing native copies of game shows, reality programs, and talk formats. In addition, the United States successfully exports programs such as *JAG* and *ER*, but they are less popular than indigenous shows.

A professor of sociology, Stuart Hall, wrote extensively about intercultural media with an emphasis on the "haves and have-nots." Hall found—empirically—that audiences don't simply accept a TV or radio program at face value without evaluating some or all of its subtext. That is, they choose to embrace or dismiss a show's message based on filters from their personal heritage, including family values and cultural beliefs. However, the influence of popular culture programming and its profitability is often irresistible.

Technological Determinism

In terms of theory building, the digital age has drawn inspiration from *media determinism*. This is the perspective developed by a Canadian scholar of English, Marshall McLuhan, whose ideas about the medium being the message shook up the communication research world of the 1960s. Through McLuhan's eyes, the message is relevant only as "the juicy piece of meat carried by the burglar to distract the watchdog of the mind."[10] It is the communication activity itself that forms a lasting impression on attitudes and behaviors. So how does the Internet factor into this vision of a world transformed by television into a *global village*? Would McLuhan speak now of the "global village" as a product of the Internet, or would he envision many global villages—virtual communities—where old distinctions based on common interests, culture, and citizenship are preserved?

TELEVISION DETERMINISM

In order to accept McLuhan's point about the "juicy meat" of content drawing people away from the media activity itself, we need to understand how his "sense ratios" or "patterns of perception" influence people's media habits. In this perspective, the age of movable type enhanced the singular sense of vision beyond hearing, taste, touch, and smell. As a result, people's thinking conformed to this sensual reemphasis on sight and the relative deprivation of other senses. Literacy moved society in logical and linear directions. In political terms, it gave rise to nations over tribes, according to McLuhan.

Cultural imperialism Term applied mainly to U.S. domination of global media production, resulting in (real or threatened) spread of consumerism and other American values and practices to other societies.

Television turned Gutenberg on his head by engaging aural and tactile senses and reordering people's *sense ratios.* This was the term McLuhan applied to the varying degree of reliance people have on their five senses (sight, hearing, taste, touch, and smell) during their communication experience. He saw the shared TV experience of electronic imagery as the opposite of the typographic experience. It had a retribalizing influence on humanity, constructing a *global village* as society began to conform to the nonlinear world of TV viewing.

The reason digital convergence has been so important may be related to the quest for *control, convenience,* and *speed* in daily communication. Interactive media give people more power over their environment, enabling them to browse websites or change channels in an instant. One can quickly exchange roles as a sender and receiver of information by voting in a web poll prompted by a sports question or a talent contest on television. Digital media are flexible and convenient, allowing access at what McLuhan called "electric speed."

INTERNET DETERMINISM

The Internet is more than a new language for the third millenium, inviting millions of global tongues into chatrooms and e-mail discourse each year. It poses a challenge to scholars because it merges forms of both interpersonal and mass media. The Internet is used for one-to-one or one-to-many communications. Web writers hold forth their ideas by merging words with pictures through audio and video channels. How should one approach this new medium, given both its traditional qualities of media and its unique facets of interactivity with layers of visual and audio messages?

Interactivity defines the Internet to the degree that users easily change roles and control the conversation from their computers. Clicking on hyperlinks invites viewers to pursue their curiosity by selecting highlighted terms and phrases, and by visiting virtual pages of new information. This style of web reading embraces intuition over logic, still, whether the web viewer reads or scans content is based as much on the individual's personality as it is on the nature of the medium and the material.

There is one concept that has tried to broadly define the web experience. *Mediamorphosis* is a term coined by Roger Fidler, who, like McLuhan, emphasized media activity over message analysis. The Internet, said Fidler, evolved from older forms of media, replicating some of their dominant features while innovating new ones.[11] Research on the Internet has shown how it appeals to playfulness, reciprocal communication, and mastery over the environment—in other words, speed, convenience, and control. The Internet represents just the latest challenge, testing a number of earlier theories explaining how and why people communicate through electronic media.

■ Journalism Theories

For years, media scholars have tracked reporters and editors to see how they shape politics and society. What professional norms and practices do they follow, and why? Researchers have asked why certain stories seem so important while others are barely noticed. They have wondered what impact news media have on politics and the outcome of elections.

A research perspective summoning the ancient idea of a walled city where a **gatekeeper** stands watch, choosing which visitors should gain entry, has produced a variety of studies involving electronic media. Acting as gatekeepers, news editors, producers, and writers allow certain stories to flow through gates, creating the agenda of entertainment and information. Is this agenda shaped, then, by *personal*

Gatekeeper Individual who has the power to control message flow from source(s) to receiver(s) through the media.

Archives of the History
of American Psychology—
The University of Akron

One beloved scholar who established key concepts for mass communication research was a traditional European professor, from his thick accent to his coffeehouse seminars. Kurt Lewin was born in 1890 to Jewish farmers in Mogilno, East Prussia, in what is today Poland. The deepening chill of anti-Semitism did not discourage him from earning a Ph.D. in psychology at the University of Berlin in 1914.

In the fashion of German professors, Lewin met informally with his students to discuss their ideas. At one Sunday-morning session in a Berlin cafe,[12] he noticed that the waiter had an uncanny ability for recalling everyone's order and collecting the check without writing down any of the information. Lewin saw this as an opportunity to test his "field theory," based on the idea

that personal intention to carry out a task creates a psychological field that is unbalanced until the task is completed. After the waiter collected everyone's payments for the coffee and pastry, the professor asked him to recollect what each student had ordered, but the indignant waiter could not do it. The field of mutually dependent facts had evaporated. Lewin maintained it was because the waiter's task had been completed and the equilibrium of his field restored.

Prior to World War II, Lewin was so eager to leave Germany that he accepted a two-year appointment in home economics at Cornell University in New York. That meant an unmistakable drop in status for this rising star among European scholars. Lewin later moved to the University of Iowa to conduct psychological experiments with the Child Welfare Research Station. There, he coined two key terms, *gatekeeper* and *channel,* to describe the selection and preparation of food menus for the family. His students adapted the concepts to the study of mass communication. Lewin is perhaps best remembered for his observation, "There is nothing so practical as a good theory."

choice, professional training, or *organizational routines?* Additional questions probe how journalists *frame* their reports. Framing deals with editorial decisions about how a problem or issue is defined by its news coverage.

GATEKEEPING RESEARCH

The idea of gatekeeping in journalism first appeared in the work of Lewin's research assistant, David Manning White, at the University of Iowa. In 1950, White interviewed a wire editor for a small-town daily newspaper to uncover the thinking behind story selection. White called his editor "Mr. Gates," borrowing the *gatekeeping* metaphor of his mentor, as he took note of the journalist's compliance with accepted news routines, jotting down his comments about each story's merits.

After White's landmark study, gatekeeping evolved as a vital area of research. Scholars measured how the mix of news items carried on national wires tends to be replicated in daily newspapers. If the Associated Press devoted about half the day's news to politics, the local newspaper's portion would be about the same. In television, gatekeeping studies have considered the visual, dramatic, and other values by which producers choose stories for their lineups. So, where do their criteria come from?

Scholars have identified a hierarchy of gatekeeping influences starting at the microlevel and working up to the macrolevel: *Individual, social, organizational, institutional,* and *ideological* forces all play a part. The news routines of the organization tend to influence story decisions more than do the personal traits of the gatekeeper, including personal ideology. So how does this *gatekeeping* activity influence the audience?

AGENDA-SETTING STUDIES

Powerful effects Perspective of media effects suggesting media hold a strong influence over individuals.

The idea of agenda setting arrived at a time when scholars were questioning the **powerful effects** notion of media. The notion came from one of Paul Lazarsfeld's colleagues, Joseph Klapper, who suggested that individual differences and intervening circumstances moderate media influences.[13] Klapper believed that reading newspapers or listening to the radio does not produce any singular effects, but acts in concert with other "mediating factors and influences,"[14] which would include some prior knowledge of the content topic.

Agenda-setting studies emerged as new evidence that the media are powerful; issues are either brought to prominence by news attention, relegated to lesser importance, or discounted as irrelevant. Two quotes appear as trademarks in this early line of research.[15] Walter Lippmann's title for the first chapter in his book *Public Opinion*, "The World Outside and the Pictures in Our Heads," resonated with scholars. Lippmann believed that "pictures in our heads" are created by journalists who assemble facts to entertain as much as to inform.

The second quote comes from a University of Wisconsin political scientist who felt that the press "may not be successful much of the time in telling people what to think, but it is stunningly successful in telling its readers what to think about." Bernard C. Cohen's subsequent explanation is frequently overlooked: "And it follows from this that the world looks different to different people, depending not only on their personal interests, but also on the map that is drawn for them by the writers, editors, and publishers of the papers they read."[16]

AMERICAN COLUMNIST AND AUTHOR WALTER LIPPMANN.

Limited versus Powerful Effects

Cohen's reference to individual differences and mitigating circumstances seems to support Klapper's conclusion about limited effects. The primary challenge to Klapper's thesis came from Professors Maxwell McCombs and Donald Shaw, who held that the priorities of the news media's agenda are detected and adopted by the audience. McCombs and Shaw drew a yardstick on issue salience by comparing media and audience priorities in the news. Their survey in Chapel Hill, North Carolina, produced path-breaking results.

PRINCIPLE 4

Agenda setting is how news media influence the public's attention to particular issues.

The two researchers counted 15 issues ranking high in news coverage of the 1968 presidential race between Vice President Richard M. Nixon and Senator Hubert Humphrey. They ranked the items on a scale of important to not-so-important. The rank order was calculated according to each story's share of time or space and its prominence in placement—whether it was in the front or back of the newspaper or newscast. Researchers then interviewed undecided voters to determine their opinions about what was important. The result was a good match between news media and audience priorities, lending support to the agenda-setting hypothesis and the case for powerful-media effects.

Agenda-Setting Evolution

The agenda-setting model helped reveal much about *salience* and its relationship to audience priorities. However, it also raised another question: Do some important stories languish in obscurity until dramatic events make them "marketable" and significant to the news agenda? The second level of agenda-setting research analyzes such issues to determine how they contribute to their ranking.

Information Need

Agenda-setting research has taken another direction in assessing how an individual's need for orientation within society influences his or her reliance on news in the first place. A person's motivation for seeking information—based on personal relevance and need for knowledge—may intensify the agenda-setting effect. One reason presidential elections generate so much scholarship is that researchers know many voters lack personal knowledge of the candidates and have a need for orientation about them. The agenda-setting phenomenon then becomes more visible.

FRAMING ANALYSIS

A good deal of study has been devoted to the question of bias in the news, such as the liberal-versus-conservative slanting of stories. The Pew Center discovered in 2004 that as many citizens believe the news media to be biased in terms of its political coverage as those who think it is not. An unprecedented 39 percent of voters polled perceived the news media to be biased, but only 38 percent thought the coverage was impartial. Whether the viewer is watching Fox News or CBS, bias is generally a reflection of personal agreement or disagreement. As a result, scholars have proceeded in a different direction by probing how issues may be changed in the audience's mind by their **framing.**

Todd Gitlin, a UC-Berkeley political scientist, said framing shows "persistent patterns of cognition, interpretation, and presentation, of selection, emphasis and exclusion, by which symbol-handlers routinely organize discourse."[17] Sociologist Erving Goffman defined *frames* as a **schema** of interpretation to help the audience "locate, perceive, identify and label" issues they have heard, read, or watched.[18] Gitlin focused on the "symbol-handlers," whereas Goffman stressed the *audience's* frames for interpreting news.

The greatest power of framing is not in defining problems but in suggesting who or what is to blame for them. That inevitably leads to questions about what must be done to remedy a situation and who must be held accountable. It is possible that framing can even influence public policy. For example, a U.S. Surgeon General's report framed the debate over lung cancer by linking it to smoking. Congress acted in 1971 to ban cigarette commercials from television.

▐ Functions and Dysfunctions

So far, this chapter has explored what the media try to do and what happens after they do it, but what of the audience's intentions? In **uses and gratifications** research, scholars emphasize the idea of an *active* audience. The underlying premise is that the media do *not* have the ultimate control in the flow of information and entertainment.

What are the principal social functions of media use? To alert the audience to the world around them is defined as *surveillance of the environment;* news provides this type of information. To formulate opinion is described as *correlation of events,* and *transmission of the social heritage* is how television, radio, and other media translate cultural values. Charles Wright, a sociologist and student of Harold Lasswell's, added the function of *entertainment,* but also indicated there were *dysfunctions* arising from media use.

USES AND GRATIFICATIONS

Research into uses of radio, television, the Internet, and other media was an outgrowth of the debate over limited versus powerful effects. Professor Bernard Berelson, a member of Paul Lazarsfeld's Bureau of Applied Social Research, wrote in 1959 that the well of thought in media research had gone dry. Elihu Katz, another Lazarsfeld colleague, disagreed and countered that only persuasion theory was spent; the functionalist paradigm was flowing with new ideas. To support his point, Katz recommended one of Berelson's own studies. His research had revealed how readers cope with a newspaper strike and respond to missing their daily newspaper.[19] Katz reinforced his argument by referencing other active-audience findings, such as the importance of adventure heroes to children involved in role playing.

Framing Organizing structure of news defining parameters of an issue, including causes and effects of a problem and the people responsible.

Schema Attitudinal basis for explaining how an individual approaches a communication experience.

Uses and gratifications Perspective based on an active audience seeking to satisfy personal needs through media rewards; involves the idea of functions and dysfunctions of media use.

Would the emergence of *uses and gratifications* then be classified as a new theory of communication? Katz did not think so. He called it a *perspective* for discovering the social and psychological *needs* that generate *media exposure* that result in *gratifications* or other *consequences*.

PRINCIPLE 5

Personal needs motivate media exposure and limit media effects.

Gratifications Obtained

Studies in *uses and gratifications* have produced some interesting results. One study asked how a person's mood might affect his or her choice of TV programs. It found that "stressed subjects watched nearly six times as much relaxing television as did bored subjects [while] bored subjects watched nearly twice as much exciting fare as did stressed subjects."[20] British scholars wondered why voters tuned in to political broadcasts on the eve of a general election. Most of the viewers simply wanted to know more about the candidates and their platforms, but one-third wished to identify with a political party.[21]

Uses researchers typically rely on self-reports prompted by asking audiences to reveal personal motives and rewards for choosing what they watch, read, and browse. The difficulty that communication scholars have encountered is not only getting the audience to identify personal motives but also determining how successful they were in obtaining gratifications.

Unintended Consequences

One theme of *uses* research is that people choose to scratch a real or imagined "media *itch*" in ways that can produce unintended consequences. One dysfunction is called *media dependency*, which explores how much people get hooked on their favorite media fare: talk shows, soap operas, violence, and even pornography. A higher level of exposure, preoccupation, and disregard for harmful effects are the markers of media dependency.

Researchers also have turned their attention to the dependency effects of video games. A British authority, Dr. Mark Griffiths, observed in 2000 how children are drawn into video games at around the age of 7, and by their early teens, at least 7 percent will be addicts, playing the games for at least 30 hours a week. In 1982, U.S. Surgeon General C. Everett Koop noted the first wave of video game addicts in the United States, noting that children "are into the games body and soul—everything is zapping the enemy."[22]

One feature of dependency theory is called *parasocial interaction*. This phenomenon was introduced in a 1956 edition of the journal *Psychiatry*. That study found audience members developing imagined relationships with real and fictional characters on TV. Television viewers fancied strong ties with glamorous personas. They imagined understanding the characters and their motives better than real-life relationships.[23] A motion picture, *Nurse Betty*, played on an extreme case of parasocial interaction involving a soap-opera devotee. A love-struck waitress fantasized about becoming a nurse for a handsome doctor in a soap opera, and drove to Hollywood to fulfill her fantasy. Media dependency shares some theoretical territory with ideas about the *cultivating* effects of heavy television

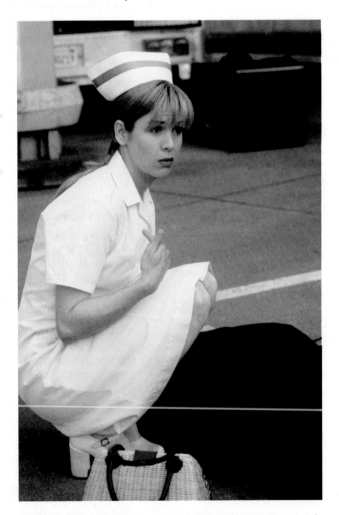

RENEE ZELLWEGER AS "NURSE BETTY." Parasocial interaction studies describe how people invest personal emotions in imagined relationships with media characters. *Nurse Betty* dramatized this concept, showing a waitress who no longer distinguished her media-inspired fantasy from reality.

viewing. (This term describes how media can "cultivate" a view of reality that the audience will accept as accurate if that is its dominant source of information.)

MEDIA VIOLENCE

Violence is familiar to both fiction and nonfiction forms of entertainment. Scholars have defined *television violence* as scenes of "intentional physical harm, compelling force against one's will, or the infliction of pain." In metropolitan areas across the nation, anchors on television narrate video of murders and other violent crimes. Reality crime shows such as *Cops* and *America's Most Wanted* play back videotaped scenes of aggressive acts, punctuated by pleas from law-enforcement officers to bring fugitives to justice. It is easy to understand how violence has become an attractive scenario for screenwriters spinning spectacular narratives. Television producers claim that the pervasiveness is due to its popularity. But "We-just-give-'em-what-they-want" rationalizations defending the use of violence to meet audience expectations do not always square with ratings success. The economic argument is easier to accept, since violence in television costs less to produce.

Researchers have pursued questions about media violence since the 1920s, when the Payne Study and Experimental Fund sponsored a report on the impact violent movies had on children. Media violence has come to be associated with a number of contradictory theories predicting that it can relieve tension, stimulate acts of aggression, desensitize viewers, or engender an exaggerated fear of reality.

Cultivation or Catharsis

The effects of media violence are *emotional*, *behavioral*, and *cognitive*, theorists have concluded. The most familiar cognitive theory dealing with media violence is the *mean-and-scary-world hypothesis* advanced by **cultivation** research. It suggests that heavy exposure to television violence cultivates an exaggerated perception of danger in the real world.

The primary emotional theory is based on the *catharsis* perspective, which holds that therapeutic effects may be derived from watching actors engage in violent and hostile activities on screen—that this somehow purges the audience of hostility toward others. This theory has little support in experiments. Similarly, the *excitation transfer theory* proposes that media excitement may be imparted by neural transmission—in other words, that TV can arouse excitement in viewers' brains. The responses need not be hostile or violent but can be warm and friendly, depending on the motivation of the viewer.

Imitative Effects

In the 1960s, social psychologists conducted experiments by watching children after they had viewed violence on television. They saw children beat up a large bobo doll (an inflatable, egg-shaped balloon creature with a weight in the bottom that makes it bob back up when you knock it down and that rocks back and forth after being punched). *Imitative* effects or modeling theory was used to explain the children's aggression. *Modeling theory* is based on four prerequisites: first the viewer remembers the violence *(retention)*. Then, if given an opportunity, and with sufficient cause, the viewer becomes *motivated*. Finally, he or she will *act out* based on his or her imitation of TV violence.

Disinhibition or Desensitization

Another theory is defined as *disinhibition*, which predicts that people feel freer to express aggression after viewing harmful acts on television. It suggests that audiences relax their control by rationalizing their tendencies toward aggression using media portrayals as a subtext.

Cultivation Theory suggesting heavy television viewing tends to distort people's perceptions of reality.

Desensitization supposes that a lack of empathy or concern results from viewing violence. The viewer can become so callous to acts of brutality on TV that she or he shrugs and turns away when confronted with a victimizing act in reality.

Violence Assessment

A 2003 study conducted by the Kaiser Family Foundation found that nearly half of U.S. parents (47 percent) with children between the ages of 4 and 6 reported that imitations of aggressive behavior had prompted similar behavior in their children by watching TV. It should be noted that children are even more likely to imitate positive behaviors viewed on television—87 percent.[24]

Two professors from Iowa State University conducted a thorough review of video game research, and reported in 2001 that children who play violent video games will become more aggressive in their real-life play. In 2003, Craig Anderson and his associates at Iowa State reported that violent music lyrics increased violent thoughts and hostile feelings among college students. This study brought violent music under the same umbrella of research that has documented the strong links between violent film viewing, television watching, and video games with aggressive and sometimes hostile behaviors. There is ample support for a positive but weak link between exposure to television violence and aggressive behavior.[25]

▌ Social Psychological Approaches

Communication has been described as the fundamental social process. Because interpersonal factors affect media processes (as the two-step flow established), social psychologists look at how people behave in particular groups. Theorists have studied how some types of people suffer in silence on certain subjects if they have been convinced that their opinions will isolate them. Other theorists have traced the spread or diffusion of new technologies to different social groups. Varying levels of knowledge have been analyzed according to age, education, income level, gender, and ethnicity.

SPIRAL OF SILENCE

Viewers may experience *feelings* of aggression or fear through repeated exposure to violence, but there is a theory suggesting that television, radio, and newspapers also may inhibit verbal expressions. Public opinion is reflected by the media's lens, but can these snapshots stifle *unpopular* views? Elisabeth Noelle-Neumann, a German professor from the University of Mainz, believed that is the case. She defined *public opinion* as a constraining force that allows people to express only acceptable feelings if they wish to avoid isolating themselves.

Basic Tenets

Noelle-Neumann's *spiral of silence* had its origin in the conformist cults of pre–World War II Germany. She began as a publicist writing for *Das Reich*, a German magazine supporting Hitler's reign. The basic tenets of her theory are fairly easy to follow: Even though people harbor private opinions, they generally want to share them in public—that is, until their "quasi-statistical organ or sense" zips their lips by telling them they're about to let loose an unpopular idea. The desire to be accepted by the dominant group as well as the fear of being shunned are incentives to this form of self-censorship.

The spiral-of-silence theory holds that electronic media play an important role in self-regulation. Their omnipresence, or **ubiquity,** and narrow range of opinion, or **consonance,** actually discourage free expression. Media consumers gather essentially the same type of information and opinions on the same subjects regardless of where they turn—newspapers, magazines, television, or the Web. The *cumulative effect* of this consonant exposure is to stifle free expression, according to the theory.

Noelle-Neumann agrees that some people are exceptions to the rule—able to withstand social pressure and go against the grain—people she describes as "hard cores." There is, in addition, the possibility that unpopular opinions may reemerge as acceptable, depending on their revised treatment in the media. Cues given by characters, actors, and story lines about cultural values—including moral codes, social roles, and other customs—encourage viewers to adopt similar attitudes.

Research Evidence

Testing the *spiral of silence* has not been easy, given the need to first identify ideas quashed by the media and then relate them to people's reports of being stifled. Historical evidence offers some clues, though. For example, during the 1930s, it was acceptable to defend U.S. proposals for socialism as a remedy to the ills of the Depression. During the 1950s, however, this defense became tantamount to treason. Did media play a role in stifling socialists from speaking out? Critics doubt it, and accuse this theory of exaggerating media's power over dissent. Yet, the power of social sanctions over unpopular expressions is obvious. The upshot of this theory is that only expressed opinions can be translated into public policy. Otherwise, they are invisible, and, as far as the public agenda is concerned, nonexistent.

DIFFUSION OF INNOVATIONS

For years, historians and anthropologists have charted the spread of new ideas through literary and cultural exchanges wrought by war and commerce. At the dawn of the twentieth century, French lawyer and sociologist Gabriel Tarde observed how his society was ordered by paths of personal influence linked to education and rank in society.

In the study of communication, diffusion theory follows the flow of ideas through media and social institutions. Its special emphasis is on the adoption of new ideas and technologies. In Iowa, agricultural scientists tested diffusion theory to find out how farmers chose to adopt hybrid seed corn.[26] The choice was attributed to both media and social relations. Diffusion's initial premise is that audiences find news of innovation stimulating and, given their heightened need for orientation, will find out more before deciding whether to adopt something new.

As in persuasion theory, diffusion follows several steps: *awareness, then attitude formation,* and *decision making.* Researchers gauge the length of time it takes for audience members to move through those phases in the process, and investigate the rationale that they use. The decision to adopt is based on five factors: the innovation's relative *advantage,* its *compatibility* with existing conditions, its *simplicity,* its *accessibility* to experimentation, and the *benefits* it reveals to others.

Adopter Categories

Diffusion theorists divide people according to their rates of adoption. First, there are the *innovators,* adventurous and savvy people who look for new ideas to try them out. They take seriously the invitation to be "the first ones on their block." Coming in a close second are the *early adopters.* They are adventurous but more cautious. The curve then thickens to form two groups: *early majority* and *late majority.* They will try an innovation only after it becomes clear that it has been tested and accepted. Finally come the *laggards,* who adopt an innovation only after tried-and-true methods become obsolete.

Ubiquity Key element in spiral of silence theory suggesting that media are available to everyone almost all of the time.

Consonance Presentation of a consistent image or message, usually concerning public issues or current events.

The diffusion of innovations in electronic media has seen transitions from film to videotape, typewriters to computers, pay phones to cell phones, broadcast to cable and satellite, and the list goes on. Interest in the theory grew from Everett Rogers's *Diffusion of Innovations,* published in 1962. Columbia University's Bureau of Applied Social Research gave its seal of approval to diffusion theory in 1966. That's when Pfizer Pharmaceuticals needed a study to find out how its advertising campaign for tetracycline, an antibiotic for skin disorders, was faring with physicians. The Columbia researchers agreed to do the study by applying diffusion theory. They established a link between doctors who adopted tetracycline and their level of social activism. Physicians who were involved in professional networking adopted the medicine sooner than socially isolated ones did.[27]

Diffusion theory suggests that the media's influence will be strongest during the initial awareness phase, after which social contacts become more relevant. An individual decides *not* to innovate if he or she concludes that the innovation's costs exceed its benefits, or that it poses unfavorable consequences.

Media Diffusion

Diffusion research in media has revealed the level at which a crucial point of saturation is achieved among audience members. Studies of e-mail and fax diffusion focused on the take-off point when a critical mass of knowledge was formed and created momentum for those two innovations to become self-sustaining.[28] In 1960, researchers began charting the diffusion of news stories. News of public affairs did not always reach a critical mass, but filtered rapidly to citizens in leadership positions and bypassing other strata of society. This made researchers wonder if education and income combined to create a gap between those in the know and those left outside the loop.

KNOWLEDGE-GAP HYPOTHESIS

Communication is essential to acquiring the basic needs of shelter, clothing, food, and most human essentials. Social scientists find that information empowers people by giving them the means of satisfying those needs. Those with personal influence enjoy a safer and easier existence than those without. British Prime Minister Benjamin Disreali once observed, "As a general rule the most successful man in life is the man who has the best information."

Social divisions founded on socioeconomic status (SES) have distinguished the line between the information-rich and information-poor. This is a way of describing people who have access to the facts they need for making good decisions versus those who do not. Studies have found that people are denied the blessings of prosperity simply because they do not know where to turn for that information.[29] These blessings include day care, welfare assistance, medical aid, and other services. Why would anyone accept information poverty? University of Minnesota researchers discovered how levels of literacy and access to channels of communication made the difference. In 1970, the researchers published their study of the socioeconomic status of population groups, linking status deficits to knowledge deficits. The knowledge gap is especially evident in news of general interest, such as public affairs. The gap is less noticeable in areas of personal interest, such as hobbies or sports.

Status and Participation

The **knowledge-gap hypothesis** predicts that people of higher socioeconomic status will have more opportunity for education and greater access to sources of quality information. They also tend to associate with people of similar backgrounds who offer useful tips for solving problems. To further widen the knowledge gap, media tend to target upscale audiences for advertising based on buying power. Targeting neglects impoverished viewers, since they are regarded as "wasted circulation."

Knowledge-gap hypothesis Prediction that people with more education and higher incomes tend to acquire better information on public issues and other necessities than those of lower socioeconomic status.

A representative democracy depends on an informed electorate to choose candidates and participate in civic affairs. Swedish scholars found there is not only a knowledge gap for lower SES citizens but an "influence gap" as well. In Sweden, well-educated citizens gain access to policymakers through social channels denied to those of lower social status.[30]

Digital Divide

Knowledge-gap theory has been applied to the use of the Internet and researchers have found that some of the same disparities have worked their way into the information age. The differences are most apparent across income levels and ethnic lines where the "have-nots" are less likely to have access to computers and on-line resources. The Pew Internet & American Life Project reported in 2003 that although the on-line population is growing, "clear demographic groups remain." Internet literacy continues to favor the young, the wealthy, the well educated, and the white. Digital-divide researchers have been criticized for appealing to notions about racial and class divisions, and for failing to regard variables such as personal and professional interests. The Pew study took into account those factors, and found that beyond the cost of a computer and Internet access, concerns about credit card theft, pornography, and fraud kept the digital divide from closing.[31]

◤ Research Methods

The scientific method of scholarship can unlock the mysteries of attitudes and behaviors by collecting evidence and processing it in a systematic manner. A variety of data-gathering techniques have gained acceptance and are broadly classified as **quantitative** and **qualitative research**. The obvious difference between them is that quantitative research draws on statistical tools, whereas qualitative scholarship collects facts and opinions from other sources before extrapolating to achieve greater understanding. Both approaches are founded on problem-solving techniques and logical reasoning.

It's tempting to distinguish quantitative research as *inductive* rather than *deductive*, because it first gathers data in an open-ended quest to solve a problem or answer a question. Deductive reasoning, on the other hand, is often linked with qualitative research, since it may begin with a stated premise and then seek evidence to support it. Such distinctions are not entirely accurate; both types of reasoning play roles in quantitative and qualitative research.

Three principal methods are involved in quantitative research of electronic media: surveys, experiments, and content analysis. Whenever tables or charts are furnished and statistical analysis is given, the methodology must be explained and defended. Specifically, an accounting must be given for the research method's **validity** and **reliability**. In other words, did the research method—like a good set of scales—give the correct weight (validity), and will it be able to give it again and again over time (reliability)? Just how did the observers extract the specific data desired from their sample, and will other researchers come to the same conclusion if they use those methods?

SURVEY RESEARCH

In the quantitative approach, one of the most familiar methods is **survey research**, which draws data from a **sample** of respondents. A day rarely goes by when the news media do not present a poll claiming to show how the American people feel about a particular issue or event. For example, when President George W. Bush first used the phrase describing an "evil axis" of countries in his 2002 State of the Union

Qualitative research Methods including interviews, study of documents, and observation; developed in social sciences to study social-cultural phenomena.

Quantitative research Methods including surveys, experiments, and numerical methods; developed in natural sciences to study natural phenomena.

Validity Quality-control check researchers use to ensure that methods and means are correctly measuring the phenomenon intended for observation.

Reliability Quality control of research observation indicating the method used will consistently give over time accurate information.

Survey research Sampling opinions from part of a population by questionnaires, telephone calls, interviews, or electronic polls.

Sample Portion of total population based on their representative or unique qualities chosen to generalize about the larger group.

Address, *Newsweek* surveyed the American people about his imagery. The polling data showed 64 percent of the American people felt the president's speech was a warning that the United States was watching Iraq, Iran, and North Korea closely, while 10 percent said it was just tough talk.[32] To document its survey snapshot of American opinion, the magazine cited the source of its polling data, the firm, its methodology, sample size, and margin of error in this fashion: Princeton Survey Research Associates interviewed 1,008 adults by telephone between January 31 and February 1, 2002. The degree to which the statistics varied from the population at large was plus or minus 3 percent.

Surveys permit comparisons based on the demographic elements discussed in previous chapters on advertising and the audience, such as age, education, and income level. Surveys can be administered by telephone, mail, in person, or over the Internet, but researchers need to follow scientific safeguards before generalizing from the data. Obviously, interactive polls conducted via televised questions, such as those on the screen during the Super Bowl, do not draw a representative sample of the audience.

In selecting a sample to be surveyed, researchers work to assure that it reflects the makeup of the whole population of interest. They ask either open-ended or closed questions, allowing respondents to originate personal answers or give multiple-choice responses. In either case, percentages are compared to determine which answers are most favored, whether the question involves a candidate, an issue, or a course of action.

The main advantage of sample surveys is that it's possible to generalize to a larger group by interviewing representative members. This is helpful but poses a problem, given the tendency for certain types of people to be near their phones and available to answer questions while others are more elusive (occupied with children and careers). The remedy often is to give additional weight to data collected from hard-to-interview types. What surveys do not reveal are cause-and-effect relationships. Social scientists have designed another method to deal with the problem of causality.

EXPERIMENTAL DESIGNS

Experimental design research originated in agriculture as a means for determining what factors—such as seeds, sun, rain, and fertilizer—would make crops grow. It became a valuable tool in the field of social science for determining how media exposure may influence different people in a variety of ways.

In such experiments, **key variables** are defined as *independent* or *dependent*, which means they contribute either to the cause (independent) or to the effect (dependent) of a particular event. For example, studies in television violence expose audiences to the *independent variable* of knifings and shootings on the TV screen, to see how that exposure might produce a reaction in the *dependent variable* of attitudes or acts of aggression.

Experiments involve two or more groups, one of which serves as a *control* for reasons of comparison. The control group is *not* exposed to the independent variable, so that researchers may compare the group with those subjects who received the stimulus. The groups that are exposed to the independent variable are called *treatment groups,* and there are often multiple levels of the treatment condition.

Experiments can be conducted either in the field or the laboratory. The laboratory gives the researcher more control so that he or she can prevent subjects from being exposed to what are considered *confounding* or **intervening variables** that can interfere with the outcome. The field experiment, however, eliminates the artificiality of placing subjects in front of a TV monitor in a classroom with strangers. Instead, it allows the researcher to measure media effects in a natural environment.

Experimental design Controlled means for varying a condition or treatment on experimental subjects in order to determine the effects such exposure has on their status.

Key variables Measurable influences responsible for stimulating or causing particular effects as part of a phenomenon.

Intervening variables Influences that serve to mitigate and interfere with what is believed to be a cause-effect relationship between variables.

protalk
Bob Papper
POLLSTER OF THE NEWSROOM

There is something about human nature that seeks to gain a better understanding of itself—even at the professional level. Broadcasters need to know how they fare among their peers in terms of the issues confronting their business. When broadcast journalists want a bird's-eye view of the newsroom, they typically turn to surveys conducted for the Radio-Television News Directors Association (RTNDA) by Bob Papper of Ball State University.

Papper leads a research team that draws up annual questionnaires compiling statistics on such varied subjects as newsroom salaries, diversity profiles in the work force, and Web use. One statistical chart that is popular among news professionals is the annual salary scale for broadcast journalists. It often appears on a newsroom's bulletin board, posted either by management to show how well everyone is doing or by employees to show how far they're lagging behind the curve.

Papper's review of minority groups in the newsroom has charted progress around the country. In television, he discovered an increase of more than 3 percent from 2003 to 2004, with the growth evenly split between Hispanic Americans and African Americans. The percentage of minority TV news directors nearly doubled, from 6.6 percent to 12.5 percent during that time period. Years ago, women were less likely to be hired as

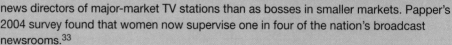

"Researchers who fudge their findings hurt the credibility of survey research."

news directors of major-market TV stations than as bosses in smaller markets. Papper's 2004 survey found that women now supervise one in four of the nation's broadcast newsrooms.[33]

The RTNDA researcher found on the Internet news directors struggling to turn a profit from their station websites. Even though more than 91 percent of the TV stations produced a site, fewer than 10 percent were turning a profit. Successful sites mostly gained an edge by breaking the news first, before the stations' TV anchors had a chance to report it. Local weather and local news ranked high on the list of what web users want, according to his research.

Like any researcher, Papper must be careful about what data he uses to make his projections, and says he will discard information if it is flawed. Data that are questionable in terms of validity or reliability will not get into his reports, he says. Researchers who "fudge" their findings, Papper adds, hurt the credibility of survey research.

CONTENT ANALYSIS

Just as a survey researcher or experimenter draws conclusions from one set of data that represents a larger group, content analysts infer findings from media materials—video- and audio-tapes, scripts of commercials and so on—by organizing them into meaningful blocks of data. Each piece of tape or paragraph of copy is coded according to definitions described by the category choices. For example, a researcher might be interested in the way women are portrayed in commercials as wives, mothers, professionals, or combinations of those roles. The coder will classify by operational definition each portrayal per spot and count it. Rather than sending out questionnaires or exposing subjects to stimuli, the researcher draws inferences from samples of those ads or other programming.

Content analysis is a systematic technique for examining elements of a message to draw some inferences either about the message's nature or its source. Before counts are taken and comparisons are made through statistical analysis, researchers check the *reliability* of the coding method. For example, if one coder found that an actress in a commercial was portraying a professional woman, would a second coder share that opinion? The level to which they agree is called *intercoder reliability*, and it is the standard required for scientific content analysis.

Content analysis has become popular as a means for studying television and other electronic media content—whether the question concerns the prevalence of gender and ethnic stereotypes in TV shows or the political opinions of talk radio. So popular has content analysis become as a tool for researching media messages that an entire edition of the *Journal of Broadcasting and Electronic Media* was devoted to articles using this research method. Scholars measured the extent to which ethnic diversity had become a part of network TV programming, as well as the degree of emphasis given sensational content.[34]

TV VIOLENCE. One of the most extensively researched questions in electronic media concerns how depictions of violence have affected both children and adults. Social psychologists believe intense viewing of media violence produces aggressive and imitative behavior among children, and cultivates distorted ideas about the threat of violence among adults.

HISTORICAL AND QUALITATIVE METHODS

Whenever a researcher employs qualitative techniques, the methodology is justified in much the same way as in quantitative research. That means the scholar must make a good-faith showing that the evidence addresses a larger question, and that the study's generalities are valid because they include a complete and representative set of data. It also means that the qualitative technique for reasoning from the evidence can be proved valid and reliable.

Qualitative studies include historical analysis, a kind of cultural research called *ethnography*, and case studies. Historical research is often a mix of words and numbers, but its primary mission is to extrapolate ideas by collecting enough incidents from the historical record to distinguish an emerging pattern. The historical scholar must consider alternative explanations, especially when attempting to draw cause-and-effect relationships. Historical research is most carefully judged by its overall literature review, and whether the scholar used comprehensive references in assembling primary and secondary sources to answer the research questions. The

Content analysis Research method for objectively and systematically studying elements of a message in order to draw inferences about the nature of the content and/or its source(s).

protalk
Norm Hecht
NORMAN HECHT RESEARCH

Television news is usually the primary budget item for local TV stations, so ratings success is critical to their profits and financial stability. That means that the news competition within each market is a key concern of station managers, and it's the reason they turn to consultants to find out how to beat their rivals for ratings.

Norman Hecht says his research team evaluates methods to assess their effectiveness for client stations and to keep pace with the dynamic media landscape. When Hecht entered the field of news measurement about 10 years ago, TV stations and research firms employed focus groups as one of their leading measures for gathering data. He has since shown clients newer methods to inform and to guide their winning strategies.

The challenges facing focus-group research include drawing a representative picture of the audience. This is complicated by any failure to achieve random selection and by a group's vulnerability to dominant members. Hecht believes other methods produce more valid and usable results. He cites, for example, the "recruit-to-view" method, in which samples of about 10 to 15 viewers are recruited each night over a given period of time—usually one or two weeks. The subjects are screened for their viewing behavior, and then are asked to watch one of the client's newscasts on a specific evening. Fans

> **"Qualitative data can provide very actionable insight into promotion and programming strategies."**

of both competitive stations and the client station are recruited. That way, a variety of newscasts are included in the sample, and the findings are more representative of the larger scheme of things. Viewers participate in the privacy of their own homes—the environment in which they would normally watch TV. This kind of "qualitative data can provide very actionable insight into promotion and programming strategies," enabling the station to serve its audience and attract new viewers, says Hecht.

Most leading stations in the top 50 television markets hire research analysts to work in their sales departments. These researchers analyze and compile ratings data as well as developing sales materials based on Nielsen and other data sources. There are hundreds, if not thousands, of attractive jobs involving primary and secondary media research, according to Hecht.

MEDIA RESEARCH AND WRITING

Job-seekers can find research positions in both the private and public sectors. There is work to be done—or for—advertising agencies, cable systems, cable networks and services, radio stations and networks, television stations and networks, station sales representative firms, and the syndicated data companies themselves, such as Arbitron and Nielsen. The positions involve both primary and secondary media research. Primary researchers harvest their own data from which to draw conclusions, whereas secondary analysts draw on statistics compiled by others, such as rating companies.

The top 50 markets usually have positions for **research directors** at local TV stations. These lucrative positions go to people who know how to compile statistics, write presentations (sometimes with the help of staff), and present their findings. In some cases, research directors "bid out" data collection to other companies, which conduct the field interviews.

Research directors need a thorough understanding of Nielsen data, as well as the statistics other firms supply. For example, they may have to interpret reports and surveys from such services as TVScan or Scarborough, which offer stations specialized data on their performance in sales, advertising, programming, web traffic, and the audience. Researchers usually have at least a master's degree in communication or a marketing degree with relevant professional experience. Research directors demonstrate their broader skills in management and budgeting by supervising staffs assigned to help them.

Researchers track a station's ratings success, make audience estimates for future program schedules, and prepare station position papers involving rating patterns for advertisers and for show producers, including newswriters and reporters. They provide data to help stations promote their news and community work. Researchers advise sales managers and account executives on ways to increase revenue.

In addition, researchers adopt a teaching role by training station employees (and their own assistants) in research methods.

Researchers must be able to explain in print and personal presentations what a station or network audience is watching, and why. They must write effectively and present their findings in meaningful ways. They are also asked to monitor the competition and its sponsors; advise their employers on copyright and contract issues; formulate market strategies; and analyze the growth potential of the station, cable system, or network. Researchers for media polling firms are hired for survey design, methodology, and report writing.

In public broadcasting, consultants often enhance fund raising and membership drives as well as public relations and internal communications.

Writers/Researchers work on network or documentary projects, on topics as diverse as global business and nature/science series. Production studios and network news organizations use researchers on creative projects and programming scripts.

Academic Positions

Colleges hire researchers as instructors and professors at the assistant, associate, and full-professor ranks. Virtually all such faculty jobs require research, publication, and service to the college and community. Professors teach theory and research in the classroom and work in collaboration with other scholars. A school that advertises for a professor to teach media theory and research gives special attention to a candidate's record of publications and scholarship.

Assistant professors in broadcasting and electronic media often have experience in various aspects of radio and television. They establish research track records by investigating the theories discussed in this chapter and by applying new theories to their fields of expertise. Professors at all ranks serve as reviewers and editors on publication boards where their knowledge of theory and research is essential to the task of screening journal articles. In addition to writing for journals and books in their fields, professors may consult professional organizations.

literature review is critical because it establishes the depth of information that has already been gathered relevant to the question at hand.

Case Studies

Qualitative research may involve participant observations or case studies. Daniel Berkowitz at the University of Iowa, for example, decided to undertake a case study of gatekeeping by spending more than 200 hours in a local TV newsroom. He spent the time observing work and conducting interviews to discover how TV news producers decide which story ideas to pursue and which ones to discard. He discovered that the criteria of importance and visual impact, as well as the producer's personal instincts, were brought to bear on the story selection process. Berkowitz concluded that the driving purpose of TV news gatekeepers was to achieve a balanced mix of stories, similar to a healthful meal.[35]

Summary

Communication research relies on sound theoretical explanations of media activities, including social effects—how media teach values, influence views, and shape personal behavior. The drive to build practical theories to explain the nature of communication involves academic scholars and professional researchers who have helped to decode and explain the way media change lives. This is accomplished through approaching the process from a structural perspective before the specific media messages are considered.

Whether conducted by independent firms or academic institutions, media research is fundamental to decision making by programmers, advertisers, producers, and writers. It may be used "after the fact" to justify personal decisions or, more appropriately, as a building tool for strategies in media planning. In any event, it is vital in shaping global channels for the future. Successful channels of radio, television, and Internet content secure the audience's trust by building on perceptions of expertise and honesty.

From the early days of the bullet theory to recent research in framing, the nature of media study has served to address key questions. Theories of persuasion asked what types of emotional and logical appeals motivate audiences. Theories of cognition have asked how people learn and apply information. Social theorists have asked how group norms and values influence media content and create multi-step flows of information. Critical theories and approaches veer from the empirical tradition of data analysis, but they still address key questions: How is media power consolidated in wealthy conglomerates? How do media depict members of the minority community? What happens when western culture collides with foreign television?

The process and effects of research have produced theories on how media cultivate aggressive behavior and fearful perceptions of the world. Research has shown how news agendas and frames are established, and who controls the gates through which news and programming pass. Digital convergence has brought about a need for new theories and methods to explain the principles of electronic media. In the future, scholars will have an opportunity to see how smaller segments of society engage in digital media. While more channels and choices vie for scattered and shrinking audiences, the traditional debate of limited-versus-powerful effects has given way to a third perspective. It defines audiences in all types of communities, "de-massified" but still connected.

Food for Thought

1. Propaganda studies focus on persuading people during times of war. If you were to conduct a study of U.S. propaganda in the Middle East, what research questions would you ask?

2. The two-step flow theory suggests that people get their news from other people's accounts of what they heard or saw in the news media. When was the last time you heard news from a friend or family member before checking it out for yourself? How did your friend or relative influence your attitude toward the news?

3. News reporters are sometimes charged with exaggerating, or "hyping," the facts of their stories. If that's the case, do you think it harms the credibility of the reporter, the news story, the station or network, or all of the above? Explain your answers.

4. Media researchers deal with problems involving stereotyping in the media. Do you consider the stereotyping of women and minorities to be a problem, and if so, what should be done about it?

5. Gatekeeping, agenda setting, and framing research have examined news media's decisions about what stories to cover, how much importance to attach to them, and how the stories define the issues. Can you think of a case where the news media covered a story or framed the issue in such a way that did not seem fair to you? Explain.

6. The diffusion-of-innovation model suggests that social networks play a greater role in making an innovation viable than media do. Are there times when media would be more powerful than friends or family in persuading you to adopt an innovation? Explain.

7. The knowledge gap hypothesis predicts that people with more education and higher socio-economic status will learn more about political affairs than people at the lower end of the spectrum. Do you agree with that assumption? Why or why not?

Public Broadcasting

Chapter 14

Imagine the concept of public broadcasting as an umbrella, under which the first two items below certainly fit:

▶ In Chicago, WBFZ-FM *public radio*—which claims more than half a million listeners—holds its fall fund-raising "pledge drive." Some 15,000 pledges come in. They add up to a voluntary commitment of $1.6 million from station fans.

▶ In New York, the country's largest *public TV* station, WNET, not only produces local shows but also cranks out *Nature, Great Performances, American Masters, EGG the Arts Show,* and other programs that reach national audiences regularly.

Those first two cases clearly involve the two basic branches of public broadcasting as it is generally defined. But what if we open our metaphorical umbrella wider so that it covers the following circumstances as well?

▶ In Hawthorne, Nevada, WKI-TV was born in 1997—in a single-wide trailer. Elevation: 4,330 feet. Power: 10 watts (this is *low power TV*). Staff: Bob and Ginny Becker. Programming:

nostalgia, religion. Purpose: Do good, stay busy, break even.[1]

▶ In a Washington, DC, suburb, at *public access* radio channel WEBR, *The Shock Jock Hair Rock Show* on Friday night is followed immediately by the serious local talk program *Focal Point.* Moodwise, it's a sharp turn, but WEBR shows are created by volunteers—not, they say, "by corporate radio robots."

▶ In Berkeley, California, KPFA, the first *community radio* station in the country—part of what might be called *alternative* radio—squabbles with its foundation bosses; at one point they lock out the workers and spend half a million dollars keeping them out.[2]

▶ In Bridgeport, Connecticut, students at Fairfield University have rocked the *college broadcasting* scene by not only airing music but also by "streaming" it through the Internet and around the world; one DJ won big awards in California for his long-distance western-swing programming.

All of the six sets of circumstances summarized here are real. And indeed, as you may have concluded by now, the old term *public broadcasting* simply may not be a large enough umbrella to cover them.

However, please bear with us. Recent decades have reinforced and expanded an old and powerful idea: Society needs many outlets for the impulse to reach people through electronic media—not to woo "audiences" as sales targets but to pursue other visions. Widespread demand for outreach through newer, inexpensive channels seems likely to draw many people to innovative tributaries of "public" broadcasting rather than to its more tightly controlled mainstream.

Central to this is the founding notion that public broadcasting first of all was to be *noncommercial*. Profit was taken off the table, removing the prospect of corporate ownership. In its place, the government would supply some tax dollars, nonprofit foundations would be asked for support, for-profit companies could buy "underwriting" messages, and the

PRINCIPLE 1

Public broadcasting's funding basis—noncommercial public support—works against its survival in a marketplace economy.

public itself would have to pick up the rest of the tab directly. This added up to a shaky footing for a broadcasting system that would need to offer high-quality programs before anybody would be willing to donate to it.

On the positive side, this organizing basis did set public broadcasting apart from what a famous regulator called the "vast wasteland" that commercial broadcasters had brought to American culture by the 1960s. However, it also placed the new media realm in a position of supplication before its prospective supporters. And as more recent years brought new technologies and social trends, members of the public have not just been handed a catering bill; increasingly, they have found ways to crash the party.

Consequently, in this chapter, we'll explore first the evolving nature and condition of mainstream public radio and TV and then also other areas of public noncommercial broadcasting—some large and influential, some small and intimate. All of these media contribute to local and national conversations today, and are sending interesting signals about the future.

◤ Traditional Roles

Public television has long been airing artistic, historical, and journalistic documentaries; offbeat comedies and dramas produced in the United States and abroad; service programs on cooking, home building, antiques, and other topics of practical viewing; and public-affairs shows from the local to the national level.

Public radio has kept up a hectic pace of its own, offering quiz-and-comedy programs; newsmaker interviews; long and thorough daily newscasts; presentations by playwrights, essayists, and critics; live coverage of momentous events; and music to feed eclectic appetites.

These sorts of programming don't appeal to everyone, no matter how fervently they are intended to do so. Studies have shown that public broadcasting fans tend to be better educated, more affluent, and much more likely to own stock or take foreign vacations than most people are. Many of them build their broadcast diets around what public stations offer. Where a commercial station or network's first role is to earn money for its owners and investors, a public station can move more directly toward the *social* roles it hopes to play.

PUBLIC SERVICE

To the dismay of public broadcasters, most media consumers primarily recall only the biggest national names that appear on their screens: ABC, Fox, HBO, ESPN, and other commercial brands. If a person remembers PBS, that may be because he or she spent long hours before *Sesame Street* as a child—or parent—rather than because the person is now glued to *NewsHour with Jim Lehrer* every evening.

Programs such as *NewsHour* and National Public Radio's *All Things Considered* occur almost exclusively in public broadcasting. They provide information at a pace and depth that seems likely to create understanding of each story or issue—or at least likelier than the wham-bam tempo of commercial newscasts. This constitutes public service to viewers and listeners who wish also to be informed citizens. *Sesame Street,* too, performs a service found almost nowhere else on television today; parents and some researchers credit the program with developing skills and a wider worldview in young children as a daily companion and playmate.

Commercial broadcasters sometimes complain that. their profit-making status makes it difficult to air much besides entertainment. They say public broadcasting, which leans on payments from "underwriters" for low-key institutional messages rather than high-energy advertising spots, is merely disguising commercialism, not rejecting it On the other hand, for-profit broadcasters make few claims of their own to a large or altruistic public-service role.

EDUCATION

Education was the very first concern of public broadcasting, dating to 1917, and has been a vibrant component of the field ever since. It arose from desires to make conventional schooling more effective and to extend it to Americans who could not make their way to classrooms. Those goals still are in place in many parts of the United States today.

Most early programming was *educational* broadcasting because it was produced by schools and colleges and often took the form of instruction—classes delivered over radio waves. However, as entrepreneurs worked to turn radio into a business, the term *educational* became broader. It came to suggest broadcasting done primarily in the public interest and not for private gain. In practice, as the landscape of topics and programs expanded, this has meant noncommercial as opposed to commercial broadcasting. Furthermore, because of its dependency on direct public support, what used to be called only educational broadcasting is generally called simply *public* broadcasting today.

It's easy to underestimate the importance of public broadcasting amid the clatter and dazzle of commercial radio and TV. However, there's more to this alternative field of electronic media than many people may realize. First, even though there are *fewer* public stations than private ones, FCC figures (for March 2004) show that both categories are substantial:

Radio	Television
AM Stations 4,781	Commercial 1,362
FM Commercial 6,224	Public 382
FM Noncommercial	
Educational 2,471	

Note that radio stations are categorized differently from TV stations. In radio, some AM and all FM educational stations are noncommercial. However, religious and some other formats are not part of *public* broadcasting. As of 2004, about 2,500 radio stations and 382 TV stations, as listed here, were doing what is known as *public broadcasting.*

sidebar

BEYOND MAKING MONEY

Public broadcasting is in most U.S. cities and in many sizable towns, especially those with college-run stations. Newcomers may have to search the spectrum to discover public broadcasting, but it's usually there to be found.

Public broadcasting deserves close examination for at least two reasons: It draws on tax dollars and donations, demanding enough from its audience to warrant scrutiny; and it pursues programming well beyond the chosen range of most commercial stations. It thus challenges the capitalist dynamic underlying our society. There are signs that money-pinched public broadcasting is mimicking some commercial practices, but so far, it remains a national experiment in *differentness.*

Shows now vary widely, from the radio comedy-quiz *Whad'Ya Know?* to TV wild-life specials to gavel-to-gavel congressional hearings coverage to *Masterpiece Theatre*, a series of dramatic performances. At its core, much public-broadcasting fare still seems intended to educate in a broad sense—to feed, stretch, and stimulate the mind. However, hot controversies over federal funding can erupt when broadcasts "go too far," offending audiences' (or at least politicians') values and tastes.

Today, some programming flows through closed systems run by schools or colleges that serve their students or wider communities with daily lessons and not much more. Other public-broadcasting fare arrives in homes via standard broadcast channels and is nationally celebrated. For example, researchers have found the long-running series *Sesame Street* to be a good preparation for prekindergarten children. Other studies, however, indicate that television can inhibit early learning and benefits affluent children more than poor ones. Still, *Sesame Street* is one of public broadcasting's greatest success stories.

INCLUSION

Commercial networks primarily have pursued mainstream-white-middle-class audiences, only gradually making room on their airwaves (and staffs) for diverse ethnic groups and perspectives. Public broadcasting, free of commercial pressures, has been able to work at inclusion more consistently—but it too has produced inconsistent results.

A quarter-century ago, the **Carnegie Commission** on the Future of Public Broadcasting declared, "Public broadcasting is a major cultural institution that can play a decisive role in bringing together the pluralistic voices and interests of the American community." The commission noted, however, that many minority-interest groups found public broadcasting to be "closed, unwilling to change, and afraid of criticism and controversy."[3] Years later, in 1992, James T. Yee of the National Asian American Telecommunications Association said public broadcasters had adopted "multiculturalism" as a policy but still ignored the nation's past: "Such things as racism, discrimination, and oppression are assumed not to exist or to have never existed"[4]

By 2002, the Corporation for Public Broadcasting (CPB) could show returns from a new focus on multicultural programming. Ethnic-minority characters or concerns were prominent in more and more PBS and NPR programs. *American Family* had become the first prime-time Latino drama series on broadcast TV. Documentaries and miniseries explored current and historical issues of immigration and integration. Commemorations of the 9/11 terrorist attacks examined Arab American viewpoints. NPR launched a talk show featuring prominent African American broadcaster Tavis Smiley (who left *NPR* in 2004, complaining of slow progress toward diversity), and public radio's African American and Hispanic American audiences were growing. Money passing from taxpayers through Congress to the CPB was funding minority projects in "new media" (i.e., Internet programs).

The minority infrastructure also was expanding: Three new minority-controlled radio stations began serving Native American and black listeners in 2002. But this drew attention to some negatives in the system: Jobs were tight overall, and although minority employment in public broadcasting had continued edging up, its share of the total workforce hadn't. More than 80 percent of public radio and TV employees were white. Just 55 radio stations and six TV stations were minority-controlled, and they produced 60 percent of all minority-oriented programming during 2001. The only boom was in Hispanic American programming as the Latino population of the United States mushroomed.[5]

PRINCIPLE 2

All media that hope to reflect or influence American values must seek diversity in their content and on their payrolls.

Carnegie Commissions Series of blue-ribbon public-broadcasting study groups funded by Carnegie Corporation of New York.

protalk
Margaret Drain
**VICE PRESIDENT FOR NATIONAL PROGRAMS,
WGBH-TV, BOSTON, MASSACHUSETTS**

Margaret Drain is one of the nation's best-regarded developers of public-television programs. She supervised the creation of probing studies of American history—the landmark PBS series, *The American Experience*—and oversaw the show for seven years. In spring 2004, Drain stepped up to a broad new level of influence over what viewers see on PBS: the job of vice president for national programming for a leading program generator, WGBH, in Boston.

Now, Drain oversees "all nonfiction and fiction programming, plus lifestyle—*Antiques Roadshow, This Old House,* cooking programs—anything that goes out on a national schedule. . . . Also, I see more pieces of the WGBH pie than in my old job—children's shows, the Internet, radio, the local station, underwriting. . . . History was my portfolio before; now I've opened up to other areas."

Drain has witnessed the launch of two digital channels through the local Comcast cable system; they carry older shows of lasting value to which WGBH still holds the rights. Thus digital paths bring high-quality series and documentaries into forms that are increasingly popular but that leave Drain joking: "Just because you can see a little picture on a telephone now, you get to watch one of our shows on a screen as big as

> "In an era when we're really worried about civic participation and civic responsibility, I think public broadcasting is more important than ever."

a ravioli." Like other public-TV executives, she must rise to meet new technologies and pursue new ways to use expensive programs.

Drain's ambitions for public broadcasting are tied to its freedom from the commercial world. A former CBS News producer, she says the networks' Iraq war coverage has typified commercial TV's failures. "It's reflexive; there's no analysis. Embedded journalists give us up-to-the-minute live reports, but in terms of any real analysis, you never see it. Newsmagazines have turned themselves over to celebrity news. . . . Nobody's taking on these stories. They've abandoned news, their *responsibility.*

"In an era when we're really worried about civic participation and civic responsibility, I think public broadcasting is more important than ever."

Her advice to students: "If people want to do quality work in a journalistic enterprise, public TV is still a place to be. You work with smart people who can bring you along . . . in a way that's not permeated with political direction and still allows you to do good work. Otherwise, you're just in a grind."

The Dollar Dilemma

Public broadcasting faces an abiding dilemma: If it sells ads or accepts too much corporate money, it will lose its special freedoms; but if it stays mostly independent of commercial forces, it will be doomed to struggle for money. In some countries, economic survival is a less daunting struggle for public broadcasters. Some systems that are at least partly tax funded have almost universal popular support and thus manage to establish long-term stability. This is not the case in the United States, where philosophies and interests contend vigorously in the political arena. American public broadcasters cling to their noncommercial status with a stubborn if sometimes tenuous grip.

COMPETITION

One aggravating factor in the public-broadcasting struggle is the wide spectrum of program choices available to most audiences today. The digitally tuned car radio and the ever-more-versatile TV "zapper" have made precise selection of programming so easy that many people, like migratory birds, tend to visit certain nesting spots. Thus, they become *niche* audiences, flocking repeatedly to the same channels.

That's fine with public broadcasters if those digital tuners are locked onto public stations. Things aren't always that simple, though. Commercial radio, dominated by popular music and political talk, may appear to program mainly for demographic groups that tolerate advertising and wouldn't choose public radio on a bet. Still, in some respects, public and commercial radio compete head to head.

For commercial radio, larger audiences mean more support for a station's advertisers; in public radio, they mean a greater field of listeners from whom to seek donations and whose presence may attract corporate underwriters.

Today's radio dial is jammed with commercial choices, some of which outshout the more genteel tones of public stations. In the public-radio universe—more than 2,000 stations across all the 50 states—a spare dollar is as rare as a pair of "shock jocks." Meanwhile, public television has suffered even more serious blows from the explosion of cable and satellite TV since the 1980s. Public stations used to be almost alone in airing tough, edgy documentaries and offbeat cultural fare. Now, with about 230 national cable networks operating and the number growing, viewers find many of the same features on specialized cable channels, virtually around the clock.

POLITICS

Largely due to political controversy over programming, government support for public broadcasting has fluctuated. Donors and underwriters could take its place, but most people who listen to public radio never contribute a dime to it. To pay their bills, public radio stations might like to sell advertising, but are not allowed to do so. These distinctions, among others, sharply separate public radio and TV stations from the business model of broadcasting described elsewhere in this book. That is why, overall, this unique electronic enterprise survives with fewer resources than commercial broadcasting.

With this relative poverty, though, comes one common advantage tied into public broadcasting's goals: Most public stations are locally owned and don't have to send money flowing up a corporate food chain to distant investors on Wall Street. The downside—a serious concern to some broadcasters—is that public broadcasting has little hope of ever getting off the fund-raising treadmill or out from under politicians' thumbs.

Birth and Growth

Broadcasting has two histories—one of them primary and prominent, the other a far less flashy narrative. Although public broadcasting evolved at about the same time and from the same technologies that built commercial stations, it always has played the smart but outcast cousin. Its programming is precious to its habitual consumers, but they are relatively few.

It's a mass-media sector that sometimes seems to reject the mass. Its programming often is labeled *elitist;* taking on tough societal issues to which activists suggest solutions, it's also often branded as *liberal.* Sometimes, too, this outcast cousin becomes a political whipping boy. The history of public broadcasting is mostly an epic of high hopes, low ratings, and more high hopes.

PRINCIPLE 3

Public broadcasting's roots in education give it a unique mission that helps to separate it from commercial media.

FOUNDATIONS

Public broadcasting was born in a classroom. Its history is that of an educational project that "morphed" into a broader cultural phenomenon. Millions of families still know the network mainly because of *Sesame Street,* a show that attempts to embed basic learning skills through entertainment for children.

Going to College

The first educational radio station, known as 9XM, was established experimentally at the University of Wisconsin in 1917. That was the year the United States entered World War I—the year after broadcast pioneer Lee De Forest broke ground by transmitting national election results from his primitive New York station. In the infancy of broadcasting as a popular medium, public radio's precursors were just as significant as their commercial counterparts.

Soon, college lectures turned into radio programs. Educators—starting in 1921 with Latter Day Saints University in Salt Lake City, Utah—received many of the first radio licenses from the government. In 1922, University of Wisconsin station 9XM became WHA. Soon, business interests and entrepreneurs eager to control the new mass medium began moving to sweep aside nonprofit visions of broadcasting.

A Double-Edged Law

The Radio Act of 1927, setting up a commission to regulate radio, borrowed a phrase from a piece of railroad legislation that still is quoted today when the purposes of broadcasting are discussed: Stations were to be licensed to use the airwaves according to the **"public interest, convenience, and necessity."** This seemed to position **public service** as a commanding goal. However, the other edge of the Radio Act disappointed educational broadcasters. To win political support, the Federal Radio Commission offered the choicest **frequencies** to commercial broadcasters—who had, as it happens, virtually designed key elements of the act.

The frequency problem would hamper educational stations. Many were trying to reach listeners not just with high-level instruction in useful subjects but with unique cultural and entertainment shows. Others sought to fulfill important missions—for instance, transmitting agricultural information, as no other medium was. By 1930, the State University of Iowa's station WOI would be broadcasting agricultural market data across a state in which half the farmers had radio sets. Government regulators and politicians did encourage commercial stations to air more educational programming. Because they were already hampered by the cost of

Public interest, convenience, and necessity Phrase in Radio Act of 1927 setting out purposes of broadcasting, used at the time mainly in a *commercial*-broadcasting context.

Public service Ill-defined concept cited as primary goal and value of public broadcasting; generally suggests not-for-profit enterprises or efforts to serve civic goals.

Frequencies Specific "spots" on the electromagnetic spectrum (airwaves) on which stations transmit their programs, as assigned by the FCC.

UNIVERSITY OF WISCONSIN 9XM/WHA. The *X* in its original name marked it as an experiment in 1917, but the University of Wisconsin station (later named WHA) inaugurated educational radio.

broadcast facilities and operations, educators were being marginalized. Without strong positions on the radio dial, many soon gave up their licenses, disappointed that they had neither attracted crowds of open-minded listeners nor generated much publicity for their colleges.

In the early 1930s, the growth of unions and other interest groups prompted Congress to focus on a fundamental question: Who should be permitted to use the airwaves "owned" by—that is, controlled by the government of—the people of the United States? It was clear that to avoid technical chaos, access would have to be limited, but to *whom?* In the end, the Communications Act of 1934 largely protected the interests of commercial broadcasters. An amendment that would have reserved good frequencies for public broadcasting was defeated. But in 1938, after a series of hearings over what to do with unassigned radio frequencies, the Federal Communications Commission (FCC) created a new class of "noncommercial educational" stations. It set aside **AM (amplitude modulation)** airspace for them in 1940 and promised that **FM (frequency modulation)** channels—with better sound quality—also might be available soon.

FM Promise

The new FM bands did emerge as cheaper and thus favored routes for education. Major public school districts and universities in the 1940s began building FM stations, newly eager to provide learning to children and adults. Development—tied mainly to money supplied by schools and colleges, rarely a rich set of backers—proceeded slowly, however. Technically, FM was a weak stepchild; it was still inaccessible to most potential listeners because few homes had FM sets. Commercial broadcasters boosted AM radio into a dominant position in American radio that lasted into the 1970s.

In this environment, most people didn't mind hearing radio ads, which, after all, kept introducing new wonders into everyday life. So it's unsurprising that a spasm of well-intended efforts to promote public service over private gain in the nation's media drew little attention from the American people. Some of these efforts came from a government-appointed group called the Hutchins Commission. This blue-ribbon panel studied the media, including radio specifically, and in 1947 urged them to follow a "social responsibility" standard in all things. Meanwhile, the FCC's own staff produced a report (known as the "Blue Book" for the color of its cover) that could have led to a blueprint for regulation that would encourage programming in the public interest. These outcomes were largely ignored, however. Despite mounting evidence to the contrary, one scholar has noted, the notion that commercial broadcasters would answer important public needs was by now deep-seated. After all, broadcasters would be regulated. Meanwhile, they were delivering exciting radio shows, some of them straight from Hollywood.

The FCC set aside frequencies for educational radio, but it languished as a colorless, sparsely available, mostly low-power medium through the 1940s and into the 1950s. Its stations, when they could be found on the local dial at all, offered a short menu of programs. National attention was shifting toward a newer, more glamorous medium—television. It seemed likely to create tremendous opportunities for public broadcasting in an open society.

AM (Amplitude modulation) Method of impressing a signal on a radio carrier wave by varying its amplitude.

FM (Frequency modulation) Method of impressing a signal on a radio carrier wave by varying its frequency.

ACCELERATION

By the early 1950s, the FCC was holding hearings on **spectrum** space for television stations nationally. One commissioner, Frieda Hennock, earned an honored place in public-broadcasting history by insisting that a fourth of the channels be reserved for noncommercial stations. In the end, the stations won fewer than 10 percent of the available channels, but that was enough to launch educational TV. A Houston, Texas, university station received the first license in 1953.

The Lure of Television

The **Ford Foundation**—a legacy of the great Ford automaking fortune—now had started supplying educational TV with its first regular outside funding. The foundation, which would become public broadcasting's most faithful and generous friend, wanted to put useful cultural and intellectual material before audiences that they wouldn't find elsewhere.

At this point, public broadcasting should have been gaining a splashy new position in modern society. After all, **test patterns** (on-screen graphics used to tune transmissions) and still-primitive programming looked pretty much the same, whether they appeared on commercial or public TV channels. To most families of that day, TV was an almost unbelievably attractive novelty. In their journeys around the dial, they should have discovered and watched many interesting public broadcasting shows.

There was one big problem: Public TV was nowhere to be found on most TV sets. In the largest U.S. cities, every spot on the standard **VHF (very high frequency)** band, from channels 2 through 13, had been granted to commercial broadcasters. To see public TV, viewers in New York, Los Angeles, and other huge **markets** would have to buy **UHF (ultrahigh frequency)** equipped TV tuners, since it wasn't accessible through existing models. Many educational station projects simply stalled as a result.

Such obstacles typified the early years of public television—but so did its advocates' stubborn commitment to public goals. Aided by Ford Foundation construction grants, people did build stations, aggressively on the VHF band, though less so on UHF. There was enough political support for reserving future space for educational channels to set off a vehement debate in the U.S. Senate in the mid-1950s. Before long, the government was starting to provide funding for public-TV programming. The FCC worked with broadcasters and manufacturers to design TV-set technology that would bring in all channels. In 1962, President Kennedy signed a law stating that this requirement must be met in all new television sets. By this time, major educational stations were on the air several hours a day—ambitious at that time, if unimaginably limited today. They had begun early to pool their resources; a *consortium*—or alliance—that in 1963 would become **National Educational Television (NET)** accepted programs from stations, made copies, and distributed them to other stations.

Nudging the Culture

President John F. Kennedy took office in 1961 and urged citizens to throw off Eisenhower-era complacency and pursue higher arts and ambitions. Even with respect to popular culture, the young president wanted action and activism on a national

FRIEDA HENNOCK. Businesses wanted television to themselves, but FCC commissioner Hennock made history by holding out successfully for educational channels.

Spectrum The electromagnetic spectrum, which is used to carry radio signals.

Ford Foundation Major financial backer of public television from its earliest years onward.

Test pattern Graphic pattern of fine and dark lines, usually printed on card or sheet, with which technicians focus and calibrate television cameras.

VHF (Very High Frequency) Refers to any frequency between 30 and 300 megahertz; denotes TV channels 2–13.

Markets Term for communities or metropolitan areas served by broadcast stations.

UHF (Ultrahigh Frequency) Refers to any frequency between 300 and 3,000 megahertz; denotes TV channels 14–69.

National Educational Television Consortium of broadcasters that became major program producer for public TV; assets acquired in 1970 by owner of New York station, renamed WNET.

38

NEWTON MINOW. President Kennedy's FCC chairman became famous by labeling TV programming "a vast wasteland" and pressing the commercial broadcasters for higher public-service standards.

scale. He found inclinations toward both in a still younger man, a lawyer named Newton N. Minow. As the new Democratic chairman of the FCC (and a strong backer of educational TV), Minow made it clear that prime-time TV programs did not square with his view of public service. He complained memorably to commercial broadcasters at their national convention: "Sit down in front of your television when your station goes on the air . . . and keep your eyes glued to that set until the station signs off. I can assure you that you will observe a **vast wasteland.**"

Minow threatened not to renew licenses of broadcasters who failed to meet public-service needs. *Vast wasteland* soon would become a popular epithet for television, despite (or, perhaps, because of) the support TV was winning from millions of viewers. With an FCC chair denouncing commercial TV, educational broadcasting suddenly looked even better. Moreover, there clearly was growing demand for more varied public-television programming beyond mere classroom instruction.

National Educational Television soon switched its emphasis from the classroom to the living room. The Ford Foundation provided money, much of it funneled to a two-year experimental program called **Public Broadcasting Laboratory (PBL),** in which NET generated cutting-edge content. Some of it was high-culture entertainment, including dramas as socially challenging in the sixties as the TV stage plays aired "live" by commercial networks had been in the fifties. There also were nonfiction programs that boldly showcased the nation's ethnic diversity; some went so far as to examine the lives of black Americans during a period of violent civil-rights confrontations in the South. Other programs delved into volatile national issues, including key government actions in Vietnam. Public broadcasting was displaying its power to tackle important and controversial issues of American life and culture.

Carnegie's Gift

Under Kennedy and his successor—Lyndon B. Johnson, a Texas commercial broadcaster—the government was encouraging the growth of this alternative medium. A 1964 meeting between broadcasters and the U.S. Office of Education created the Carnegie Commission on Educational Television, charged with charting its future path. The Carnegie recommendations, announced in 1967, would come to be viewed as historically significant.

The commission declared noncommercial broadcasting to be a vitally important communication asset of the American people. It held that public stations "should be individually responsive to the needs of the local communities and collectively strong enough to meet the needs of a national audience. Each must be a product of local initiative and local support." Moving from the general to the specific, the commission members called for greatly increased federal support for public broadcasting—specifically for funding to be drawn from a new 2 to 5 percent excise tax on the sale of television sets. The commission asked for the creation of a semipublic corporation to administer the funds.

President Johnson was listening: He introduced a bill to that effect in Congress. Hearings and debate soon brought radio into the process as well, and on November 7, 1967, the **Public Broadcasting Act** became law. It retired the term *educational* from popular use. Besides committing federal financial aid, the act established the

Vast wasteland Term coined by FCC Chairman Newton N. Minow to describe commercial television programming in 1961 speech to broadcasters.

Public Broadcasting Laboratory Two-year experimental program (in 1960s) to generate innovative programming for public television.

Public Broadcasting Act Law enacted in 1967 committing federal money to public broadcasting and establishing Corporation for Public Broadcasting to develop system.

39

Corporation for Public Broadcasting (CPB), a 15-member board drawn from civic and cultural organizations. It would use tax dollars to help create high-quality radio and TV programs, set up interconnections, develop stations, and generally promote the growth of noncommercial broadcasting.

The corporation could not govern stations; the FCC licensed them individually, so power in public broadcasting remained formally at the local level. (Even NET, a vital program supplier to stations, never had managed to bring stations together.) So there was less to CPB's formal authority than its title might suggest. It was not to be the hub of a national network. Still, some local stations and regional public-TV networks objected to the idea of a central power interconnecting them all; they insisted on autonomy.

PRINCIPLE 4

If centralized funding = centralized power, then government funding = the threat of government interference.

Moreover, there was the specter of political influence from Washington—not at all welcome in many communities. An earlier plan under which the U.S. Department of Health, Education and Welfare would issue directly all public-broadcasting grants—keeping the purse strings out of CPB's hands—had given way to full grant-making power for the corporation's board. One of its central tasks was to shield stations from direct government pressure—but would they need to be shielded from CPB itself?

The CPB Factor

Money from the new corporation seemed certain to make stations attentive to the leanings of its board members. They would, of course, have leanings: The board was to be appointed by the president—initially, at least, in a brutal political climate. The Vietnam War polarized national politics; it battered Kennedy's successor, President Johnson, into announcing that he would not run for reelection, and brought Republican Richard Nixon back to the forefront. Once elected, in 1968, Nixon quickly appointed a long-time political supporter and Johnson critic, Albert L. Cole, to the CPB board. In this political environment, many local station managers became nervous about possible "top-down" pressure on them to accept certain types of programming.

Pressure certainly would come in the years ahead, especially from members of Congress, goaded by constituents and special-interest groups, and from the White House. In the turbulent late sixties, with the Public Broadcasting Act taking effect, the main questions about public broadcasting had to do with how independently and how rapidly the newly elevated field would grow to serve public needs. Speed was important partly because of how television in particular was permeating American life. Fewer than 1 in 10 homes had TV sets in 1950, but one decade later, in 1960, 87 percent of homes had acquired "the tube." A few years after that, the Carnegie Commission recognized that only with large quantities of federal and other noncommercial dollars could public broadcasting be seen and heard through a growing mass of commercial programming.

The Corporation for Public Broadcasting began dispersing funds to pay American Telephone & Telegraph (AT&T)—at that time still the country's dominant "Bell" telephone system—to interconnect the public TV stations. It wasn't as if they hadn't been exchanging programs for many years, however; their primary collective, the **National Association of Educational Broadcasters (NAEB),** had organized interstation shipping by mail back in 1950. (The process was known as *bicycling tapes.*) But universal movement of programs *through phone lines* (and, very soon, satellites) would standardize operations and help move public TV into the big leagues.

After a six-month test to prove the distribution web would work, the next job was to set up a guiding organization to push programs through it. With support from the older entities NET and NAEB (soon to begin receding from their leadership roles), CPB in 1969 created the Public Broadcasting Service (PBS). The Cor-

National Association of Educational Broadcasters Main organization and program supplier of public broadcasting until demise in 1981.

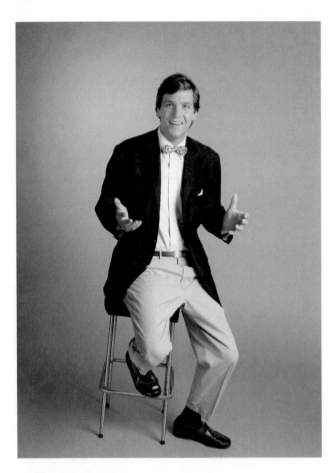

TUCKER CARLSON. This conservative commentator moved from CNN's *Crossfire* to PBS, a rare switch from commercial to public broadcasting—prompting claims that PBS was using Carlson to woo right-wing support.

Localism Regulatory principle declaring that the broadcast system is based on local stations meeting local needs.

poration for Public Broadcasting would become a central force in both public television and public radio, facilitating their growth. It would also help them evolve into a sort of counterindustry to commercial broadcasting, and would become a bull's-eye for critics.

TRIUMPHS AND TROUBLES

Public Broadcasting Service (PBS) arrived on the scene at a historical moment that in some ways would symbolize its future. In 1969, the country was torn by dissent. The Nixon White House seemed suspicious of all mass media. Young people marched in the streets; older people argued at work about the war and politics, and many found it difficult to get through the day without ideological headaches. Most got home from work and chose one among their few channels of *commercial* television. So the creation of a potentially mind-expanding broadcast service that would feed on tax dollars, replacing "educational" TV with "public" TV, invited close attention as an alternative.

Localism's Pull

In late 1970, PBS launched its live-feed web among local stations. The Nixon White House—whose Office of Telecommunications Policy kept a wary eye on perceived liberal bias—resisted the prospect of a fourth network that might use federal money to infiltrate America's communities with antigovernment attitudes. President Nixon vetoed a CPB funding request in 1972. Meanwhile, public broadcasting had developed internal conflicts. More than its commercial counterpart, it had been rooted in **localism.** That principle held that local stations should retain power over programming, even when shows came in from a national source. Now the rising influence of CPB and PBS alarmed some local broadcasters, especially in conservative regions that were wary of Washington's ways and New York's producers. They resented even the possibility that PBS would force unified scheduling on them, requiring them to show programs that didn't fit their communities in schedule slots when other shows would do better.

The Public Broadcasting Service was revamped in 1973 to foster creation of programming, not just distribute it—but under control of a PBS board packed with local-station executives. Thus, perhaps paradoxically, a national network was being assigned to support localism. Stations would be able to influence, if not rule on, program topics and treatments. Moreover, each station would continue to be free to design its own schedule and to reject any network-distributed program.

Along the way, radio joined the network trend. Unlike TV, educational radio generally had received little federal funding and had remained generally poor, weak, campus based, and heard by few. However, more than 400 stations were on the air, a substantial base for a national system. In 1971, with CPB backing, National Public Radio was established and gave those stations—at least the 90 that initially signed on as charter members—a program service that would be their version of a network.

The Trust-Fund Option

The single most nagging problem of public broadcasting was—and is—its lack of assured long-term financing. Broadcasting is an expensive medium to run. With-

CHAPTER 14 *Public Broadcasting*

342

protalk
Tripp Sommer
NEWS DIRECTOR, KLCC-FM, EUGENE, OREGON

Tripp Sommer is a tall man—often sporting bushy hair and a gigantic, well-brushed beard—with eyes that twinkle behind his glasses. People who know only his name can't appreciate this image because Sommer's province is public *radio.* Nor does he strive for dramatic effect. In drawing distinctions between his station's news coverage and that of commercial radio, he settles for the subtle.

"We use Associated Press as a source, and some of it's going to be similar if not almost identical stories [to commercial versions]," Sommer concedes. "We try to rewrite them and not just 'rip'n' read them. But then we also try to enterprise our own stories. It might be just from a fax or a telephone call, but we're not just attached to the news wire."

Sommer was drawn in the protest-rich Vietnam War years to "underground" stations that defied the Establishment. He knew he could write but he didn't want to become a novelist or professor.

"I was a bartender," he recalls, "and I had a sense that there was a great radio story in there somewhere, so I started calling radio stations in Eugene." The only station to invite him over was KLCC, and like many public-radio workers, he started as a volun-

> "We'll try to ask different questions, get a different take on a story. . . . We're looking for different and more aspects of the story."

teer. Sommer soon found a paid home at the helm of the station's newscasts. These include brief inserts into National Public Radio's *Morning Edition* and *All Things Considered,* as well as a daily half-hour of regional news and discussion called *Northwest Passage* of which Summer is chief anchor.

He reaches for perspectives missing from the sparse, rat-a-tat news bursts heard on Eugene's commercial stations. While KLCC sounds only mildly radical compared with the protest radio of decades gone by, its programming is far more diverse and "liberal" than what commercial outlets carry. Sommer covers community news with a modest budget, a tiny paid staff, and a corps of eager volunteers. Many stories—too many, in his view—are covered entirely by telephone from the newsroom-studio at Eugene's Lane Community College.

"Part of that has to do with the time factor," he says, "specifically, volunteers who are up here four or five hours per day . . . and whether it's car or bus, and [factoring in] their time up and back, that's about all they have. I mean, we still get the story a lot of the time but we're not always *there* . . .

"We'll try to ask different questions, get a different take on a story. . . . We're looking for different and more aspects of the story."

Sommer has made KLCC's news programming work well enough to win national attention. He served two terms as president of the Public Radio News Directors Association in the early 1990s.

/ #41

PAUL GIGOT. Noted for his own conservative political analysis, Gigot, of the *Wall Street Journal,* was part of public TV's diversification moves. PBS gave him a show of his own.

out the sort of revenue streams enjoyed by commercial broadcasters, public outlets might have to live every day on the verge of literally begging for sustenance. From the beginning, educational stations had to get by on meager stipends from (most) colleges and whatever donations or "bake-sale" proceeds they could gather. With national interest in education increasing, federal and foundation grants came intermittently through the mid-century years. Even after CPB began funneling tax money to them, public broadcasters knew that every dollar could turn out to be their last from Uncle Sam.

Soon, broadcasters were giving on-air mention to commercial enterprises that made donations. Stations gave this kind of financial support a dignified name: **underwriting.** The funding problem was a key factor in the formation of still another commission, the Carnegie Commission on the Future of Public Broadcasting. Its report in 1979 pressed the government to pay about half of the estimated $1.2 billion that would be needed annually to run public broadcasting by 1985. The balance would come from state governments, businesses, audience members, and other sources. Rather than set up special taxes to supply the federal share, the commission said, the government should charge commercial broadcasters for their use of the public airwaves and put the proceeds into a **trust fund** that would protect the money for use only in public broadcasting.

That didn't happen. Among other reasons, the commercial broadcast industry had a strong, sophisticated **lobby** in Washington, the National Association of Broadcasters. The NAB once had supported public broadcasting, but came to perceive it as a threat to commercial broadcasting. The broadcast lobby had encouraged President Nixon to veto a CPB budget proposal in 1972, and now—feeling some real competition from public TV—lobbied away any hope that Congress would approve a commercially supported trust fund. Nor did other revisions proposed by "Carnegie II" bear fruit.

The Reagan Effect

In 1981, Ronald Reagan became president. During his administration, taxes were cut and federal spending increased, though generally along politically conservative lines. The CPB's budget consequently was cut by tens of millions of dollars. Throughout the 1980s, public broadcasting struggled. After a near-disastrous deficit in 1983, National Public Radio reordered its financial arrangements with NPR stations and with CPB, and then continued to expand its programming.

Public radio and TV experimented with the definition of *noncommercial,* stretching the limits of the short, restrained sponsor messages they were allowed by law to air. Ten TV stations, with special dispensation from Washington, DC, even broadcast true commercials. The FCC eventually authorized something called *enhanced* underwriting, in which companies donating to public broadcasting could receive up to 30-second plugs in return. Meanwhile, the most popular programs on public TV spun off some aggressive "tie-in" merchandising. Dolls and other replicas of characters from *Sesame Street* and other programs were up for sale across the country—often in store sections bearing public-broadcasting logos. None of this looked very "noncommercial," but it did help the stations and networks to remain solvent. It also laid the groundwork for a future of reduced, truly modest federal funding—a future that seemed almost certain.

In 1990, as competitive pressures from cable TV mounted, PBS made a radical change in its strategy that would prove controversial within the broadcast community. The network's national programming chief was given sole control of decisions over which new programs to fund for station use. Previously, stations—especially

Underwriting Usually, financial support of public broadcasting by a company or institution.

Trust fund Money set aside and held "in trust" for a specific purpose.

Lobby Lawyers or other advocates for a particular cause or industry before Congress or regulatory agencies (e.g., the "broadcast lobby").

large ones producing many PBS programs—had had a voice in that process, as had independent producers and minority representatives. It cost the stations much of their autonomy in program choice, but PBS promised that more and better shows would reach more stations under the new system.

Washington Woes

The 1990s saw conservatives mounting new attacks on the content and financing of public broadcasting. After a Republican landslide in the 1994 congressional elections, House Speaker Newt Gingrich and other top Republicans led an attack. They wanted cuts in funding to the "liberal" CPB—in fact, Gingrich said he wanted to "zero it out"—and public broadcasting wound up with a two-year funding freeze. Controversy erupted again in 1999, centering on the security of each donor's name once he or she had sent money to a public station. There were reports that some stations were turning over donor lists to political organizations. Public broadcasters long had obtained lists *from* political groups in order to tap their members for contributions. To provide unwitting donors' names *to* the political parties, however, seemed well beyond public broadcasting's mission.

The Corporation for Public Broadcasting—which had its own inspector general— investigated and reported that 53 stations had in fact exchanged the names of their contributors for other names from political fund-raisers. Minnesota Public Radio, a source of many popular programs aired by stations nationwide, handed over 10,000 donor names during the 1990s and got more than 46,000 prospective contributors' names in return (a profitable ratio, at least). When CPB reported that most donor lists had gone to Democratic organizations, the old accusations of liberal bias echoed again. Congress threatened to withhold funding unless the practice ended, and CPB cracked down on those involved.

Little had occurred to calm the stormy seas of public broadcasting by the dawn of the twenty-first century. Federal support remained touch-and-go, subject to political shifts, and the "dollar dilemma" was unresolved. Nonetheless, large foundations kept giving, businesses kept underwriting, and public broadcasters were covering more than 80 percent of their expenses with money from private sources. These included the viewers and listeners. Because of them, the **pledge break**—an interruption in programming during which staffers or volunteers repeatedly implore the

Pledge break Interruption of programming to ask viewers or listeners to donate money.

audience to donate money—entered the language. Giving by individual supporters (members) has been impressive: In a recent year, they contributed an estimated $140 million to public stations, about twice as much as the checks drafted by private underwriters.

Programming

After much deregulation and drift from lofty public-service goals, today's commercial stations have been asked to do little more than entertain the people. By contrast, the mission of public stations and their programs arouse debate over their value to the public and the increasing need to keep the whole enterprise running fairly independently. Today's managers watch Nielsen and Arbitron ratings closely, paying special attention to demographics—not how many are tuned in, but who. Some stations have begun to create large mosaics of little "niche" audiences, ready with their checkbooks, who in turn will lure richer underwriters.

TELEVISION

When *TV Guide* magazine announced its "20 top shows" of the 1980s, the fifth-ranked program was PBS's *Brideshead Revisited*. As the writer noted, "Who'd have guessed this British drama about two young Oxford men who might be lovers—one of whom carries a teddy bear—would become a cult classic, one of the most popular public television shows ever made?"

Who knew, indeed. Certainly some things are expected of public television's programs: They are expected to be high-quality exercises; to have a tone that is gentle but tough, sensitive but probing; to be rich in learning and lore; to be serious; and to be droll. A very few are expected to be just a little racy (usually in a "British" sort of way). Almost no program, though, is expected to be wildly popular—except with children. Kids were the first core audience, and not just in the classrooms served by early educational broadcasting.

Courting the Kids

Years before PBS was founded, *Sesame Street* was a gleam in the eye of Joan Ganz Cooney, a young public-TV producer. She got some foundation money, brought a puppet-making genius named Jim Henson into the fold, and tried out her new idea on youngsters in Philadelphia. The show didn't work. Cooney took its failure with the trial group as a spur to success, retooling it. *Sesame Street* hit the air nationally in November 1969. This time it worked, and well. In fact, it quickly had children reciting their letters and numbers while parents chuckled in the background at witty remarks from Henson's "Muppets."

Today, the program is a fountain of stunning statistics: It has encompassed far more than 4,000 episodes; it's the longest-running children's show on television; and it's seen in 120 other countries. In short, it's an international fixture. The Public Broadcasting Service now airs a slowly churning mix of programs for youngsters, from *Clifford the Big Red Dog* to *Reading Rainbow* to *Zoom* to *Barney and Friends* to *Teletubbies*. However, cable-TV channels, including Nickelodeon and Disney, now offer programming for preschoolers—competing directly with *Sesame Street* and prompting cutbacks in its production.

Fortunately, given its educational mission, public TV's programs for children spark enthusiasm in the classroom as well as at home. In a 2003 (PBS-commissioned) survey, more than 1,000 teachers ranked PBS as their most

PRINCIPLE 5

Even "elitist" public broadcasting can be a platform for comedy, drama, and human-interest programs that cross class lines.

STREET OF WONDER. To generations of American children and many abroad, *Sesame Street* has held magic for decades. Even this public-TV icon, however, has been weakened by media proliferation: Cable channels compete by airing children's shows of their own.

useful television source and named six of its programs among their top 10: *Reading Rainbow, National Geographic Specials, Arthur, Sesame Street, NOVA,* and *Between the Lions.*

Angling for Adults

For adults who enjoy the arts, public television offers a richly varied schedule. One showcase, *Great Performances,* ranges from opera to comedy to "performance biographies" of figures such as composers Rodgers and Hart. Another stalwart success, *Austin City Limits,* presents Texas-style concerts, often by big-name country and pop performers with baby-boomer appeal.

Public TV loved history and Americana even before producer Ken Burns became the equivalent of an auteur through his poetic, anecdote-rich documentaries *(The Civil War, Baseball, Jazz). The American Experience* also has reexamined major periods, issues, and characters, from female pilots *(Fly Girls)* to the Great Depression *(Riding the Rails)* to an antislavery crusader *(John Brown's Holy War).* Another kind of history—natural history—occupies many prime-time hours. *The Living Edens* takes great cinematographers and, through them, the viewers into enchanting places. *NOVA,* the network's most-watched ongoing documentary series, explores the sciences vividly. Although PBS gets most of the publicity and attention, stations around the country own, control, and supply its programming. The programming enterprise is a collective. Indeed, without a few large stations—WNET (New York), WGBH (Boston), WETA (Washington), and KQED (San Francisco)—producing most of the major shows, PBS might have few shows to deliver to audiences.

A timely assignment for public television early in the twenty-first century is to serve an audience that's getting older, as well as younger adults who need tips for living better. *The Perennial Gardener* is among regular visitors to homeowners, as is the entertaining home-renovation show *This Old House.* Advice-and-information programs in the long PBS lineup include *The Whole Child* (parenting), *HealthWeek* and *Body & Soul* (health), and *Religion & Ethics Newsweekly.* These, too, are produced by local stations or independent companies.

JIM LEHRER. His *NewsHour* moves deliberately through a few issues every weeknight, giving public television a prestigious presence in Washington, DC, where many public TV funding decisions are made.

The NewsHour with Jim Lehrer, a program that helps public television fulfill its imperative for news and public-affairs content, airs each weeknight. It presents an opening news summary and then typically examines no more than four or five issues. It does so at a pace that contrasts sharply with the dizzying rush of an ABC, CBS, or NBC newscast. The show's tempo may be short on excitement but is long on information, giving it the added authority and credibility needed to book important guests, who know they usually will be allowed to finish their sentences. *NewsHour* is produced in Washington, where it's seen by lawmakers who control federal funding and who often sit at its microphones.

There's much more in the public-affairs bag: Documentaries of *Frontline* highlight important yet often underreported problems (uncontrolled trade in imported guns, marketing of popular culture to teenagers, the HMOs' toll on conscientious doctors). *P. O. V.* (for "Point of View"), sometimes working with minority or activist filmmakers, focuses on rugged individualists, including a 12-year-old boy who challenged the Boy Scouts' antigay policies, and on interesting back roads of society.

Local Initiatives

Although audiences enjoy nationally distributed programs, some also favor local shows with local hosts who appeal to community tastes. Stations approach their "localism" mandate in a variety of ways. For starters, there are some utterly local public-TV programs. Among them are noteworthy news inquiries and discussions, often in large cities where commercial stations might scan complex issues quickly— if at all—but where public stations can develop topics and issues through long-form programs.

At the other extreme are vigorous small operations delivering news and public-service programming. In Alaska, *Anchorage Edition* tackles public-affairs issues and is broadcast simultaneously (simulcast) on public TV and radio. At Northern Michigan University, WNMU brings in medical specialists 17 times a year for *Ask the Doctors,* a live call-in show. Television production is expensive, however, and ambitious shows sometimes require more specialists or equipment than a small station can afford. That's why producers are often asked to bring in private-funding proposals, complete with named potential donors, before a station will commit airtime to a specific show. Programs about regional history and culture, always popular among public-TV fans, are often produced by a station and then circulated widely to diffuse the cost. Oregon Public Broadcasting airs minidocumentaries on the state's past, from immigration surges to train robberies; University of Oregon students research and produce the pieces to gain experience. Other states' public stations pursue mixed objectives: A Fargo station provides North Dakotans with kids' education programs even as its website alerts residents to flooding on the Red River.

Public television's viewers aren't always obsessed with the *serious.* Leisure pursuits make for good TV, and any state that's interesting or attractive enough to generate an outdoors or tourism show can peddle it to stations around the region or even the country. One good example is *Outdoor Idaho,* produced by Idaho Public Television and marketed widely. There are also a few big-name shows with strong local roots. *This Old House,* produced at WGBH in Boston, is viewed nationwide but displays a cultural trait rarely heard on commercial networks: rich New England accents.

For all the productivity of some public-TV stations, many smaller operations produce little beyond fund-raising shows and occasional community discussions.

RADIO

Radio has a unique ability to create worlds in listeners' heads, even as it frees us from the grip of the visual and encourages abstract thought. That's a fancy way of explaining why public radio is alive and fairly well across the country. Its use of the airwaves to broadcast effective programming has made it important to millions (about 1.8 million people a day hear *All Things Considered*). Like public-TV viewers, they endure pledge breaks—but they listen.

The Rise of NPR

Through the Public Broadcasting Act of 1967, the federal government set the stage for the creation of National Public Radio. Unlike PBS, the new radio network was intended from its inception to be a program service to local stations, and CPB helped give birth to NPR. Its debut, in April 1971, was live coverage of U.S. Senate hearings on the Vietnam War—important in terms of public broadcasting's civic mission, since stations all over the country picked up the feed. Less than a month later, the news program *All Things Considered* went on the air, its reporters covering antiwar demonstrations. The show's then-director and later co-anchor, Linda Wertheimer, would remember that day in Washington, DC, as "scented with spring flowers and tear gas."[6]

For KLCC in Eugene, Oregon, it was NPR's second important show, not its first, that made a dramatic difference. "The biggest impact NPR had on us was *Morning Edition,*" recalls program director Don Hein of the arrival of a new public-affairs show in late 1979. "We were running classical music in the mornings up till then. We just dumped the classics . . . and inserted the new show." After the arrival of *Morning Edition,* KLCC's audience grew, to the surprise of some staffers. "Inside the station, we thought everybody wanted classics," Hein says, "but *outside* the station, people wanted morning news."

Today, *Morning Edition* is public radio's top audience draw nationally, making it one of many programming "brands" that distinguish NPR. Its programs can be heard on hundreds of local stations, and other suppliers also contribute to the public-radio mix. One syndicator of programs, Public Radio International (PRI;

formerly American Public Radio), provides stations with shows including *A Prairie Home Companion,* the business report *Marketplace,* a getaway guide called *The Savvy Traveler,* and a popular Canadian public-affairs show, *As It Happens.* Public Radio International also distributes *From the Top,* hosted by pianist Christopher O'Riley, featuring classical musicians from ages 9 to 19 who hold the promise of virtuoso careers. In that respect, PRI is a direct rival to NPR—which in 1997 briefly sought to merge with it.

Most stations mix such material with local programs that reflect local tastes. An example is in Louisville, Kentucky, where WFPL-FM was launched in 1950. As the first station anywhere to be owned by a public library, the station focused on culture and learning, in contrast with the formats of nearby commercial outlets. Besides its classical music, WPFL aired lectures, language lessons, and other educational material. Later, it became "your jazz and information station." Today, the station works in partnership with two other stations and with the University of Louisville, and its schedule looks typical of most: NPR programs all morning, local issues and music in the afternoon, and a partly local- but largely national-programming menu on weekday evenings.

A Focus on People

Much public-radio programming is aimed at well-educated audiences—but not all of it. A show called *A Prairie Home Companion* hit the air in 1974, concentrated on programming for the common people, and became the most popular show in public-radio history. Host Garrison Keillor quit in exhaustion in 1987, ran the show from New York under a different title, then returned it to its birthplace of St. Paul, Minnesota, under its old name in 1993. The *Prairie* format ranges from music to comedy to Keillor's quirky, folksy tales of the imaginary town of Lake Wobegon. Other people-centered programs, such as

GARRISON KEILLOR. Minnesota-based, he became a *Prairie Home Companion* to hordes of public-radio listeners, putting Lake Wobegon on the nation's mythic map.

the jokey auto advice show *Car Talk* and comic Michael Feldman's *Whad'ya Know?*, sprang up to engage wider audiences than music or public-affairs programs might have reached.

An especially bright critical and popular success has been *This American Life*, a Chicago-based show born in 1995 and blossoming ever since. Its host, Ira Glass, addresses urban angst and other modern issues, and its storytellers often combine the funny with the bittersweet. The show is one of many that have originated at local public-radio stations, often with the help of government or foundation grants. Some of these funds come from the Corporation for Public Broadcasting. *This American Life* has given voice to people who feel disadvantaged if not disenfranchised—like the people who depend on yet a different wing of noncommercial radio.

◼ A Broadcasting Public

Working outside the sphere of "public broadcasting" as the public has understood it for many years are thousands of people who use the same tools but do somewhat different work. They do this to reach mass audiences or simply narrowly defined communities that still can be very large. And they represent some of the more truly democratic processes for communicating with audiences—noncommercially.

Many of these broadcast entities depend less on quasi-commercial "underwriting" than do conventional public radio and TV. The people in these lower-cost enterprises seem committed entirely to certain social causes, or to serving citizens at the block and neighborhood level, or simply to training themselves for electronic-media professions while informing or entertaining others.

COMMUNITY RADIO

In spirit, it's a natural-born hippie. It once was called *alternative* radio and now more often is known as *community* radio. Often fueled by "progressive" politics, it considers itself more closely aligned than any other medium with the needs of its host communities, and is loudly committed to free speech and universal access to the airwaves.

The people whom many community radio stations most hope to serve include ethnic minorities, the economically disadvantaged, political factions, aggrieved immigrant groups, and others whose lives fall outside "mainstream" middle-class society. Noncommercial radio long has aired such people's views, but in its earlier years aired only messages via small college-campus stations. A breakthrough came in 1949, when the Pacifica Foundation (as peace-oriented as its name) established KPFA-FM in Berkeley, California.

This station would become a legend—a megaphone for liberal-to-radical-left causes, giving airtime to communists, gays, and marijuana smokers in the 1950s, when they often were ostracized, arrested, censured, or at least confined to society's back rooms. KPFA broadcast rallies and rhetoric throughout the Vietnam War–protest years, some of that talk coming close to inciting rebellion. Its messages caught fire throughout the San Francisco Bay area at a time when it was a hotbed of war resistance. Yet, the station maintained that it simply was expressing the needs and wants of common people usually overlooked by mainline broadcasters.

A key to this stand was that KPFA and Pacifica had rejected not only advertising but also government funding through the Corporation for Public Broadcasting, a lifeline for many conventional public-radio stations. This enterprise was determined to operate without Establishment ties. Pacifica grew, adding stations in Los Angeles, Houston, Washington, DC, and New York. Its news, commentary, and

COMMUNITY RADIO. KBOO-FM in Portland specializes in broadcasting "unpopular, controversial, or neglected perspectives" to "diverse communities and unserved or underserved groups." Many such community stations—once called "alternative" radio—date from the turbulent 1960s and 70s.

documentaries reached largely counterculture audiences in those major cities. The station also drew listeners to the musical genres of a rainbow of ethnic groups and nationalities unheard anywhere else on radio.

At this point we should pause to distinguish community radio more clearly from other public radio. When CPB and a research firm combined on a national study, it defined *community radio* as "the five Pacifica stations and any community licensee station that generates 80% or more of its listener-hours from locally-originated programming." Only 40 stations—of more than 400 public-radio stations nationally—met that criterion.[7] In fact, to the contrary, *non*-local content today probably comprises 80 percent of what many stations broadcast. In order to gather listeners, stations have had to acquire high-quality programming from National Public Radio and other sources. They need big numbers—yes, *ratings*—to attract enough individual contributions and private-sector underwriting to stay afloat.

The feisty Pacifica arm of community radio has subsisted without some of the usual support systems in noncommercial broadcasting, including "networks" like NPR. But, like most community radio operators, it has gained only a relatively minuscule share of the national radio audience. The same CPB study that set up a definition of community radio found that it had just 6 percent of U.S. listeners at any moment. The report pointed to what might be one ironic reason: While supporting minorities and seeking diverse listeners, community radio draws mainly white, middle-class listeners. Moreover, they show less esteem for and loyalty to these stations than the fans of other forms of noncommercial radio show to theirs.

Still, Pacifica reported in 2004 that Arbitron ratings had placed its New York, Los Angeles, and Washington, DC, outlets among the top 30 public radio stations in the country. The foundation said its total listenership had grown 34 percent since 1998.[8] If so, this was a blow struck for programming that by no means is the only kind available through community radio stations. Whereas many emphasize progressive politics as Pacifica does, there is a wide array of approaches to both programming and civic relations.

Most stations belong to the National Federation of Community Broadcasters (NFCB), which has helped them become unified in some ways but by no means homogenized. Some accept money from the Corporation for Public Broadcasting; some do not, preferring to keep fund-raising as local and individual as possible. Many wage war against the gray sameness of today's commercial music formats by showcasing offbeat and indigenous styles. This helps them court both minority and white-majority listeners, as when KKCR in Kaua'i plays mostly Hawaiian music but also airs Pacifica's political-news-commentary show, *Democracy Now!* Northern California's KMUD says it has only six paid staffers but 120 volunteers; it accepts CPB grants but plays many hours a week of ethnic and "world" music.

Florida's WMNF pushes equality, peace, and social and economic justice; its programs range from NPR's *Fresh Air* interview show to an overnight block of "experimental, psychedelic, progressive" music. WUMB Folk Radio in Boston reaches listeners all over New England through a network of stations licensed to the University of Massachusetts. It's one of many outlets to combine community radio principles with those of a somewhat different stratum: college broadcasters.

COLLEGE BROADCASTING

Emerson College has the oldest noncommercial radio station in the nation's fifth-largest broadcast market. It's both an extension of the esteemed educational institution and a player in Boston community life. WERS-FM was established in 1949. Today, it's a true college station in that much of its programming is youth oriented, with students doing hands-on broadcasting tasks. It's also a *public-radio* station in that it serves a large off-campus audience and runs on contributions from listeners, local underwriters, and Emerson itself—not the sale of commercial time.

The college scene nationwide is a beehive of radio programming, on many levels of effort. From the consumer's standpoint, one reason for its success is that music long has had a high priority, personally and socially, for college-age people. Another factor is technology: Increasingly, students carry (or wear, and perhaps someday will have implanted) their own compact audio players and therefore can absorb this rich cultural currency throughout the day—whether from radio or from Internet downloads.

GOING TO SCHOOL. Broadcast curricula and college stations prepare many students for the rewarding rigors of radio and television. In some university towns, students operate the only public radio stations around, learning to work with broader communities than just their own campuses.

For student broadcasters, a major cause of passionate interest in their generally unpaid work is that the radio stations simply exist. Hundreds of colleges and universities have built, bought, or come to control FM stations and low-power AM, or have access to off-campus stations. The Broadcast Education Association, a society of educators and some students as well, lists more than 330 institutions as members, most of which have or use local radio outlets.[9] It is in these stations that many students both learn and apply the skills of broadcasting. But of course radio also reaches out: Campus groups promoting causes or events use the stations as amplifiers to other students and communities. People who never set foot on campus enjoy the usually eclectic mix of music, and sometimes news and commentary, emanating from the college stations.

Today, many college radio stations not only broadcast but also (or alternatively) *stream* their programming on the Internet. It allows far-off alumni, prospective students, and others to hear what a college station is producing. This "webcasting" is a fast-changing tangle of rules, technology, and relations with the record industry, among others, but its relatively low cost and almost infinite geographic reach have made streaming a growth area. To a lesser degree (so far), webcasting also carries college television programming.

Because TV is a far more costly enterprise than radio, relatively few colleges and universities can afford to own and operate stations; many instead rely on local cable systems. This allows them to produce programming—usually newscasts, sports, or campus-buzz shows—in their own facilities and then simply transmit or drive tapes of the shows to the cable headquarters. In Utah, Weber State University's prize-winning operation puts two newscasts a week on the AT&T TV cable that serves most TV viewers in two counties. To the south, Brigham Young University's *BYU NewsNet* has done live all-night webcasts while covering elections with one of the nation's best-equipped newsrooms. Both universities, and many more, feed their graduates into professional media work throughout the country. That career lure may be the most powerful factor in keeping college broadcasting vigorous.

LOW-POWER FM

In 2000, the FCC created a new category of radio station known as *low-power FM (LPFM)*, and in doing so lifted the hearts and hopes of thousands of citizens and activists across the United States. LPFM was intended to put broadcasting in the hands of people who had been denied access to it by high cost and red tape. But the FCC had another objective as well: to slow the proliferation of "pirate" stations (or *microbroadcasters*) run without government licenses at the risk of federal prosecution.

Pirate stations, many of which remain active and are a target of law enforcement, generally put out signals reaching a mile or less. The new LPFM class of station is a bit heftier: Still operating on 100 watts or less, it can cover roughly a three-and-a-half-mile radius around the antenna.[10] Licensees may include schools or public transportation districts, but not individuals—and not companies; the key word is still *noncommercial*.

Each new licensee might feel triumphant, because LPFM suffered through a politically difficult birth and infancy. The commercial broadcast industry had warned for years that these tiny stations would interfere with the much stronger signals on which the radio-TV economy depends. The National Association of Broadcasters and even National Public Radio persuaded Congress to stall full implementation of LPFM until interference tests were completed. In 2004, LPFM got a clean bill of health. Thousands already had applied for licenses, hundreds of licenses had been granted, and more than 300 stations were on the air. The Senate Commerce Committee backed a bill to move ahead with licensing full-bore.

The new broadcasters with their special objectives included some politically progressive groups that hoped to pressure mainstream public-radio stations to do more local programming. But it also included the Archangel Gabriel Association of Wilmington, Delaware, sending out Catholic programming; a couple of guys in Macomb, Illinois, named Tom and Darryl, campaigning for free speech and democracy; and the Multicultural Association of Southern Oregon opening its doors wide to local voices—much the way another type of noncommercial broadcasting, *public access TV*, was long ago set up to do.

PUBLIC ACCESS TV

It was in 1972 that the Federal Communications Commission started requiring cable-TV systems in the 100 largest TV markets "to provide channels for government, for educational purposes, and, most importantly, for public access." What was believed to have been the first community public-access channel already was on the air: The Junior Chamber of Commerce in a Virginia town launched it in 1968.[11] Public access arrived in Boston and New York in 1971. After the FCC order, such channels began multiplying, and by the end of the twentieth century, as many as 1,200 public-access channels were operating, by one estimate. [12]

These channels invite any member of the public—anyone—into TV production and performance settings. It then allows them, and often trains them, to use video cameras, editing equipment, and microphones to broadcast almost *whatever they wish to show or say,* free of charge. This approach answers statements in the Federal Communications Act of 1934 that the public owns the airwaves and that multiple voices should be brought to public issues—especially with commercial media avoiding some viewpoints altogether.

The new freedom given to all manner of TV producers led to controversies during the Vietnam War years, as "guerilla" broadcasters used their access to studios to attack alleged government-corporate sins. Activism endures more than three decades later, as open-to-the-public channels across the country allow all sorts of rhetoric and visual images onto the air. The system maintains equilibrium, however, because any speaker may be followed to the microphone by someone with the opposite viewpoint. Not everything proposed makes it to the TV schedule: Sexual material has been an issue at times, but local cable usually relegates spicy content to the overnight hours or otherwise keeps it away from children.

Bolstered by the Alliance for Community Media, which promotes citizen access to all sorts of public outlets, the access movement has let "sunshine" into unusual (or usually suppressed) attitudes and viewpoints more directly than any other noncommercial pipeline. In San Diego, for example, in a regular clash of ideas, a Muslim, a Jew, and two Christians regularly sit down before the public-access cameras to present *Alternate Focus.* They debate perspectives on the Middle East and the recent Iraq war. The show lasts half an hour—about 30 minutes longer than the four friends could count on receiving from commercial or even many noncommercial media.

LOW-POWER TV

This category is something of a sleeping giant. The number of LPTV stations operating in the United States is higher than the total of all other kinds of television stations—in other words, well above 1,500. Yet, although some LPTVs have existed for more than two decades, we seldom hear of the work these stations do. One reason is that, if this giant decided to roar, it would tend to roar commercially. One study suggested that two of every three LPTV outlets were set up as *businesses.*

All of these stations exist because the FCC decided a quarter-century ago to drop the barriers to low-power TV (1,000 watts or less). Formerly, the use of LPTV was restricted to *translators*—transfer systems that took in over-the-air signals from distant points and converted them for transmission through local channels. The FCC

GRASS-ROOTS TV. On a high Nevada plain, Bob and Virginia Becker have joined amateurs around the United States licensed to operate tiny, low-power TV stations—a movement only recently given government approval.

has since decided that these licensed units could begin to originate programming. The result is greater localism, particularly in rural areas that had been served poorly by mainstream television, but also in cities where people believed they could sell advertising time or subscriptions to support what amounted to hobby broadcasting.

And so, the arrival of LPTV as a programming source ignited a spirit of adventure or enterprise in many people who had not made television before. Low-power television stations often could get along with minimal investment, because one person with a VHS camera and low-power transmitter could produce and present a story or program from start to finish. As with public-access channels, the finished product might not look ready for network prime time, but it might fill some viewers' needs or at least hold their interest.

A sign of health in low-power television is the buying pressure that drives up the cost of LPTV stations. A broadcast information agency reported that in the first half of 2004, 75 stations were sold, for a combined total of about $11 million. That averages out to less than $147,000 per station—not a bad toll for entry into the world of broadcast media ownership.

Into the (Digital) Future

Public broadcasting has become one of the most closely investigated and extensively analyzed sectors of the modern mass media. Its mission, structure, and history have brought that about. Now in its ninth decade, public broadcasting as a generally altruistic enterprise sometimes soars but often displays the faltering movements of a newborn. Like the nation, it's an ongoing experiment, and its near and long-term future will be subject to some strong crosswinds.

INTERNET APPROACHES

Public broadcasting has been a technological pioneer—in satellite distribution of radio, for instance—and public TV and radio stations have pushed webcasting forward. A major reason is to stimulate interest in their programs among computer-addicted teenagers and young adults; they are among the least likely to check into conventional public broadcasting at all. PBS president and CEO Pat Mitchell began making inroads into youthful indifference when she bought the reality-documentary series *American High* after it failed on the Fox network. Rather than merely promoting and airing the program, Mitchell saw to it that youth-oriented websites began creating "buzz" about the short-run show. It drew zealous youth audiences—as much as 150 percent larger than normal—after it went on the air in spring 2001. The normally calm PBS website reportedly topped 100,000 visits a week. Certainly, the nature and quality of *American High* accounted for much of the enthusiasm, but it was clear that synergy between on-air content and the on-line habits of teenagers did its work, as well.

In another experiment with cyberspace, public TV in early 2002 began airing *American Family,* a drama about a Latino clan in East Los Angeles. One of the characters, young Cisco Gonzales, kept an Internet journal on the Web, and—in a migration from fantasy to reality—the show's viewers could enter the site and read the diary for themselves. This convergence, which once might have seemed bizarre, captured a mood of media transition that was rippling throughout the popular culture of the nation. Even though one critic called it "a sentimentalized, idealized, sanitized soap opera," *American Family* had a fair chance of bringing new viewers to a branch of television with a white, middle-aged image.

The Public Broadcasting Service also was aggressively seeking reasons to use the Internet not only to support but also to complement broadcasting. The network claimed to be putting 135,000 pages of *program content,* not just promotional "cross-plugs," at its website, and to have set up companion sites for more than 450 TV shows. Public radio also was engaging web users: National Public Radio acquired *Justice Talking,* a show that examined important court cases in front of a live audience, and made it the first public-radio program ever to have its premiere on the Internet. At the same time, many local public-radio stations already were operating websites as alternative receivers for radio listeners. With few people buying their way onto "broadband" paths that could carry clean, smooth video, the relative ease of Internet audio transmission put radio well ahead in that new medium.

Some of the greatest technical boons to public broadcasting probably still are in the future—possibly the near future. If the government lets all public broadcasters use extra digital channels indefinitely to raise money or for other purposes, the size, shape, and definition of programming could change along with the amount of incoming revenue. More revenue combined with new technology and (most important) ideas could open many doors. Of course, it also could change the *values and practices* of this noncommercial retort to the values and practices of for-profit broadcasters.

GENERAL CONCERNS

There's no neat way to sort, divide, and categorize the potential problems that public broadcasting faces; they are too tightly interwoven and too dependent on the unknown. However, it's fair to predict that the toughest challenges ahead will be recognizable extensions of the past. As William Faulkner noted, the past isn't over; it isn't even past. Sure enough, fundamental concerns loom ahead—just as they always have.

Ideals

Some public broadcasters may be forced by economic and political pressures to admit—like alcoholics entering AA—that their lives have become unmanageable. Certainly their mission is extraordinarily demanding. Over time, it may require a special breed of civic heroes.

There are signs that such people exist, people who would commit themselves to public broadcasting—perhaps as a challenge or, worse, a gamble—no matter what the odds. Even if Congress in 1967 had not ordered public broadcasters to hew selflessly to the public interest, some say they would have done so. It's not often that a person speaks openly and without apparent guile about having such noble impulses. It happens, though: With evident sincerity, former *Morning Edition* host Bob Edwards told how he felt after visiting CBS News to discuss a possible job one day in 1975, just as the Vietnam War was ending in a Communist victory:

> I felt ashamed of myself. I should have been back in my own newsroom putting together a story on the fall of Saigon . . . with lots of time to review the history, lots of time for analysis, lots of time to capture the full impact of such an important story. . . .

BOB EDWARDS.

I returned to NPR a changed man. For the first I time realized what we had there. What we were doing was a lot more important journalistically than anything Walter Cronkite or anyone else was putting out.[13]

It did not seem to bother Edwards that his moral certainty could be interpreted as condescending toward commercial broadcasters; his focus was on a mission of public service. Some public broadcasters fear that focus is getting fuzzier. True, audiences still inhabit the cozy clubbiness of high ground, as when someone insists, "I watch only PBS." In more substantive terms, however, it can be hard to winnow out the public service qualities of some noted programs. *Car Talk,* for example, often is less an advice show than an extended comedy routine by its hosts. On the other hand, no one ever said public broadcasting must avoid entertaining its listeners.

Public Needs

Scholars and philosophers have argued for ages over just what a *public* is. Small wonder, then, that for public broadcasting, the notions of *public interest* and *public service* often have proved too soft-edged to define and visualize, much less bring to life or defend.

Clearly, public service is in play when broadcasters extend their gifts to poorer, weaker, or more ethnically diverse communities than typically are served by—or seen in, or heard in—commercial broadcasting. Some improvement has occurred: One study indicates that at least 12 to 15 percent of mainstream public-radio listeners regard themselves as "other than 'White/Caucasian.'" As noted earlier, the most dramatic results occur when members of underserved communities practice public broadcasting themselves. This has happened—for example, in Anchorage, Alaska, where a comprehensive daily radio report called *National Native News* originates and is fed daily to 130 noncommercial U.S. stations. It also has happened for *Latino USA,* out of Austin, Texas, heard on more than 200 stations.

One interpretation of *public interest* would lean toward honoring that other core value we've noted, localism. That is not just a future concern but a present one. Some public-TV station managers complain that, although they are the ostensible masters of public broadcasting, their community priorities are muffled by top-down pressures on them to behave like commercial affiliates.

Money

Hope for success in fulfilling their mission hinges partly on how well public broadcasters resolve their money-and-power issues in the years ahead. They may not be *of* the world of commerce but they are undeniably *in* it—paying for buildings, electronic equipment, phone service, programs, and expertise at prices set by an aggressive commercial marketplace. That marketplace also is making it difficult for public television to guard its special content realm: Cable-TV channels now run programs so much like the prize offerings of PBS and its stations that they are struggling to hold onto their audiences.

Meanwhile, the availability of money to public broadcasters rises and falls, much as it does for commercial broadcasters. In the 2004 election period, Arbitron ratings for public radio rose just as ratings rose for commercial radio; that meant underwriters reached more listeners in the same way advertisers did.

CAR TALK. What can you say about a couple of fun-loving Massachusetts mechanics and brothers who've made a call-in show about pings and pistons into a public-radio phenomenon?

DOCUMENTARIAN. Hawaiian filmmaker Lisette Marie Flanary won wide notice with "American Aloha: Hula Beyond Hawaii," a documentary aired on PBS. Her mission—"telling stories that reflect the challenges we face today in perpetuating Hawaiian culture and traditions"—fits neatly into public broadcasting's avowed role in society.

Beyond underwriting, an inability to make large, long-term audience gains makes it all the more important for public broadcasters to tap the viewers and listeners they already have. Research tells public radio that "programming causes audience," that "public service causes giving," and that news-public affairs listeners give the most money of all. NPR's *Morning Edition* and *All Things Considered* account for almost one-third of the money donated by listeners. So, local programs are shuffled and some are eliminated as stations try to mirror audience desires. Whether this will work before some stations are forced to thin their budgets and staffs further is in question.

The researchers also say that once plenty of listeners are in place, appeals to them must be efficient and effective. For some audiences, star power helps: When WEDU in Tampa broadcast a concert by Donny Osmond, the singer himself went on the air from the studio to seek pledges of support. Viewers who pledged $250 apiece to Los Angeles station KCET heard their requests played by pianist Roger Williams. For its part, PBS is pulling out all the quasi-commercial stops. It now employs an under-40 woman, Lesli Rotenberg, as "senior vice president—brand management and promotion."

Government

Although the federal government has long paid less than a quarter of public broadcasting's costs and recently has cut the percentage lower, it still clings to its right to oversee the service and sometimes to intervene. That's what it did during the Nixon years and threatened to do again when Ronald Reagan and then George H. W. Bush held the presidency. Intervention from *state* government can afflict local and regional public broadcasters, as well. When Idaho Public Television aired a PBS-supplied documentary on children with gay and lesbian parents, outrage exploded among Republicans in the state legislature. Lawmakers required the public-TV network to air daily warnings that some acts depicted in its programs violated Idaho laws (including a ban on homosexuality—"sodomy").

Local and regional cultural issues can lead to such assaults, or threats of them, at any time; public broadcasting is especially vulnerable because of its partial

PUBLIC BROADCASTING

Careers in public broadcasting probably appeal to some job-seekers at least as much as commercial broadcasting does. Fortunately, since the disciplines and technologies of broadcasting are fairly similar everywhere, public radio and TV require little special training. Only a few positions—largely at upper levels and involving fund-raising or external relations—are likely to differ much from commercial jobs. Unfortunately, however—because they run on contributions, not on a business model, and inhabit a smaller wing of the mass media—public-broadcasting stations tend to offer fewer and lower-paying jobs than do commercial media. Despite this, thousands of applicants each year try to enter public radio or television, hoping for careers or at least rewarding experiences. Some have studied in college broadcast-production programs and are ready for work upon graduation. They may already have entered public stations, as volunteers, interns, or part-time employees—a definite plus, but not a requirement.

Because public broadcasting's mission still includes its first purpose, education, it's a good fit with schools and universities that work it into their own curricula. For example, KECG-FM, owned by a school district in El Cerrito, California, broadcasts music and news while training high school students to enter the field. Its emphasis on multicultural programming not only fills a need in the San Francisco Bay Area radio market but also, importantly, encourages minority students to aspire to public-broadcasting jobs. At Northern Michigan University, broadcast students may work for WNMU-FM and WNMU-TV, both public-broadcasting stations.

Perhaps the most critical question for prospective public broadcasters is whether they are a good fit for this demanding arm of the electronic media. Many university-based stations offer good public-employee benefits. Mostly, though, the field features low pay, relatively scant public attention, decent job security (there is no ongoing flood of glamour-crazed applicants), and a chance to spread culture and sometimes give voice to neglected sectors of society. Bottom line: If that combination suits you, try it—but study it carefully first.

Jobs

Account executive	Production coordinator
Announcer/Producer	Program director
Assistant manager	Promotions director
Development director	Reporter
Maintenance technician	Sales manager
Managing producer	Secretarial assistant
News anchor	Station manager
News writer	Systems engineer
Producer	Weekend on-air

dependence on taxpayer support. This situation shows no sign of abating. It's one reason that, again and again, friendly critics have urged that public broadcasting's finances be built and protected within a permanent fund of some sort. The second Carnegie Commission suggested a trust fund well insulated, if possible, from politics, as did a task force of the Twentieth Century Fund (now the Century Foundation) in 1992; activist Jerold Starr and his **Citizens for Independent Public Broadcasting,** an advocacy group trying to get corporate-donor influence out of the public media; the Gore Commission in 1998; and, in a 2001 book, ex-PBS-chief Lawrence Grossman and his coauthor, Newton N. Minow. (Yes, the same man who looked at commercial television more than 40 years ago and coined the term *vast wasteland*.) All advocate some form of long-range, independent financial security for public broadcasting.

Then, there are the listeners and viewers, who support the mission of public stations more fervently than anyone else does. When St. Louis's PBS station, KETC,

Citizens for Independent Public broadcasting Activist group focused mainly on freeing public broadcasting from corporate money and influence.

decided against airing a political debate, a letter-writer to the *St. Louis Post-Dispatch* noted that the station did manage to find airtime for less important material. Acidly, he cited "the repeated broadcast of a year in the life of a badger or the broadcasts, re-broadcasts and re-re-broadcasts of lectures by New Age gurus." The *Post-Dispatch* itself took the issue further, declaring editorially, "Political debates, whether high-brow, lowlife or sleep-inducing, are precisely the kind of programming that public television is created for."[14] The debates were heard on public radio, at least.

Summary

Public broadcasting has grown out of early experiments in education via radio, and has brought with it that public-service mission. Nationally, local public stations "own" and—on paper—control most of the medium. Yet, recent economic stresses and political pressures have increased the chances that top-down, commercial-style management and distribution of programming will take over as mainline public broadcasting defends its privileged territory.

The medium lives, meanwhile, with a growing mandate to welcome many voices to its work and programs as the United States becomes a more ethnically diverse country. That need and others have brought changes, such as vigorous community and low-power FM. These could draw attention from federal overseers who complicate broadcasting even in a free society. Experts agree that continued noncommercial operation is the best if not only plan for the success of public broadcasting's mission. However, this has not ruled out experiments in marketing, promotion, and ratings-oriented programming in a heavily commercialized age.

Food for Thought

1. Look up the word *public* in a dictionary and ask yourself (and some friends) if the definition you find rings true. How well does the term fit public broadcasting?
2. What are the qualities that seem to distinguish an hour of public television from an hour of commercial TV? How important are those differences, and why?
3. Why should government help to support public broadcasting as it exists today? Can you organize a persuasive argument?
4. Does public broadcasting meet the objectives of the Carnegie Commission on Educational Television's 1967 report (www.current.org/pbpb/index.html)? In what ways does it succeed or fail in meeting them?
5. In your view, what would be a "perfect" program on public radio or public TV? (Remember that it must represent an *alternative* to commercial programming.)

En flykt för livet

Chapter 15

The size of the world: In our minds, at least, electronic communication has somehow affected that over the past century. Good, bad, grotesque, uplifting, the bald extremes of human experience draw nearer to us and to others everywhere, often doing so *right now,* and *vividly.* Look again at the Picasso quotation above, then imagine these once-unimaginable highs and lows:

Crackling voice messages tell us the *Titanic* is sinking—as it begins to sink—450 ice-riddled miles from New York.

We cheer the end of World War II and hear others join us by radio, miraculously, at that glorious moment, across the country and beyond.

Rushing breathlessly to a television on September 11, 2001, we watch the Twin Towers fall and feel our hearts fall with them. Wherever else we are—we're there.

About 144 million people worldwide view at least part of the 2004 Super Bowl football game as it happens, wardrobe malfunction and all.

The rise of ocean-spanning media is no longer a dream. It's not even novel. It remains, however, wondrous. Distant peoples who once judged the United States mainly by its movies, for which they lined up outside theaters, now channel-surf at home to learn more about us than movies can show. We reciprocate by watching them back.

Today, if a nation seems headed for upheaval or war, we very likely can monitor its parliament or protesters hour by hour via Internet television. We can hear the raucous variety of the world at any time via satellite or web radio. Technology nears the point of "wiring" us (often without wires, of course) directly to every crevice of the globe.

Why does all this matter? Answers are beyond counting, but an arresting one comes from philosopher Peter Singer. He notes that media growth has made the United States more globally conspicuous than ever in its lifestyles, attitudes, and behavior. That, Singer argues, could bring a new accountability: "If the revolution in communications has created a global audience, then we might feel a need to justify our behavior to the whole world."[1]

Some may hope he's wrong, but global electronic media—as much as our "behavior" itself—have put the issue in play.

▜ A World Connected

Although they may seem vividly "American" to many, electronic media are international in origin and impact. As scholar James Carey has pointed out, the French were the first to use a preelectric telegraph—in the eighteenth century, just as the United States was coming into existence.[2] As noted earlier in this text, an Italian, Guglielmo Marconi, would pioneer "wireless" communication—radio. In Canada, in 1920, the Marconi Company's XWA (later CFCF Montreal) hit the air. In 1920 Russian Vladimir Zworykin and Briton John Logie Baird (with American Philo Farnsworth) made breakthroughs in television. More recently, it was Tim Berners-Lee, a British-born inventor working in Switzerland, who gave birth to the World Wide Web.

The role of any medium is interwoven with the customs, values, and economic-political history of the people using and operating it. What's more, there are sharp differences between rich and poor countries in their ability simply to *afford* electronic media. Expressed in millions of U.S. dollars, for example, the money required to set up and run just one television station for a substantial audience puts it virtually beyond reach for quite a few nations. Nor are skilled broadcast or computer personnel available everywhere. Today, however, electronic messages of all sorts criss-cross continents and oceans. In so doing, they express and may even be blending human communities that only recently seemed literally worlds apart.

RADIO

An American living atop a mountain or driving through a desert can use a short-wave satellite radio to pick up the programming of stations overseas. It's also easy for people with Internet access to hear overseas broadcasts through their stations' websites. International demand never seems to slack off; people think of radio as basic and, in some places, essential. Nearly every country in the world has many radio stations, with good reason: There are at least 2.5 billion radio sets in use around the planet—about one-third of them in the United States and still more in Asia.[3] This gives radio the greatest "penetration" into households and communities of any mass medium.

A Spectrum of Purposes

Advanced industrialized nations—including the United States—transmit programming to huge sections of the globe, often through special international services. A growing array of radio programs travels by satellite or the Internet from cities everywhere to listeners everywhere else. For example, just one source, Radio France Internationale (RFI), beams programs in many languages to Africa, India, and the Middle East and also reaches the United States via cable radio satellite, AM & FM, and the Web.

Some countries where broadcasters were under severe restrictions until little more than a decade ago are reaching out. In former territories and "client" states of the communist Soviet Union, pent-up urges to broadcast have been exploding onto the airwaves. The Republic of Belarus, formerly part of the Soviet Union itself—and hardly a Western household name—now sends its programs to much of Europe and to North America. Just as in the United States, some radio stations abroad are commercial, working to win listeners who will buy advertisers' products. Young people often are the primary targets. In Norway, radio has drawn youthful audiences with what U.S. programmers call the "hot adult contemporary" music genre,

protalk
María Paz Epelman
COMMUNICATIONS MANAGER, VTR GLOBALCOM, CHILE

It seems inadequate to apply the word *challenge* to a task that Maria Paz Epelman once faced: to get the Playboy Channel onto the daily viewing menu in a staunchly Catholic country in Latin America. "You don't have the media on your side, you don't have the Church on your side," says Epelman. "The Church and the government are not going to take risks. But the population is more neutral—not so conservative."

The people—the prospective customers—were Epelman's compatriots as well as her business targets. She's communications manager for VTR GlobalCom, S.A.—the Chilean arm of United Global Com, a telecommunications company operating in 26 countries.

However eager some of her countrymen and women were to view American sex kittens via satellite, important members of the Chilean establishment had their doubts. So Epelman, a journalist turned corporate spokesperson, made speeches and used her knowledge of the news industry to try to separate "skin" from "sin." "The way we did it in the press helped us to introduce [the Playboy Channel] very smoothly," she says. "After that, when anybody was against it, there was an explosion in the press, pictures of Playmates in the paper. It was very funny. Our sales went crazy."

> "You open the newspapers every day and you don't know what you'll see there ... we try cable TV, telephony, Internet—through the same connection."

Chilean individualism worked in her company's favor: "Lots of people bought that programming not really for viewing it. They did it as something symbolic, in the same way that people who choose VTR [for general TV service] choose it because they don't want to be censored."

Cultural tensions have not kept Chile from attracting a large and competitive set of foreign-based media companies over the past decade. This long, slender nation on South America's Pacific Coast had a closed economy under former military ruler Augusto Pinochet. But since a civilian government replaced him in 1990, the country has welcomed more outsiders.

United Global Com, based in Denver, Colorado, acquired VTR and with it Chile's largest cable-TV system, providing digital and telephone services, as well. It claims 60 percent of the cable-TV market nationwide and 40 percent of subscribers in the largest city, Santiago. Its customers can choose from 45 to 50 "basic" channels and 5 "premium" channels—Playboy among them.

RADIO MARTÍ. For two decades, this U.S.-government-funded radio operation has beamed anti-Castro messages into Cuba, an overt use of an electronic medium for political ends.

and other types, live on the air and on the Web. In Bangladesh, Radio Metrowave attracts younger people by sponsoring rock concerts.

Even though U.S.-style mainstream radio is broadly popular, many stations choose to pursue unique "niche" approaches, some based on geography and ethnicity. In the Hungarian capital city of Budapest, for instance, Radio C had to struggle to find start-up money. It had a social mission: To be the world's first radio outlet aimed at the Roma people—often called Gypsies, an ancient ethnic group—who comprise Hungary's largest minority. Meanwhile, Radio Terunajaya is the sole broadcaster in a poor, hilly district on the south coast of Java in Indonesia. The privately owned station broadcasts folktales and modern music originating in that region and also allows listeners to send urgent messages to neighboring villages; there's little doubt that someone they're targeting is tuned in.[4]

Political Factors

Radio in many countries still is sponsored and, sometimes, constrained by national leaders. Some governments run their own stations and networks and use them to reach out to other countries for public-relations or diplomatic reasons. Elsewhere, hybrid public-private corporations operate quite freely with only general instructions from their government. One of these, Radio Canada International, broadcasts continuously in seven languages, supplementing radio with a daily "cyberjournal" for Internet users worldwide. The British Broadcasting Corporation (BBC) says its World Service, transmitting in 43 languages, reaches 146 million listeners a week outside Britain—including increasing numbers in the United States.

Like these government-dependent organizations, three noteworthy radio operations represent the United States on the global airwaves. Some of what they broadcast is openly government-driven; other material aims to serve audiences more independently. Just how such conditions mix is a matter of continuous debate, but these channels exist, first, because they serve U.S. official interests abroad.

1. **RFE/RL (Radio Free Europe/Radio Liberty)** is a news service founded privately in the early 1950s to broadcast news into Soviet-controlled countries where media had been silenced. RFE/RL now is government financed and is a broadly useful policy tool for Washington. Based in Prague, it claims a growing audience in Middle Eastern countries, including postwar Iraq (where a companion

Radio Free Europe/Radio Liberty (RFE/RL) A semiprivate, government-funded broadcast news service that carries U.S. -oriented news and information into other countries.

operation, Radio Sawa, carries news, commentary, and cultural information, all in Arabic). Among its other Iraq initiatives, RFE/RL has serialized an Arab scholar's book on the reign and atrocities of Saddam Hussein.

2. **Radio Martí** was created by Congress in 1983 to beam radio news into Cuba in defiance of Fidel Castro's controls on Cuban media. In its two decades as a U.S. tool to promote anti-Castro sentiment among Cubans, this service has provoked criticism that it's mismanaged and circulates biased news; consequently, its listenership has plunged. President George W. Bush has backed Radio Martí, which costs about $25 million annually to run. The Cuban government answers with its own international radio broadcasts in nine languages, plus a website in four.

3. **Voice of America (VOA)** is a radio-TV-Internet service transmitting what it calls "a balanced and comprehensive projection of significant American thought and institutions"[5] to an audience of 91 million worldwide in 53 languages. Voice of America is government funded and also filters U.S. policy through news and entertainment programs. Established on radio during World War II, the service has spent decades building a close-to-neutral journalistic reputation; it has a staff of 1,200 and a $147 million budget. However, it has been prone to political interference. In late 2001, Congress moved to keep the service from broadcasting interviews with officials of terrorism-linked nations. Voice of America's director said it would keep doing appropriate interviews to maintain balance in the news.

Government entities such as VOA are not the only model under which political ideas are broadcast internationally. Hybrid public-private systems have developed around the world, especially in nations accustomed to strong central control. In Estonia, for example, Radio Tallinn must depend on government support and has been seeking more funding to expand its Internet programming. Instead, mainly to hold down spending, Estonian radio and television operations have been ordered to merge by 2008.

Conversely, there is a place in history for *anti*government broadcasting. During World War II, broadcasting systems in Europe became virtual captives of Hitler's occupying forces, which used radio to assert domination. The "resistance" in that war used radio to send demoralizing messages into the enemy's military camps. Most important to citizens of the occupied lands, clandestine radio stations kept them apprised of the war's true progress and alerted them to threatening troop movements. Meanwhile, anti-German broadcasts from abroad easily passed into the nation's airwaves, building grass-roots resistance to Hitler.

TELEVISION

At least 1.5 billion TV sets are in use worldwide—almost one set for every four people. That may seem amazing, considering the size, diversity, and uneven access to resources of all the world's cultures. An appetite for on-screen entertainment in the home now seems common to them. However, one's access to television has a lot to do with where one lives.

Gaining Access

Although by 1995 homes in ultramodern Japan had more TV sets than flush toilets,[6] there are places where—a decade later—a quiet evening before the "tube" is an unattainable luxury. Pakistan has about 135 million people—equal to nearly half the U.S. population—but fewer than 3 million TV sets, or less than 1 for every 45 people. In Africa in the late 1990s, the figure was about 60 sets per 1,000 people.

Radio Martí Organization created by Congress to transmit radio news into Cuba despite its limits on freedom of information.

Voice of America (VOA) Radio-TV-Internet network under U.S. government control that attempts to send objective news reports to other countries but sometimes suffers political interference.

OVERSEAS TV VIEWERS.
While 99 percent of U.S. homes have television, access to it abroad varies wildly, and it's still a rare luxury in much of the developing world.

(By the end of the century, perhaps 99 percent of all U.S. homes had at least one TV set.) Poverty, distance, and cost all are factors. Perhaps surprisingly to many Americans, large regions of the world, including much of South Asia, do not yet have home electricity, much less electronic media access.

By contrast, some countries have been generating television as long as, or longer than, the United States has. The United States initiated programming in 1939—three years after the British. Postwar France adopted it in 1948, and half a dozen Latin American and European nations had TV by 1951—about the time American television was establishing its first national icons on the air. Early programming in most lands conformed to their own societies' tastes (and their governments' wishes), just as the brisk Westerns, hard-bitten "teleplays," and mild white-collar comedies of the 1950s conformed to U.S. tastes. However, most of Asia had no access to television at all until the 1960s.

One huge Asian country, India, showed how politics could delay even history's most widely sought-after leisure technology. After the English pulled out at mid-twentieth century, the country's postcolonial leaders frowned on TV as a time-waster for a newly independent nation, so the medium stayed black and white until the early 1980s.[7] Since then, however, with the help of a vigorous domestic film industry, Indian TV has become rich and colorful. About 140 other nations also have their own television systems.

What TV Can Do

Programming internationally ranges far beyond the forgettable entertainment that Americans experience nightly—not that there isn't plenty of frivolous programming on every continent. Countries originate TV newscasts and nonfiction programming that suit their own purposes, including the need to promote tourism and prepare their citizens for greater global interchange. Turkish viewers can see not only their indigenous entertainment but also Turkish newscasts in several languages. In Japan, the ubiquitous **NHK** network has aired programs in English since 1925 and currently does so in seven other tongues, as well.

Like other mass media, television occasionally forces a wedge into a tight political situation. This happened during the Iraq war, when U.S.-backed Alhurra satellite TV duelled with the controversial Al-Jazeera service. Earlier, **EuroNews**—a Euro-

NHK Japan's quasigovernmental broadcast company, run by governors approved by the nation's parliament; like Britain's BBC, NHK is funded primarily by license fees paid by TV viewers.

EuroNews International satellite-TV news channel broadcasting in numerous languages.

pean satellite-news service—gained permission to send programming into Russia. It was the first time a Western media outlet had won the right to serve Russians through their own network. At last they would have the chance to see and hear outside perspectives on their country's affairs. In return, President Vladimir Putin agreed to supply Russian programming to the all-news channel.

Some television operations abroad, especially in Britain and Australia, produce first-rank entertainment of general appeal that is exported, as is, to the United States. Similarly, European networks turn out fine programs that need only translation into additional languages to attract large audiences abroad. It's U.S. programming, however, that packs the greatest marketing punch worldwide. European channels sign deals to bring American sports to eager new audiences. Some countries acquire U.S. network newscasts because of their exotic appeal, wide-ranging content, and relative low cost compared with originating newscasts. Hollywood movies, of course, are the hottest TV import in many lands.

THE INTERNET

Of all international electronic media, perhaps the hardest to encircle with a clear definition—or with limits of any kind—is the Internet. It knows no borders and, being digital, can move its messages in a variety of ways from user to user via satellite, telephone wires, fiber-optic cable, new wireless technologies, and more. In its purest sense, it's utterly anarchic; nobody's in charge. It's the product and fiefdom of its individual contributors, moment to moment.

At its simplest level, the Internet is a messaging service; hundreds of billions of e-mail messages are sent every year.[8] However, even e-mail can carry informative or entertaining content from one person to many. At higher levels of complexity, the Internet already carries radio and TV programming from Iran, Lebanon, China, and other countries to still other countries through sites on the World Wide Web. Europe is constructing a fiber-optic system that will take the Internet and many other types of electronic media across many borders. It's part of the world's projected—and to some extent already realized—information superhighway.

With complex interactivity—connections permitting interaction with electronic systems—users from around the world can sign onto computer games and play

CONNECTING CULTURES.
A sign over a labor rally in Rajasthan, India, suggests the global pervasiveness of the Internet. Increasingly, places in the world that lack on-line access are considered victims of a "digital divide."

them simultaneously with others. The success of games already has helped make *gambling* an Internet sport.

For ordinary users, the simple act of "surfing" the Internet for surprises, or pursuing flimsy leads toward some fascinating and elusive bit of information, is entertainment enough—so far. Unfortunately, most of the world's citizens can't indulge even in these simple pleasures, because of a vexing problem: They currently have little or no access to computers. The gap between larger masses of cyber-challenged people and the relatively few who do have on-line access has been termed the **digital divide.** This ephemeral boundary seems to run along the immense plains, deserts, and mountain ranges that separate one city and its electronic technologies from the next. It also distinguishes the world's urban upper classes from the urban poor, and poor countries from rich ones.

The digital divide causes knowledge inequities. For example, by one account, 80 percent of the information about Africa is generated *outside* Africa; but by 1995, only about one of every four countries on the continent had the Internet access they needed to reach much of that information.[9] Lack of money was the dominant reason.

PRINCIPLE *3*

Issues of cultural imperialism and the "digital divide" are built into electronic media.

In the face of what some fear is **cultural imperialism** perpetrated by rich countries through their media products—that is, an imposition of Western tastes and ideas on other cultures—this lack of access to broad information sources poses a serious threat of worsening global disparity. In sum, then, the Internet is a potential smash as an international mass medium, already pleasing a rapidly growing user community; yet millions remain out in the cold.

Digital divide Gap between people who have computer and Internet access and those who do not; used to describe both domestic and international rich-poor divisions.

Cultural imperialism Term applied mainly to U.S. domination of global media production, resulting in (real or threatened) spread of consumerism and other American values and practices to vulnerable societies elsewhere.

◼ Authority Over Media

Nowhere on earth, not even in the freedom-loving United States of America, are electronic media permitted to operate entirely and exactly as they wish. Usually, government gets in the way. In colonial America—long before the arrival of electronic

protalk
Veran Matic
RADIO B92, BELGRADE

A s chief editor of Belgrade's most celebrated radio station, Veran Matic has seen much excitement—sometimes a bit *too* much, perhaps. Here, he recalls some of the tactics that the free-speech movement, represented by Radio B92, and the repressive regime of Slobodan Milosevic used against each other:

"The most popular method that the regime was using against us was electronic jamming of our programs. When that failed, they would simply turn off our transmitters. In December 1996, during mass anti-regime protests [by] the citizens, they turned off our transmitter. . . . On that occasion, even Cathy Morton, the president of CPJ [Committee to Protect Journalists] back then . . . came all the way from New York to Belgrade to express her protest personally to Milosevic.

"Milosevic gave an extensive explanation, using numerous technical terms, about 'the water leakage within our cable system' which caused the disruption of our radio transmitter. Many jokes and funny stories were made about that [on the air] afterwards, and technical description of program disruption was used upon every action that the regime took against the independent media. Of course, lots of journalists and editors

"Often the people stated that our radio was the only reason for them to stay in the country."

were receiving threatening warnings; some of them were even physically attacked. We were faced with the classic methodology of police work in totalitarian regimes.

"We put on the air many 'silly' clips. Our listeners expected something like that from us. . . . In addition, we made so-called serious shows as well, which can be considered mere travesty. For instance, in 1992, we changed the whole program agenda during a single evening, along with the style and manner of speaking, so that it seemed as though the ruling party took over the radio station. We intended to continue with this for the whole 24 hours, but we were forced to stop as the listeners started smashing their radios in anxiety. . . .

"Radio was extremely important primarily for persevering . . . for the common sense of most of the people. Often, the people stated that our radio was the only reason for them to stay in the country. People . . . easily identified with our radio, and we have often been under [the] impression that we became part of the family. That fact imposes a higher level of responsibility for every single word that we put on the air."

media—mobs angered by news or agitated by politicians sometimes wanted to shut down printing presses. On more than one occasion, for instance, Benjamin Franklin's grandson, Benjamin Franklin Bache, felt the politicians' fury over something he had printed about the day's events. This is how it has been in the rest of the world, too; governments in particular have complicated life for media producers even in the age of international broadcasting and the Internet.

FORMS OF GOVERNANCE

Besides holding a stick, political authorities often hold the carrot necessary to the success of electronic media. It is public agencies that can launch and support stations and networks. Governments always like to keep their hands near the media power levers; even when they don't repress, they regulate. One scholar has identified three rationales for regulation: *technical,* under which government controls use of frequencies to prevent chaos on the airwaves; *monopoly,* which assumes that dominant media left unregulated would provide poor programming at high prices; and *political,* which holds media accountable to certain standards that range from venal to virtuous, depending on the government in charge.[10]

Authoritarian Rule

The power to broadcast is often seen by *authoritarian* leaders—who place the power of the state above individual freedoms—as central to their control of citizens. Usually, in a **coup d'etat** attempt (an effort to overthrow authorities), both rebels and government rush to occupy the broadcast stations, pushing journalists aside to reach the microphones. Fierce skirmishes often result, as in Romania and Lithuania when the Soviet Union was collapsing.

In a "pure" authoritarian regime, media have no independent power to begin with; the government owns and tightly manages the broadcast stations (while keeping newspapers on a short leash, too). Again, such controls flow organically from the political situation of the times, and certain times seem to last far longer than others. The communist revolutionary Fidel Castro took over Cuba in 1959 and still remains in charge. After more than 40 years of Soviet-style rule, some Cubans' frustration with news controls has led to an Internet "underground." Although Internet access is limited to a favored few, independent journalists manage to file antigovernment news with overseas websites. The government knows of this illegal journalism and harasses the journalists, but generally tolerates them.[11]

This sort of tension between government domination and breakaway broadcast media—often led by one station—is part of the unofficial political calculus in authoritarian countries. In some, such as sub-Saharan Africa's fragile nations, governments hold only tenuous power. Perhaps in compensation, they tend to crack down when broadcasters get out of line (rarely, as a rule) in order to restrict citizens' access to "destabilizing" ideas. One of the more severe examples is North Korea. With just a handful of domestic channels operating, the politically isolated country's roughly 23 million people are forbidden by their government to watch foreign TV. Next door in South Korea, by contrast, nearly all homes have television sets and people use electronic media avidly. Westward across Asia, the Taliban sect of ultraconservative Muslims that once ruled Afghanistan was burning television sets in the streets in 2001—until U.S.-aided Afghan rebels drove the clerics from power.

Only the least fortunate societies live under such iron-fisted rule, of course; government pressure on free speech and media in democracies is usually subtle. Since the disintegration of Soviet communism, which had sponsored authoritarian regimes on several continents, many nations have edged toward "western" models. For them, broadcasting and the Internet are used not merely to wage political warfare but also to develop well-rounded entertainment and information programs and content.

Coup d'etat Sudden political action, usually resulting in change of government by force.

Six weeks after the World Trade Center toppled in 2001—with countless TV cameras trained on it—the *Iran News* reported a government raid on citizens' satellite dishes. Iranian police confiscated about 1,000 dishes in 48 hours, with most people giving them up quietly. The radical Islamic regime explained the move as a way of heading off "elements abroad" that were plotting to use TV to mount a political challenge in Iran.

This wasn't new; the country once run by Persian royalty had outlawed the use of satellite dishes in the 1990s. However, they generally were tolerated after the election of reform-minded President Seyed Mohammad Khatami in 1997. Now, however, the international journalism group *Reporters Sans Frontieres (RSF)* (Reporters without Borders) protested that some 7,000 dishes had been confiscated since March 2001, and that "satellite dishes are, along with radios, one of the rare means for Iranians to have access to foreign information." That, of course, seemed to be the point of the confiscations. The RSF said that most Iranians routinely hide their dishes under tarps or in air conditioning units.

"AN ARROW IN THE DARK"

One government official spoke out against the confiscations. "It does not make sense in this day and age to block information, because ultimately citizens, using various means and methods, will gain access to the information they seek," said Mohammad Reza Saidee, a parliament deputy from Tehran. "Trying to negate information is like shooting an arrow in the dark." Predictably, Iran's government hadn't switched off the power to its own satellite broadcasters, who stuck to their global schedule. Soon after, citizens began finding their way to new satellite frequencies, as dishes gradually became available again to careful, media-starved Iranians.

Sources: "Iran Crackdown," *Global News Wire,* 1 February 2002; "Iran: Hardliners Target Internet, Satellite Dishes," *Inter Press Service,* 27 December 2001; Islamic Republic of Iran Broadcasting (www.irib.com).

Moderate Rule

Many nonauthoritarian countries tend to have **paternalistic** policies—applying moderate controls to the media while supporting their growth. Radio and television can spend less time battling repression and more time providing news and entertainment. An example of such progress is South Africa, a predominantly black nation once run by a white minority. One popular TV entertainment show is *Madam and Eve,* in which Madam is a wealthy white woman and Eve her black maid. Their struggle to shed old prejudices and inhibitions in order to get along together has seemed to help South African viewers work through their own post-**apartheid** (white-run discrimination) anxieties. The program would have been suppressed little more than a decade ago, when the country's old political system still was in place.[12]

Apartheid had other effects on broadcasting, at least one of them memorably bizarre. The government-controlled South African Broadcasting Corporation aired newscasts in English, Afrikaans (the unique tongue of the country's white Dutch settlers), and the indigenous languages Xhosa and Zulu. However, the few black anchors who were permitted to deliver news in English and Afrikaans reportedly had to do so utterly free of "black" accents, to avoid offending white viewers. Even in this subtle way, the media under apartheid were forced to acknowledge publicly who was in charge of the country.[13]

Laissez-Faire Rule

A number of economically advanced countries have *laissez-faire* governance, which entails neither sponsoring media nor getting in their way at every turn. The media in these countries usually are run as private concerns and have a large measure of freedom in what they transmit. This approach prevails in the United States, where business goals generally hold sway. Political appointees to the Federal Communications Commission (FCC) have power to thwart broadcasters' moves—to buy up

Paternalistic Fatherlike—in both the kind and the stern sense—in dealing with people, groups, or nations; some governments tend to behave paternalistically toward business.

Apartheid Racially based political system in which white rulers of South Africa discriminated against black citizens in most aspects of life; this included forcing them into all-black "homelands."

Laissez-faire Noninterference in the affairs of others; in governance, tolerance of autonomous action by citizens or organizations.

competitors, for example—but in recent years have tended to facilitate and even encourage such actions. While Congress can pass restrictive laws, it must work within the bounds of the media-shielding First Amendment to the Constitution; anyway, lawmakers depend on broadcasting to transmit their political messages.

Still, even progressive countries find they must impose some rules: U.S. law prohibits the broadcast of material that is judged to be obscene, for example. Also, in every country with multiple stations, there's a traffic-directing function for the government, since broadcast signals occupy the publicly "owned" airwaves and can interfere with one another. This has led to regulation, including licenses that specify the conditions under which radio and TV stations may operate. Licenses usually do not set very specific limits or requirements on programming content, instead obligating broadcasters to honor general principles such as "the public interest." Indeed, truly daring adventures rarely happen in commercial radio and television, because broadcasters want most to keep operating without government meddling; it's a basic urge that mainly reflects economic interests.

Government Funding

In the United States, most broadcasters meet expenses and make profits by charging advertisers money for using the airwaves to push their products. However, lightly regulated commercial broadcasting is but one economic model. Another, generally known globally as *public-service broadcasting*, draws substantial funding (and, with it, close scrutiny) from the government. When governments demand total control of their country's media—which have the built-in power to reach and possibly arouse the "masses"—the most direct approach is to run them. A government that wants to *be* a broadcaster can use a routine tax or whatever other revenues it has to underwrite the costs.

Iraq, a former monarchy that gave way to a dictatorship, is an example of this. Half a century before military ruler Saddam Hussein challenged the Western powers, Iraq's king launched a national radio station. The revolution that overthrew the royalty in the 1950s expanded radio and launched television—supported by funds from the Soviet Union.

SHOWING THE WORLD. A TV journalist reports from the "spider hole" in which Iraqi dictator Saddam Hussein was hiding on December 14, 2003, when U.S. troops captured him. International electronic media send such powerful images into homes worldwide.

Britain's approach to financing broadcasting is a function of its political and economic history. The famous BBC (British Broadcasting Corporation) was set up in 1922 as a cooperative within the radio industry, which already was growing on the commercial model. It was the *government,* however, that pressured private companies to accept the BBC in their midst; it was the government that permitted it to thrive on license fees paid by radio-set owners. The BBC (affectionately called "the Beeb") thus became a sort of public-service monopoly with a clear, officially sanctioned edge over its commercial rivals.[14] This approach spread across Western Europe, tying the fortunes of broadcasters to license fees.

More recently, the BBC has strayed from its roots as a noncommercial public-service entity that draws sustenance directly from the people. Like U.S. public broadcasters, it has suffered mushrooming costs and increasing competition, and there

are limits on its chief revenue source (license fees). The result is that the Beeb is selling more and more of its programs and many "tie-in" products on the international market in an effort to create a wide and reliable income stream. This is controversial among many BBC fans and employees who have resisted commercial infiltration, but the British government has acquiesced and leading politicians support the Beeb's new strategy. It includes alliances with foreign broadcasters and draws on commercial tools; in the United States, for example, the cable/satellite service known as BBC America carries advertising.[15] In January 2002—to the chagrin of traditionalists—the BBC announced the creation of its second commercial-TV division.

The government used broadcasting mainly to transmit messages and to reinforce cultural, religious, and political norms that would strengthen the regime's hold on power. Later, Saddam Hussein poured oil money into the purchase of powerful transmitters to give Iraq a louder radio voice in the Middle East. Most of its programming remained political in nature and thus of little appeal beyond the Iraqi people. Until the United States deposed him, Saddam retained control of both radio and television as government-run propaganda tools.[16]

PRINCIPLE 4

Mixed Money Sources

Broadcast economics around the world is a patchwork quilt, varying from one country to the next. Government often seeks both to influence and to support the growth of broadcasting by **subsidizing**—contributing money to—private companies. Then, often, the companies may sell advertising to gain other revenues. With government blessings, they also may charge citizens license fees on their home receivers. Japan's vast NHK network subsists on such fees; it's a public-service network like the BBC, barred from amassing profits.

Different forces are at work in every country, however. Consider, for example, Argentina. When TV arrived in the 1950s, three private channels tried to do business in a country which, in one researcher's words, has had "a jagged history of military intervention and populist dictatorship."[17] Dictator Juan Perón **nationalized** broadcast outlets in the 1970s (turned them into government property) and then they were **privatized** again in the 1980s (returned to private hands). Since then, aided greatly by U.S. and other foreign investors, Argentine TV and radio have expanded and grown affluent on advertising dollars—but remain politically cautious.

In France, as in other European countries, advertising does not bear the full financing burden. The information ministry subsidizes broadcasters, providing part of their income while at the same time practicing paternalism by controlling their programming and limiting their competition from foreign broadcasters.[18] Many other nations grant economic protection to their broadcasters. This helps domestic operators monopolize the advertising marketplace. European **unification** moves—lowering economic barriers among generally friendly countries—plus international

> Authoritarian rulers see popular media as avenues to control the people.

Subsidizing Financially assisting persons or organizations; for example, governments sometimes subsidize farmers to help them keep producing food, or subsidize broadcasters to help them stay on the air.

Nationalize To convert private property (often companies or industries) to government or public property.

Privatize To convert public property to private property, as when a state or school system turns over some of its responsibilities to private business.

Unification The uniting of different forces or groups; a number of European countries have moved toward unification by agreeing to use a common currency.

media mergers since the fall of communism have helped to expand consumer markets and boost expenditures on advertising. Robust demand for products and services can feed enough money into a radio or TV system to reduce its reliance on government funds.

The world broadcasting map shows both progress and stagnation. Most of Western Europe has pushed privatization in recent years as Europe's unification advanced. The trend extended to postcommunist countries: Hungary began privatizing broadcasting in 1995 and now has nine commercial TV stations that run advertising.

Croatia, part of the former Yugoslavia, is still tense over its region's future, and has yet to set broadcasting free. State-owned broadcasting does accept advertising and does acquire satellite-fed programs from other countries. The operating key is that Croatia's three TV channels have a 95 percent market share in their country—that is, they command 95 percent of the viewership and the ad market.[19] A country that so effectively discourages foreign competition can avoid taking the controversial step of banning it.

▗ Global Reach

When astronaut Neil Armstrong walked on the moon in July 1969, 600 million people watched on earth via live television. They could do so because a huge satellite dish had been placed in a small Australian town named Parkes. Pictures and sound sent by *Apollo 11* were bounced from that dish to other, smaller dishes that pulled the signals into TV networks everywhere. The moon walk was new then, but this technology wasn't. The United States had launched its first communications satellite nine years earlier. Live TV pictures from the United States were reaching England via satellite by 1962. So, by the sixties, television had become extensively international, spanning oceans with live signals sent into space and relayed back to the home planet.

Decades later, humans work in space stations, and thousands of satellites have been launched, many for communications use. Media companies increasingly are leaping across continents, racing one another to connect the right services to the right populations and maximize their profits.

TAKING CONTROL

Those who own or control media in many lands around the globe—or would like to—see revenues at many levels. They come from advertising (of products known and purchased globally); from subscription fees paid by consumers; from side deals with other companies, often in other countries; and from many other sources.

Going it alone has proved difficult for even long-established businesses trying to reach millions of new customers for broadcast programming and high-tech communications services. Instead, a media company—often led by a bold chief executive—often moves to execute a merger or acquisition that can help the company assemble resources needed to ring the planet. Undoubtedly, the most prominent (and prominently aggressive) global-media magnate to follow this course has been Rupert Murdoch, the Australian-turned-U.S.-citizen whose family controls News Corporation (see Spanning the World).

However, several U.S.-based giants could compete with Murdoch for global reach. They include Time Warner, already the world's largest media company; Viacom; Walt Disney Company; and the gigantic conglomerate NBC Universal, which houses the European broadcast-satellite powerhouse Vivendi (and the ultra-

RUPERT MURDOCH. His name became synony-mous with global reach as the Australian-born media baron gained footholds in more and more countries.

The most aggressive media globalizer, Rupert Murdoch, began life in Australia but has succeeded beyond any borders. He reigns over a vast media empire and is one of the world's very richest men, with a personal fortune estimated at $7.8 billion. He also, at this writing, is in his mid-seventies, with two sons and a daughter already in major roles in the family firm, News Corporation. Adding to the air of corporate royalty, Murdoch became a father again in 2001, when his third wife gave birth to a girl—fifth in line to his throne.

Murdoch has built his emerging dynasty with passport in hand. Starting with a newspaper chain in Australia, he moved on to England and assembled a tabloid empire, getting into broadcasting with a "superstation" sending programs around Europe.

Murdoch migrated to New York in the 1970s and by 1985 was becoming a U.S. citizen in order to buy the six Metromedia TV stations for $1.55 billion. He then created Twentieth-Century Fox Film Corporation, acquiring what would become the Fox TV stations.

By 2005, Murdoch's News Corporation owned 35 U.S. television stations. The corporation also con-trolled British Sky Broadcasting, a leading European satellite-TV carrier. It had stakes in pay-TV services covering Germany and Italy. News Corp. owned or had large stakes in DirecTV in the United States, plus supplied programming to Asia—a reported 300 mil-lion viewers in 53 countries—through STAR TV, with Murdoch's son James at the helm. Its joint venture in China had bundled U.S. cable channels like Fox Sports and Fox News and claimed to have become China's most-watched foreign source. News Corpo-ration had moved into programming vacuums from Mexico to Europe to its Australian homeland.

Adding the company's interests in cable networks and other enterprises, it seemed possible that an aging Australian could snap his fingers and darken many screens around the world. Murdoch had one more angle: a growing corporate arm called NDS, which was exploiting new technologies. It sold electronic gateways that denied viewers access to pay-TV channels unless they'd paid, and also had launched a business allowing Brazilians to do interactive banking and helping users in the Chinese province of Sichuan trade stocks via satellite.

high-valued General Electric Company). These corporations and a few others work to keep themselves constantly in position to acquire media holdings—worldwide.

Some acquisitions are aimed at making money off audiences within individual countries by running their broadcast outlets, or by supplying programming to them. But, increasingly, an eye for "synergy" prompts corporate leaders to seek interconnection between what they already own and the tempting media properties across the sea.

Paths to Audiences

Regardless of who controls them, it's the hardware and software of far-flung com-munications systems that keep media programming flowing to and from points around the globe. Broadcast content can be sent in its entirety through satellite relays to distant hub cities or to individual customers at home. Wired, land-tethered systems also are coming into heavy use as planned extensions of satellite networks. Many of these space-to-land pathways have been constructed by U.S.-based media companies working with in-country partners overseas.

PRINCIPLE 5

The urge to merge and to dominate media crosses boundaries as easily as signals do.

TV IN SPAIN. Most economically advanced countries have developed home-grown media environments. Although they create much of the programming popular with citizens, there remains a brisk traffic in shows imported from elsewhere, especially the United States.

Satellites and Synergy

Communications satellites have been around since the 1960s, born as one of the U.S. responses to the Soviet launch of *Sputnik* in 1957. Predictably, the Department of Defense had an edge in that era, and the best early satellites were developed for military use. But soon, some of their technologies were spun off for the great communications marketplace. RCA, American Telephone and Telegraph (AT&T), and other companies began building their own "comsats." One of them picked up video from the 1964 Tokyo Olympics and passed it on to distant receivers. Within a few years, it became routine for at least part of a satellite's relay capacity to be reserved for use by U.S. TV networks, among others who paid for the service.

Never have such orbiting links been employed more broadly than they are on the global-media scene today. EuroNews, the pan-European television consortium, began using satellites in 2001 to serve Russian television audiences; in 2002, it added subscribers in Spain, the Federal Republic of Yugoslavia, Macedonia, and the Czech Republic. All that brought the enterprise several million viewers. In 2003, it gained 27 million more by having its sky-fed "channel" passed through Russian cable systems in 89 regions—more than half of that vast country. Now, the All-Russian State Television and Radio Company (VGTRK) has become the fourth major stockholder in EuroNews.

Thus, the reach and influence of a single news source may be multiplied many times through a combination of satellite and land-based distribution. Similar networks have been or are being established in other regions of the world. The typical net is likely to be limited to a particular region by the maximum number of languages and cultural differences it can juggle. Some experts see such issues as much harder to handle than the technical issues involved in maintaining satellites.

The Internet eventually may fit into space-land networks, and might even help break down region-to-region boundaries. In a broadband future, Internet hubs might be able to superimpose local languages on some content (or, if necessary, "gatekeep" incoming languages to avoid cultural offenses) and move it into new territory. This could help companies extend their programming into fresh markets— for better or worse. The growth of already huge global enterprises such as News

For years, broadcast networks based in Latin America—and some that have sprung up in the United States—have aired news and other programming in Spanish. When one such company, Telemundo, joined in broadcasting a live U.S. telethon for survivors of the 2001 terrorist attack on the World Trade Center, it reflected the presence of new immigrants as well as millions of Hispanic natives of the United States. The network translated the event into Spanish for 7.7 million Hispanic households.

Telemundo, founded in 1986 (and later reborn after a bout with bankruptcy), is the nation's second-largest Spanish-language broadcaster, claiming to reach 85 percent of all Hispanic Americans. The good news for advertisers is that total buying power for Hispanics increased by 84 percent to $383 billion during the 1990s. The Latino boom in broadcasting has raised

an issue common to both Telemundo and its larger rival, Univision. In general, the question concerns assimilation; in particular, a soap-opera genre called *telenovelas* that has thrived, in Spanish, on both networks. It reminds viewers of their foreign heritage. In the view of Hispanic critics, however, including some in the media, such programming may be delaying immigrants' full transition into American life while excluding many of those who have completed that transition. The end result is a cultural divide within the U.S. Latin community. TV advertising reaches all, however, which helps explain why, in fall 2001, NBC paid $2.6 billion to acquire Telemundo.

Corporation and Disney has been somewhat unpopular in their homelands but less so in countries waiting to be served more programming.

NEWS THAT TRAVELS

The power of electronic media can be detected almost everywhere humans live today, and at no time more dramatically than when news breaks. CNN war reporter Christiane Amanpour told a national audience of news directors that she feels it as she travels: "I am so identified over the world—because CNN is seen all over the world . . . that wherever I go, people say jokingly, or maybe not so jokingly, that they shudder when they see me: 'Oh my god. Amanpour is coming. Is something bad going to happen to us?' "[20]

Filling the Skies

CNN is based in the United States, but has TV outlets in more than 200 countries and claims to reach more than a *billion* viewers. This path-breaking enterprise sprang from the fertile mind of a media visionary, Ted Turner, whose story by now is well known. He launched the cable network in 1980, used foreign journalists to supply news from abroad, added his own staffers, and dominated coverage of the 1990–1991 Persian Gulf War. In the process, Turner forced national leaders around the world to watch Atlanta-based CNN. This was because the network often acquired important information before their governments did and fed it via satellite to their own constituents. Satellites have been used to cover the biggest international TV stories—and radio stories, too, generating pictures in our heads that reach beyond mere "news."

Picture one memorable moment: A young Chinese man standing bravely in front of an oncoming tank in Beijing's

CHRISTIANE AMANPOUR. CNN's most conspicuous globe-trotter is recognizable—and, she says, a little unnerving to the locals—wherever she shows up around the world.

THE WORST. When militants took over a school in Beslan, Russia, they shot video of their captives, and it quickly was aired on Russian television. When such stark force and mortal fear become visible, viewers worldwide may find reason to take sides.

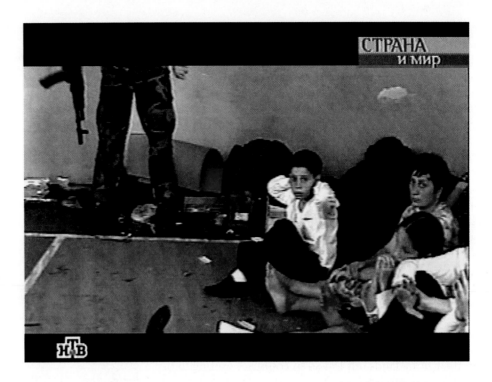

Tienanmen Square. CNN transmitted that image via satellite in August 1989 as a pro-democracy revolution bubbled over. It turned out that the strength—or brutality—suggested by a large piece of military equipment in proximity to a human would shake China's world image, spurring outrage. Lewis A. Friedland has noted that the subsequent live coverage made CNN temporarily "the primary source of information for much of the U.S. government" in an international moment of tension.[21]

One of the more important effects of those events in China was to demonstrate the danger of taking television lightly in an increasingly connected world. At least during crises, global leaders no longer could afford to wait for their aides to prepare a neat summary of the day's stories (as President Ronald Reagan preferred in the 1980s). More than a dozen years later, news that's happening now can be captured and transmitted from almost anywhere.

Now other sources of TV news—many of them based outside the United States—populate the global marketplace. A sample of the smorgasbord: Via satellite, DirecTV carries news from the U.S. networks ABC, CBS, and NBC; the United Kingdom's BBC; CNN International; Canadian-run News World International (offering foreign newscasts in their own languages or English); Galavision, broadcasting from Mexico; Univision, from Mexico and Venezuela; the nonprofit Worldlink TV; and other programming sources that at least occasionally provide international news. A world as closely linked by news as this needs coverage that is shared quite openly and fluidly. On the contrary, U.S. journalists working abroad constantly rediscover that their country's free-press model is the exception to the international rule.

Industrialized countries routinely send reporters, cameras, and microphones into the world's news "hot spots." Any globetrotting journalist who walks into one of these hot zones is potentially in danger. Journalists—whether foreign or domestic—sometimes are killed or captured while reporting in developing countries torn by political turmoil or civil war. A nonprofit support organization called the Committee To Protect Journalists (CPJ) compiles the annual death toll. In 2003, it added up to 35 journalists killed worldwide. Among them were nine journalists killed in

PRINCIPLE 6

Global media for entertainment—valuable; global media for news—priceless.

Afghanistan by Taliban forces or hired gunmen as the Western "war on terrorism" moved into gear.[22] The dead also included four radio journalists—in Costa Rica, Thailand, and the Philippines—who had criticized powerful forces or commented on political scandals and were killed by unknown assailants. In 2002, Islamic militants in Pakistan captured Wall Street Journal reporter Daniel Pearl; officials announced a month later that he had died in his captors' hands. His murderers later released a video to document their atrocity.

Technical Wizardry

Overall, newsgathering technology has played a huge role in international journalism. In fact, without the development of increasingly sophisticated gear for sending pictures and words over land and through the sky, reporters would have transmitted far less news to us over the past 150 years. The telegraph, telephone, and wireless communication helped supply us with foreign dispatches from the mid-nineteenth century through the mid-twentieth century. During the Persian Gulf War of 1990–1991, small network news teams with truckloads of technology and portable satellite dishes transmitted live pictures to the United States from embattled Kuwait before Allied liberation troops arrived. As the legendary broadcast journalist Daniel Schorr noted, such techniques could confound government censors on both sides by literally flying over them.[23] Veteran CNN correspondent Jim Bittermann put it this way: "Where earthquakes, coups and crises happen, the satellite dish is there, for better for worse."[24]

Critical to coverage of the Kosovo conflict in the late 1990s was the satellite phone. It predated the videophone and handled only voice communication, but it was a ramp into space, where no official was standing ready and eager to intercept and censor digital signals.[25]

Foreign correspondents expect discomfort, and modern advances have made that likelier than ever. Bittermann said that the portability of today's satellite gear means that, rather than hustling back to safe hotels after a day's action, reporters "are very often eating and sleeping in the disaster they are covering."[26] When he made that comment, he hadn't even worked with the **videophone.** It was April 2001 when this device—actually a data-handling package the size of a small suitcase—gave CNN a big international news "exclusive." It came when an enterprising TV crew furtively photographed a U.S. plane being held on a Chinese island, using a digital camera and the videophone to feed live pictures and sound to the world via satellite. Much cheaper and smaller than previous video-feed systems, the videophone came to be useful at especially sensitive news scenes.

Today, besides the videophone, a whole array of electronic tools comes into play, especially in television. The "miniaturization" of field television gear is an unceasing process as news companies seek to get difficult work done with as few workers as possible. In early 2003, awaiting the impending U.S. war on Iraq, NBC correspondent Kerry Sanders told an interviewer: "There has been an amazing, quantum leap in broadcast technology. A year ago in Afghanistan, the gear we needed to go 'live' filled 75 to 100 cases. With the technology improvements we've seen since then, we can now fit everything we need in five or six."[27]

This phenomenon is evident among the press corps of many developed countries where new-media tools, however costly, are affordable to networks. Correspondents can make wireless connections to the Internet, from compact but high-powered laptop computers. They can do extensive research from the same computers, then transmit finished broadcast stories or segments as video and/or audio files. Improved broadband applications make it likely that the videophone's shaky images will settle down, restoring the production gloss to which viewers worldwide are accustomed (or becoming so).

But technical advances in television news have long been a two-edged sword: What they facilitate—speed, dexterity, visual flash—is what some news managers

Videophone Field transmission kit that allows TV correspondents to send video and voice via satellite from remote news locales to their home countries.

inevitably come to *require*—what they promote and showcase. This, say some critics, has endangered the journalistic value of the enterprise, which presumably is its reason for existing.

A British scholar put it harshly enough: "For the media, war is now about transmitting information fast and furiously in order to feed the plethora of 24/7 broadcasting stations that gulp down unverified reports that are often without context."[28]

That, no doubt, should be a topic of concern and debate for all of us as the high-tech globalization of news continues into the future.

▌ Into the Future

Some of the most intense concerns for the future of international electronic media are directed at the growth of the corporations that increasingly shape what people view and hear around the world. Another issue addresses the quality of journalism and the influence of news technology.

CONTROL, CULTURES, AND ECONOMIES

The world map of media control is changing slowly, but clearly most of the power over programming is in American, European, and Australian hands. This leaves several continents relatively weak in influence over media content and thus—some argue—exposed to whatever cultural messages the richest nations wish to force on them.

This argument includes the cultural-imperialism claim that global media will tend (or attempt) to convert every listener, viewer, and Internet consumer to Western values, including a desire for consumer products. As media scholar Robert McChesney points out, "infomercials" and TV shopping channels are among the hottest-selling programming on the world market.

As competition grows, Western media are cutting deals that bring local distributors and producers into the revenue stream. This begins to deposit money in other countries, not just send it back to the producers' own bank accounts and shareholders. Among other results, if this trend grows, it could encourage the "localism" that some believe is the key to global media that will satisfy all. In the meantime, Western-generated entertainment and information continue to have negative effects on much of the world, says McChesney. He says that, because they see consumerism as equal to democracy, young people in postauthoritarian countries such as Chile are buying more and voting less.[29]

Another analyst, Dietrich Berwanger, argues that cultural imperialism through media has done little to impose the will or values of rich nations on the world's less affluent societies—at least "compared with the effects of Western religion, economy, and weaponry" throughout history.[30] Recently, important media producers have begun to emerge in some of the global media's client countries: India's Kishore Lulla sold his subscription TV channel featuring Bombay-made movies and music to 170,000 South Asian-immigrant subscribers on six continents—a sort of reverse cultural imperialism.[31]

JOURNALISM

News, if narrowly defined as "breaking" events, can burst out anywhere, anytime. But seen more broadly, news is occurring continuously; it's the endless recording and analysis of issues, the "first rough draft of history," as the late *Washington Post* publisher Philip Graham described it. Will future journalism explain the world to

INTERNET USER (WORLD).
Still incalculable are the effects of spreading Internet use on nations and cultures that have lacked U.S.-style access to the fruits of electronic media.

itself consistently and conscientiously? If not, how will new international relationships and entanglements—economic, cultural, political, even military—be understood and evaluated by world citizens?

Some answers may be found close to home—anyone's home. A country's own journalists often are best able to tell its stories fairly, accurately, and thoroughly. Indeed, when they've been silenced by their own governments, reporters who know "where the bodies are buried" sometimes serve as invaluable tipsters to arriving foreign correspondents. After all, if local journalists can't report what they know by conventional means, why not give it to the world any way they can?

Unfortunately, many serious developments go unreported by either domestic broadcasters or the sparse and busy corps of globe-trotting journalists. Globalization may begin to lift some government pressures off the shoulders of indigenous journalists. Until such progress appears, however, our common knowledge of many regions and nations is bound to remain sketchy and spotty—unless media corporations begin to invest in a larger and more experienced international news corps.

That's improbable, given the U.S. experience: Guided by shrinking budgets and by audience research—some of it from what CNN's Christiane Amanpour has called "hocus-pocus focus groups"[32]—TV networks since the 1980s have shifted increasingly toward human-interest and consumer news. The networks have cut back on government coverage (except during crises or scandals), slashed overseas staffs, and reduced the amount of foreign news on TV by more than half.[33] The September 2001 terrorist attacks and the 2003 Iraq war boosted the coverage again—but for how long?

Summary

Electronic media are the products of more than a century of international invention—and are inherently international, thanks to their ability to hurl messages over borders and oceans. Radio is as pervasive on other continents as it is in the United States. It has connected citizens at the village level and across cultural boundaries, and has served governments and social movements. Private money, public funds, and sometimes a blend of the two have helped broadcasting thrive.

Television is now beamed into almost every corner of the earth. Propagated via satellite, animated by Hollywood movies and tailored to regional cultures, TV is at the heart of a global commercial enterprise. Corporations have grown fabulously wealthy transmitting U.S. media products around the world. Although some of their work draws criticism on intercultural grounds, the infrastructure and technologies created and used by global electronic media help to move news and entertainment from almost anywhere to almost anywhere else. Internet initiatives have given millions of world citizens a new tool with which to acquire news and information and exchange culture, in more self-tailored ways than the traditional media provide.

Food for Thought

1. Is *cultural imperialism* by media-producing countries a serious threat to less-developed nations and communities? How? Give examples.
2. How important is *news* in most of the world's societies? How does it compare with entertainment as a driving force in the spread of global media?
3. Why do Americans, in general, show little interest in TV and radio programs from other lands—when those countries show great interest in U.S. media?
4. Is the global transmission of government propaganda—including deliberately deceptive reports from the Pentagon and State Department in their "war on terrorism"—a legitimate use of electronic media? Explain.
5. If you had the opportunity to begin a dialogue with an African, Asian, or European person via radio or the Internet, would you do so? If you would, then what cultural materials would you wish to exchange?

Notes

Chapter 1

1. For a profound discussion of this process, see Walter Lippmann, *Public Opinion* (New York: Harcourt Brace, 1922).
2. An exception is the low-power FM radio stations that in 2001 were granted permission to seek FCC licenses; their coverage areas can be as small as a single neighborhood.
3. *Random House Dictionary*, 647.
4. Pamela J. Shoemaker, "Hardwired for News: Using Biological and Cultural Evolution to Explain the Surveillance Function," *Journal of Communication* 46/3 (Summer 1996): 32.
5. For more insights on news and fear, see David L. Altheide, "The News Media, the Problem Frame, and the Production of Fear," *Sociological Quarterly* 38/4 (Fall 1997): 647.
6. Nielsen Media Research, via Television Bureau of Advertising website, 30 October 2001, www.tvb.org/tvfacts/tvbasics/basics32.html.

Chapter 2

1. For more on Morse, see Carlton Mabee, *The American Leonardo: A Life of Samuel F. B. Morse.* (New York: Alfred A. Knopf, 1944), and Edward Lind Morse (Ed.), *Samuel F. B. Morse: His Letters and Journals* (Boston: Houghton Mifflin, 1914).
2. See Orrin E. Dunlap, Jr., *Radios 100 Men of Science* (New York: Harper and Brothers Publishers, 1944). See also M. D. Fagen (Ed.), *A History of Engineering and Science in the Bell System: The Early Years (1875–1925)* (New York: Bell Telephone Laboratories, 1975); J. Brooks, *Telephone: The First Hundred Years* (New York: Harper and Row, 1975); and I. De Sola Pool, *Forecasting the Telephone: A Retrospective Technology Assessment* (Norwood, NJ: Ablex, 1983).
3. G. R. M. Garratt, "Hertz," in *The Early History of Radio from Faraday to Marconi* (London: The Institution of Electrical Engineers, 1994), 34–50. See also Oliver J. Lodge, *Signaling through Space without Wires: The Work of Hertz and His Successors,* 3rd ed. (New York: Van Nostrand, 1900; reprint New York: Arno Press, 1974).
4. See Giancarlo Masini, *Marconi* (New York: Marsilio Publishers, 1995); Degna Marconi *My Father, Marconi* (New York: McGraw-Hill, 1962); and Orrin E. Dunlap, *Marconi: The Man and His Wireless* (New York: Macmillan, 1937; reprint New York: Arno Press, 1971).
5. Lawrence W. Lichty and Malachi Topping (Eds.), *Part Two/Stations* in *American Broadcasting: A Source Book on the History of Radio and Television* (New York: Hastings, 1975), 12–19.
6. Louise M. Benjamin, "In Search of the 'Radio Music Box' Memo," *Journal of Broadcasting and Electronic Media* (Summer 1993): 325–335.

7. Lewis J. Paper, *Empire: William S. Paley and the Making of CBS* (New York: St. Martin's Press, 1987). See also William S. Paley, *As It Happened: A Memoir* (Garden City, NY: Doubleday, 1979).
8. Sonia Williams, "Black Drama" and "Black Formats" in D. Godfrey and F. Leigh, eds. *Historical Dictionary of American Radio* (Westport, CT: Greenwood, 1998) 46–49.
9. Frank Buxton and Bill Owen, *The Big Broadcast, 1920–1950* (New York: Viking, 1972) and Vincent Terrace, *Radio's Golden Years: The Encyclopedia of Radio Programs, 1930–1960* (San Diego: A. S. Barnes and Co., 1981).
10. See Howard Koch, *The Panic Broadcast* (New York: Avon Books, 1971).
11. Reuven Frank, *Out of Thin Air: The Brief Wonderful Life of Network News* (New York: Simon & Schuster, 1991).
12. Jay Perkins, "Television Covers the 1952 Political Convention in Chicago; An Oral History Interview with Sig Mickelson," *Historical Journal of Film, Radio and Television* 18, no. 1 (March 1998): 95.
13. Melvin Patrick Ely, *The Adventures of Amos 'n' Andy—A Social History of an American Phenomenon* (New York: Free Press, 1991).
14. Ben Fong-Torres, "Like a Rolling Stone Richard Fatherly Knows Best," Special to the Repository of Reel Radio Presents a Special Report: Todd Storz and Radio's Revolution, www.reelradio.com/storz/index.html.
15. As cited by David Weinstein in a related entry on Todd Storz in *Historical Dictionary of American Radio*, ed. D. G Godfrey and F. A. Leigh (Westport, CT: Greenwood, 1998), 373–374; "King of Giveaway," *Time* (4 June 1956): 100–102. See also David T. MacFarland, *The Development of the Top 40 Radio Format* (New York: Arno Press, 1979).
16. "America's Long Vigil," *TV Guide* (25 January 1964).
17. Mark J. Braun, "FM Radio (Frequency Modulation)" in *Historical Dictionary of American Radio*, ed. Godfrey and Leigh, 164–166.
18. "Reality Bubble Bursts—Agency Finds TV Genre Losing Steam," *Media Daily News*, MediaPost, 14 October 2004, www.mediapost.com.
19. Norman Pattiz, telephone interview by author, 11 November 2000.
20. "Local Broadcasters Offer Cheaper Premium Services," *The New York Times*, 3 May 2004.

Chapter 3

1. Spot-beam satellites will allow limited coverage of local markets. See "Sats Pitch 'Must Carry,'" *Broadcasting & Cable* (31 July 2000): 38.
2. See "Pioneers," at Cable Center web page www.cable center.org.

3. Mary Alice Mayer Phillips, *CATV—A History of Community Antenna Television* (Evanston, IL: Northwestern University Press, 1972), 7–8.
4. Ibid., 9.
5. George Mair, *Inside HBO—The Billion Dollar War between HBO, Hollywood, and the Home Video Revolution* (New York: Dodd, Mead & Company, 1988).
6. Howard J. Barr, "Commission Releases Report on 2001 Cable Industry Prices," Womble Carlyle web page, 11 April 2002, www.wcsr.com/FSL5CS/telecommunication-memos/telecommunicationmemos1252.asp.
7. Ibid., 30.
8. See David H. Waterman and Andrew A. Weiss, *Vertical Integration in Cable Television* (Washington, DC: American Enterprise Institute, 1993), 9.
9. *Cable Communications Policy Act of 1984, U.S. Code,* vol. 47, sec. 531 (1984).
10. William R. Davie and Jung-Sook Lee, "Handling Hate Speech on Public Access Television," *Feedback* 40, no. 3 (1999): 33–41. See also Mark D. Harmon, "Hate Groups and Cable Public Access," *Journal of Mass Media Ethics* 6, no. 3 (1991): 149, 153.
11. James Roman, *Love, Light, and a Dream* (Westport, CT: Praeger, 1996), 241, 248.
12. See Robert W. Crandall and Harold Furchtgott-Roth, *Cable TV: Regulation or Competition?* (Washington, DC: Brookings Institute, 1996).
13. 26 F.C.C. 403 (1959).
14. First Report and Order, 38 F.C.C. 683 (1965).
15. Second Report and Order, 2 F.C.C. 2d 725 (1966).
16. *United States* v. *Southwestern Cable Co.,* 392 U.S. 157 (1968).
17. *Quincy Cable TV, Inc.* v. *F.C.C.* (D.C. Cir. No. 83-1283, 1985).
18. "Alternative Delivery Systems," *Electronic Media* (31 July 2000): 14.
19. FCC, CC Docket 87-266, July 16, 1992, Local Telephone Companies to be Allowed to Offer Video Dialtone Services; Repeal of Statutory Congress.
20. F. Leslie Smith, John W. Wright II, and David H. Ostroff, *Perspectives on Radio and Television,* 4th ed. (Mahwah, NJ: Lawrence Erlbaum, 1998), 139–140.
21. "Western Show Attendance Down," *Multichannel News International,* 30 November 2001. www.tvinsite.com/multiinternational/index.asp?layout+story&doc_id=58879&display=breaking News.
22. See "Industry Statistics," National Cable & Telecommunications Association, www.ncta.co . . . _overview/indStats.cfm?statID=7.
23. See "Career Opportunities with Cable Systems," National Cable & Telecommunications Association, www.ncta.com/careers/careers.cfm?careerID=2.

Chapter 4

1. Joel Brinkley, *Defining Vision—The Battle for the Future of Television* (San Diego: Harcourt Brace, 1997), 146–148.
2. Ralph Donald and Thomas Spann, *Fundamentals of Television Production* (Ames: Iowa State University Press, 2000), 138–139.
3. In video, RGB colors are kept isolated and delivered from their source to the display device over separate wires, resulting in higher-quality pictures.
4. Originally, UHF was channels 14 to 83 after the freeze was lifted in 1952 on new station licenses. This UHF plan provided for more than 2,000 stations in about 1,300 communities, including 242 noncommercial and educational stations. Since the UHF band represented much higher frequencies than the original VHF channels, it presented a problem—most existing TV sets could not receive UHF channels. The FCC then mandated that all new TV sets must be able to receive both VHS and UHF channels.
5. Stephen Labaton, "255 Licenses Are Awarded for Low-Power FM Radio," *The New York Times* (22 December 2000): C5.
6. "What Is iBiquity Digital?" http://www.ibiquity.com/01content.html.
7. Joan Van Tassel, "Digital Video Compressions," in *Communication Technology Update,* 3rd ed., ed. August E. Grant (Newton, MA: Butterworth Heinemann, 1994), 9.
8. NTSC also refers to the television system called *composite video,* wherein sync, luminance, and color are combined into a single analog signal.
9. Closed captioning text for the hearing impaired takes only one line of the VBI. Broadcasters use other lines of the VBI to communicate with local stations and cable systems, for purposes such as clock signals and commercial timing.

Chapter 5

1. Gary Wolf, "How the Internet Invented Howard Dean," *Wired Magazine,* www.wiredcom/wired/archive/12.01/dean.html.
2. Irwin Lebow, *The Digital Connection* (New York: Computer Science Press, 1991), 167, 170–174. Also "Harry Nyquist," www.geocities.com/bioelectrochemistry/nyquist.htm.
3. Steven Lubar, " 'Do not fold, spindle or mutilate': A cultural history of the punch card," May 1991, ccat.sas.upenn.edu/slubar/fsm.
4. Internet Society, "A Brief History of the Internet," *All about the Internet.* http://www.isoc.org/internet-history/brief.html.
5. Preston Gralla, *How the Internet Works,* 4th ed. (Indianapolis: Que–Macmillan, 1998).
6. M. Mitchell Waldrop, "No, This Man Invented the Internet," Forbes.com, www.forbes.com/asap/2000/1127/105.html. Also "Part I: The History of ARPA Leading Up to the ARPANET," *History of the ARPANET,* www.dei.isep.ipp.pt/does/arpa—1.html
7. Tim Berners-Lee, with Mark Fischetti, *Weaving the Web: The Original Design and Ultimate Destiny of the World Wide Web by Its Inventor* (San Francisco: Harper, 1999).
8. Adam Cohen, "Coffee with Pierre—A Better World—That's the Dream of eBay Founder Pierre Omidyar," *Time.com: 1999 Persona of the Year,* 27 December 1999, www.time.com/time/poy/pierre.html. Also, Susan Moran. "The Pro," *Business 2.0 Magazine Indepth,* Auction Watch

Daily—Viewpoint—The Insider, www.auctionwatch.com/
awdaily/viewpoint/inside/3-082399.html.

9. Saul Hansell, "Demand Grows for Net Service at High
Speed," *The New York Times* (22 December 2001): C1, C4.

10. The Editors of *The New Atlantis*, "Life Is Just a Game," *The
New Atlantis*, 4 (Wiunter 2004): 105–108.

Chapter 6

1. "Internet as Unique News Source," Pew Internet and
American Life Project, July 8, 2004, www.pewinternet.
org/

2. Telephone interview by author, October 11, 2000.

3. Willard Sterne Randall, *A Little Revenge: Benjamin Franklin
at War with His Son* (New York: Quill/William Morrow,
1984), 42.

4. Disney news release on *Business Wire*, January 5, 2000.

5. www.disney.go.com/investors/, January 2000.

6. Ken Auletta, *Three Blind Mice: How the TV Networks Lost
their Way* (New York: Random House, 1991), 4.

7. "Who Owns the Airwaves? Ownership Ranks Rapidly
Thinned by Consolidation," *Electronic Media* (18 May
1998): 1A.

8. Corporate history at Citadel website: www.citadelcommu-
nications.com/about/history.html.

9. "Sinclair Seeks to Refinance Loan to Avoid Default," *Electronic Media Online*, April 3, 2001, www.emonline.com/.

10. "The Myth of Media Concentration: Why the FCC's Media
Ownerhsip Rules Are Unnecessary," James Gattuso, Heritage Foundation. WebMemo #284, May 29, 2003, www.
heritage.org/Research/InternetandTechnology/wm284.
cfm.

11. Hearing of the Senate Commerce, Science, and Transportation Committee on competition in the telecommunications industry, 253 Russell Senate Office Building,
Washington, DC, January 14, 2003.

12. "Independent Production Companies," Museum of
Broadcast Communications, www.mbcnet.org/ETV/I/
htmlI/independentp/independentp.htm.

13. U.S. Bureau of Labor Statistics, 2004, www.bls.gov/oco/
cg/cgs017.htm#addinfo.

14. Ibid.

15. Herbert H. Howard, "TV Station Group and Cross-Media
Ownership: A 1995 Update." *Journalism & Mass Communication Quarterly*, 72, no. 2 (Summer 1995): 390–401.

16. Minority Meda & Telecom Council, www.mmtconline.
org/FAQ_s/rtb/index.shtml.

17. "Forecast Predicts Nearly 60 Million U.S. Homes with
HDTV by 2008," The Yankee Group web site, July 28,
2004, www.yankeegroup.com/public/home/daily_view-
point.jsp?ID=11772.

Chapter 7

1. John Giglione, "Suddenly, 'Reality' TV Is Too . . . Real,"
Christian Science Monitor online, 11 October 2001.

2. Chris Carter, closed-circuit television interview, National
Association of Television Program Executives Educational
Foundation, 1997.

3. David E. Kelley, closed-circuit television interview,
National Association of Television Program Executives
Educational Foundation, 1998.

4. "Kelley Puts Himself on Personal Hiatus," *Television Week*
(3 May 2004): 6.

5. *Success* refers to several cable news channels' ability to
prosper while drawing audiences, sometimes topping 2
million viewers nationally but usually less—for some, far
less. By contrast, nightly newscasts on the old "Big Three"
broadcast networks are attracting, at this writing, between 6 million and 9 million viewers. That's still a small
percentage of the total TV audience, which in recent years
has been scattered among many more outlets.

6. "Rumble in the Morning: *Today Show* Stumbles as Rivals
Sharpen Up," *Broadcasting and Cable* (26 July 2004): 8.

7. Telephone interviews by author, August 2001.

8. "Langley Collars Fox on *Cops* Deals," The *Hollywood Reporter*, 11 July 2000.

9. "Network Affiliation and Programming," E. W.
Scripps annual report for 2000, www.scripps.com/
2000annualreport/financials/08.html

10. Peter Maroney, conversation with author, 29 June 2000.

11. Show syndicators typically base their prices on a station's
audience size; thus, New York and Los Angeles stations
would be charged much more for the same show than
Portland stations would; small-city stations would pay
much less.

12. Maroney, conversation with author.

13. "Genre Looks for Third Act," *Daily Variety* (8 June 2004):
A2.

14. Marilyn Lavin, "Creating Consumers in the 1930s: Irna
Phillips and the Radio Soap Opera," *Journal of Consumer
Research* (June 1995): 75.

15. "WCLV to Play Full Pieces on Monday Marathons," *The
Plain Dealer* (Cleveland, Ohio) (2 August 1998): 31.

16. "Tuning in to Hispanic Music," *Billboard* (8 December
2001): LM-1

17. "Vying for Listeners: KBIG Tunes in to Likes, Needs,
Habits of Women," *Los Angeles Times* (On the Air
column) (30 May 2000): B4.

18. "'The Basket' Shoots. . . . Will It Score?" *Spokane* (Wash.)
Spokesman-Review (30 April 2000): F3.

19. "Citadel Yanks *Don & Mike*; Replacement Show Sought,"
Albuquerque Tribune (28 August 1999): D5.

20. Young viewers may not have realized that TV "reality"
arguably began half a century earlier, when CBS's *Candid
Camera* began capturing ordinary people in mildly embarrassing situations. Often adapted before Fox launched
Cops and similar shows in the 1980s, reality also brought
us *When Animals Attack* and similar thrillfests.

21. Dyan Machan, "Barry Diller's Next Course," Forbes (9
March 1998): 122.

22. "Little to Head RuffNation; Radio Vet Assumes President's
Post Jan. 15," *Billboard*, 30 December 2000.

23. Michael Stroud, "*Felicity* Voted Most Likely to Succeed,"
Broadcasting & Cable, 7 (September 1998): 22.

24. "FCC Inquiry Focuses on Impact of Violent TV Shows,"
Communications Daily, 29 (July 2004).

25. "TV Kids Become Smokers, Eaters," 15 July 2004, Associated Press.
26. "Transfixed by the Tube," *Cleveland Plain Dealer*, 22 April 2004.

Chapter 8

1. Ted Koppel, telephone interview with Al Tompkins, Poynter Institute for Media Studies, 29 April 2004.
2. The prior publicity evidently helped. *The Fallen* scored 22 ratings points higher nationally than the previous Friday's *Nightline* and 29 points above the previous week's average—despite the Sinclair stations' pullout. However, because no commercials were shown during the program, its ratings were not included in the May statistics by which future advertising rates were calculated.
3. "On *Nightline*, a Grim Sweeps Roll Call," *Washington Post* 28 April 2004, C01.
4. Bureau of Labor Statistics, U.S. Department of Labor. http://stats.bls.gov/oes/current/oes273020.htm.
5. A. M. Rubin, E. M. Perse, R. A. Powell, "Loneliness, Parasocial Interaction, and Local Television News Viewing," *Human Communication Research*, 12, no. 2, 155–180.
6. Deborah Tannen, *The Argument Culture: Moving from Debate to Dialogue* (New York: Random House, 1998).
7. "Popular Policies and Unpopular Press Lift Clinton Ratings," report by the Pew Research Center on the People and the Press, 6 February 1998. www.people-press.org/content.htm.
8. *The State of the News Media 2004: An Annual Report on American Journalism*, Project for Excellence in Journalism (2004). www.stateofthenewsmedia.org/.
9. Peter Fornatale and Joshua E. Mills, *Radio in the Television Age* (Woodstock, NY: Overlook Press, 1980).
10. "On the Road to Recovery," RTNDA/BSU local news survey, 2003. www.bsu.edu/web/rpapper/staff%2003%20v6.htm.
11. "Why the Overhaul of TV News Shows?" *U.S. News and World Report* (20 November 1978): 51.
12. Ken Auletta, *Three Blind Mice* (New York: Random House, 1991), 341.
13. "On the Road."
14. "On the Road."
15. Deborah Potter, "Getting What You Pay For," *American Journalism Review* (October 2000): 94.
16. "Female Anchors on Local TV Paid 28% Less," *Los Angeles Times* (1 June 2000): A1.
17. "Fox Trots Ahead of the Rest," *Houston Chronicle* (8 March 2001): 4.
18. "Diverse Auds Just Want to Be Shown the Money," *Variety* (26 April–2 May 1999): 36.
19. "New Comcast Channel to Air Regional Sports," *Detroit Free Press*, 18 August 2004. www.freep.com/money/business/comcast18e_20040818.htm
20. *A Nation Online: How Americans Are Expanding Their Use Of The Internet*. National Telecommunications and Information Administration, February 2002.
21. *Internet Sapping Broadcast News Audience; Investors Now Go Online for Quotes, Advice*, Pew Research Center for People and the Press, 11 June 2000.
22. Ad published in *ShopTalk* (Internet newsletter for TV-news industry), July 12, 2000.
23. "Aired Live but Was It News?" *Baltimore Sun* (7 March 2001): 1E.
24. *Local TV News Project—2001*, Project for Excellence in Journalism, 15 November 2001.
25. "KDKA Caught Cheating on Spots—Tip of the Iceberg?" *Electronic Media*, 7 November 2001.

Chapter 9

1. "Nominations Reflect HBO's Advantage over Broadcast Networks," *Tampa Tribune* (19 July 2002): 4.
2. Matthew Zelkind personal conversation, July 11, 2001.
3. "Newsman Shaw Signing Off CNN," *Hollywood Reporter*, 13 November 2000.
4. "AT&T Hands Up on $110 Mil; Ma Bell, Shifting Focus, Cancels Upfront," *Television Week* (9 August 2004): 1.
5. Karen S. Buzzard, *Chains of Gold: Marketing the Ratings and Rating the Markets* (Metuchen, NJ: Scarecrow Press, 1990), 15.
6. A. C. Nielsen Co., Museum of Broadcast Communications.
7. "Response-Rate Drop Stopped," *Mediaweek.com*, published 25 June 2001, www.mediaweek.com.
8. "Who Needs the Sweeps? TV's Periodic Race for Ratings Seems to Have Lost Its Purpose," *New York Times* (24 April 2000): C1.
9. Tad Friend, "The Next Big Bet: Is a Family of Depressed Morticians HBO's Best Hope for Life after *The Sopranos*"? *New Yorker* (14 May 2001): 80.
10. For example, in 2002, Nike introduced the *Air Jordan XVII*, a $200 pair of basketball shoes that came in a metal briefcase-style box with an interactive CD-ROM showing how the shoes were made. The target audience was teenaged boys, who, said a retail analyst, "basically keep it under their bed and, when friends come over, show it to them." (From "Nike's New Air Jordans: $200 a Pair," *Register-Guard* (Eugene, Oregon)(2 February 2002): 5B.
11. "A Multicultural Affair: If Your Target Is the Traditional American Family, Make Sure Your Message Is in Many Languages," *American Demographics* Forecast (June 2001): 4.
12. Questionnaire posted on SRI Consulting Business Intelligence website, July 2001, http://future.sri.com/.
13. "'Grain' Pleads for Nielsen Families' Help," *Chicago Sun-Times*, (22 December 1993): 51.
14. Joseph M. Kayany and Paul Yelsma, "Displacement Effects on Online Media in the Socio-Technical Contexts of Households," *Journal of Broadcasting and Electronic Media*, 44, no. 2 (Spring 2000): 215–229.
15. "Advertisers Join Fans in Protesting Dumping of KIRO-FM's Cashman," Seattle *Post-Intelligencer* (8 April 1999): Entertainment, 1.
16. *Lou Dobbs Moneyline*, CNN, 17 July 2001.
17. "One Big Happy Channel," *Salon.com*, 28 June 2001.
18. "A Dim View of the Ratings: Broadcasters Say the Nielsen Numbers Don't Add Up," *Washington Post* (11 April 1996): D9.

19. National Association of Broadcasters, *Survey of Nielsen Ratings Service Quality Issues, Nov. 2000.* Stations queried: 1,069. Responding: 506 (47.3 percent). Sample was all full-power commercial TV stations with known fax numbers, not a random sample; NAB stresses that views of nonrespondents cannot be known.

20. Marc Gunther, *The House That Roone Built: The Inside Story of ABC News* (Boston: Little, Brown, 1994), 31.

21. Interview with Richard Wald, in *Inside the TV Business* (New York: Sterling Publishing, 1979), 215.

22. "Ch. 29 News Chief 'Had a Great Run,' Not So Great Ratings," *Philadelphia Inquirer,* 28 June 2001.

23. "Nielsens Schmielsens: TV Ratings System Is More Unpopular than Ever, But Executives and Advertisers Have Nowhere Else to Turn," *Minneapolis Star Tribune* (20 April 1997): 1F.

24. Posted at Nielsen//NetRatings website, July 2001, www.nielsen-netratings.com/. The PTC also hands out a grade card for the TV networks by charting the "influx of adult-themed programming infiltrating the 'family hour.'" There were only two networks (CBS and WB) with passing scores in 2000. ABC, UPN, and NBC received unsatisfactory grades, and the Fox network flunked PTC's criteria for wholesome entertainment.

25. http://www.jnm.com, 8/16/01.

Chapter 10

1. *Printers' Ink* (April 12, 1922) in Juliann Sivulka, *Soap, Sex, and Cigarettes* (Belmont, CA: Wadsworth, 1997), 183.

2. Sivulka, *Soap, Sex, and Cigarettes,* p. 201.

3. An early version of AdBusters, today's Media Foundation publication, is dedicated to counteradvertising.

4. Sivulka, *Soap, Sex, and Cigarettes,* p. 222.

5. Joe Flint, "Networks Face Off Over Ad Dollars," *Wall Street Journal* (3 May 2004): B8; John M. Higgins, "Ad nauseum," *Broadcasting & Cable* (12 July 2004): 3; "Broadcast Took 99 of Top 100 Programs in July," and "Wired Cable Penetration Hits 9-Year Low in July," TVB website: www.tvb.org/nav/build_frameset.asp?url=/docs/homepage.asp; and "Cable Levels Playing Field On 'Reach-Ability,'" Cable Television Advertising Bureau website: www.onetvworld.org.

6. Charles Warner and Joseph Buchman, *Broadcast and Cable Selling* (Belmont, CA: Wadsworth, 1993), 306.

7. Warner and Buchman, *Broadcast and Cable Selling,* 96–130.

8. Rosser Reeves, *Reality in Advertising* (New York: Knopf, 1960), 34.

9. "Home Shopping Network," 6 July 2001, www.hsn.com/content/article.

10. Jim Sterne, *World Wide Web Marketing,* 2nd ed. (New York: Wiley Computer Publishing, 1999), p. 259.

11. "Pop-Under Ads Fuel Negative Perception for Internet," www.unicast.com.

12. "Cookies," www.illumintus.com/cookie/.

13. Sterne, *World Wide Web Marketing,* p. 259.

14. David Lieberman, "Cable, Satellite, Net Grab Chunk of Election Ad Bucks," *USA Today* (18 August 2004): B1.

15. Ed Fulginiti, "Television Collapsing under Clutter's Weight," *Electronic Media* (31 December 2001): 9.

16. Jean Folkerts, Stephen Lacy, and Lucinda Davenport, *The Media in Your Life: An Introduction to Mass Communication* (Boston: Allyn and Bacon, 1998), 362.

17. Sec. 52 Stat. 111 (1938) and 38 Stat. 717 (1914). F.T.C. Policy Statement on Deception, Appended to Cliffdale Associates, Inc. 103 F.T.C. at 174.

18. *Warner-Lambert* v. *FTC,* 562 F.2nd 749, cert. denied, 435 U.S. 950 (1978).

19. Cigarette Advertising, 9 FCC 2d 921, 949 (1967), aff'd *Banzhaf* v. *FCC,* 405 F.2d 1082, 1091 (DC Cir. 1968), cert. denied, 396 U.S. 842 (1969). Congress enacted the Public Health Cigarette Smoking Act of 1969 banning tobacco ads from television and radio. Because of congressional ban on cigarette commercials, the FCC ruled in 1970 that counterads were no longer required. Mothers Against Drunk Driving appealed to the FCC in 1997 to have it apply to alcohol spots, as well. MADD argued before the FCC for countercommercials to drinking based on the extraordinary impact on health and safety.

Chapter 11

1. "Regulation," *American Broadcasting—A Source Book on the History of Radio and Television,* 2nd ed., ed. Lawrence Lichty and Malachi Topping (New York: Hastings House, 1976), p. 527.

2. Andrew F. Inglis, *Behind the Tube—A History of Broadcasting Technology and Business* (Stoneham, MA: Butterworth, 1990), p. 84.

3. *Hoover* v. *Intercity Radio Co.,* 286 F. 1003 (1923).

4. Marvin R. Bensman, "Regulation of Broadcasting by the Department of Commerce, 1921–1927," *American Broadcasting—A Source Book on the History of Radio and Television,* 2nd ed., ed. Lawrence Lichty and Malachi Topping (New York: Hastings House, 1976), p. 554.

5. *Hoover* v. *Intercity Radio Co., Inc.,* 286 F. 1003 (D.C. Cir.), February 25, 1923, and *United States* v. *Zenith Radio Corporation et al.,* 12 F. 2d 614 (N.D. Ill.), April 16, 1926.

6. Maurice E. Shelby, Jr., "John R. Brinkley: His Contribution to Broadcasting," and Thomas W. Hoffer, "TNT Baker: Radio Quack," *American Broadcasting—A Source Book on the History of Radio and Television,* 2nd ed., ed. Lawrence Lichty and Malachi Topping (New York: Hastings House, 1975), 560–577.

7. *Trinity Methodist Church, South* v. *Fed. Radio Comm.,* 62 F.2d. 850.

8. Telegram from Aimee Semple McPherson, KFSG Radio, 1925, cited by Erik Barnouw, *A Tower in Babel: A History of Broadcasting in the United States to 1933* (New York: Oxford University Press, 1966), p. 180.

9. As cited by Richard J. Meyer, "Reaction to the 'Blue Book,'" *American Broadcasting—A Source Book on the History of Radio and Television,* 2nd ed., ed. Lawrence Lichty and Malachi Topping (New York: Hastings House, 1976), 590.

10. Frank Ahrens. "Michael Powell: The Great Deregulator," *Washtech News* 18 June 2001. www.washtech.com/news/regulation/10574-1.html.

11. FCC statements, 8 January 2002, www.fcc.gov/Speeches/Copps/Statements/2002/stmjc201.html.

12. PCS is the abbreviation for *personal communications ser-vices,* which is regarded as the next generation of two-way wireless after cellular phones functioning digitally at a different frequency (1900 MHz) with data as well as voice transmission.

13. *The Federal Register* is also found on the Web at http://fr.cos.com.

14. Until the Federal Radio Commission was formed in 1927, the Commerce Department issued radio licenses for two-year terms, then extended them to three-year terms with the creation of the FRC. The extension to eight-year terms came with the 1996 Telecommunications Act. See W. Jefferson Davis, "The Radio Act of 1927," *American Broadcasting—A Source Book on the History of Radio and Television,* 2nd ed., ed. Lawrence Lichty and Malachi Topping (New York: Hastings House, 1976), 556.

15. *Bechtel* v. *FCC,* 10 F. 3d 875, 1993.

16. Online News Hour Update (PBS). "Congress Approves Massive Spending Bill." 22 January 2004, www.PBS.org/newshow/update/congress_01-22-04.html.

17. *Associated Press* v. *United States,* 326 U.S. 1, 20 (1945) as cited by Philip M. Napoli, "Deconstructing the Diversity Principle," *Journal of Communication,* 7, no. 4 (1999): 7–34.

18. *Fox Television Stations, Inc.* v. *Federal Communications Commission,* 280 F. 3d 1027, 2002.

19. Kalpana Srinivasan, "Entertainment: Number of Minority-Owned TV Stations Drops," 16 January 2001, www.nandotimes.com.

20. Ivy Planning Group LLC. *Historical Study of Market Entry Barriers, Discrimination and Changes in Broadcast and Wireless Licensing 1950 to Present.* Document prepared for the Office of General Counsel, Federal Communications Commission, December 2000.

21. D.C. Cir. No. 97-1116, April 14, 1998.

22. *DC/MD/DE Broadcasters Association* v. *FCC,* 236 F.3d 13 (D.C. Cir. 2001).

23. Rob Puglisi, "How and When Will HDTV Affect Television News?" *RTNDA Communicator* (May 1988): 12, 14–15.

24. "Advanced Television Systems and Their Impact upon the Existing Television Broadcast Service," MM Docket No. 87-286, FCC 97-115 Orel. April 21, 1997 (Sixth Report and Order).

25. Telephone Interview, January 4, 2005.

26. *Turner Broadcasting Inc.* v. *FCC,* 819 F. Supp.32 (D.D.C. 1993) and *Turner Broadcasting System, Inc.* v. *FCC,* 512 U.S. 622, 114 S. Ct. 2445 129 L. Ed. 2d 497 (1994).

27. *City of Dallas* v. *FCC,* 165 F 3d 341, 5th Cir., 1999.

28. In Re-Application of Great Lakes Broadcasting Co., FRC Docket 4900, 3 F.R.C. Ann. Rep. 32 (1929).

29. Federal Radio Commission, Third Annual Report 33 (1929).

30. 13 FCC 1246 (1949).

31. See *Syracuse Peace Council* v. *FCC,* 867 F. 2d 654 (D.C. Cir. 1989) cert denied, 493 U.S. 1019 (1990) and *Arkansas AFL-CIO* v. *FCC,* 11 F.3d 1430 (8th Cir. 1993).

32. *Radio-Television News Directors Association and National Association of Broadcasters* v. *FCC et al.,* United States Court of Appeals for the District of Columbia Circuit October 11, 2000 No. 98–1305, consolidated with No. 98-1334 On Motion to Recall the Mandate or for an Order Pursuant to 47 U.S.C._402(h).

33. Title III, Part I, Sec. 315.

34. 47 U.S.C. Sec. 312 (a) (7).

35. David Broder, "A Word from Our TV Stations," *Washington Post* syndicate, *Fort Worth Star-Telegram* (24 February 2002): 5E.

36. *Atlanta NAACP,* 36 FCC 2d 635 (1972).

37. *Gillett Communications of Atlanta Inc. (WAGA-TV5)* v. *Becker 21.* (DC N.Ga), Med. L. Rptr. 702 (1992) cited by Kenneth Creech, *Electronic Media Law and Regulation,* 3rd ed. (Boston: Focal Press, 2000), 60–61.

38. See 47 U.S.C.A. Sec. 315(a)(1)-(4).

39. *Aspen Institute Program on Communications and Society Petition,* 35 R.R. 2d 49 (1975).

40. Barry Cole and Mal Oettinger, *Reluctant Regulators: The FCC and the Broadcast Audience* (Reading, MA: Addison-Wesley, 1988), 248–250.

41. See "Children's Television Programming and Advertising Practices," 75 FCC 2d 138 (1979).

42. See *Washington Association for Television and Children* v. *FCC,* 712 F. 2d 677 [D.C. Cir. 1983] and *ACT* v. *FCC,* 821 F. 2d 741 (D.C. Cir. 1987).

43. 47 U.S.C. Sec. 303(b), *Consideration of Children's Television Service in Broadcast License Renewal.*

44. Report and Order in the Matter of Policies and Rules Concerning Children's Television Programming; MM Docket No. 98-48, FCC 96-355, released Aug. 8, 1996.

45. See *Midler* v. *Ford Motor Company,* 849 F.2d 460 (9th Cir., 1988) and *Vanna White* v. *Samsung Electronics America Inc.,* 971 F.2d 1395, 20 Med.L.Rptr.1457 (9th Cir. 1992), cert. denied, 508 U.S. 951, 113 S.Ct. 2443, 124 L.Ed.2d 660 (1993).

46. *Eldred* v. *Ashcroft,* v. 537 U.S. 186 (2003).

47. *Luther R. Campbell a.k.a. Luke Skywalker* v. *Acuff-Rose Music, Inc.,* 510 U.S. 569, 114 S.Ct. 1164, 127 L.Ed.2d 500, 22 Med. L. Rptr. 1353 (1994).

48. *Tin Pan Apple* v. *Miller Brewing* (DC So NY, 1990) 17 Med. L. Rptr. 2273.

49. *Sony Corporation of America* v. *Universal Studios,* 464 U.S. 417, 1984.

50. *Metro-Goldwyn-Mayer Studios* v. *Grokster, Ltd.,* 2003 U.S. Dist. LEXIS 6994 (D.C. Cal. April 25, 2003.

51. Restatement (Second) of Torts Section 568A (1977), cited in Roy L. Moore, *Mass Communication Law and Ethics,* 2nd ed. (Mahwah, NJ: Lawrence Erlbaum, 1999), 322.

52. *Sagan* v. *Apple Inc.,* 22 Media Law Rptr. 2141, 874 F. Supp.1072 (D.C.C.Cal. 1994).

53. 485 U.S. 46, 1987.

54. 376 U.S. 279-80 (1964).

55. *Newton* v. *NBC* (D.C. Nev., 1987) 114 Med. L. Rptr. 1914.

56. S. D. Warren and L. D. Brandeis, "The Right to Privacy," 4 *Harv. L. Rev.* (1890), 193.

57. Ibid.

58. Ibid.

59. *American Civil Liberties Union of Georgia* v. *Miller,* 977 F. Supp. 1228 (N.D.Ga.1997).

60. *McVeigh* v. *Cohen*, 983 F.Supp. 215 (D.C. Cir. 1998).

61. *Chandler* v. *Florida*, 449 U.S. 560 (1981).

62. *Miller* v. *California*, 413 U.S. 15 (1973).

63. Federal law prohibits obscene content on broadcast television (18 USC 1464) cable (18 USC 1468; 47 USC 559) and satellite television (18 USC 1468). The provision of federal law prohibiting "indecent" material on broadcast TV is enforced only between the hours of 6:00 A.M. and 10:00 P.M. Congress passed a law in 1996 (47 USC 561) requiring cable operators to scramble the signals for channels dedicated to sexually oriented programming.

64. The FCC issued a new policy statement on indecency in April 2001—one that had been promised since 1994, when the FCC found itself at a stalemate with a Chicago radio station, WLUP-AM. The station's owner, Evergreen Media Corporation, responded to an indecency complaint from the FCC by challenging its process regarding indecency on constitutional grounds.

65. *FCC* v. *Pacifica Foundation*, 438 U.S. 726, 98 S. Ct. 3026, 57 L. Ed. 2d 1073, 3 Med. L. Reporter 2553 (1978).

66. 18 U.S. C. 1464 and 47 CFR Section 73.3999.

67. *Cruz* v. *Ferre*, 571 F. Supp. 125 (S.D. Fla. 1983).

68. *Reno* v. *ACLU* (CDA), 521 U.S. 844 (1997) and *ACLU* v. *Reno* (CDA), 929 F. Supp. 824 E.D.Pa. 1996).

69. *ACLU* v. *Reno* (COPA), 217 F. 3d 162 (2000).

Chapter 12

1. Philip Patterson and Lee Wilkins, *Media Ethics: Issues & Cases*, 4th ed. (New York: McGraw-Hill Higher Education, 1997).

2. Peter Fuss, *The Moral Philosophy of Josiah Royce.* (Cambridge, MA: Harvard University Press, 1965), as cited by Philip Patterson and Lee Wilkins, *Media Ethics—Issues & Cases*, 4th ed. (New York: McGraw-Hill, 1997), 71–80.

3. Philip Patterson and Lee Wilkins, *Media Ethics*, 10.

4. Louis A. Day, *Ethics in Media Communications: Cases and Controversies*, 2nd ed. (Belmont, CA: Wadsworth, 1997).

5. Joseph A. Mirando, "Lessons on Ethics in News Reporting Textbooks, 1867–1997," *Journal of Mass Media Ethics*, 13, no. 1 (1998): 26–39.

6. John Rawls, *A Theory of Justice* (Cambridge, MA: Harvard University Press, 1971).

7. Val Limburg, *Electronic Media Ethics* (Newton, MA: Butterworth-Heinemann, 1994), 49–59.

8. Vernon Stone, *Evolution of the RTNDA Code of Ethics*, 7 October 2001.

9. Jennifer Sinco-Kelleher. "Story from Mosque Inspired Me to Be a Voice for the Voiceless," Newsroom Diversity Freedom Forum.org, 25 September 2001, www.freedom forum.org/templates/document. asp?document=149m.

10. See *Advertising Principles of American Business* at www.aaf. org/about/principles.html.

11. Alexander Kendrick, *Prime Time: The Life of Edward R. Murrow* (Boston: Little, Brown, 1969), 466.

12. "Geraldo's Story: Truth or Consequences?" *More Talk About Ethics: Commentary, Analysis, & Advice from the Director of Poynter's Ethics Program*, 21 December 2001, www. poynter.org/column.asp?id=36&aid=887.

13. "News for Robert Greenwald—Ailes Attempts to Outfox Critics," *IMDB*, 27 July 2004, www.imdb.com/name/nm0339254/news.

14. Maureen Dowd, "Lights, Camera, Wartime TV Action," *The New York Times in the Fort Worth Star-Telegram* (26 February 2002): 11B.

15. Molly Ivins, "A Bad Idea and Some Better Ones," *The Fort Worth Star-Telegram* (24 February 2002): 5E.

16. L. Brent Bozell III, "No *Pride* at NBC," Parents Television Council, 17 September 2004, www.parentstv.org/PTC/publications/lbbcolumns/2004/0917.asp.

17. See Press Release: "RAND Study Finds Adolescents Who Watch a Lot of TV with Sexual Content Have Sex Sooner," 25 September 2004, www.rand.org/news/press.04/09.07.html.

18. Joe Scholosser, "Family-Hour Feud," and "More Sound and Fury," *Broadcasting & Cable*, (New York: Cahners, 6 August 2001): 8, 46.

19. Children & The Media, *Fall Colors: Prime Time Diversity Report—2003*, twelve-page report printed May 2004 provides a five-year analysis of the networks' coverage of minorities, www.childrennow.org/media/fc2003/fc-2003-highlights.cfm.

20. Natalie Cortes, "Latinos Woefully Underrepresented in U.S. Media, Panelists Say," Freedom Forum Online, 8 October 2001, www.freedom forum.org.

21. Lawrence Soley, "The Power of the Press Has A Price," *Extra!* July/August 1997, www.fair.org/extra/9707/ad-survey.htm.

22. Bob Steele and Al Tompkins, *Newsroom Ethics: Decision-Making for Quality Coverage*, 2nd ed., The Radio and Television News Directors Foundation, 2000, 42.

23. "We Paid $3 Billion for the Stations. We'll Decide What the News Is." *Extra!* Update, June 1998, www.foxbghsuit.com and www.monitor.net/rachel/r593.html.

24. Lawrie Mifflin, "ABC News Reporter Discovers the Limits of Investigating Disney," *The New York Times*, 19 October 1998, www.corpwatch.org/trac/corner/worldnews/other/226.htm.

25. Sam Roberts, "Infomercials Disguised as News." 9 September 2001, rtvj.l@server2.umt.edu.

26. Allison Romano, "Grist from CNN More Subtle than Some," *Broadcasting & Cable Online*. 30 July 2001, http://broadcastingcable.com.

27. "Embedded Ads in TV Stories," *CSMonitor.com*, 01 October 2003, www.csmonitor.com/2003/1001/p08s05-comv.htmi.

28. "Forbes.com Links Words to Ads," *The Media Center at the American Press Institute. Cyberjournalist.net*, www.cyber-journalist.net/ethics_and_credibility/.

29. Brian Krebs, "News Web Sites Court Campaign Ads," *Washingtonpost.com* 30 July 2004, www.washington post.com/wp-dyn/articles/A26819-.html.

30. Jay Black, Bob Steele, and Ralph Barney, *Doing Ethics in Journalism—A Handbook with Case Studies* (Boston: Allyn and Bacon, 1999), 145, 159.

31. *Editor and Publisher Online*, "Anchor Fired for Giving Political Money," 14 November 2000, www.media into. com/ephomo/news/newshtm/stories/20400 or 2.htm.

32. Vernon Stone, *Evolution of the RTNDA Code of Ethics*, 9.

33. Tom Maurstad, *"Shock Value: Horror of Terrorism Playing on Home TV Screens,"* 10 July 2004, www.dallasnews.com.

34. "Americans Divided over Graphic War Images," Pew Internet and American Life Project via the Center for Media Research, 27 July 2004, www.typepad.com/t/track-back/961249.

35. *NPPA Code of Ethics*, www.nppa.org/professional_development/business_practices/ethics.

36. Kelly McBride, "Hiding the Paper: A First for a Journalist," *Poynteronline*, 14 May 2004, www.poynter.org, Quicklink: A65963.

37. Shirley Biagi, "Special Perspectives from Judy Woodruff," *News Talk II* (Belmont, CA: Wadsworth, 1987), 41.

38. John McKay, "Survivor's Mark Burnett Offers Success Tips at Motivational Speaker Event, *National Post*, 24 September 2004, www.cp.org.

39. Ted Koppel, *Nightline*, 10.

40. Dan Weaver and Jason Siegel, *Breaking into Television* (Princeton, NJ: Peterson's, 1998), 9–10.

41. Weaver and Siegel, *Breaking into Television*, 189.

42. Biagi, "Special Perspectives from Susan Spencer," *News Talk II*, 5–6.

43. Lucinda Watson, *How They Achieved* (New York: Wiley and Sons, 2001), 35–43.

44. Watson, *How They Achieved*, 54.

45. Sumner Redstone (with Peter Knobler), *A Passion to Win* (New York: Simon and Schuster, 2001), 20.

Chapter 13

1. "Statement Regarding Polling on Terrorism Prior to 9/11/01," 6 August, 2004. Survey by Pew Research Center. Methodology: Conducted by Princeton Survey Research Associated, May 11–May 20, 2001, and based on telephone interviews with a national adult sample of 1,587—this item was asked of form 2 of half sample. AAPOR website accessed August 28, 2004, www.aapor.org.

2. Albert Einstein (1933), cited by Don W. Stacks and Michael B. Salwen, *An Integrated Approach to Communication Theory and Research* (Mahwah, NJ: Lawrence Erlbaum, 1996), p. 3.

3. John Fiske, *Introduction to Communication Studies* (London: Routledge, 1982), 6.

4. Central Command website, accessed 28 August, 2004, www.centcom.mil/galleries/lealet.images.uzd-8104.jpg.

5. Wilbur Schramm, *The Beginnings of Communication Study in America—A Personal Memoir* (Thousand Oaks, CA: Sage, 1997), 111.

6. Harold D. Lasswell, *Propaganda Technique in the World War* (New York: Knopf, 1971)(Original work published 1927).

7. F. R. Dulles, Review of *Propaganda Technique in the World War: The Bookman*, 1968, cited by Wilbur Schramm, *The Beginnings of Communication Study in America—A Personal Memoir* (Thousand Oaks, CA: Sage, 1997), p. 35.

8. Harold D. Laswell, "The Structure and Function of Communication in Society," *The Communication of Ideas*, ed. L. Bryson (New York: Harper & Brothers), reported in Wilbur Schramm (Ed.), *Mass Communications* (Urbana: University of Illinois Press), 117–130.

9. Carl Hovland, Irving L. Janis, and H. H. Kelley, *Communication and Persuasion* (New Haven, CT: Yale University Press, 1953) and M. Pfau, "Designing Messages for Behavioral Inoculation," *Designing Health Messages: Approaches for Communication Theory and Public Health Practice*, ed. E. Maibach and R. L. Parrott (Thousand Oaks, CA: Sage, 1995), 99–113.

10. Marshall McLuhan, *Understanding Media: The Extensions of Man* (New York: McGraw-Hill, 1965), 12.

11. Roger Fidler, *Mediamorphosis: Understanding New Media* (Thousand Oaks, CA: Pine Forge, 1997).

12. Everett M. Rogers, *A History of Communication Study: A Biographical Approach* (New York: MacMillan, 1994), p. 321.

13. Joseph Klapper, *The Effects of Mass Communication* (New York: Free Press, 1967).

14. Klapper, *The Effects of Mass Communication*, 8.

15. Professors Werner J. Severin and James W. Tankard have noted other scholars referencing this phenomenon. In 1958, Norton Long, for example, referred to the newspaper's role in "setting the territorial agenda," which influences "what most people will be talking about, what most people will think the facts are, and what most people will regard as the way problems are to be dealt with." Quote taken from *American Journal of Sociology*, 64: 260. The husband and wife team of Kurt and Gladys Lang the following year discussed how "mass media force attention to certain issues," and "are constantly suggesting what individuals in the mass should think about, know about, have feelings about." Quote taken from Kurt Lang and Gladys E. Lang, "The Mass Media and Voting," *American Voting Behavior*, ed. E. Burdick and A. J. Brodbeck (Glencoe, IL: Free Press, 1959), 232, cited by Werner J. Severin and James W. Tankard, Jr., *Communication Theories—Origins, Methods, and Uses in the Mass Media*, 3rd ed. (New York: Longman, 2001), 221–22.

16. Bernard C. Cohen, *The Press and Foreign Policy* (Princeton, NJ: Princeton University Press, 1963), p. 13 and Walter Lippmann, *Public Opinion* (New York, Macmillan, 1922).

17. Todd Gitlin. *The Whole World Is Watching: Mass Media in the Making and Unmaking of the New Left* (Berkeley: University of California Press, 1980), 7.

18. Erving Goffman, *Frame Analysis* (New York: Harper & Row, 1974), 21.

19. Bernard Berelson, "What 'Missing the Newspaper' Means," *The Process and Effects of Mass Communication*, ed. Wilbur Schramm (Urbana, University of Illinois, 1965), 36–47.

20. Jennings Bryant and Dolf Zillman, "Using Television to Alleviate Boredom and Stress: Selective Exposure as a Function of Induced Excitational States," *Journal of Broadcasting*, 28 (1984): 1–20.

21. Jay G. Blumler and Denis McQuail, *Television in Politics: Its Uses and Influence* (Chicago: University of Chicago Press, 1964).

22. "Video Games: Cause for concern?" BBC News, 26 November 2000, accessed at www.bbc.org. Also see James

D. Ivory, "Video Games and the Elusive Search for Their Effects on Children," presented to the Mass Communication and Society Division at the Association for Education in Journalism and Mass Communication's Annual Convention, Washington, DC, August 2001.

23. D. Horton and R. R. Wohl. "Mass Communication and Parasocial Interaction: Observation on Intimacy at a Distance," *Psychiatry*, 19(1956): 216.

24. Kaiser Family Foundation, Children and Violence study cited in "Research on the Effects of Media Violence," Media Awareness Network, www.media-awareness.ca/english/issues/violence/effects_media_violence.cfm.

25. Craig Anderson et al., cited in "Research on the Effects of Media Violence," Media Awareness Network, www.media-awareness.ca/english/issues/violence/effects_media_violence.cfm.

26. Bryce Ryan and Neal C. Gross. "The Diffusion of Hybrid Seed Corn in Two Iowa Communities," *Rural Sociology*, 8 (1943): 15–24.

27. Everett M. Rogers, and Arvind Singhal, "Diffusion on Innovations," *An Integrated Approach to Communication Theory and Research*, ed. Michael B. Salwen and Don W. Stacks (Mahwah, NJ: Lawrence Erlbaum, 1996), 417.

28. Rogers and Arvind, "Diffusion of Innovation," 418.

29. T. Childers and J. Post, *The Information-Poor in America* (Metuchen, NJ: Scarecrow Press, 1975).

30. Cecilie Gaziano and Emanual Gaziano, "Theories and Methods in Knowledge Gap Research Since 1970," *An Integrated Approach to Communication Theory and Research*, ed. Michael B. Salwen and Don W. Stacks (Mahwah, NJ: Lawrence Erlbaum, 1996), 136.

31. Amanda Lenhart, John Horrigan, Lee Rainie, Katherine Allen, Angie Boyce, Mary Madden, and Erin O'Grady, "The Ever-Shifting Internet Population—A New Look at Internet Access and the Digital Divide," Pew Internet & American Life Project, 16 April 2003, www.pewinternet.org/pdfs/PIP_Shifting_net_Pop_Reprt.pdf).

32. Michael Hirsh and Roy Gutman, "Powell's New War," *Newsweek* (11 February 2002): 25, www.nydailynews.com_/06/New York_Now/Television/a-140420.

33. Bob Papper, "Recovering Lost Ground—Minorities Gain Ground and Women Make Management Strides in Radio and TV Newsrooms in 2004," www.rtndf.org/diversity/Diversity2004.pdf.

34. Bradley S. Greenberg and Larry Collette, "The Changing Faces on TV: A Demographic Analysis of Network Television's New Seasons, 1966–1992," *Journal of Broadcasting and Electronic Media*, 41, no. 1 (1997): 14–24.

35. Daniel A. Berkowitz, "Refining the Gatekeeping Metaphor for Local Television News," *Journal of Broadcasting & Electronic Media*, 34 (1990): 55–68.

Chapter 14

1. "Nothing is 'Big Time' about Channel 13," *Out West*, 39 (July 1997).

2. "Nearly $500,000 Spent During KPFA Lockout," *San Francisco Chronicle*, 8 September 1999, A21.

3. *A Public Trust: The Landmark Report of the Carnegie Commission on the Future of Public Broadcasting* (New York: Bantam Books, 1979), pp. 281–282.

4. James T. Yee, "Background Statement on Minority Needs," in *Assessing the Public Broadcasting Needs of Minority and Diverse Audiences* (Queenstown, MMMD: The Aspen Institute, 1992), p. 39.

5. *Public Broadcasting's Services to Minorities and Diverse Audiences: A Report to the 107th Congress and the American People*, Corporation for Public Broadcasting, December, 2002.

6. National Public Radio Time Line, 30th Anniversary Celebration, www.npr.org/programs/atc/atc30.

7. *Audience 98: The Importance of Community Radio*, report of Audience Research Analysis and the Corporation for Public Broadcasting, www.aranet.com/a98/reports/a98-r16.htm.

8. "Pacifica Radio Stations Break into Nation's Top 30," press release from Pacific Radio Foundation, 9 August 2004.

9. 2002 statistics, Broadcast Education Association, www.beaweb.org/.

10. "Low Power FM Broadcast Radio Stations," FCC Audio Division, www.fcc.gov/mb/audio/1pfm/.

11. Gilbert Gillespie, *Public Access Cable Television in the United States and Canada* (New York: Praeger Publishers, 1975) 35–36.

12. "Public Access Television," Museum of Broadcast Communications, www.museum.tv/archives/.

13. Bob Edwards, *News and Views from National Public Radio: The M. L. Seidman Memorial Town Hall Lecture Series*. Memphis, TN: Rhodes College 1987, 7.

14. *St. Louis Post-Dispatch*, Editorial, 11 October 2000, F8.

Chapter 15

1. Peter Singer, *One World: The Ethics of Globalization*. (New Haven, CT: Yale University Press, 2002).

2. James Carey, "Time, Space and the Telegraph," in *Communication in History: Technology, Culture and Society*, 3rd ed., ed. David Crowley and Paul Heyer (New York: Longman), 135.

3. *UNESCO Statistical Yearbook 1999*, Institute for Statistics, United Nations Organization for Education, Science and Culture.

4. Krishna Sen and David T. Hill, *Media, Culture and Politics in Indonesia* (Melbourne, Australia: Oxford University Press, 2000), 92.

5. Language from Voice of America charter, at VOA website, www.ibb.gov/pubaff/voacharter.html.

6. Maynard Parker, *Mixed Signals: The Prospects for Global Television News* (A Twentieth Century Fund Report) (New York: Twentieth Century Fund Press, 1995), 5.

7. "India's Television History," Indiantelevision.com, www.indiantelevision.com/indianbrodcast/history/historyoftele.htm.

8. "How Much Information?" (research report), School of Information Management and Systems, University of California at Berkeley, 2001. http://info.berkeley.edu/research/projects.

9. Carla Brooks Johnston, *Global News Access: The Impact of New Communications Technologies* (Westport, CT: Praeger, 1998), 81.

10. Peter J. Humphreys, *Mass Media and Media Policy in Western Europe* (European Policy Research Unit Series) (Manchester, England: Manchester University Press, 1996), 112–116.

11. "Faint Voices Rise from Cuba," *WIRED.COM,* 29 May 2001.

12. "New, Multiracial Beginning in Story of 'Madam and Eve,'" *Los Angeles Times* (7 March 2001): F6.

13. Report from correspondent Phyllis Crockett on *All Things Considered,* National Public Radio, 27 November 1993.

14. Humphreys, *Mass Media and Media Policy,* 112.

15. Lucy Küng-Shankleman, *Inside the BBC and CNN: Managing Media Organisations* (London: Routledge, 2000).

16. For excellent overviews of broadcasting in Iraq and neighboring countries, see Douglas Boyd, *Broadcasting in the Arab World: A Survey of the Electronic Media in the Middle East,* 3rd ed. (Ames: Iowa State University Press, 1999).

17. John Sinclair, *Latin American Television: A Global View* (Oxford, England: Oxford University Press, 1999), 84.

18. Priscilla Parkhurst Ferguson (translator's note) in Pierre Bourdieu, *On Television and Journalism* (London: Pluto Press, 1998), 83.

19. "Croatia Country Profile," Quest Economics Database, *Europe Review World of Information,* 23 August 2001.

20. Speech to Radio-Television News Directors Association annual convention, Minneapolis, MN, 15 September 2000.

21. Lewis A. Friedland, *Covering the World: International Television News Services* (Perspectives on the News series) (New York: Twentieth Century Fund, 1992), 6.

22. Committee to Protect Journalists website, www.cpj.org.

23. Daniel Schorr, "On National Security, Five Ways to Respond to Restraints," *Nieman Reports,* 52, no. 1 (Spring 1998): 42.

24. Bittermann, Jim, in *Live from the Trenches: The Changing Role of the Television News Correspondent,* ed. Joe S. Foote (Carbondale: Southern Illinois University Press, 1998), 123.

25. Kevin McAuliffe, "Kosovo: A Special Report," *Columbia Journalism Review* (May/June 1999): 28.

26. Bitterman, "Live from the Trenches," 123.

27. "From ENG to SNG: TV Technology for Covering the Conflict with Iraq," by Mike Wendland, Poynter Institute for Media Studies website, 6 March 2003, www.poynter. org/.

28. "New Media in Modern War," by Philip Taylor, from website of Channel 4 (British TV), April 2003.

29. Robert W. McChesney, *Rich Media Poor Democracy: Communication Politics in Dubious Times* (Urbana: University of Illinois Press, 1999), 100–112.

30. Dietrich Berwanger, "The Third World," in *Television: An International History* (Oxford, England: Oxford University Press, 1998), 192.

31. "A Passage from India," *The Economist,* U.S. edition, 21 October 2000.

32. Speech to Radio-Television News Directors.

33. Garrick Utley, "The Shrinking of Foreign News," *Foreign Affairs* (March/April 1997): 2.

Photo Credits

KTXA-TV; p. 230: Corbis/Bettmann; p. 233: National Recreational Properties/AP Wide World Photos; p. 234: Jeffery Allan Salter/Corbis/SABA Press Photos, Inc.; p. 236: Courtesy of ZZ Mylar, MySanAntonio.com; p. 238: Courtesy Kerry Campaign via Getty Images/Getty Images; p. 214: 20TH CENTURY FOX/THE KOBAL COLLECTION/Picture Desk, Inc./Kobal Collection; p. 241: Jose Luis Pelaez/Corbis/Stock Market; p. 247: Neal Peters Collection; p. 249: FCC; p. 250: Kevin Lamarque/Corbis/Reuters America LLC; p. 258: Courtesy of Richard Wiley and DuPont Photographers, Inc.; p. 263: Wally McNamee/CORBIS-NY; p. 264: Steve Liss/Getty Images/Time Life Pictures; p. 266: AP Wide World Photos; p. 267a: Neal Preston/Corbis/Bettmann; p. 267b: AP Wide World Photos; p. 269: Courtesy of Heidi Constantine; p. 271: Corbis/Bettmann; p. 272: Corbis/Sygma; p. 274: AP Wide World Photos; p. 275: Picturequest–Royalty Free; p. 278: Texas National Guard/CORBIS-NY; p. 281: Lyrl Ahern; p. 282a: Library of Congress; p. 282b: Library of Congress; p. 287: HO/AFP/Getty Images Inc.–Agence France Presse; p. 288: Tony Esparza/CBS/Photofest; p. 289: Spencer Platt/Getty Images; p. 290: Fred Prouser/Getty Images Inc.–Liaison; p. 291: Courtesy of Ernest Sotomayor, Newsday.com; p. 296: Courtesy of Robert Steele, Poynter Institute; p. 297: Courtesy of Deborah Potter, NewsLab; p. 299: HO/AFP/WASHINGTON POST/Getty Images Inc.–Agence France Presse; p. 300: Vivian Ronay/Getty Images Inc.–Liaison; p. 304: AP Wide World Photos; p. 306: AP Wide World Photos; p. 312: Tony Esparza/CBS/Everett Collection; p. 314: Archives of the History of American Psychology; p. 315: Getty Images Inc.–Hulton Archive Photos; p. 317: Getty Images Inc.–

Liaison; p. 324: Courtesy of Robert Papper, Ball State University; p. 325: Edouard Berne/Getty Images Inc.–Stone Allstock; p. 326: Courtesy of Norman Hecht, Norman Hecht Research; p. 327: AP Wide World Photos; p. 330: Tony Nagelmann/NPR; p. 335: Courtesy of American Experience; p. 338: Courtesy of University of Wisconsin Archives; p. 339: Courtesy of the Scheslinger Library, Radcliffe Institute, Harvard University; p. 340: Courtesy of Newton Minow; p. 342: © Herrmann + Starke 2004, Courtesy of WETA, NPR; p. 343: Courtesy of Tripp Sommer, KLCC-FM, Eugene, Oregon; p. 344: Courtesy of Paul Gigot/The Wall Street Journal; p. 345: Courtesy of Maryland Public Television; p. 347: Everett Collection; p. 348: Michael Dibari/AP Wide World Photos; p. 350: Philip Gould/Corbis/Bettmann; p. 352: Courtesy of KPFA-FM, Berkeley, California; p. 353: Courtesy of WERS-FM, Emerson College, Boston, Massachusetts; p. 356: Courtesy of Bob and Virginia Becker, low-power television operators; p. 357: Courtesy of WBUR/National Public Radio; p. 358: Courtesy of WBUR/National Public Radio; p. 359: Courtesy of Lisette Marie Flanary; p. 360: Lon C. Diehl/PhotoEdit; p. 362: SVEN NACKSTRAND/Getty Images Inc.–Agence France Presse; p. 365: Courtesy of Maria Paz Epelman; p. 366: AP Wide World Photos; p. 368: John Ali/Getty Images Inc.–Liaison; p. 369: Annebicque Bernard/Corbis/Sygma; p. 370: Margot Granitsas/The Image Works; p. 371: Courtesy of Veran Matic; p. 374: Shamil Zhumatov/CORBIS-NY; p. 377: AP Wide World Photos; p. 378: Quin Llenas/COVER/The Image Works; p. 379: Koren Ziv/Corbis/Sygma; p. 380: NTV/Via Reuters TV/Corbis/Reuters America LLC; p. 383: Phillippe Lopez/Corbis/Bettmann